STUDENT SOLUTIONS MANUAL

GEX PUBLISHING SERVICES

INTERMEDIATE ALGEBRA
FOURTH EDITION

Tom Carson
Franklin Classical School

Bill E. Jordan
Seminole State College of Florida

PEARSON

Boston   Columbus   Indianapolis   New York   San Francisco   Upper Saddle River
Amsterdam   Cape Town   Dubai   London   Madrid   Milan   Munich   Paris   Montreal   Toronto
Delhi   Mexico City   São Paulo   Sydney   Hong Kong   Seoul   Singapore   Taipei   Tokyo

ISBN-13: 978-0-321-91228-2
ISBN-10: 0-321-91228-4

www.pearsonhighered.com

PEARSON

CONTENTS

Chapter 1 Real Numbers and Expressions... 1

Chapter 2 Linear Equations and Inequalities in One Variable 11

Chapter 3 Equations and Inequalities in Two Variables and Functions.. 44

Chapter 4 Systems of Linear Equations and Inequalities 81

Chapter 5 Exponents, Polynomials, and Polynomial Functions............ 123

Chapter 6 Factoring... 148

Chapter 7 Rational Expressions and Equations 170

Chapter 8 Rational Exponents, Radicals, and Complex Numbers 219

Chapter 9 Quadratic Equations and Functions 254

Chapter 10 Exponential and Logarithmic Functions............................... 306

Chapter 11 Conic Sections... 338

Chapter 1
Real Numbers and Expressions

Exercise Set 1.1

1. {Saturday, Sunday}

3. {January, June, July}

5. {North Carolina, North Dakota}

7. {0, 1, 2, 3, 4}

9. {3, 6, 9, ...}

11. {−1, 0, 1}

13. { } or ∅

15. $\{x | x$ is an integer$\}$

17. $\{x | x$ is a letter of the English alphabet$\}$

19. $\{x | x$ is a day of the week$\}$

21. $\{x | x$ is a natural-number multiple of 5$\}$

23.

25.

27.

29.

31. False, because "n" is not a vowel.

33. True, because "go" is a verb.

35. True, because 4 is a rational number.

37. True, because "James" is not the last name of any U.S. president.

39. False, because −0.6 is a real number.

41. True, because each of the vowels is a letter of the English alphabet

43. True, because each number listed is an integer.

45. True, because the set of integers is a subset of the rational numbers.

47. True, because the set of irrational numbers is a subset of the real numbers.

49. False, because a real number can be either rational or irrational but not both.

51. True, because the whole numbers consists of the natural numbers plus 0.

53. True, because the definition of rational number is "a real number that can be expressed in the form $\frac{a}{b}$, where a and b are integers and $b \neq 0$."

55. True, because each natural number can be expressed in the form $\frac{a}{1}$, where a and 1 are integers and $1 \neq 0$, which is a rational number.

57. $|2.6| = 2.6$

59. $\left|-1\frac{2}{5}\right| = 1\frac{2}{5}$

61. $|-1| = 1$

63. $|-8.75| = 8.75$

65. $-6 > -8$ because -6 is farther right on a number line than -8.

67. $0 > -1.8$ because 0 is farther right on a number line than -1.8.

69. $3\frac{4}{5} > 3\frac{3}{4}$ because $3\frac{4}{5}$ is farther right on a number line than $3\frac{3}{4}$.

71. $|-3| = |3|$ because the absolute value of -3 is 3, which is the same as the absolute value of 3.

73. $6.7 = |6.7|$ because the absolute value of $|6.7|$ is the same as 6.7.

75. $\left|-\frac{2}{3}\right| < \left|-\frac{4}{3}\right|$ because $\left|-\frac{4}{3}\right| = \frac{4}{3}$, which is farther right on a number line than $\left|-\frac{2}{3}\right| = \frac{2}{3}$.

77. $-0.6, -0.44, 0, |-0.02|, 0.4, \left|1\frac{2}{3}\right|, 3\frac{1}{4}$

79. $-12.6, -9.6, 1, |-1.3|, \left|-2\frac{3}{4}\right|, 2.9$

81. {2005, 2006, 2008}

83. {2006, 2008}

85. {Internet Explorer, Firefox, Chrome}

87. {Internet Explorer, Firefox}

89. {Justin Bieber, Katy Perry, Lady Gaga}

91. {Justin Bieber, Katy Perry, Lady Gaga, Taylor Swift}

Exercise Set 1.2

1. Additive inverse

3. Additive identity

5. Multiplicative identity

7. Multiplicative inverse

9. Multiplicative identity

11. Additive inverse

13. The additive inverse of 8 is -8 because $-8 + 8 = 0$.

   The multiplicative inverse of 8 is $\dfrac{1}{8}$ because $8 \cdot \dfrac{1}{8} = 1$.

15. The additive inverse of -7 is 7 because $-7 + 7 = 0$.

   The multiplicative inverse of -7 is $-\dfrac{1}{7}$ because $-7 \cdot \left(-\dfrac{1}{7}\right) = 1$.

17. The additive inverse of $-\dfrac{5}{8}$ is $\dfrac{5}{8}$ because $-\dfrac{5}{8} + \dfrac{5}{8} = 0$.

   The multiplicative inverse of $-\dfrac{5}{8}$ is $-\dfrac{8}{5}$ because $-\dfrac{5}{8} \cdot \left(-\dfrac{8}{5}\right) = 1$.

19. The additive inverse of 0.3 is -0.3 because $-0.3 + 0.3 = 0$.

   The multiplicative inverse of $0.3 = \dfrac{3}{10}$ is $\dfrac{10}{3}$ because $0.3 \cdot \dfrac{10}{3} = 1$.

21. Commutative property of addition

23. Distributive property

25. Associative property of multiplication

27. Associative property of addition

29. Commutative property of addition

31. Commutative property of multiplication

33. $9 + (-16) = -7$

35. $-27 + (-13) = -40$

37. $-15 + 9 = -6$

39. $14 + (-19) = -5$

41. $\begin{aligned}
-\dfrac{3}{4} + \dfrac{1}{6} &= -\dfrac{3(3)}{4(3)} + \dfrac{1(2)}{6(2)} \\
&= -\dfrac{9}{12} + \dfrac{2}{12} \\
&= \dfrac{-9+2}{12} \\
&= -\dfrac{7}{12}
\end{aligned}$

43. $\begin{aligned}
-\dfrac{1}{8} + \left(-\dfrac{2}{3}\right) &= -\dfrac{1(3)}{8(3)} + \left(-\dfrac{2(8)}{3(8)}\right) \\
&= -\dfrac{3}{24} + \left(-\dfrac{16}{24}\right) \\
&= \dfrac{-3 + (-16)}{24} \\
&= -\dfrac{19}{24}
\end{aligned}$

45. $-0.18 + 6.7 = 6.52$

47. $-3.28 + (-4.1) = -7.38$

49. $-(-7) = 7$

51. $-(-(-2.7)) = -(2.7) = -2.7$

53. $-|12| = -12$

55. $-\left|-\dfrac{3}{4}\right| = -\left(\dfrac{3}{4}\right) = -\dfrac{3}{4}$

57. $2 - (-3) = 2 + 3 = 5$

59. $10 - (-2) = 10 + 2 = 12$

61. $7 - 11 = 7 + (-11) = -4$

63. $8 - 3 = 8 + (-3) = 5$

65. $\dfrac{7}{10}-\left(-\dfrac{3}{5}\right)=\dfrac{7}{10}+\dfrac{3}{5}=\dfrac{7}{10}+\dfrac{6}{10}=\dfrac{13}{10}$

67. $-\dfrac{1}{5}-\left(-\dfrac{1}{5}\right)=-\dfrac{1}{5}+\dfrac{1}{5}=0$

69. $6.2-3.65=6.2+(-3.65)=2.55$

71. $-6.1-(-4.5)=-6.1+4.5=-1.6$

73. $4(-3)=-12$

75. $(-2)(-1)=2$

77. $-2\cdot\dfrac{1}{4}=-\dfrac{2}{4}=-\dfrac{1}{2}$

79. $-25\div(-5)=5$

81. $-\dfrac{1}{4}\div\dfrac{2}{3}=-\dfrac{1}{4}\cdot\dfrac{3}{2}=-\dfrac{3}{8}$

83. $-12\div0.3=-40$

85. $-1(-2)(-3)=-6$

87. $2.7(-0.1)(-2)=0.54$

89. $-2(5)(-2)(-3)=-60$

91. $(-1)(-3)(-5)(2)(3)=-90$

93. $2261.18-(-19.65)=2261.18+19.65=2280.83$

95. $-1475.84+1200+(-124.75)+(-12.50)$
    $+(-225.65)+(-175.92)=-\$814.66$

97. $560.5+(-2402.5)+560.5=-1281.5$ N

99. $4.8(-0.035)=-0.168$
    $4.8+(-0.168)=\$4.632$ million

101. a) Earth: $-32.2(5.5)=-177.1$ lb.
     The Moon: $-5.5(5.5)=-30.25$ lb.
     Mars: $-12.3(5.5)=-67.65$ lb.
     b) Earth

103. $-6.4(8)=-51.2$ V

Review Exercises

1. {Washington, Adams, Jefferson, Madison}

2. No, because an element, 6, is in B but not in A.

3. infinite

4. -6 can be written as $-\dfrac{6}{1}$

5. $|-25|=25$

6. $-\dfrac{5}{6}=-\dfrac{15}{18}$ because $-\dfrac{15}{18}=-\dfrac{\cancel{3}\cdot5}{\cancel{3}\cdot6}=-\dfrac{5}{6}$

Exercise Set 1.3

1. $(-4)^3$
   base = -4; exponent = 3; negative four to the third power, or negative four cubed

3. -1^7
   base = 1; exponent = 7; the additive inverse of one raised to the seventh power

5. $5^4=5\cdot5\cdot5\cdot5=625$

7. $(-3)^4=(-3)\cdot(-3)\cdot(-3)\cdot(-3)=81$

9. $(-4)^3=(-4)\cdot(-4)\cdot(-4)=-64$

11. $-6^2=-6\cdot6=-36$

13. $(-1)^{10}$
    $=(-1)(-1)(-1)(-1)(-1)(-1)(-1)(-1)(-1)(-1)$
    $=1$

15. $\left(-\dfrac{3}{8}\right)^2=\left(-\dfrac{3}{8}\right)\left(-\dfrac{3}{8}\right)=\dfrac{9}{64}$

17. $\left(-\dfrac{5}{6}\right)^3=\left(-\dfrac{5}{6}\right)\left(-\dfrac{5}{6}\right)\left(-\dfrac{5}{6}\right)=-\dfrac{125}{216}$

19. $(0.4)^3=(0.4)(0.4)(0.4)=0.064$

21. $(-2.1)^3=(-2.1)(-2.1)(-2.1)=-9.261$

23. ±15

25. ±16

27. $\pm\dfrac{2}{3}$

29. No real-number roots exist.

31. $\sqrt{25}=5$

33. $\sqrt[5]{32}=2$

35. $\sqrt{0.36} = 0.6$

37. $\sqrt[3]{-27} = -3$

39. $\sqrt[3]{\dfrac{8}{27}} = \dfrac{2}{3}$

41. $\sqrt{-36}$ is not a real number.

43. $\sqrt[3]{\dfrac{24}{3}} = \sqrt[3]{8} = 2$

45. $-4 + 3(-1)^4 + 18 \div 3 = -4 + 3(1) + 18 \div 3$
$$= -4 + 3 + 6$$
$$= -1 + 6$$
$$= 5$$

47. $-8^2 + 36 \div (9 - 5) = -64 + 36 \div 4$
$$= -64 + 9$$
$$= -55$$

49. $12^2 \div \sqrt{45 - 9} - 8 = 12^2 \div \sqrt{36} - 8$
$$= 144 \div 6 - 8$$
$$= 24 - 8$$
$$= 16$$

51. $-2|8 - 2| \div (-5)(2) = -2|-10| \div (-5)(2)$
$$= -2 \cdot 10 \div (-5)(2)$$
$$= -20 \div (-5)(2)$$
$$= 4(2)$$
$$= 8$$

53. $-24 \div (-6)(2) + \sqrt{169 - 25}$
$$= -24 \div (-6)(2) + \sqrt{144}$$
$$= -24 \div (-6)(2) + 12$$
$$= 4(2) + 12$$
$$= 8 + 12$$
$$= 20$$

55. $-18 \cdot \dfrac{2}{9} \div (-2) + |9 - 5(-2)|$
$$= -18 \cdot \dfrac{2}{9} \div (-2) + |9 + 10|$$
$$= -18 \cdot \dfrac{2}{9} \div (-2) + |19|$$
$$= -4 \div (-2) + 19$$
$$= 2 + 19$$
$$= 21$$

57. $13.02 \div (-3.1) + 6^2 - \sqrt{25}$
$$= 13.02 \div (-3.1) + 36 - 5$$
$$= -4.2 + 36 - 5$$
$$= 26.8$$

59. $(1 - 0.8)^2 + 2.4 \div (0.3)(-0.5)$
$$= (0.2)^2 + 2.4 \div (0.3)(-0.5)$$
$$= 0.04 + 2.4 \div (0.3)(-0.5)$$
$$= 0.04 + 8(-0.5)$$
$$= 0.04 - 4$$
$$= -3.96$$

61. $\dfrac{4}{5} \div \left(-\dfrac{1}{10}\right) \cdot (-2) + \sqrt[5]{16 + 16}$
$$= \dfrac{4}{5} \div \left(-\dfrac{1}{10}\right) \cdot (-2) + \sqrt[5]{32}$$
$$= \dfrac{4}{5} \cdot \left(-\dfrac{10}{1}\right) \cdot (-2) + 2$$
$$= 16 + 2$$
$$= 18$$

63. $\dfrac{9}{8} \cdot \left(-\dfrac{2}{3}\right) + \left(\dfrac{1}{5} - \dfrac{2}{3}\right) \div \sqrt{\dfrac{125}{5}}$
$$= \dfrac{9}{8} \cdot \left(-\dfrac{2}{3}\right) - \dfrac{7}{15} \div \sqrt{25}$$
$$= -\dfrac{3}{4} - \dfrac{7}{15} \div 5$$
$$= -\dfrac{3}{4} - \dfrac{7}{75}$$
$$= -\dfrac{253}{300}$$

65. $\dfrac{12 - 2^3}{5 - 3 \cdot 2} = \dfrac{12 - 8}{5 - 6}$
$$= \dfrac{4}{-1}$$
$$= -4$$

67. $\dfrac{6^2 - 3(4 + 2^5)}{5 + 20 - (2 + 3)^2} = \dfrac{6^2 - 3(4 + 32)}{5 + 20 - 5^2}$
$$= \dfrac{6^2 - 3(36)}{5 + 20 - 25}$$
$$= \dfrac{36 - 3(36)}{0}$$
$$= \text{undefined}$$

69. Associative property of multiplication is used to multiply $3 \cdot 3$ instead of multiplying $1 \cdot 3$ from left to right.

71. Distributive property was applied instead of adding $-1 + 36$ in the parentheses.

73. Mistake: Multiplied before division.
Correct: $12 \div 4 \cdot 3 - 11 = 3 \cdot 3 - 11$
$$= 9 - 11$$
$$= -2$$

75. Mistake: Found the square root of the addends 16 and 9 instead of their sum.
Correct: $30 \div 2 + \sqrt{16 + 9} = 30 \div 2 + \sqrt{25}$
$$= 30 \div 2 + 5$$
$$= 15 + 5$$
$$= 20$$

77. a) $\dfrac{(76.5 + 74.5 + 71.4 + 69.2)}{4} = \dfrac{291.6}{4} = 72.9$ min.
b) No, her average is greater than 72 minutes.

79. $\dfrac{2470 + 2400 + 2370 + 2210}{4}$
$$= \dfrac{9450}{4}$$
$$= 2362.5$$

81. $\dfrac{(1385.30 + 1379.32 + 1375.32 + 1365.00 + 1390.99)}{5}$
$$= \dfrac{6895.93}{5}$$
$$= 1379.186$$

83. $\dfrac{4(3) + 4(4) + 3(2) + 3(1)}{14} = \dfrac{37}{14} \approx 2.64$

85. $35.5 + 0.10(658 - 500) + 0.12(45)$
$$= 35.5 + 0.10(158) + 0.12(45)$$
$$= 35.5 + 15.8 + 5.4$$
$$= \$56.70$$

87. $0.35(814) + 54.50 + 3(89.90) + 112.45$
$$= 284.9 + 54.50 + 269.70 + 112.45$$
$$= \$721.55$$

89. $\left[2(8.95) + 2(10.95) + 6.95 + 2(1.45)\right] \div 5$
$$= (17.90 + 21.90 + 6.95 + 2.90) \div 5$$
$$= 49.65 \div 5$$
$$= \$9.93$$

91. $1200 + 349(3)(12) + 0.15(52,000 - 45,000)$
$$= 1200 + 349(3)(12) + 0.15(7000)$$
$$= 1200 + 12,564 + 1050$$
$$= \$14,814$$

93. $8^4 = 4096$

95. $2^7 = 128$

97. $3^3 \cdot 10^4 = 27 \cdot 10,000 = 270,000$

Review Exercises

1. $\{0, 1, 2, 3, 4, 5, 6, 7, 8\}$

2. $19 - (-54) = 19 + 54 = 73$

3. $-2(-5)(-6) = -60$

4. $\dfrac{-38}{2} = -19$

5. It is an expression because it has no equal sign.

6. Commutative property of multiplication

Exercise Set 1.4

1. $5n$ 3. $3n - 2$ 5. $5 - p$

7. $\dfrac{n^4}{8}$ 9. $2n - 20$ 11. $x^4 y^2$

13. $\dfrac{p}{q} - \dfrac{1}{2}$ 15. $m - 3(n + 5)$

17. $(4 - t)^5$ 19. $\dfrac{6}{7}x + 7$

21. $(m - n) - (x + y)$

23. Mistake: Incorrect order.
Correct: $y - 6$

25. Mistake: Sum means use parentheses.
Correct: $4(r + 7)$

27. $5w$ 29. $2 - 3w$ 31. $2r$

33. $17 - n$ 35. $\left(t + \dfrac{1}{2}\right)$ hr.   37. πd

39. $\dfrac{1}{2}h(a + b)$ 41. $\dfrac{1}{3}\pi r^2 h$ 43. $\dfrac{1}{2}mv^2$

45. $\dfrac{Mm}{d^2}$ 47. $\sqrt{1-\dfrac{v^2}{c^2}}$

49. $-0.4(x+2)-5 = -0.4(3+2)-5$
$$= -0.4(5)-5$$
$$= -2-5$$
$$= -7$$

51. $-3m^2+5m+1 = -3\left(-\dfrac{2}{3}\right)^2+5\left(-\dfrac{2}{3}\right)+1$
$$= -3\left(\dfrac{4}{9}\right)+5\left(-\dfrac{2}{3}\right)+1$$
$$= -\dfrac{4}{3}-\dfrac{10}{3}+\dfrac{3}{3}$$
$$= -\dfrac{11}{3}$$

53. $\left|2x^2-3y\right|+x = \left|2(5)^2-3(-1)\right|+5$
$$= \left|2(25)+3\right|+5$$
$$= \left|50+3\right|+5$$
$$= \left|53\right|+5$$
$$= 53+5$$
$$= 58$$

55. $\sqrt[3]{c}-2ab^2 = \sqrt[3]{8}-2(-1)(-2)^2$
$$= 2-2(-1)(4)$$
$$= 2+8$$
$$= 10$$

57. a) $ad-bc = 5(7)-(0.2)(-3) = 35+0.6 = 35.6$

 b) $ad-bc = -8\left(-\dfrac{5}{6}\right)-\left(\dfrac{2}{3}\right)(2) = \dfrac{20}{3}-\dfrac{4}{3} = \dfrac{16}{3}$

59. a) $\dfrac{y_2-y_1}{x_2-x_1} = \dfrac{-7-(-1)}{5-3} = \dfrac{-7+1}{2} = \dfrac{-6}{2} = -3$

 b) $\dfrac{y_2-y_1}{x_2-x_1} = \dfrac{-2-(-1)}{-1-3} = \dfrac{-2+1}{-4} = \dfrac{-1}{-4} = \dfrac{1}{4}$

61. If $x = 0$, this expression is undefined because the denominator is 0.

63. $y-6 = 0$
$$y = 6$$
If $y = 6$, this expression is undefined because the denominator is 0.

65. $y+5 = 0$ $y-1 = 0$
$$y = -5 \qquad\qquad y = 1$$
If $y = -5$ or $y = 1$, this expression is undefined because the denominator is 0.

67. $4x+1 = 0$
$$4x = -1$$
$$x = -\dfrac{1}{4}$$
If $x = -\dfrac{1}{4}$, this expression is undefined because the denominator is 0.

69. $9(3x-5) = 9\cdot 3x - 9\cdot 5 = 27x - 45$

71. $-5(m+2) = -5\cdot m - 5\cdot 2 = -5m - 10$

73. $\dfrac{3}{8}\left(\dfrac{2}{9}x-24\right) = \dfrac{3}{8}\cdot\dfrac{2}{9}x - \dfrac{3}{8}\cdot 24 = \dfrac{1}{12}x - 9$

75. $-2.1(3x+2.4) = -2.1\cdot 3x - 2.1(2.4) = -6.3x - 5.04$

77. $2x-13x = -11x$

79. $\dfrac{6}{7}b^2 - \dfrac{8}{7}b^2 = -\dfrac{2}{7}b^2$

81. $4x-9y-12+y+3x = 4x+3x-9y+y-12$
$$= 7x - 8y - 12$$

83. $1.5x+y-2.8x+0.3-y-0.7$
$$= 1.5x-2.8x+y-y+0.3-0.7$$
$$= -1.3x - 0.4$$

85. $2.6h^2+\dfrac{5}{3}h-\dfrac{2}{5}h^2+h+7$
$$= 2.6h^2-\dfrac{2}{5}h^2+\dfrac{5}{3}h+h+7$$
$$= 2.2h^2+\dfrac{8}{3}h+7$$

87. $2(5n-6)+4(n+1)-8 = 10n-12+4n+4-8$
$$= 10n+4n-12+4-8$$
$$= 14n-16$$

89. $7a+5b-3(4a+2b)-12+8b$
$$= 7a+5b-12a-6b-12+8b$$
$$= 7a-12a+5b-6b+8b-12$$
$$= -5a+7b-12$$

91. a) $14+(6x-8x)$ b) $14-2x$

 c) $14-2(-3) = 14+6 = 20$

Review Exercises

1. $\{x | x \text{ is an integer and } x \geq -2\}$

2. $-5^2 + 3(6-8) - \sqrt{16} = -25 + 3(-2) - 4$
$$= -25 - 6 - 4$$
$$= -35$$

3. $(20-24)^3 - 4|3-8| = (-4)^3 - 4|-5|$
$$= -64 - 4(5)$$
$$= -64 - 20$$
$$= -84$$

4. $-7^2 = -49$

5. Commutative property of addition

6. Distributive property

Chapter 1 Review Exercises

1. $\{$Alaska, Hawaii$\}$

2. $\{\ldots, -3, -1, 1, 3, 5, \ldots\}$

3. $\{5, 10, 15, \ldots\}$

4. $\{$s, i, m, p, l, f, y$\}$

5. $\{x | x \text{ is a natural-number multiple of } 3\}$

6. $\{x | x \text{ is a whole number}\}$

7. $\{x | x \text{ is a prime number}\}$

8. $\{x | x \text{ is a day of the week}\}$

9. True, because March is not one of the days of the week.

10. False, because the numbers in the set A are not all contained in the set B.

11. False, because $\dfrac{2}{3}$ is not an irrational number.

12. True, because $\sqrt{2}$ is not an integer.

13. $-(-5) = |-5|$

14. $\left|-3\dfrac{1}{4}\right| > -3\dfrac{1}{4}$

15. $0.5 = \dfrac{1}{2}$

16. $-|-3| > -4$

17. Additive inverse

18. Multiplicative inverse

19. Additive identity

20. Multiplicative identity

21. Distributive property

22. Associative property of multiplication

23. Commutative property of addition

24. Commutative property of multiplication

25. Associative property of addition

26. Distributive property

27. $6 + (-7) = -1$

28. $-4 + 9 = 5$

29. $-15 + (-2) = -17$

30. $-2 + (-5) = -7$

31. $7 - 9 = 7 + (-9) = -2$

32. $-2 - 8 = -2 + (-8) = -10$

33. $15 - (-2) = 15 + 2 = 17$

34. $-8 - (-1) = -8 + 1 = -7$

35. $-2(4) = -8$

36. $-3(-5) = 15$

37. $7(-8) = -56$

38. $25 \div (-5) = -5$

39. $-10 \div (-5) = 2$

40. $-8 \div 4 = -2$

41. $-1(-2)(-3) = -6$

42. $-50 \div [4 \div (-2)] = -50 \div (-2)$
$$= 25$$

43. $-245.85 + 125.00 + (-72.34) + (-12.50)$
$$+ (-14.75) = -\$220.44$$

44. -2^7
base = 2; exponent = 7; the additive inverse of two raised to the seventh power; -128

45. $(-1)^4$

  base = -1; exponent = 4; negative one raised to the fourth power; 1

46. $-3^2 = -(3 \cdot 3) = -9$

47. $-2^3 = -(2 \cdot 2 \cdot 2) = -8$

48. $(-4)^2 = (-4)(-4) = 16$

49. $\left(-\dfrac{2}{5}\right)^3 = \left(-\dfrac{2}{5}\right)\left(-\dfrac{2}{5}\right)\left(-\dfrac{2}{5}\right) = -\dfrac{8}{125}$

50. $\sqrt{121} = 11$

51. $\sqrt[4]{81} = 3$

52. $\sqrt[3]{-8} = -2$

53. $\sqrt[8]{1} = 1$

54. $-7\left|8 - 2^4\right| + 7 - 3^2 = -7|8 - 16| + 7 - 3^2$
$$= -7|-8| + 7 - 3^2$$
$$= -7 \cdot 8 + 7 - 9$$
$$= -56 + 7 - 9$$
$$= -58$$

55. $8\left(1 - 3^2\right) + \sqrt{16 - 7} = 8(1 - 9) + \sqrt{9}$
$$= 8(-8) + 3$$
$$= -64 + 3$$
$$= -61$$

56. $\sqrt[3]{8} - 7 \cdot 3^2 = 2 - 7 \cdot 3^2$
$$= 2 - 7 \cdot 9$$
$$= 2 - 63$$
$$= -61$$

57. $\sqrt{16} + \sqrt{9} - 3(2 - 7) = 4 + 3 - 3(-5)$
$$= 4 + 3 + 15$$
$$= 22$$

58. $5^2 (3 - 8)^2 = 5^2 (-5)^2 = 25 \cdot 25 = 625$

59. $\dfrac{3^4 - 5\left[3 - 4(-2)\right]}{16 \div 8(-5)} = \dfrac{3^4 - 5[3 + 8]}{2(-5)}$
$$= \dfrac{3^4 - 5 \cdot 11}{-10}$$
$$= \dfrac{81 - 55}{-10}$$
$$= \dfrac{26}{-10} = -2.6$$

60. $\dfrac{2(3) + 4(4) + 3(3) + 3(1)}{12} = \dfrac{6 + 16 + 9 + 3}{12}$
$$= \dfrac{34}{12}$$
$$\approx 2.83$$

61. $59 + 0.15(37) = 59 + 5.55 = \$64.55$

62. $5^3 \cdot 10^5 = 125 \cdot 100,000 = 12,500,000$ codes

63. $14 - 8n$

64. $2(n - 2)$

65. $n + \dfrac{1}{3}(n - 4)$

66. $\dfrac{m}{\sqrt{n}}$

67. $\dfrac{1}{2}(n - 8) - 16$

68. $(n + 5) + 20$

69. $2w$

70. $\left(\dfrac{1}{3} + t\right)$ hr

71. $3a^2 - 4a + 2 = 3(-1)^2 - 4(-1) + 2$
$$= 3(1) + 4 + 2$$
$$= 3 + 4 + 2$$
$$= 9$$

72. $-4|-b + 2ac| = -4|-(-2) + 2(1)(-1)|$
$$= -4|2 - 2|$$
$$= -4|0|$$
$$= -4(0)$$
$$= 0$$

73. $15(x+y)^2 + x^2 = 15(1+(-2))^2 + 1^2$
$$= 15(-1)^2 + 1^2$$
$$= 15 \cdot 1 + 1$$
$$= 15 + 1$$
$$= 16$$

74. $\sqrt{a-b} + 3^0 - a = \sqrt{25-16} + 3^0 - 25$
$$= \sqrt{9} + 3^0 - 25$$
$$= 3 + 1 - 25$$
$$= -21$$

75. $x - 3 = 0$
$$x = 3$$
If $x = 3$, this expression is undefined because the denominator is 0.

76. $2x - 1 = 0$
$$2x = 1$$
$$x = \frac{1}{2}$$
If $x = \frac{1}{2}$, this expression is undefined because the denominator is 0.

77. $-2(5x+1) = -2 \cdot 5x - 2 \cdot 1 = -10x - 2$

78. $4(2a+3b-4) = 4 \cdot 2a + 4 \cdot 3b + 4 \cdot (-4)$
$$= 8a + 12b - 16$$

79. $3x + 2x^2 - 4x - x - 3x^2$
$$= 2x^2 - 3x^2 + 3x - 4x - x$$
$$= -x^2 - 2x$$

80. $5m^5 + 3mn - 2mn^2 - mn - 4m^5$
$$= 5m^5 - 4m^5 - 2mn^2 + 3mn - mn$$
$$= m^5 - 2mn^2 + 2mn$$

81. $-7ab + 3ab^2 + 2ab + 3a - 7a^2 - 8$
$$= -7a^2 + 3ab^2 - 7ab + 2ab + 3a - 8$$
$$= -7a^2 + 3ab^2 - 5ab + 3a - 8$$

82. $6r - 3 - r - 2r - 7$
$$= 6r - r - 2r - 3 - 7$$
$$= 3r - 10$$

Chapter 1 Practice Test

1. $|8.1| = 8.1$

2. $-\left|-\frac{11}{4}\right| = -\frac{11}{4}$

3. $\sqrt{169} = 13$

4. $\sqrt[3]{125} = 5$

5. $\sqrt[5]{\frac{1}{32}} = \frac{1}{2}$

6. Commutative property of addition, because the order of the addends is changed.

7. Associative property of multiplication, because the grouping of the factors is changed.

8. $9 + (-1) = 8$

9. $\frac{2}{3} - \left(-\frac{1}{4}\right) = \frac{2}{3} + \frac{1}{4}$
$$= \frac{2(4)}{3(4)} + \frac{1(3)}{4(3)}$$
$$= \frac{8}{12} + \frac{3}{12}$$
$$= \frac{11}{12}$$

10. $(-3)(2.5) = -7.5$

11. $(-5)^2 = (-5)(-5) = 25$

12. $-\frac{2}{5} \div \frac{5}{2} = -\frac{2}{5} \cdot \frac{2}{5} = -\frac{4}{25}$

13. $\sqrt[4]{16} = 2$

14. $8 \div 4 \cdot 2 = 2 \cdot 2 = 4$

15. $-3^2 + 7 - 2(5-1) = -3^2 + 7 - 2(4)$
$$= -9 + 7 - 2(4)$$
$$= -9 + 7 - 8$$
$$= -10$$

16. $\sqrt[4]{16} + 8 - (3+1)^2 = \sqrt[4]{16} + 8 - 4^2$
$$= 2 + 8 - 16$$
$$= -6$$

17. $4 \div |8-6| + 2^4 = 4 \div |2| + 2^4$
$$= 4 \div 2 + 16$$
$$= 2 + 16$$
$$= 18$$

18. $(5-4)^5 + (2-3)^3 = 1^5 + (-1)^3$
$$= 1 - 1$$
$$= 0$$

19. $\sqrt{9+16} + \left[-2^2 + 3(2-5) \right]$
$$= \sqrt{25} + \left[-2^2 + 3(-3) \right]$$
$$= 5 + (-4 - 9)$$
$$= 5 + (-13)$$
$$= -8$$

20. $-423.75 + (-84.50) + (-24.80) + 500$
$+ (-356.45) = -\$389.50$

21. $\dfrac{23.1 + 6.2 + 5.9 + 3.9 + 3.1 + 2.7 + 2.2 + 2.1 + 1.4 + 1.3}{10}$
$= 5.19$ million

22. Let $x = 2$ and $y = -3$.
$$-2\left|3 - 4xy^2\right| = -2\left|3 - 4(2)(-3)^2\right|$$
$$= -2\left|3 - 4(2)(9)\right|$$
$$= -2\left|3 - 72\right|$$
$$= -2\left|-69\right|$$
$$= -2(69)$$
$$= -138$$

23. Let $a = 16$ and $b = 4$.
$$\frac{a}{b} - \sqrt{a} + \sqrt{b} = \frac{16}{4} - \sqrt{16} + \sqrt{4}$$
$$= 4 - 4 + 2$$
$$= 2$$

24. $-7(3x+5) = -7 \cdot 3x + (-7) \cdot 5$
$$= -21x + (-35)$$
$$= -21x - 35$$

25. $\dfrac{2}{5}x + \dfrac{3}{4}y - 5x + 6y + 2.7$
$$= \frac{2}{5}x - 5x + \frac{3}{4}y + 6y + 2.7$$
$$= \frac{2}{5}x - \frac{25}{5}x + \frac{3}{4}y + \frac{24}{4}y + 2.7$$
$$= -\frac{23}{5}x + \frac{27}{4}y + 2.7$$

Chapter 2
Linear Equations and Inequalities in One Variable

Exercise Set 2.1

1.
$$2x - 2 = 10$$
$$2x - 2 + 2 = 10 + 2$$
$$2x = 12$$
$$\frac{2x}{2} = \frac{12}{2}$$
$$x = 6$$

Check: $2(6) - 2 \overset{?}{=} 10$

$$12 - 2 \overset{?}{=} 10$$
$$10 = 10$$

3.
$$9 - 3m = 12$$
$$9 - 9 - 3m = 12 - 9$$
$$-3m = 3$$
$$\frac{-3m}{-3} = \frac{3}{-3}$$
$$m = -1$$

Check: $9 - 3(-1) \overset{?}{=} 12$

$$9 - (-3) \overset{?}{=} 12$$
$$9 + 3 \overset{?}{=} 12$$
$$12 = 12$$

5.
$$2x + 2 = x + 3$$
$$2x - x + 2 = x - x + 3$$
$$x + 2 = 3$$
$$x + 2 - 2 = 3 - 2$$
$$x = 1$$

Check: $2(1) + 2 \overset{?}{=} 1 + 3$

$$2 + 2 \overset{?}{=} 4$$
$$4 = 4$$

7.
$$4t + 3 = 2t + 9$$
$$4t - 2t + 3 = 2t - 2t + 9$$
$$2t + 3 = 9$$
$$2t + 3 - 3 = 9 - 3$$
$$2t = 6$$
$$\frac{2t}{2} = \frac{6}{2}$$
$$t = 3$$

Check: $4(3) + 3 \overset{?}{=} 2(3) + 9$

$$12 + 3 \overset{?}{=} 6 + 9$$
$$15 = 15$$

9.
$$2(3z + 1) = 8$$
$$6z + 2 = 8$$
$$6z + 2 - 2 = 8 - 2$$
$$6z = 6$$
$$\frac{6z}{6} = \frac{6}{6}$$
$$z = 1$$

Check: $2(3(1) + 1) \overset{?}{=} 8$

$$2(3 + 1) \overset{?}{=} 8$$
$$2(4) \overset{?}{=} 8$$
$$8 = 8$$

11.
$$n + 2(3n + 1) = 9$$
$$n + 6n + 2 = 9$$
$$7n + 2 = 9$$
$$7n + 2 - 2 = 9 - 2$$
$$7n = 7$$
$$\frac{7n}{7} = \frac{7}{7}$$
$$n = 1$$

Check: $1 + 2(3(1) + 1) \overset{?}{=} 9$

$$1 + 2(3 + 1) \overset{?}{=} 9$$
$$1 + 2(4) \overset{?}{=} 9$$
$$1 + 8 \overset{?}{=} 9$$
$$9 = 9$$

13. $6u - 17 = 4(u + 3) - 3$

$6u - 17 = 4u + 12 - 3$

$6u - 17 = 4u + 9$

$6u - 4u - 17 = 4u - 4u + 9$

$2u - 17 = 9$

$2u - 17 + 17 = 9 + 17$

$2u = 26$

$\dfrac{2u}{2} = \dfrac{26}{2}$

$u = 13$

Check: $6(13) - 17 \overset{?}{=} 4(13 + 3) - 3$

$78 - 17 \overset{?}{=} 4(16) - 3$

$61 \overset{?}{=} 64 - 3$

$61 = 61$

15. $\dfrac{1}{2}z - 1 = 4z - 3 - 3z$

$\dfrac{1}{2}z - 1 = z - 3$

$2\left(\dfrac{1}{2}z - 1\right) = 2(z - 3)$

$z - 2 = 2z - 6$

$z - z - 2 = 2z - z - 6$

$-2 = z - 6$

$-2 + 6 = z - 6 + 6$

$4 = z$

Check: $\dfrac{1}{2}(4) - 1 \overset{?}{=} 4(4) - 3 - 3(4)$

$2 - 1 \overset{?}{=} 16 - 3 - 12$

$1 = 1$

17. Multiply both sides by the LCD, 12.

$\dfrac{1}{4}x + \dfrac{1}{3}x + \dfrac{5}{6} = \dfrac{15}{4}$

$12\left(\dfrac{1}{4}x + \dfrac{1}{3}x + \dfrac{5}{6}\right) = 12\left(\dfrac{15}{4}\right)$

$12 \cdot \dfrac{1}{4}x + 12 \cdot \dfrac{1}{3}x + 12 \cdot \dfrac{5}{6} = 12 \cdot \dfrac{15}{4}$

$3x + 4x + 10 = 45$

$7x + 10 = 45$

$7x + 10 - 10 = 45 - 10$

$7x = 35$

$\dfrac{7x}{7} = \dfrac{35}{7}$

$x = 5$

Check:

$\dfrac{1}{4}x + \dfrac{1}{3}x + \dfrac{5}{6} \overset{?}{=} \dfrac{15}{4}$

$\dfrac{1}{4}(5) + \dfrac{1}{3}(5) + \dfrac{5}{6} \overset{?}{=} \dfrac{15}{4}$

$\dfrac{5}{4} + \dfrac{5}{3} + \dfrac{5}{6} \overset{?}{=} \dfrac{15}{4}$

$\dfrac{15}{12} + \dfrac{20}{12} + \dfrac{10}{12} \overset{?}{=} \dfrac{15}{4}$

$\dfrac{45}{12} \overset{?}{=} \dfrac{15}{4}$

$\dfrac{15}{4} = \dfrac{15}{4}$

19. Multiply both sides by 100 to clear decimals.

$0.5x + 0.95 = 0.2x - 1$

$100(0.5x + 0.95) = 100(0.2x - 1)$

$100 \cdot 0.5x + 100 \cdot 0.95 = 100 \cdot 0.2x - 100 \cdot 1$

$50x + 95 = 20x - 100$

$50x - 20x + 95 = 20x - 20x - 100$

$30x + 95 = -100$

$30x + 95 - 95 = -100 - 95$

$30x = -195$

$\dfrac{30x}{30} = \dfrac{-195}{30}$

$x = -6.5$

Check:

$0.5(-6.5) + 0.95 \overset{?}{=} 0.2(-6.5) - 1$

$-3.25 + 0.95 \overset{?}{=} -1.3 - 1$

$-2.3 = -2.3$

21. Multiply both sides by 10 to clear decimals.

$$1.6x - 18 + 0.9x = 0.1x + 6$$

$$2.5x - 18 = 0.1x + 6$$

$$10(2.5x - 18) = 10(0.1x + 6)$$

$$25x - 180 = x + 60$$

$$25x - x - 180 = x - x + 60$$

$$24x - 180 = 60$$

$$24x - 180 + 180 = 60 + 180$$

$$24x = 240$$

$$\frac{24x}{24} = \frac{240}{24}$$

$$x = 10$$

Check:

$$1.6(10) - 18 + 0.9(10) \overset{?}{=} 0.1(10) + 6$$

$$16 - 18 + 9 \overset{?}{=} 1 + 6$$

$$7 = 7$$

23. $17a - 5 - 7a = 2a + 19$

$$10a - 5 = 2a + 19$$

$$10a - 2a - 5 = 2a - 2a + 19$$

$$8a - 5 = 19$$

$$8a - 5 + 5 = 19 + 5$$

$$8a = 24$$

$$\frac{8a}{8} = \frac{24}{8}$$

$$a = 3$$

Check:

$$17(3) - 5 - 7(3) \overset{?}{=} 2(3) + 19$$

$$51 - 5 - 21 \overset{?}{=} 6 + 19$$

$$25 = 25$$

25. $8r - 2(r + 5) = 5(2r - 6) - 8$

$$8r - 2r - 10 = 10r - 30 - 8$$

$$6r - 10 = 10r - 38$$

$$6r - 6r - 10 = 10r - 6r - 38$$

$$-10 = 4r - 38$$

$$-10 + 38 = 4r - 38 + 38$$

$$28 = 4r$$

$$\frac{28}{4} = \frac{4r}{4}$$

$$7 = r$$

Check:

$$8(7) - 2(7 + 5) \overset{?}{=} 5(2(7) - 6) - 8$$

$$56 - 2(12) \overset{?}{=} 5(14 - 6) - 8$$

$$56 - 24 \overset{?}{=} 5(8) - 8$$

$$32 \overset{?}{=} 40 - 8$$

$$32 = 32$$

27. $5 - 5(3x - 2) = 38 - 2(x - 8)$

$$5 - 15x + 10 = 38 - 2x + 16$$

$$15 - 15x = 54 - 2x$$

$$15 - 15x + 2x = 54 - 2x + 2x$$

$$15 - 13x = 54$$

$$15 - 15 - 13x = 54 - 15$$

$$-13x = 39$$

$$\frac{-13x}{-13} = \frac{39}{-13}$$

$$x = -3$$

Check:

$$5 - 5(3(-3) - 2) \overset{?}{=} 38 - 2(-3 - 8)$$

$$5 - 5(-9 - 2) \overset{?}{=} 38 - 2(-11)$$

$$5 - 5(-11) \overset{?}{=} 38 - (-22)$$

$$5 - (-55) \overset{?}{=} 38 + 22$$

$$5 + 55 \overset{?}{=} 60$$

$$60 = 60$$

29. $8h - 27 + 5(2h - 3) = 62 - (3h + 8)$

$$8h - 27 + 10h - 15 = 62 - 3h - 8$$

$$18h - 42 = 54 - 3h$$

$$18h + 3h - 42 = 54 - 3h + 3h$$

$$21h - 42 = 54$$

$$21h - 42 + 42 = 54 + 42$$

$$21h = 96$$

$$\frac{21h}{21} = \frac{96}{21}$$

$$h = \frac{32}{7}$$

Check:

$$8\left(\frac{32}{7}\right) - 27 + 5\left(2\left(\frac{32}{7}\right) - 3\right) \overset{?}{=} 62 - \left(3\left(\frac{32}{7}\right) + 8\right)$$

$$\frac{256}{7} - 27 + 5\left(\frac{64}{7} - 3\right) \overset{?}{=} 62 - \left(\frac{96}{7} + 8\right)$$

$$\frac{256}{7} - 27 + 5\left(\frac{43}{7}\right) \overset{?}{=} 62 - \frac{152}{7}$$

$$\frac{256}{7} - 27 + \frac{215}{7} \overset{?}{=} 62 - \frac{152}{7}$$

$$\frac{282}{7} = \frac{282}{7}$$

31. Multiply both sides by the LCD, 15.

$$\frac{2}{3}q - 4 = \frac{1}{5}q + 10$$

$$15\left(\frac{2}{3}q - 4\right) = 15\left(\frac{1}{5}q + 10\right)$$

$$10q - 60 = 3q + 150$$

$$10q - 3q - 60 = 3q - 3q + 150$$

$$7q - 60 = 150$$

$$7q - 60 + 60 = 150 + 60$$

$$7q = 210$$

$$\frac{7q}{7} = \frac{210}{7}$$

$$q = 30$$

Check:

$$\frac{2}{3}(30) - 4 \overset{?}{=} \frac{1}{5}(30) + 10$$

$$20 - 4 \overset{?}{=} 6 + 10$$

$$16 = 16$$

33. Multiply both sides by the LCD, 35.

$$\frac{2}{7}(x - 5) = -2 + \frac{2}{5}x$$

$$35\left(\frac{2}{7}(x - 5)\right) = 35\left(-2 + \frac{2}{5}x\right)$$

$$10(x - 5) = -70 + 14x$$

$$10x - 50 = -70 + 14x$$

$$10x - 14x - 50 = -70 + 14x - 14x$$

$$-4x - 50 = -70$$

$$-4x - 50 + 50 = -70 + 50$$

$$-4x = -20$$

$$\frac{-4x}{-4} = \frac{-20}{-4}$$

$$x = 5$$

Check:

$$\frac{2}{7}(5 - 5) \overset{?}{=} -2 + \frac{2}{5}(5)$$

$$\frac{2}{7}(0) \overset{?}{=} -2 + 2$$

$$0 = 0$$

35. Multiply both sides by 100 to clear decimals.

$$0.5(x - 2) + 1.76 = 0.3x + 0.8$$

$$0.5x - 1 + 1.76 = 0.3x + 0.8$$

$$100(0.5x + 0.76) = 100(0.3x + 0.8)$$

$$50x + 76 = 30x + 80$$

$$50x - 30x + 76 = 30x - 30x + 80$$

$$20x + 76 = 80$$

$$20x + 76 - 76 = 80 - 76$$

$$20x = 4$$

$$\frac{20x}{20} = \frac{4}{20}$$

$$x = 0.2$$

Check:

$$0.5(0.2 - 2) + 1.76 \overset{?}{=} 0.3(0.2) + 0.8$$

$$0.5(-1.8) + 1.76 \overset{?}{=} 0.06 + 0.8$$

$$-0.9 + 1.76 \overset{?}{=} 0.06 + 0.8$$

$$0.86 = 0.86$$

37. Multiply both sides by 100 to clear decimals.

$$2x - 3.24 + 2.4x = 6.2 + 0.08x + 3.52$$
$$4.4x - 3.24 = 0.08x + 9.72$$
$$100(4.4x - 3.24) = 100(0.08x + 9.72)$$
$$440x - 324 = 8x + 972$$
$$440x - 8x - 324 = 8x - 8x + 972$$
$$432x - 324 = 972$$
$$432x - 324 + 324 = 972 + 324$$
$$432x = 1296$$
$$\frac{432x}{432} = \frac{1296}{432}$$
$$x = 3$$

Check:

$$2(3) - 3.24 + 2.4(3) \overset{?}{=} 6.2 + 0.08(3) + 3.52$$
$$6 - 3.24 + 7.2 \overset{?}{=} 6.2 + 0.24 + 3.52$$
$$9.96 = 9.96$$

39. $$6(m+3) - 5 + 2m = 3(3m+1) - m$$
$$6m + 18 - 5 + 2m = 9m + 3 - m$$
$$8m + 13 = 8m + 3$$
$$8m - 8m + 13 = 8m - 8m + 3$$
$$13 \neq 3$$

    no solution

41. $$7(n+2) - 3n = 4 + 4n + 10$$
$$7n + 14 - 3n = 14 + 4n$$
$$4n + 14 = 14 + 4n$$
$$4n - 4n + 14 = 14 + 4n - 4n$$
$$14 = 14$$

    all real numbers

43. Mistake: The distributive property was not used correctly.
Correct:

$$2(x+3) = 7x - 1$$
$$2x + 6 = 7x - 1$$
$$2x - 2x + 6 = 7x - 2x - 1$$
$$6 = 5x - 1$$
$$6 + 1 = 5x - 1 + 1$$
$$7 = 5x$$
$$\frac{7}{5} = \frac{5x}{5}$$
$$\frac{7}{5} = x$$

45. Mistake: Subtracted before distributing into the parentheses.
Correct: $$4 - 5(x+1) + 4x = 7 - 11$$
$$4 - 5x - 5 + 4x = -4$$
$$-1 - x = -4$$
$$-1 + 1 - x = -4 + 1$$
$$-x = -3$$
$$x = 3$$

47. Mistake: Did not distribute the $-$ with the 2.
Correct:

$$-2(x+4) + 6x = 2x + 6$$
$$-2x - 8 + 6x = 2x + 6$$
$$4x - 8 = 2x + 6$$
$$4x - 2x - 8 = 2x - 2x + 6$$
$$2x - 8 = 6$$
$$2x - 8 + 8 = 6 + 8$$
$$2x = 14$$
$$\frac{2x}{2} = \frac{14}{2}$$
$$x = 7$$

49. $$P = R - C$$
$$P + C = R - C + C$$
$$P + C = R$$
$$P - P + C = R - P$$
$$C = R - P$$

51. $$A = bh$$
$$\frac{A}{h} = \frac{bh}{h}$$
$$\frac{A}{h} = b$$

53. $$A = 2\pi pw$$
$$\frac{A}{2\pi w} = \frac{2\pi pw}{2\pi w}$$
$$\frac{A}{2\pi w} = p$$

55. $$A = \frac{1}{2}\theta r^2$$

$$2 \cdot A = 2 \cdot \frac{1}{2}\theta r^2$$

$$2A = \theta r^2$$

$$\frac{2A}{r^2} = \frac{\theta r^2}{r^2}$$

$$\frac{2A}{r^2} = \theta$$

57. $$F = \frac{kMm}{d^2}$$

$$d^2 \cdot F = d^2 \cdot \frac{kMm}{d^2}$$

$$Fd^2 = kMm$$

$$\frac{Fd^2}{km} = \frac{kMm}{km}$$

$$\frac{Fd^2}{km} = M$$

59. $$A = \pi s (R + r)$$

$$\frac{A}{\pi(R+r)} = \frac{\pi s(R+r)}{\pi(R+r)}$$

$$\frac{A}{\pi(R+r)} = s$$

61. $$P = 2l + 2w$$

$$P - 2w = 2l + 2w - 2w$$

$$P - 2w = 2l$$

$$\frac{P-2w}{2} = \frac{2l}{2}$$

$$\frac{P-2w}{2} = l$$

63. $$3x + 2y = 6$$

$$3x - 3x + 2y = 6 - 3x$$

$$2y = 6 - 3x$$

$$\frac{2y}{2} = \frac{6-3x}{2}$$

$$y = \frac{6-3x}{2}$$

65. $$F = \frac{9}{5}C + 32$$

$$F - 32 = \frac{9}{5}C + 32 - 32$$

$$F - 32 = \frac{9}{5}C$$

$$\frac{5}{9}(F - 32) = \frac{5}{9} \cdot \frac{9}{5}C$$

$$\frac{5}{9}(F - 32) = C$$

67. $$x = vt + \frac{1}{2}at^2$$

$$x - vt = vt - vt + \frac{1}{2}at^2$$

$$x - vt = \frac{1}{2}at^2$$

$$2(x - vt) = 2 \cdot \frac{1}{2}at^2$$

$$2(x - vt) = at^2$$

$$\frac{2(x-vt)}{t^2} = \frac{at^2}{t^2}$$

$$\frac{2(x-vt)}{t^2} = a$$

69. Mistake: Subtracted lw instead of dividing lw.

Correct: $$V = lwh$$

$$\frac{V}{lw} = \frac{lwh}{lw}$$

$$\frac{V}{lw} = h$$

71. Mistake: Subtracted the coefficient of l instead of dividing by the coefficient.

Correct: $$P = 2l + 2w$$

$$P - 2w = 2l + 2w - 2w$$

$$P - 2w = 2l$$

$$\frac{P-2w}{2} = \frac{2l}{2}$$

$$\frac{P-2w}{2} = l$$

Review Exercises

1. $\{1, 3, 5, 7, 9, 11, 13\}$

2. $14 - 12\left[6 - 8\left(3 + 2^5\right)\right] + \sqrt{9 \cdot 16}$

    $= 14 - 12\left[6 - 8\left(3 + 32\right)\right] + \sqrt{144}$

    $= 14 - 12\left[6 - 8 \cdot 35\right] + 12$

    $= 14 - 12\left[6 - 280\right] + 12$

    $= 14 - 12\left[-274\right] + 12$

    $= 14 + 3288 + 12$

    $= 3314$

3. $7n - 9$

4. $-3\left(n + 8\right)$

5. $-2x - 6y + 5$

6. $-54m + 24$

Exercise Set 2.2

1. $F = \dfrac{9}{5}C + 32$

    $-109.3 = \dfrac{9}{5}C + 32$

    $-141.3 = \dfrac{9}{5}C$

    $\dfrac{5}{9}(-141.3) = \dfrac{5}{9} \cdot \dfrac{9}{5}C$

    $-78.5^{\circ} = C$

3. Find the height of the box.

    $A = 2lw + 2lh + 2wh$

    $2280 = 2 \cdot 24 \cdot 15 + 2 \cdot 24 \cdot h + 2 \cdot 15 \cdot h$

    $1140 = 360 + (24 + 15)h$

    $780 = 39h$

    $20 = h$

    The height of the box is 20 in.

5. Find the length of the side labeled a.

    $A = \dfrac{1}{2}h\left(a + b\right)$

    $370 = \dfrac{1}{2} \cdot 20\left(a + 25\right)$

    $370 = 10\left(a + 25\right)$

    $37 = a + 25$

    $12 = a$

    The length of the side is 12 cm.

7. Find the dimensions of the room.

    $P = 2l + 2w$

    $84 = 2(w + 6) + 2w$

    $42 = w + 6 + w$

    $42 = 2w + 6$

    $36 = 2w$

    $18 = w$

    $l = w + 6$

    $l = 18 + 6$

    $l = 24$

    The length of the room is 24 ft. and the width of the room is 18 ft.

9. Find the area of the wall: $A = 10 \cdot 22$

    $A = 220 \text{ ft.}^2$

    Find the area of one window: $A = 5 \cdot 4$

    $A = 20 \text{ ft.}^2$

    Subtract the area of the 2 windows from the area of the wall to find the area covered by paint.

    $220 - 2(20) = 220 - 40 = 180 \text{ ft.}^2$

11. Find the area of the smaller box:

    $A = \left(20\right) \cdot \left(x + 10 - x\right)$

    $A = 20 \cdot 10$

    $A = 200$

    Find the area of the larger box:

    $A = x \cdot \left(32 + 20\right)$

    $A = 52x$

    Add the two boxes together and set equal to the given area.

    $2020 = 200 + 52x$

    $1820 = 52x$

    $35 = x$

    $x = 35\text{ft.}, x + 10 = 45\text{ft.}$

13. Set the expressions equal to each other.

    $50n - 1200 = 24n + 750$

    $26n = 1950$

    $n = 75$

    75 units must be sold to break even.

15. $B = P + Prt$

$$3330 = P + P(0.055)(2)$$
$$3330 = P + 0.11P$$
$$3330 = 1.11P$$
$$\$3000 = P$$

17. Use the formula $F = ma$ and solve for m.

$$F = ma$$
$$-135 = m \cdot -32.2$$
$$4.2 \approx m$$

The woman's mass rounded to the nearest tenth of a slug is 4.2 slugs.

19. Use the formula $F = ma$ and solve for m.

$$F = ma$$
$$-4512.9 = m \cdot -9.8$$
$$460.5 = m$$

The mass the man lifted rounded to the nearest tenth of a kilogram is 460.5 kg.

21. Use the formula $d = v_i t + \dfrac{1}{2}at^2$ and solve for v_i.

$$d = v_i t + \frac{1}{2}at^2$$
$$260 = v_i \cdot 4 + \frac{1}{2} \cdot 25 \cdot 4^2$$
$$60 = 4v_i + 200$$
$$60 = 4v_i$$
$$15 = v_i$$

The object's initial velocity is 15 m/sec.

23. Use the formula $d = v_i t + \dfrac{1}{2}at^2$ and solve for a.

$$d = v_i t + \frac{1}{2}at^2$$
$$525 = 50 \cdot 6 + \frac{1}{2} \cdot a \cdot 6^2$$
$$525 = 300 + 18a$$
$$225 = 18a$$
$$12.5 = a$$

The car's acceleration is 12.5 ft./sec.^{2}.

25. Let x be the number. Translate to an equation and solve for x.

$$5x - 3 = 37$$
$$5x - 3 + 3 = 37 + 3$$
$$5x = 40$$
$$\frac{5x}{5} = \frac{40}{5}$$
$$x = 8$$

27. Let y be the number. Translate to an equation and solve for y.

$$5y + 8 = 3y - 10$$
$$2y = -18$$
$$y = -9$$

29. Let x be the number. Translate to an equation and solve for x.

$$3(x + 4) = -6$$
$$3x + 12 = -6$$
$$3x + 12 - 12 = -6 - 12$$
$$3x = -18$$
$$\frac{3x}{3} = \frac{-18}{3}$$
$$x = -6$$

31. Let x be the number. Translate to an equation and solve for x.

$$5(x - 2) - 8 = 2x$$
$$5x - 10 - 8 = 2x$$
$$5x - 18 = 2x$$
$$5x - 5x - 18 = 2x - 5x$$
$$-18 = -3x$$
$$\frac{-18}{-3} = \frac{-3x}{-3}$$
$$6 = x$$

33. Let p represent the wholesale price. Because $135.68 is the retail price, it is the result of adding 25% of the wholesale price to the wholesale price. Translate to an equation and solve for p.

$$0.25p + p = 135.68$$
$$1.25p = 135.68$$
$$\frac{1.25p}{1.25} = \frac{135.68}{1.25}$$
$$p \approx 108.54$$

The wholesale price is approximately $108.54.

35. Let p represent the original price. Because \$699.95 is the discounted price, it is the result of subtracting 30% of the original price from the original price. Translate to an equation and solve for p.

$$p - 0.30p = 699.95$$
$$0.70p = 699.95$$
$$\frac{0.70p}{0.70} = \frac{699.95}{0.70}$$
$$p \approx 999.93$$

The original price is approximately \$999.93.

37. Let x represent the number of units sold in the previous year. Because the 45,000 units sold this year is a 20% increase over the previous year, it is the result of adding 20% of x to x. Translate to an equation and solve.

$$0.20x + x = 45,000$$
$$1.20x = 45,000$$
$$\frac{1.20x}{1.20} = \frac{45,000}{1.20}$$
$$x = 37,500$$

The number of units sold the previous year was 37,500 units.

39. Create a table.

Categories	Rate	Time	Distance
car 1	65	t	$65t$
car 2	50	t	$50t$

Translate the information in the table to an equation and solve.

$$65t + 50t = 230$$
$$115t = 230$$
$$\frac{115t}{115} = \frac{230}{115}$$
$$t = 2$$

The cars will meet in 2 hours.

41. Create a table.

Categories	Rate	Time	Distance
northbound	9	x	$9x$
southbound	15	x	$15x$

Translate the information in the table to an equation and solve.

$$9x + 15x = 42$$
$$24x = 42$$
$$\frac{24x}{24} = \frac{42}{24}$$
$$x = 1.75$$

The boats will be 42 miles apart in 1.75 hours.

Review Exercises

1. $-5\frac{3}{8}, -\frac{1}{6}, 0.02, 4.5\%, \sqrt{48}, |-15.8|$

2. 2 is the smallest prime number.

3. $-|5.8| = -|-5.8|$

4. $-(-6) > -(-(-8))$

5.
$$6x - 19 = 4x - 31$$
$$6x - 4x - 19 = 4x - 4x - 31$$
$$2x - 19 = -31$$
$$2x - 19 + 19 = -31 + 19$$
$$2x = -12$$
$$\frac{2x}{2} = \frac{-12}{2}$$
$$x = -6$$

6.
$$-\frac{4}{9}n = \frac{5}{6}$$
$$-\frac{9}{4} \cdot -\frac{4}{9}n = -\frac{9}{4} \cdot \frac{5}{6}$$
$$n = -\frac{15}{8}$$

Exercise Set 2.3

1. a) $\{x | x \geq 5\}$ b) $[5, \infty)$

    c)

3. a) $\{q | q < -1\}$ b) $(-\infty, -1)$

    c)

5. a) $\left\{ p \,\middle|\, p < \frac{1}{5} \right\}$ b) $\left(-\infty, \frac{1}{5} \right)$

    c)

7. a) $\{r | r \leq 1.9\}$ b) $(-\infty, 1.9]$

    c)

9. $r - 6 < -12$

     $r - 6 + 6 < -12 + 6$

         $r < -6$

a) $\{r \mid r < -6\}$ b) $(-\infty, -6)$

c)

           $-10\,-9\,-8\,-7\,-6\,-5\,-4\,-3\,-2\,-1\;0\;\;1\;\;2$

11. $-4y \le -16$

    $\dfrac{-4y}{-4} \ge \dfrac{-16}{-4}$

      $y \ge 4$

a) $\{y \mid y \ge 4\}$ b) $[4, \infty)$

c)

         $-2\,-1\;0\;\;1\;\;2\;\;3\;\;4\;\;5\;\;6\;\;7\;\;8\;\;9\;\;10$

13. $3p + 9 < 21$

     $3p + 9 - 9 < 21 - 9$

         $3p < 12$

         $\dfrac{3p}{3} < \dfrac{12}{3}$

          $p < 4$

a) $\{p \mid p < 4\}$ b) $(-\infty, 4)$

c)

         $-2\,-1\;0\;\;1\;\;2\;\;3\;\;4\;\;5\;\;6\;\;7\;\;8\;\;9\;\;10$

15. $6 - 5x > 29$

    $6 - 6 - 5x > 29 - 6$

        $-5x > 23$

       $\dfrac{-5x}{-5} < \dfrac{23}{-5}$

         $x < -\dfrac{23}{5}$

a) $\left\{x \mid x < -\dfrac{23}{5}\right\}$ b) $\left(-\infty, -\dfrac{23}{5}\right)$

c)

             $-\frac{23}{5}$

        -6 -5 -4 -3

17. $\dfrac{a}{8} + 1 < \dfrac{3}{8}$

    $8 \cdot \left(\dfrac{a}{8} + 1\right) < 8 \cdot \left(\dfrac{3}{8}\right)$

        $a + 8 < 3$

    $a + 8 - 8 < 3 - 8$

         $a < -5$

a) $\{a \mid a < -5\}$ b) $(-\infty, -5)$

c)

          $-10\,-9\,-8\,-7\,-6\,-5\,-4\,-3\,-2\,-1\;0\;\;1\;\;2$

19. $7 + 2x < -2 + x$

    $7 + 2x - x < -2 + x - x$

       $7 + x < -2$

    $7 - 7 + x < -2 - 7$

         $x < -9$

a) $\{x \mid x < -9\}$ b) $(-\infty, -9)$

c)

       $-11\;\;-10\;\;-9\;\;-8\;\;-7\;\;-6$

21. $-11k - 8 > -16 - 9k$

   $-11k + 9k - 8 > -16 - 9k + 9k$

     $-2k - 8 > -16$

    $-2k - 8 + 8 > -16 + 8$

       $-2k > -8$

       $\dfrac{-2k}{-2} < \dfrac{-8}{-2}$

         $k < 4$

a) $\{k \mid k < 4\}$ b) $(-\infty, 4)$

c)

        $-3\,-2\,-1\;0\;\;1\;\;2\;\;3\;\;4\;\;5\;\;6\;\;7\;\;8\;\;9$

23. $2(3w - 4) - 5 \le 17$

      $6w - 8 - 5 \le 17$

       $6w - 13 \le 17$

   $6w - 13 + 13 \le 17 + 13$

        $6w \le 30$

        $\dfrac{6w}{6} \le \dfrac{30}{6}$

         $w \le 5$

a) $\{w \mid w \le 5\}$ b) $(-\infty, 5]$

c)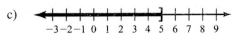

        $-3\,-2\,-1\;0\;\;1\;\;2\;\;3\;\;4\;\;5\;\;6\;\;7\;\;8\;\;9$

25. $4(y+2)+5 > 3(2y-1)$

$4y+8+5 > 6y-3$

$4y+13 > 6y-3$

$4y-6y+13 > 6y-6y-3$

$-2y+13 > -3$

$-2y+13-13 > -3-13$

$-2y > -16$

$\dfrac{-2y}{-2} < \dfrac{-16}{-2}$

$y < 8$

a) $\{y \mid y < 8\}$		b) $(-\infty, 8)$

c)

```
<---|---|---|---|---|---|---|---|---|---|---|-)-|--->
   -3 -2 -1  0  1  2  3  4  5  6  7  8  9
```

27. $\dfrac{1}{6}(5x+1) < -\dfrac{1}{6}x - \dfrac{5}{3}$

$6 \cdot \dfrac{1}{6}(5x+1) < 6 \cdot -\dfrac{1}{6}x - 6 \cdot \dfrac{5}{3}$

$5x+1 < -x-10$

$5x+x+1 < -x+x-10$

$6x+1 < -10$

$6x+1-1 < -10-1$

$6x < -11$

$\dfrac{6x}{6} < \dfrac{-11}{6}$

$x < -\dfrac{11}{6}$

a) $\left\{x \mid x < -\dfrac{11}{6}\right\}$		b) $\left(-\infty, -\dfrac{11}{6}\right)$

c)

```
         -11/6
<--------------)---|---|---|---------|--->
   -3        -2       -1         0
```

29. $\dfrac{1}{6}(7m+2) - \dfrac{1}{6}(11m-7) \geq 0$

$6 \cdot \dfrac{1}{6}(7m+2) - 6 \cdot \dfrac{1}{6}(11m-7) \geq 6 \cdot 0$

$7m+2 - (11m-7) \geq 0$

$7m+2-11m+7 \geq 0$

$-4m+9 \geq 0$

$-4m+9-9 \geq 0-9$

$-4m \geq -9$

$\dfrac{-4m}{-4} \leq \dfrac{-9}{-4}$

$m \leq \dfrac{9}{4}$

a) $\left\{m \mid m \leq \dfrac{9}{4}\right\}$		b) $\left(-\infty, \dfrac{9}{4}\right]$

c)

```
                  9/4
<---|---|---[---|---|--->
    0   1   2   3   4
```

31. $0.7x-0.3 \leq 0.8x+0.7$

$0.7x-0.8x-0.3 \leq 0.8x-0.8x+0.7$

$-0.1x-0.3 \leq 0.7$

$-0.1x-0.3+0.3 \leq 0.7+0.3$

$-0.1x \leq 1$

$\dfrac{-0.1x}{-0.1} \geq \dfrac{1}{-0.1}$

$x \geq -10$

a) $\{x \mid x \geq -10\}$		b) $[-10, \infty)$

c)

```
<---|---[---|---|---|---|---|---|---|---|---|---|--->
  -12 -11 -10 -9 -8 -7 -6 -5 -4 -3 -2 -1  0
```

33. $0.09z+20.34 < 3(1.4z-1.5)+2.1z$

$0.09z+20.34 < 4.2z-4.5+2.1z$

$0.09z+20.34 < 6.3z-4.5$

$0.09z-6.3z+20.34 < 6.3z-6.3z-4.5$

$-6.21z+20.34 < -4.5$

$-6.21z+20.34-20.34 < -4.5-20.34$

$-6.21z < -24.84$

$\dfrac{-6.21z}{-6.21} > \dfrac{-24.84}{-6.21}$

$z > 4$

a) $\{z \mid z > 4\}$		b) $(4, \infty)$

c)

35. $3(x+2)-4 < x+5+2x$

$$3x+6-4 < 3x+5$$
$$3x+2 < 3x+5$$
$$3x-3x+2 < 3x-3x+5$$
$$2 < 5$$

Because there are no variable terms left in the inequality and the inequality is true, every real number is a solution for the original inequality.

a) $\{x \mid x \text{ is a real number}\}$

b) $(-\infty, \infty)$

c)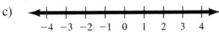

37. $4b-3(b+5) \geq 2(2b+3)-3(b+5)$

$$4b-3b-15 \geq 4b+6-3b-15$$
$$b-15 \geq b-9$$
$$b-b-15 \geq b-b-9$$
$$-15 \geq -9$$

Because there are no variable terms left in the inequality and the inequality is false, there are no solutions for the original inequality.

a) $\{\ \}$ or $\varnothing$

b) No interval notation

c) ![number line from -4 to 4]

39. $\dfrac{3}{4}x < -6$

$$\frac{4}{3} \cdot \frac{3}{4}x < \frac{4}{3} \cdot (-6)$$
$$x < -8$$

41. $5x-1 > 14$

$$5x-1+1 > 14+1$$
$$5x > 15$$
$$\frac{5x}{5} > \frac{15}{5}$$
$$x > 3$$

43. $1-4x \leq 25$

$$1-1-4x \leq 25-1$$
$$-4x \leq 24$$
$$\frac{-4x}{-4} \geq \frac{24}{-4}$$
$$x \geq -6$$

45. $6+2(x-5) \leq 12$

$$6+2x-10 \leq 12$$
$$2x-4 \leq 12$$
$$2x-4+4 \leq 12+4$$
$$2x \leq 16$$
$$\frac{2x}{2} \leq \frac{16}{2}$$
$$x \leq 8$$

47. Let x represent Yvonne's score on the fifth test.

$$\frac{92+100+83+81+x}{5} \geq 90$$
$$\frac{356+x}{5} \geq 90$$
$$356+x \geq 450$$
$$356-356+x \geq 450-356$$
$$x \geq 94$$

Yvonne must score 94 or higher.

49. Let x represent Blake's score on his fourth round.

$$\frac{86+82+88+x}{4} \leq 84$$
$$\frac{256+x}{4} \leq 84$$
$$256+x \leq 336$$
$$256-256+x \leq 336-256$$
$$x \leq 80$$

Blake must score 80 or less.

51. Let w represent the width of the bedroom.

$$18w \leq 234$$
$$\frac{18w}{18} \leq \frac{234}{18}$$
$$w \leq 13$$

The width must be 13 ft. or less.

53. Let r represent the radius of the pot.

$$150 \geq 2\pi r$$
$$150 \geq 2(3.14)r$$
$$150 \geq 6.28r$$
$$\frac{150}{6.28} \geq \frac{6.28r}{6.28}$$
$$23.9 \geq r$$

The radius must be approximately 23.9 in. or less.

55. Let r represent Juan's average rate.

$$180 \leq r(3)$$
$$\frac{180}{3} \leq \frac{3r}{3}$$
$$60 \leq r$$

He must drive at least 60 mph.

57. To find profit, we subtract costs from revenue.

$$(12.5n + 3000) - (8n + 21,000) \geq 0$$
$$12.5n + 3000 - 8n - 21,000 \geq 0$$
$$4.5n - 18,000 \geq 0$$
$$4.5n - 18,000 + 18,000 \geq 18,000$$
$$\frac{4.5n}{4.5} \geq \frac{18,000}{4.5}$$
$$n \geq 4000$$

4000 or more lamps must be sold to break even or make a profit.

59. $F = \frac{9}{5}C + 32$

$$F = \frac{9}{5}(1064.58) + 32$$
$$F = 1948.244$$

a) $t < 1948.244°F$ b) $t \geq 1948.244°F$

61. current × resistance = voltage

$$x \cdot 6 \leq 15$$
$$\frac{6x}{6} \leq \frac{15}{6}$$
$$x \leq 2.5$$

The current can be 2.5 amps or less

Review Exercises

1. Yes, because each element of A is a member of B.

2. Commutative property of addition

3. $0.5r^2 - t = 0.5(-6)^2 - 8$
$$= 0.5(36) - 8$$
$$= 18 - 8$$
$$= 10$$

4. $\left| x^3 - y \right| = \left| (-4)^3 - (-19) \right|$
$$= \left| -64 + 19 \right|$$
$$= \left| -45 \right|$$
$$= 45$$

5.
$$6x + 9 = 13 + 4(3x + 2)$$
$$6x + 9 = 13 + 12x + 8$$
$$6x + 9 = 21 + 12x$$
$$6x + 9 - 21 = 21 - 21 + 12x$$
$$6x - 12 = 12x$$
$$6x - 6x - 12 = 12x - 6x$$
$$-12 = 6x$$
$$\frac{-12}{6} = \frac{6x}{6}$$
$$-2 = x$$

6.
$$\frac{3}{4}n - 1 = \frac{1}{5}(2n + 3)$$
$$20\left(\frac{3}{4}n - 1\right) = 20 \cdot \frac{1}{5}(2n + 3)$$
$$15n - 20 = 4(2n + 3)$$
$$15n - 20 = 8n + 12$$
$$15n - 8n - 20 = 8n - 8n + 12$$
$$7n - 20 = 12$$
$$7n - 20 + 20 = 12 + 20$$
$$7n = 32$$
$$n = \frac{32}{7}$$

Exercise Set 2.4

1. $\{1, 3, 5\}$ 3. $\{7, 8\}$

5. $-4 < x < 5$ 7. $0 < y \leq 2$

9. $-7 < w < 3$ 11. $0 \leq u \leq 2$

13. $2 < x < 7$

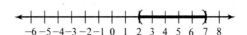

15. $-1 < x \le 5$

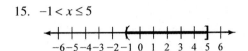

17. $1 \le x \le 10$

19. $-3 \le x < 4$

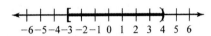

21. $-3 < x < -1$

a)

b) $\left\{x \mid -3 < x < -1\right\}$ c) $(-3, -1)$

23. $x + 2 > 5$ and $x - 4 \le 2$

    $x > 3$ $x \le 6$

a)

b) $\left\{x \mid 3 < x \le 6\right\}$ c) $(3, 6]$

25. $-x > 1$ and $-2x \le -10$

    $x < -1$ $x \ge 5$

a)

b) $\varnothing$ c) no interval notation

27. $3x + 8 \ge -7$ and $4x - 7 < 5$

    $3x \ge -15$ $4x < 12$

    $x \ge -5$ $x < 3$

a)

b) $\left\{x \mid -5 \le x < 3\right\}$ c) $[-5, 3)$

29. $-3x - 8 > 1$ and $-4x + 5 \le -3$

    $-3x > 9$ $-4x \le -8$

    $x < -3$ $x \ge 2$

a)

b) $\varnothing$ c) no interval notation

31. $-3 < x + 4 < 1$

    $-7 < x < -3$

a)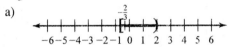

b) $\left\{x \mid -7 < x < -3\right\}$ c) $(-7, -3)$

33. $-7 \le 4x - 3 \le 5$

    $-4 \le 4x \le 8$

    $-1 \le x \le 2$

a)

b) $\left\{x \mid -1 \le x \le 2\right\}$ c) $[-1, 2]$

35. $0 \le 2 + 3x < 8$

    $-2 \le 3x < 6$

    $-\dfrac{2}{3} \le x < 2$

a)

b) $\left\{x \mid -\dfrac{2}{3} \le x < 2\right\}$ c) $\left[-\dfrac{2}{3}, 2\right)$

37. $-1 < -2x + 5 \le 5$

    $-6 < -2x \le 0$

    $3 > x \ge 0$

a)

b) $\left\{x \mid 0 \le x < 3\right\}$ c) $[0, 3)$

39. $3 \le 6 - x \le 6$

    $-3 \le -x \le 0$

    $3 \ge x \ge 0$

a)

b) $\left\{x \mid 0 \le x \le 3\right\}$ c) $[0, 3]$

41. $\left\{a, c, d, g, o, t\right\}$ 43. $\left\{w, x, y, z\right\}$

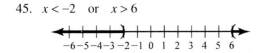

45. $x < -2$ or $x > 6$

47. $y < -3$ or $y \geq 0$

49. $w > -3$ or $w > 2$

51. $u \geq 0$ or $u \leq 2$

53. $y + 2 < -7$ or $y + 2 > 7$

$y < -9$ $y > 5$

a)

b) $\{y \mid y < -9 \text{ or } y > 5\}$

c) $(-\infty, -9) \cup (5, \infty)$

55. $4r - 3 < -11$ or $2r - 3 > -1$

$4r < -8$ $2r > 2$

$r < -2$ $r > 1$

a)

b) $\{r \mid r < -2 \text{ or } r > 1\}$

c) $(-\infty, -2) \cup (1, \infty)$

57. $-w + 2 \leq -5$ or $-w + 2 \geq 3$

$-w \leq -7$ $-w \geq 1$

$w \geq 7$ $w \leq -1$

a)

b) $\{w \mid w \leq -1 \text{ or } w \geq 7\}$

c) $(-\infty, -1] \cup [7, \infty)$

59. $7 - 4k \leq -5$ or $6 - 2k \geq 2$

$-4k \leq -12$ $-2k \geq -4$

$k \geq 3$ $k \leq 2$

a)

b) $\{k \mid k \leq 2 \text{ or } k \geq 3\}$

c) $(-\infty, 2] \cup [3, \infty)$

61. $\dfrac{2}{3}x - 4 \geq -2$ or $\dfrac{2}{5}x - 2 \leq -4$

$\dfrac{2}{3}x \geq 2$ $\dfrac{2}{5}x \leq -2$

$2x \geq 6$ $2x \leq -10$

$x \geq 3$ $x \leq -5$

a)

b) $\{x \mid x \leq -5 \text{ or } x \geq 3\}$

c) $(-\infty, -5] \cup [3, \infty)$

63. $-2(c - 1) < -4$ or $-2(c - 2) < -6$

$-2c + 2 < -4$ $-2c + 4 < -6$

$-2c < -6$ $-2c < -10$

$c > 3$ $c > 5$

a)

b) $\{c \mid c > 3\}$ c) $(3, \infty)$

65. $2(x + 3) + 3 \leq 1$ or $2(x + 1) + 7 \leq -3$

$2x + 6 + 3 \leq 1$ $2x + 2 + 7 \leq -3$

$2x + 9 \leq 1$ $2x + 9 \leq -3$

$2x \leq -8$ $2x \leq -12$

$x \leq -4$ $x \leq -6$

a)

b) $\{x \mid x \leq -4\}$

c) $(-\infty, -4]$

67. $-3(x - 1) - 1 \leq -1$ or $-3(x + 1) + 5 \geq -2$

$-3x + 3 - 1 \leq -1$ $-3x - 3 + 5 \geq -2$

$-3x + 2 \leq -1$ $-3x + 2 \geq -2$

$-3x \leq -3$ $-3x \geq -4$

$x \geq 1$ $x \leq \dfrac{4}{3}$

a)

b) $\{x \mid x \text{ is a real number}\}$ or $\mathbb{R}$

c) $(-\infty, \infty)$

69. $-3 < x+4$ and $x+4 < 7$

  $-7 < x$ $x < 3$

  a)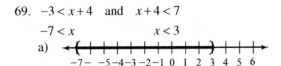
  $\begin{array}{c} -7- \;\; -5\,-4\,-3\,-2\,-1 \;\; 0 \;\; 1 \;\; 2 \;\; 3 \;\; 4 \;\; 5 \;\; 6 \end{array}$

  b) $\{x|-7 < x < 3\}$ c) $(-7, 3)$

71. $4x+3 < -5$ or $3x+2 > 8$

  $4x < -8$ $3x > 6$

  $x < -2$ $x > 2$

  a)
  $\begin{array}{c} -6\,-5\,-4\,-3\,-2\,-1 \;\; 0 \;\; 1 \;\; 2 \;\; 3 \;\; 4 \;\; 5 \;\; 6 \end{array}$

  b) $\{x|x < -2 \text{ or } x > 2\}$

  c) $(-\infty, -2)\cup(2, \infty)$

73. $4 < -2x < 6$

  $-2 > x > -3$

  a)
  $\begin{array}{c} -6\,-5\,-4\,-3\,-2\,-1 \;\; 0 \;\; 1 \;\; 2 \;\; 3 \;\; 4 \;\; 5 \;\; 6 \end{array}$

  b) $\{x|-3 < x < -2\}$ c) $(-3, -2)$

75. $7 \le 4x-5 \le 19$

  $12 \le 4x \le 24$

  $3 \le x \le 6$

  a)
  $\begin{array}{c} -6\,-5\,-4\,-3\,-2\,-1 \;\; 0 \;\; 1 \;\; 2 \;\; 3 \;\; 4 \;\; 5 \;\; 6 \end{array}$

  b) $\{x|3 \le x \le 6\}$ c) $[3, 6]$

77. $-5 < 1-2x < -1$

  $-6 < -2x < -2$

  $3 > x > 1$

  a)
  $\begin{array}{c} -6\,-5\,-4\,-3\,-2\,-1 \;\; 0 \;\; 1 \;\; 2 \;\; 3 \;\; 4 \;\; 5 \;\; 6 \end{array}$

  b) $\{x|1 < x < 3\}$ c) $(1, 3)$

79. $x-3 < 2x+1$ and $2x+1 < 3x$

  $-x < 4$ $-x < -1$

  $x > -4$ $x > 1$

  a)
  $\begin{array}{c} -6\,-5\,-4\,-3\,-2\,-1 \;\; 0 \;\; 1 \;\; 2 \;\; 3 \;\; 4 \;\; 5 \;\; 6 \end{array}$

  b) $\{x|x > 1\}$ c) $(1, \infty)$

81. $36 \le 0.09x \le 45$

  $400 \le x \le 500$

  a)
  $\begin{array}{c} 250 \;\; 300 \;\; 350 \;\; 400 \;\; 450 \;\; 500 \;\; 550 \end{array}$

  b) $\{x|400 \le x \le 500\}$

  c) $[400, 500]$

83. $80 \le \dfrac{95+80+82+88+x}{5} < 90$

  $400 \le 345 + x < 450$

  $55 \le x < 105$

  a)
  $\begin{array}{c} 40 \;\; 50 \;\; 60 \;\; 70 \;\; 80 \;\; 90 \;\; 100 \;\; 110 \end{array}$

  b) $\{x|55 \le x < 105\}$ c) $[55, 105)$

85. a)
  $\begin{array}{c} 66 \;\; 68 \;\; 70 \;\; 72 \;\; 74 \;\; 76 \;\; 78 \end{array}$

  b) $\{x|68° \le x \le 78°\}$ c) $[68, 78]$

87. a)
  $\begin{array}{c} 68 \;\; 70 \;\; 72 \;\; 74 \;\; 76 \;\; 78 \;\; 80 \;\; 82 \end{array}$

  b) $\{x|72° \le x \le 80°\}$ c) $[72, 80]$

89. $6000 \le \dfrac{1}{2}\cdot 50(60+x) \le 8000$

  $6000 \le 25(60+x) \le 8000$

  $6000 \le 1500 + 25x \le 8000$

  $4500 \le 25x \le 6500$

  $180 \le x \le 260$

  a)
  $\begin{array}{c} 170\,180\,190\,200\,210\,220\,230\,240\,250\,260\,270 \end{array}$

  b) $\{x|180 \text{ ft.} \le x \le 260 \text{ ft.}\}$

  c) $[180, 260]$

Review Exercises

1. No, the absolute value of zero is zero and zero is neither positive nor negative.

2. $-|2-3^2|-|-4| = -|2-9|-|-4|$

  $= -|-7|-4$

  $= -7-4$

  $= -11$

3. $-|-|-16|| = -|-16| = -16$

4. $3x - 14x = 17 - 12x - 11$
$$-11x = 6 - 12x$$
$$x = 6$$

5. $\dfrac{3}{5}(25 - 5x) = 15 - \dfrac{3}{5}$
$$15 - 3x = \dfrac{72}{5}$$
$$5 \cdot 15 - 5 \cdot 3x = 5 \cdot \dfrac{72}{5}$$
$$75 - 15x = 72$$
$$-15x = -3$$
$$x = \dfrac{1}{5}$$

6. $\dfrac{1}{2}(x + 2) = 0$
$$2 \cdot \dfrac{1}{2}(x + 2) = 2 \cdot 0$$
$$x + 2 = 0$$
$$x = -2$$

Exercise Set 2.5

1. $x = 2$ or $x = -2$

3. $|a| = -4$

    Because the absolute value of every real number is a positive number or zero, this equation has no solution.

5. $x + 3 = 8$ or $x + 3 = -8$
$$x = 5 \qquad\qquad x = -11$$

7. $2m - 5 = 1$ or $2m - 5 = -1$
$$2m = 6 \qquad\qquad 2m = 4$$
$$m = 3 \qquad\qquad m = 2$$

9. $6 - 5x = 1$ or $6 - 5x = -1$
$$-5x = -5 \qquad\qquad -5x = -7$$
$$x = 1 \qquad\qquad x = \dfrac{7}{5}$$

11. $4 - 3w = -6$ or $4 - 3w = 6$
$$-3w = -10 \qquad\qquad -3w = 2$$
$$w = \dfrac{10}{3} \qquad\qquad w = -\dfrac{2}{3}$$

13. $|4m - 2| = -5$

    Because the absolute value of every real number is a positive number or zero, this equation has no solution.

15. $4w - 3 = 0$
$$4w = 3$$
$$w = \dfrac{3}{4}$$

17. $|2y| - 3 = 5$
$$|2y| = 8$$
$$2y = 8 \quad \text{or} \quad 2y = -8$$
$$y = 4 \qquad\qquad y = -4$$

19. $|y + 1| + 2 = 4$
$$|y + 1| = 2$$
$$y + 1 = 2 \quad \text{or} \quad y + 1 = -2$$
$$y = 1 \qquad\qquad y = -3$$

21. $|b - 4| - 6 = 2$
$$|b - 4| = 8$$
$$b - 4 = 8 \quad \text{or} \quad b - 4 = -8$$
$$b = 12 \qquad\qquad b = -4$$

23. $3 + |5x - 1| = 7$
$$|5x - 1| = 4$$
$$5x - 1 = 4 \quad \text{or} \quad 5x - 1 = -4$$
$$5x = 5 \qquad\qquad 5x = -3$$
$$x = 1 \qquad\qquad x = -\dfrac{3}{5}$$

25. $1 - |2k + 3| = -4$
$$-|2k + 3| = -5$$
$$|2k + 3| = 5$$
$$2k + 3 = 5 \quad \text{or} \quad 2k + 3 = -5$$
$$2k = 2 \qquad\qquad 2k = -8$$
$$k = 1 \qquad\qquad k = -4$$

27. $4 - 3|z - 2| = -8$
$$-3|z - 2| = -12$$
$$|z - 2| = 4$$
$$z - 2 = 4 \quad \text{or} \quad z - 2 = -4$$
$$z = 6 \qquad\qquad z = -2$$

29. $6 - 2|3 - 2w| = -18$

$-2|3 - 2w| = -24$

$|3 - 2w| = 12$

$3 - 2w = 12$ or $3 - 2w = -12$

$-2w = 9$ $-2w = -15$

$w = -\dfrac{9}{2}$ $w = \dfrac{15}{2}$

31. $|3x - 2(x + 5)| = 10$

$|3x - 2x - 10| = 10$

$|x - 10| = 10$

$x - 10 = 10$ or $x - 10 = -10$

$x = 20$ $x = 0$

33. $2x + 1 = x + 5$ or $2x + 1 = -(x + 5)$

$x = 4$ $2x + 1 = -x - 5$

$3x = -6$

$x = -2$

35. $x + 3 = 2x - 4$ or $x + 3 = -(2x - 4)$

$7 = x$ $x + 3 = -2x + 4$

$3x = 1$

$x = \dfrac{1}{3}$

37. $3v + 4 = 1 - 2v$ or $3v + 4 = -(1 - 2v)$

$5v = -3$ $3v + 4 = -1 + 2v$

$v = -\dfrac{3}{5}$ $v = -5$

39. $2n + 3 = 3 + 2n$ or $2n + 3 = -(3 + 2n)$

$3 = 3$ $2n + 3 = -3 - 2n$

$4n = -6$

$n = -\dfrac{3}{2}$

One equation leads to a solution with no variables yet is true. The solution to this absolute value equation is all real numbers.

41. $2k + 1 = 2k - 5$ or $2k + 1 = -(2k - 5)$

$0 = -6$ $2k + 1 = -2k + 5$

no solution $4k = 4$

$k = 1$

This absolute value equation has only one solution, which is 1.

43. $|10 - (5 - h)| = 8$

$|10 - 5 + h| = 8$

$|5 + h| = 8$

$5 + h = 8$ or $5 + h = -8$

$h = 3$ $h = -13$

45. $\dfrac{b}{2} - 1 = 4$ or $\dfrac{b}{2} - 1 = -4$

$\dfrac{b}{2} = 5$ $\dfrac{b}{2} = -3$

$b = 10$ $b = -6$

47. $\dfrac{4 - 3x}{2} = \dfrac{3}{4}$ or $\dfrac{4 - 3x}{2} = -\dfrac{3}{4}$

$4(4 - 3x) = 2 \cdot 3$ $4(4 - 3x) = -3 \cdot 2$

$16 - 12x = 6$ $16 - 12x = -6$

$-12x = -10$ $-12x = -22$

$x = \dfrac{5}{6}$ $x = \dfrac{11}{6}$

49. $\left|2y + \dfrac{3}{2}\right| - 2 = 5$

$\left|2y + \dfrac{3}{2}\right| = 7$

$2y + \dfrac{3}{2} = 7$ or $2y + \dfrac{3}{2} = -7$

$2y = \dfrac{11}{2}$ $2y = -\dfrac{17}{2}$

$y = \dfrac{11}{4}$ $y = -\dfrac{17}{4}$

Review Exercises

1. $6 - 4x > -5x - 1$ 2. $-4x \le 12$

$x > -7$ $x \ge -3$

3. $n - 2 < 5$

$n - 2 + 2 < 5 + 2$

$n < 7$

4. $(-2, 0]$

5.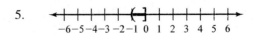

6. $-1 < \dfrac{3x+2}{4} < 4$

$-4 < 3x+2 < 16$

$-6 < 3x < 14$

$-2 < x < \dfrac{14}{3}$

Exercise Set 2.6

1.   a)

   b) $\{x \mid -5 < x < 5\}$ c) $(-5, 5)$

3. $-7 \le x+3 \le 7$

$-10 \le x \le 4$

   a)

   b) $\{x \mid -10 \le x \le 4\}$ c) $[-10, 4]$

5. $|s+3| + 6 < 9$

$|s+3| < 3$

$-3 < s+3 < 3$

$-6 < s < 0$

   a)

   b) $\{s \mid -6 < s < 0\}$ c) $(-6, 0)$

7. $|2m-5| - 3 < 6$

$|2m-5| < 9$

$-9 < 2m-5 < 9$

$-4 < 2m < 14$

$-2 < m < 7$

   a)

   b) $\{m \mid -2 < m < 7\}$ c) $(-2, 7)$

9. $|-3k+5| + 7 \le 8$

$|-3k+5| \le 1$

$-1 \le -3k+5 \le 1$

$-6 \le -3k \le -4$

$2 \ge k \ge \dfrac{4}{3}$

   a)

   b) $\left\{k \mid \dfrac{4}{3} \le k \le 2\right\}$ c) $\left[\dfrac{4}{3}, 2\right]$

11. $2|x| + 7 \le 3$

$2|x| \le -4$

$|x| \le -2$

Because absolute values cannot be negative, this inequality has no solution.

   a)

   b) $\{\ \}$ or $\varnothing$ c) no interval notation

13. $2|w-3| + 4 < 10$

$2|w-3| < 6$

$|w-3| < 3$

$-3 < w-3 < 3$

$0 < w < 6$

   a)

   b) $\{w \mid 0 < w < 6\}$ c) $(0, 6)$

15.   a)

   b) $\{c \mid c < -12 \text{ or } c > 12\}$

   c) $(-\infty, -12) \cup (12, \infty)$

17. $y+2 \le -7$ or $y+2 \ge 7$

$y \le -9$ $y \ge 5$

   a)

   b) $\{y \mid y \le -9 \text{ or } y \ge 5\}$

   c) $(-\infty, -9] \cup [5, \infty)$

19. $|p-6|-3>5$

$|p-6|>8$

$p-6<-8$ or $p-6>8$

$p<-2$ $p>14$

a)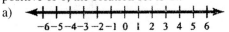

b) $\{p|p<-2 \text{ or } p>14\}$

c) $(-\infty,-2)\cup(14,\infty)$

21. $|3x+6|-3\geq 9$

$|3x+6|\geq 12$

$3x+6\leq -12$ or $3x+6\geq 12$

$3x\leq -18$ $3x\geq 6$

$x\leq -6$ $x\geq 2$

a)

b) $\{x|x\leq -6 \text{ or } x\geq 2\}$

c) $(-\infty,-6]\cup[2,\infty)$

23. $|-4n-5|+3>8$

$|-4n-5|>5$

$-4n-5<-5$ or $-4n-5>5$

$-4n<0$ $-4n>10$

$n>0$ $n<-\dfrac{5}{2}$

a)

b) $\left\{n\left|n<-\dfrac{5}{2} \text{ or } n>0\right.\right\}$

c) $\left(-\infty,-\dfrac{5}{2}\right)\cup(0,\infty)$

25. $4|v|+3\geq 7$

$4|v|\geq 4$

$|v|\geq 1$

$v\leq -1$ or $v\geq 1$

a)

b) $\{v|v\leq -1 \text{ or } v\geq 1\}$

c) $(-\infty,-1]\cup[1,\infty)$

27. $4|y+2|-1>3$

$4|y+2|>4$

$|y+2|>1$

$y+2<-1$ or $y+2>1$

$y<-3$ $y>-1$

a)

b) $\{y|y<-3 \text{ or } y>-1\}$

c) $(-\infty,-3)\cup(-1,\infty)$

29. $|4m+8|-2>10$

$|4m+8|>12$

$4m+8<-12$ or $4m+8>12$

$4m<-20$ $4m>4$

$m<-5$ $m>1$

a)

b) $\{m|m<-5 \text{ or } m>1\}$

c) $(-\infty,-5)\cup(1,\infty)$

31. $-6<-3x+6<6$

$-12<-3x<0$

$4>x>0$

$0<x<4$

a)

b) $\{x|0<x<4\}$ c) $(0,4)$

33. $|2r-3|>-3$

This inequality indicates that the absolute value is greater than a negative number. Because the absolute value of every real number is either positive or 0, the solution set is $\mathbb{R}$.

a)

b) $\{r|r \text{ is a real number}\}$

c) $(-\infty,\infty)$

35. $4 - 2|x+3| > 2$

$$-2|x+3| > -2$$

$$|x+3| < 1$$

$$-1 < x+3 < 1$$

$$-4 < x < -2$$

a)
-6 -5 -4 -3 -2 -1 0 1 2 3 4 5 6

b) $\{x \mid -4 < x < -2\}$ c) $(-4, -2)$

37. $6|2x-1| - 3 < 3$

$$6|2x-1| < 6$$

$$|2x-1| < 1$$

$$-1 < 2x-1 < 1$$

$$0 < 2x < 2$$

$$0 < x < 1$$

a)
-6 -5 -4 -3 -2 -1 0 1 2 3 4 5 6

b) $\{x \mid 0 < x < 1\}$ c) $(0, 1)$

39. $5 - |w+4| > 10$

$$-|w+4| > 5$$

$$|w+4| < -5$$

Since absolute values cannot be negative, this inequality has no solution.

a)
-6 -5 -4 -3 -2 -1 0 1 2 3 4 5 6

b) $\varnothing$ c) no interval notation

41. $\left| 2 - \dfrac{3}{2}k \right| \le 5$

$$-5 \le 2 - \dfrac{3}{2}k \le 5$$

$$-10 \le 4 - 3k \le 10$$

$$-14 \le -3k \le 6$$

$$\dfrac{14}{3} \ge k \ge -2$$

a)
-6 -5 -4 -3 -2 -1 0 1 2 3 4 5 6

b) $\left\{ k \mid -2 \le k \le \dfrac{14}{3} \right\}$ c) $\left[-2, \dfrac{14}{3} \right]$

43. $|0.25x - 3| + 2 > 4$

$$|0.25x - 3| > 2$$

$$0.25x - 3 < -2 \quad \text{or} \quad 0.25x - 3 > 2$$

$$0.25x < 1 \qquad\qquad 0.25x > 5$$

$$x < 4 \qquad\qquad\qquad x > 20$$

a)
0 4 8 12 16 20 24 28

b) $\{x \mid x < 4 \text{ or } x > 20\}$

c) $(-\infty, 4) \cup (20, \infty)$

45. $\left| 2.4 - \dfrac{3}{4}y \right| \le 7.2$

$$-7.2 \le 2.4 - \dfrac{3}{4}y \le 7.2$$

$$-9.6 \le -\dfrac{3}{4}y \le 4.8$$

$$12.8 \ge y \ge -6.4$$

a)
-8 -6 -4 -2 0 2 4 6 8 10 12 14

b) $\{y \mid -6.4 \le y \le 12.8\}$

c) $[-6.4, 12.8]$

47. $|2p - 8| + 5 > 1$

$$|2p - 8| > -4$$

This inequality indicates that the absolute value is greater than a negative number. Because the absolute value of every real number is either positive or 0, the solution set is $\mathbb{R}$.

a)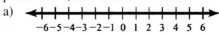
-6 -5 -4 -3 -2 -1 0 1 2 3 4 5 6

b) $\{p \mid p \text{ is a real number}\}$

c) $(-\infty, \infty)$

49. $|x| < 3$

51. $|x| \ge 4$

53. $|x+1| > 1$

55. $|x-3| \le 2$

57. $|x| >$ any negative number

Review Exercises

1. true

2. true

3. $23 - 5(2 - 3x) = 12x - (4x + 1)$

    $23 - 10 + 15x = 12x - 4x - 1$

    $13 + 15x = 8x - 1$

    $13 + 7x = -1$

    $7x = -14$

    $x = -2$

4. $\dfrac{5}{6}x - \dfrac{3}{4} = \dfrac{2}{3}x + \dfrac{1}{2}$

    $\dfrac{1}{6}x - \dfrac{3}{4} = \dfrac{1}{2}$

    $\dfrac{1}{6}x = \dfrac{5}{4}$

    $6 \cdot \dfrac{1}{6}x = 6 \cdot \dfrac{5}{4}$

    $x = \dfrac{15}{2}$

5. $Ax + By = C$

    $By = C - Ax$

    $\dfrac{By}{B} = \dfrac{C - Ax}{B}$

    $y = \dfrac{C - Ax}{B}$

6. Let x be the number. Translate to an equation and solve for x.

    $\dfrac{2}{5}(x + 5) = \dfrac{3}{4}x - 5$

    $\dfrac{2}{5}x + 2 = \dfrac{3}{4}x - 5$

    $\dfrac{8}{20}x + 2 = \dfrac{15}{20}x - 5$

    $7 = \dfrac{7}{20}x$

    $20 = x$

Chapter 2 Review Exercises

1. $3x - 9 = 12$

    $3x = 21$

    $x = 7$

    Check:

    $3(7) - 9 \overset{?}{=} 12$

    $21 - 9 \overset{?}{=} 12$

    $12 = 12$

2. $7m - 3 = 4m + 9$

    $3m - 3 = 9$

    $3m = 12$

    $m = 4$

    Check:

    $7(4) - 3 \overset{?}{=} 4(4) + 9$

    $28 - 3 \overset{?}{=} 16 + 9$

    $25 = 25$

3. $6(3a - 4) + (3a + 4) = 5(a - 2) - 10$

    $18a - 24 + 3a + 4 = 5a - 10 - 10$

    $21a - 20 = 5a - 20$

    $16a - 20 = -20$

    $16a = 0$

    $a = 0$

    Check:

    $6(3 \cdot 0 - 4) + (3 \cdot 0 + 4) \overset{?}{=} 5(0 - 2) - 10$

    $6(0 - 4) + (0 + 4) \overset{?}{=} 5(-2) - 10$

    $6(-4) + (4) \overset{?}{=} 5(-2) - 10$

    $-24 + 4 \overset{?}{=} -10 - 10$

    $-20 = -20$

4. $2 - 2y - 6(y + 3) = 5 - 7y$

$2 - 2y - 6y - 18 = 5 - 7y$

$-16 - 8y = 5 - 7y$

$-16 - y = 5$

$-y = 21$

$y = -21$

Check:

$2 - 2(-21) - 6(-21 + 3) \overset{?}{=} 5 - 7(-21)$

$2 - (-42) - 6(-18) \overset{?}{=} 5 - (-147)$

$2 + 42 + 108 \overset{?}{=} 5 + 147$

$152 = 152$

5. $\dfrac{1}{5}d - 2 = -3 + d$

$-\dfrac{4}{5}d - 2 = -3$

$-\dfrac{4}{5}d = -1$

$-5\left(-\dfrac{4}{5}d\right) = -5(-1)$

$4d = 5$

$d = \dfrac{5}{4}$

Check:

$\dfrac{1}{5}\left(\dfrac{5}{4}\right) - 2 \overset{?}{=} -3 + \dfrac{5}{4}$

$\dfrac{1}{4} - 2 \overset{?}{=} -\dfrac{7}{4}$

$-\dfrac{7}{4} = -\dfrac{7}{4}$

6. $\dfrac{1}{2}k + \dfrac{5}{8} = \dfrac{1}{4}$

$8\left(\dfrac{1}{2}k + \dfrac{5}{8}\right) = 8\left(\dfrac{1}{4}\right)$

$4k + 5 = 2$

$4k = -3$

$k = -\dfrac{3}{4}$

Check:

$\dfrac{1}{2}\left(-\dfrac{3}{4}\right) + \dfrac{5}{8} \overset{?}{=} \dfrac{1}{4}$

$-\dfrac{3}{8} + \dfrac{5}{8} \overset{?}{=} \dfrac{1}{4}$

$\dfrac{2}{8} \overset{?}{=} \dfrac{1}{4}$

$\dfrac{1}{4} = \dfrac{1}{4}$

7. $2w + 2(7 - w) = 2(5w + 10) + 4$

$2w + 14 - 2w = 10w + 20 + 4$

$14 = 10w + 24$

$-10 = 10w$

$-1 = w$

Check:

$2(-1) + 2(7 - (-1)) \overset{?}{=} 2(5(-1) + 10) + 4$

$-2 + 2(7 + 1) \overset{?}{=} 2(-5 + 10) + 4$

$-2 + 2(8) \overset{?}{=} 2(5) + 4$

$-2 + 16 \overset{?}{=} 10 + 4$

$14 = 14$

8. $5r - 2(5r - 3) = 6 - 5r$

$5r - 10r + 6 = 6 - 5r$

$-5r + 6 = 6 - 5r$

$0 + 6 = 6$

$6 = 6$

All real numbers

9. $0.53 - 0.2z = 0.2(2z - 13) - 0.47$

$0.53 - 0.2z = 0.4z - 2.6 - 0.47$

$0.53 - 0.2z = 0.4z - 3.07$

$0.53 - 0.6z = -3.07$

$-0.6z = -3.6$

$z = 6$

Check:

$0.53 - 0.2(6) \overset{?}{=} 0.2(2(6) - 13) - 0.47$

$0.53 - 1.2 \overset{?}{=} 0.2(12 - 13) - 0.47$

$0.53 - 1.2 \overset{?}{=} 0.2(-1) - 0.47$

$0.53 - 1.2 \overset{?}{=} -0.2 - 0.47$

$-0.67 = -0.67$

10. $18(v + 2) - 6v = 30 + 12v - 2$

$18v + 36 - 6v = 28 + 12v$

$12v + 36 = 28 + 12v$

$36 \neq 28$

no solution

11. $\dfrac{2}{5}p - \dfrac{3}{20} = \dfrac{13}{20} - \dfrac{3}{10}(6 - 2p)$

$20\left(\dfrac{2}{5}p - \dfrac{3}{20}\right) = 20\left(\dfrac{13}{20} - \dfrac{3}{10}(6 - 2p)\right)$

$8p - 3 = 13 - 6(6 - 2p)$

$8p - 3 = 13 - 36 + 12p$

$8p - 3 = -23 + 12p$

$-4p - 3 = -23$

$-4p = -20$

$p = 5$

Check:

$\dfrac{2}{5}(5) - \dfrac{3}{20} \overset{?}{=} \dfrac{13}{20} - \dfrac{3}{10}(6 - 2(5))$

$2 - \dfrac{3}{20} \overset{?}{=} \dfrac{13}{20} - \dfrac{3}{10}(6 - 10)$

$2 - \dfrac{3}{20} \overset{?}{=} \dfrac{13}{20} - \dfrac{3}{10}(-4)$

$2 - \dfrac{3}{20} \overset{?}{=} \dfrac{13}{20} + \dfrac{12}{10}$

$\dfrac{37}{20} = \dfrac{37}{20}$

12. $0.5(l + 4) = 0.3(3l - 1) - 0.9$

$0.5l + 2 = 0.9l - 0.3 - 0.9$

$0.5l + 2 = 0.9l - 1.2$

$2 = 0.4l - 1.2$

$3.2 = 0.4l$

$8 = l$

Check:

$0.5(8 + 4) \overset{?}{=} 0.3(3(8) - 1) - 0.9$

$0.5(12) \overset{?}{=} 0.3(24 - 1) - 0.9$

$6 \overset{?}{=} 0.3(23) - 0.9$

$6 \overset{?}{=} 6.9 - 0.9$

$6 = 6$

13. $I = Prt$

$\dfrac{I}{Pr} = \dfrac{Prt}{Pr}$

$\dfrac{I}{Pr} = t$

14. $P = 2l + 2w$

$P - 2l = 2w$

$\dfrac{P - 2l}{2} = \dfrac{2w}{2}$

$\dfrac{P - 2l}{2} = w$

15. $A = \dfrac{1}{2}bh$

$2A = bh$

$\dfrac{2A}{b} = h$

16. $P = a + b + c$

$P - b - c = a$

17. $C = \dfrac{5}{9}(F - 32)$

$-196 = \dfrac{5}{9}(F - 32)$

$-\dfrac{1764}{5} = F - 32$

$-352.8 = F - 32$

$-320.8° = F$

18.
$$B = P + Prt$$
$$4368 = P + P(0.04)(1)$$
$$4368 = P + 0.04P$$
$$4368 = 1.04P$$
$$\$4200 = P$$

19. Let n represent the number.
$$\frac{2}{3}n = 2 + \frac{1}{2}n$$
$$\frac{1}{6}n = 2$$
$$n = 12$$

20. Let n represent the number.
$$2(n-7) = -6$$
$$2n - 14 = -6$$
$$2n = 8$$
$$n = 4$$

21. Let p represent the original price. Because $44.94 is the discounted price, it is the result of subtracting 30% of the original price from the original price. Translate to an equation and solve for p.
$$p - 0.30p = 44.94$$
$$0.70p = 44.94$$
$$\frac{0.70p}{0.70} = \frac{44.94}{0.70}$$
$$p = 64.20$$
The original price is $64.20.

22. Let x represent the number of units sold in the previous year. Because the 37,791 units sold this year is a 10.5% increase over the previous year, it is the result of adding 10.5% of x to x. Translate to an equation and solve.
$$0.105x + x = 37,791$$
$$1.105x = 37,791$$
$$\frac{1.105x}{1.105} = \frac{37,791}{1.105}$$
$$x = 34,200$$
The number of units sold the previous year was 34,200 units.

23. Create a table.

	rate	time	distance
car 1	50	x	$50x$
car 2	55	x	$55x$

Translate into an equation and solve.
$$50x + 55x = 315$$
$$105x = 315$$
$$x = 3$$
They will meet in 3 hours.

24.
$$-8n \geq -32$$
$$n \leq 4$$

a) $\{n | n \leq 4\}$ 　　 b) $(-\infty, 4]$

c)

25.
$$3x + 5 > -1$$
$$3x > -6$$
$$x > -2$$

a) $\{x | x > -2\}$ 　　 b) $(-2, \infty)$

c)

26.
$$9 - 2m > 15$$
$$-2m > 6$$
$$m < -3$$

a) $\{m | m < -3\}$ 　　 b) $(-\infty, -3)$

c)

27.
$$5h - 11 \geq 9h + 9$$
$$-4h - 11 \geq 9$$
$$-4h \geq 20$$
$$h \leq -5$$

a) $\{h | h \leq -5\}$ 　　 b) $(-\infty, -5]$

c)

28. $\dfrac{2}{3}t - 1 \le \dfrac{1}{4}t + \dfrac{1}{2}$

$12\left(\dfrac{2}{3}t - 1\right) \le 12\left(\dfrac{1}{4}t + \dfrac{1}{2}\right)$

$8t - 12 \le 3t + 6$

$5t - 12 \le 6$

$5t \le 18$

$t \le \dfrac{18}{5}$

a) $\left\{t \mid t \le \dfrac{18}{5}\right\}$ b) $\left(-\infty, \dfrac{18}{5}\right]$

c)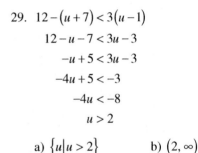

29. $12 - (u + 7) < 3(u - 1)$

$12 - u - 7 < 3u - 3$

$-u + 5 < 3u - 3$

$-4u + 5 < -3$

$-4u < -8$

$u > 2$

a) $\{u \mid u > 2\}$ b) $(2, \infty)$

c)

30. Let l be the length of the pool.

$12l \le 300$

$l \le 25$

The length of the pool can be 25 ft. or less.

31. Intersection: $A \cap B = \{1, 5, 9\}$

Union: $A \cup B = \{1, 5, 7, 9\}$

32. Intersection: $A \cap B = \{\ \}$ or $\varnothing$

Union: $A \cup B = \{1, 2, 3, 4, 5, 6, 7\}$

33. $4 < -2x < 6$

$-2 > x > -3$

a)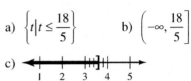

b) $\{x \mid -3 < x < -2\}$ c) $(-3, -2)$

34. $9 \le -3x \le 15$

$-3 \ge x \ge -5$

a)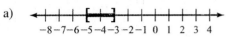

b) $\{x \mid -5 \le x \le -3\}$ c) $[-5, -3]$

35. $-3 < x + 4 < 7$

$-7 < x < 3$

a)

b) $\{x \mid -7 < x < 3\}$ c) $(-7, 3)$

36. $0 < x - 1 \le 3$

$1 < x \le 4$

a)

b) $\{x \mid 1 < x \le 4\}$ c) $(1, 4]$

37. $-1 \le 2x + 3 < 3$

$-4 \le 2x < 0$

$-2 \le x < 0$

a)

b) $\{x \mid -2 \le x < 0\}$ c) $[-2, 0)$

38. $3x - 2 > 1$ and $3x - 2 < -8$

$3x > 3$ $3x < -6$

$x > 1$ $x < -2$

a)

b) $\{\ \}$ or $\varnothing$ c) no interval notation

39. Let x represent the fourth test score.

$80 \le \dfrac{80 + 89 + 83 + x}{4} < 90$

$80 \le \dfrac{252 + x}{4} < 90$

$320 \le 252 + x < 360$

$68 \le x < 108$

a)

b) $\{x \mid 68 \le x < 108\}$

c) $[68, 108)$

40. Let x represent the length of the rectangular building.

$12,000 \le 80x \le 16,000$

$150 \le x \le 200$

a)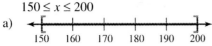

b) $\{x \mid 150 \le x \le 200\}$

c) $[150, 200]$

41. $w + 4 \le -2$ or $w + 4 \ge 2$

$\quad w \le -6 \qquad w \ge -2$

a)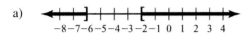

b) $\{w \mid w \le -6 \text{ or } w \ge -2\}$

c) $(-\infty, -6] \cup [-2, \infty)$

42. $4w - 3 < 1$ or $4w - 3 > 0$

$\quad 4w < 4 \qquad 4w > 3$

$\quad w < 1 \qquad w > \dfrac{3}{4}$

a) ![number line from -6 to 6 with arrows both directions]

b) $\{w \mid w \text{ is a real number}\}$

c) $(-\infty, \infty)$

43. $2m - 5 < 0$ or $2m - 5 > 5$

$\quad 2m < 5 \qquad 2m > 10$

$\quad m < \dfrac{5}{2} \qquad m > 5$

a) ![number line from -6 to 6 with open circle at 5/2 and open paren at 5]

b) $\left\{ m \mid m < \dfrac{5}{2} \text{ or } m > 5 \right\}$

c) $\left(-\infty, \dfrac{5}{2} \right) \cup (5, \infty)$

44. $3x + 2 \le -2$ or $3x + 2 \ge 8$

$\quad 3x \le -4 \qquad 3x \ge 6$

$\quad x \le -\dfrac{4}{3} \qquad x \ge 2$

a) ![number line from -6 to 6 with bracket at -4/3 and bracket at 2]

b) $\left\{ x \mid x \le -\dfrac{4}{3} \text{ or } x \ge 2 \right\}$

c) $\left(-\infty, -\dfrac{4}{3} \right] \cup [2, \infty)$

45. $-x - 6 \le -2$ or $-x - 6 \ge 3$

$\quad -x \le 4 \qquad -x \ge 9$

$\quad x \ge -4 \qquad x \le -9$

a) ![number line from -12 to 0 with bracket at -9 and bracket at -4]

b) $\{x \mid x \le -9 \text{ or } x \ge -4\}$

c) $(-\infty, -9] \cup [-4, \infty)$

46. $-4w + 1 \le -3$ or $-4w + 1 \ge 5$

$\quad -4w \le -4 \qquad -4w \ge 4$

$\quad w \ge 1 \qquad w \le -1$

a) ![number line from -6 to 6 with bracket at -1 and bracket at 1]

b) $\{w \mid w \le -1 \text{ or } w \ge 1\}$

c) $(-\infty, -1] \cup [1, \infty)$

47. $-4, 4$

48. $x - 4 = -7$ or $x - 4 = 7$

$\quad x = -3 \qquad x = 11$

49. $2w - 1 = -3$ or $2w - 1 = 3$

$\quad 2w = -2 \qquad 2w = 4$

$\quad w = -1 \qquad w = 2$

50. $|5r + 8| = -3$

Because the absolute value of every real number is a positive number or zero, this equation has no solution.

51. $|q - 4| - 3 = 8$

$\quad |q - 4| = 11$

$\quad q - 4 = -11$ or $q - 4 = 11$

$\quad q = -7 \qquad q = 15$

52. $|5w| - 2 = 13$

$|5w| = 15$

$5w = -15$ or $5w = 15$

$w = -3$ $w = 3$

53. $2|3x - 4| = 8$

$|3x - 4| = 4$

$3x - 4 = -4$ or $3x - 4 = 4$

$3x = 0$ $3x = 8$

$x = 0$ $x = \dfrac{8}{3}$

54. $4 - 2|r - 5| = -8$

$-2|r - 5| = -12$

$|r - 5| = 6$

$r - 5 = -6$ or $r - 5 = 6$

$r = -1$ $r = 11$

55. $|3x - 2| = |x + 2|$

$3x - 2 = x + 2$ or $3x - 2 = -(x + 2)$

$2x = 4$ $3x - 2 = -x - 2$

$x = 2$ $4x = 0$

$x = 0$

56. $|2x - 1| = |3x + 2|$

$2x - 1 = 3x + 2$ or $2x - 1 = -(3x + 2)$

$-3 = x$ $2x - 1 = -3x - 2$

$5x = -1$

$x = -\dfrac{1}{5}$

57. $|-3x + 1| = |3 - 2x|$

$-3x + 1 = 3 - 2x$ or $-3x + 1 = -(3 - 2x)$

$-2 = x$ $-3x + 1 = -3 + 2x$

$4 = 5x$

$\dfrac{4}{5} = x$

58. $|9 - 4x| = |7 - 2x|$

$9 - 4x = 7 - 2x$ or $9 - 4x = -(7 - 2x)$

$2 = 2x$ $9 - 4x = -7 + 2x$

$1 = x$ $16 = 6x$

$\dfrac{8}{3} = x$

59. $|x| < 5$

a)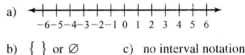

b) $\{x \mid -5 < x < 5\}$ c) $(-5, 5)$

60. $-4 < 2m + 6 < 4$

$-10 < 2m < -2$

$-5 < m < -1$

a)

b) $\{m \mid -5 < m < -1\}$ c) $(-5, -1)$

61. $|3s - 1| \le -2$

Because absolute values cannot be negative, this inequality has no solution.

a)

b) $\{\ \}$ or $\varnothing$ c)   no interval notation

62. $7|m + 3| \le 21$

$|m + 3| \le 3$

$-3 \le m + 3 \le 3$

$-6 \le m \le 0$

a)

b) $\{m \mid -6 \le m \le 0\}$ c) $[-6, 0]$

63. $|p| \ge 4$

a)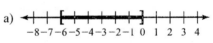

b) $\{p \mid p \le -4 \text{ or } p \ge 4\}$

c) $(-\infty, -4] \cup [4, \infty)$

64. $x - 3 < -7$ or $x - 3 > 7$

$x < -4$ $x > 10$

a)

b) $\{x \mid x < -4 \text{ or } x > 10\}$

c) $(-\infty, -4) \cup (10, \infty)$

65. $5|b| - 2 > 3$

$$5|b| > 5$$

$$|b| > 1$$

$b < -1$ or $b > 1$

a)

b) $\{b \mid b < -1 \text{ or } b > 1\}$

c) $(-\infty, -1) \cup (1, \infty)$

66. $-2|t - 5| < -10$

$$|t - 5| > 5$$

$t - 5 < -5$ or $t - 5 > 5$

$t < 0$ $\qquad$ $t > 10$

a)

b) $\{t \mid t < 0 \text{ or } t > 10\}$

c) $(-\infty, 0) \cup (10, \infty)$

67. $5 - 2|2k - 3| \le -15$

$$-2|2k - 3| \le -20$$

$$|2k - 3| \ge 10$$

$2k - 3 \le -10$ or $2k - 3 \ge 10$

$2k \le -7$ $\qquad$ $2k \ge 13$

$k \le -\dfrac{7}{2}$ $\qquad$ $k \ge \dfrac{13}{2}$

a)

b) $\left\{ k \mid k \le -\dfrac{7}{2} \text{ or } k \ge \dfrac{13}{2} \right\}$

c) $\left(-\infty, -\dfrac{7}{2} \right] \cup \left[\dfrac{13}{2}, \infty \right)$

68. $3 - 7|2p + 4| \le 24$

$$-7|2p + 4| \le 21$$

$$|2p + 4| \ge -3$$

a)

b) $\{p \mid p \text{ is a real number}\}$ $\quad$ c) $(-\infty, \infty)$

Chapter 2 Practice Test

1. Let $A = \{h, o, m, e\}$ and $B = \{h, o, u, s, e\}$

$A \cap B = \{e, h, o\}$ because e, h, and o are the only elements in both A and B.

$A \cup B = \{e, h, m, o, s, u\}$ because these are the elements that are either in A or B or both.

2. $2(w + 2) - 3 = 3(w - 1)$

$$2w + 4 - 3 = 3w - 3$$

$$2w + 1 = 3w - 3$$

$$2w - 2w + 1 = 3w - 2w - 3$$

$$1 = w - 3$$

$$1 + 3 = w - 3 + 3$$

$$4 = w$$

3. Multiply both sides by the LCD, 15.

$$\frac{2}{3}q - 4 = \frac{1}{5}q + 10$$

$$15\left(\frac{2}{3}q - 4 \right) = 15\left(\frac{1}{5}q + 10 \right)$$

$$10q - 60 = 3q + 150$$

$$10q - 3q - 60 = 3q - 3q + 150$$

$$7q - 60 = 150$$

$$7q - 60 + 60 = 150 + 60$$

$$7q = 210$$

$$\frac{7q}{7} = \frac{210}{7}$$

$$q = 30$$

4. $|x+3| = 5$

   $x+3 = -5$ or $x+3 = 5$

   $x = -8$ $x = 2$

5. $3 - |2x-3| = -6$

   $-|2x-3| = -9$

   $|2x-3| = 9$

   $2x-3 = -9$ or $2x-3 = 9$

   $2x = -6$ $2x = 12$

   $x = -3$ $x = 6$

6. $|2x+3| = |x-5|$

   $2x+3 = x-5$ or $2x+3 = -(x-5)$

   $x+3 = -5$ $2x+3 = -x+5$

   $x = -8$ $3x+3 = 5$

                                $3x = 2$

                                $x = \dfrac{2}{3}$

7. $|5x-4| = -3$

   This equation has the absolute value equal to a
   negative number. Because the absolute value of
   every real number is a positive number or zero,
   this equation has no solution.

8. $A = \dfrac{1}{2}h(b+B)$

       $2A = h(b+B)$

       $\dfrac{2A}{h} = b + B$

   $\dfrac{2A}{h} - B = b$

9. $-3 < x+4 \le 7$

   $-7 < x \le 3$

   a)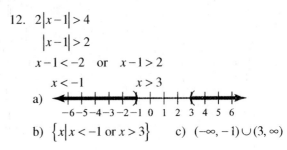

   b) $\{x| -7 < x \le 3\}$ c) $(-7, 3]$

10. $4 < -2x \le 6$

   $-2 > x \ge -3$

   a)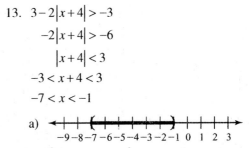

   b) $\{x| -3 \le x < -2\}$ c) $[-3, -2)$

11. $|x+4| < 9$

   $-9 < x+4 < 9$

   $-13 < x < 5$

   a)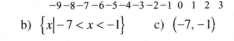

   b) $\{x| -13 < x < 5\}$ c) $(-13, 5)$

12. $2|x-1| > 4$

   $|x-1| > 2$

   $x-1 < -2$ or $x-1 > 2$

   $x < -1$ $x > 3$

   a) ![number line]

   b) $\{x| x < -1 \text{ or } x > 3\}$ c) $(-\infty, -1) \cup (3, \infty)$

13. $3 - 2|x+4| > -3$

   $-2|x+4| > -6$

   $|x+4| < 3$

   $-3 < x+4 < 3$

   $-7 < x < -1$

   a) ![number line]

   b) $\{x| -7 < x < -1\}$ c) $(-7, -1)$

14. $|3y-2| < -2$

   Because the absolute value of every real number is
   a positive number or zero, this inequality has no
   solution.

   a) ![number line]

   b) $\{\ \}$ or $\varnothing$ c) no interval notation

15. $|8t+4| \ge -12$

   Because the absolute value of every real number is
   greater than or equal to -12, this inequality is true
   for all real numbers.

   a) ![number line]

   b) $\{t| t \text{ is a real number}\}$ c) $(-\infty, \infty)$

16. $2|3x-4| \le 10$

$|3x-4| \le 5$

$-5 \le 3x-4 \le 5$

$-1 \le 3x \le 9$

$-\dfrac{1}{3} \le x \le 3$

a)

b) $\left\{ x \middle| -\dfrac{1}{3} \le x \le 3 \right\}$ c) $\left[-\dfrac{1}{3}, 3 \right]$

17. Find the width of the box.

Use the formula for surface area.

$A = 2lw + 2lh + 2wh$

$815 = 2 \cdot 15 \cdot w + 2 \cdot 15 \cdot 8 + 2 \cdot w \cdot 8$

$815 = 30w + 240 + 16w$

$575 = 46w$

$12.5 = w$

The height of the box is 12.5 in.

18. Let x represent the original price. Translate to an equation and solve.

$31.36 = x - 0.30x$

$31.36 = 0.70x$

$\dfrac{31.36}{0.70} = \dfrac{0.70x}{0.70}$

$44.8 = x$

The original price of the dress is $44.80.

19. Using the relationship rate $\times$ time = distance, complete a table.

Categories	Rate	Time	Distance
car 1	65	t	$65t$
car 2	70	t	$70t$

Translate to an equation.

$60t + 75t = 54$

$135t = 54$

$\dfrac{135t}{135} = \dfrac{54}{135}$

$t = 0.4$

If the cars are 54 miles apart, it will take 0.4 hours for the cars to meet.

20. Let x represent the number of books Johnson orders.

$40 < 8x < 50$

$5 < x < 6.25$

Since he would have to order between 5 and 6.25 books, he needs to order 6 books.

Chapters 1 and 2 Cumulative Review

1. False; $\dfrac{3}{2}$ is a rational number, but not an integer.

2. False; subtraction is not commutative.

3. False; $-a < x < a$ is not the same as $x > a$ and $x < -a$.

4. identity

5. real number

6. Reverse the inequality symbol.

7. $-\left| -\dfrac{5}{9} \right| = -\dfrac{5}{9}$

8. $\sqrt[3]{-125} = -5$

9. $-16 + 9 = -7$

10. $(-4)(-5)(4) = 80$

11. $-9^2 = -(9) \cdot (9) = -81$

12. $14 - 4(-2)^2 = 14 - 4(4)$

$= 14 - 16$

$= -2$

13. $\left| 7 - 3(8-2) \div 6 \right| + \left[4\left(2 - 4^2\right) \div 2 \right]$

$= \left| 7 - 3(6) \div 6 \right| + \left[4(2 - 16) \div 2 \right]$

$= \left| 7 - 18 \div 6 \right| + \left[4(-14) \div 2 \right]$

$= \left| 7 - 3 \right| + (-56 \div 2)$

$= \left| 4 \right| + (-28)$

$= 4 + (-28)$

$= -24$

14. $\left[-3^2-2(-6+2)\right]+\sqrt{25+144}$

$=\left[-3^2-2(-4)\right]+\sqrt{169}$

$=\left[-9-2(-4)\right]+13$

$=(-9+8)+13$

$=-1+13$

$=12$

15. $\dfrac{4y-5x^2}{3y^3-z^2}=\dfrac{4(2)-5(-3)^2}{3(2)^3-(-4)^2}$

$=\dfrac{4(2)-5(9)}{3(8)-(16)}$

$=\dfrac{8-45}{24-16}$

$=\dfrac{-37}{8}$

16. $-3\left(2x^2-4x+3\right)=-3\cdot 2x^2-(-3)\cdot 4x+(-3)\cdot 3$

$=-6x^2+12x-9$

17. $\dfrac{x-5}{x+4}$ is undefined when the denominator is equal
to zero, $x+4=0$. This occurs when $x=-4$.

18. $4d^2-4+4d+4d^2-9-3d$

$=4d^2+4d^2+4d-3d-4-9$

$=8d^2+d-13$

19. $4(2y-5)-2(3y-6)+6y$

$=8y-20-6y+12+6y$

$=8y-8$

20. $12-4(5-x)=12-20+4x=4x-8$

21. $8(2z-3)-z=5z-34$

$16z-24-z=5z-34$

$15z-24=5z-34$

$10z-24=-34$

$10z=-10$

$z=-1$

22. $1.4u-0.5(44-6u)=66+2.4u$

$10(1.4u-0.5(44-6u))=10(66+2.4u)$

$14u-5(44-6u)=660+24u$

$14u-220+30u=660+24u$

$44u-220=660+24u$

$20u-220=660$

$20u=880$

$u=44$

23. $\dfrac{3x}{4}-\dfrac{2}{5}=\dfrac{3x}{10}+\dfrac{1}{2}$

$20\left(\dfrac{3x}{4}-\dfrac{2}{5}\right)=20\left(\dfrac{3x}{10}+\dfrac{1}{2}\right)$

$15x-8=6x+10$

$9x-8=10$

$9x=18$

$x=2$

24. $7u-10(2u+3)\geq 22$

$7u-20u-30\geq 22$

$-13u-30\geq 22$

$-13u\geq 52$

$u\leq -4$

a) $\{u\mid u\leq -4\}$

b) $(-\infty,-4]$

c)
$$\xleftarrow{\quad}\overset{\textstyle]}{+\!+\!+\!+\!+\!+}\rightarrow$$
$$-7\ -6\ -5\ -4\ -3\ -2$$

25. $|2x-5|-1\leq 4$

$|2x-5|\leq 5$

$-5\leq 2x-5\leq 5$

$0\leq 2x\leq 10$

$0\leq x\leq 5$

a) $\{x\mid 0\leq x\leq 5\}$

b) $[0,5]$

c)
$$\xleftarrow{\quad}\overset{\textstyle[\quad\quad\quad]}{+\!+\!+\!+\!+\!+\!+\!+\!+}\rightarrow$$
$$-2\ -1\ 0\ 1\ 2\ 3\ 4\ 5\ 6\ 7$$

26. $3 < -2y + 5 \le 7$

    $-2 < -2y \le 2$

     $1 > y \ge -1$

    $-1 \le y < 1$

   a) $\{y \mid -1 \le y < 1\}$

   b) $[-1, 1)$

   c)
                $-3\ -2\ -1\ \ 0\ \ 1\ \ 2\ \ 3$

27. $|3x - 9| + 6 = 12$

       $|3x - 9| = 6$

   $3x - 9 = -6$ or $3x - 9 = 6$

      $3x = 3$ $3x = 15$

        $x = 1$ $x = 5$

28.   Let n represent the number.

   $3 - 2(n - 6) = 3n - 5$

   $3 - 2n + 12 = 3n - 5$

     $15 - 2n = 3n - 5$

        $15 = 5n - 5$

        $20 = 5n$

          $4 = n$

   The number is 4.

29. Set expressions equal to each other and solve for n.

   $7n - 250 = 2n + 1200$

        $5n = 1450$

         $n = 290$

The company must sell 290 units to break even.

30. Create a table.

Categories	Rate	Time	Distance
passenger jet	560	t	$560t$
private jet	440	t	$440t$

Translate the information in the table to an equation and solve.

   $560t + 440t = 3000$

       $1000t = 3000$

          $t = 3$

The jets will meet in 3 hours.

Chapter 3

Equations and Inequalities in Two Variables and Functions

Exercise Set 3.1

1. A: (2, 4), B: (–2, 1), C: (0, –3),
    D: (4, –5)

3. a) I   b) IV   c) III   d) y-axis

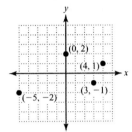

5. a) x-axis   b) IV   c) III   d) I

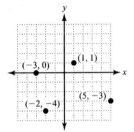

7. a) The figure is shown below.

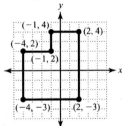

    b) The perimeter of the figure is found by adding
       up the lengths of each side.
       $3 + 7 + 6 + 5 + 3 + 2 = 26$ units
    c) The area of the figure is found by counting the
       grid squares included within the figure: 36
       square units.

9. Replace x with 4 and y with 2 and see if the
    equation is true.
    $$2x + 3y \;=\; 14$$
    $$2(4) + 3(2) \;\overset{?}{=}\; 14$$
    $$8 + 6 \;\overset{?}{=}\; 14$$
    $$14 \;=\; 14$$
    Yes, $(4, 2)$ is a solution.

11. Replace x with $-\dfrac{1}{3}$ and y with $\dfrac{2}{3}$ and see if the
    equation is true.
    $$x - y \;=\; -1$$
    $$-\frac{1}{3} - \frac{2}{3} \;\overset{?}{=}\; -1$$
    $$-\frac{3}{3} \;\overset{?}{=}\; -1$$
    $$-1 \;=\; -1$$
    Yes, $\left(-\dfrac{1}{3}, \dfrac{2}{3}\right)$ is a solution.

13. Replace x with 5 and y with -1 and see if the
    equation is true.
    $$\frac{2}{3}x - y \;=\; 8$$
    $$\frac{2}{3}(5) - (-1) \;\overset{?}{=}\; 8$$
    $$\frac{10}{3} + 1 \;\overset{?}{=}\; 8$$
    $$\frac{13}{3} \;\neq\; 8$$
    No, $(5, -1)$ is not a solution.

15. Replace x with 2.5 and y with -2.1 and see if the
    equation is true.
    $$5.2x \;=\; 6.7 - 3y$$
    $$5.2(2.5) \;\overset{?}{=}\; 6.7 - 3(-2.1)$$
    $$13 \;\overset{?}{=}\; 6.7 + 6.3$$
    $$13 \;=\; 13$$
    Yes, $(2.5, -2.1)$ is a solution.

17. $x - y = 4$

x	y	Ordered Pair
0	-4	$(0, -4)$
4	0	$(4, 0)$
2	-2	$(2, -2)$

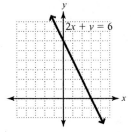

23. $2x - 5y = 10$

x	y	Ordered Pair
0	-2	$(0, -2)$
5	0	$(5, 0)$
-5	-4	$(-5, -4)$

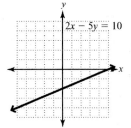

19. $2x + y = 6$

x	y	Ordered Pair
0	6	$(0, 6)$
3	0	$(3, 0)$
2	2	$(2, 2)$

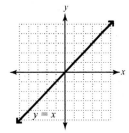

25. $y = x$

x	y	Ordered Pair
0	0	$(0, 0)$
3	3	$(3, 3)$
-2	-2	$(-2, -2)$

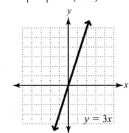

21. $x - 2y = 10$

x	y	Ordered Pair
0	-5	$(0, -5)$
10	0	$(10, 0)$
2	-4	$(2, -4)$

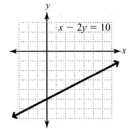

27. $y = 3x$

x	y	Ordered Pair
0	0	$(0, 0)$
1	3	$(1, 3)$
2	6	$(2, 6)$

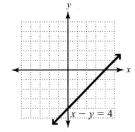

29. $y = -x$

x	y	Ordered Pair
0	0	$(0, 0)$
−2	2	$(-2, 2)$
3	−3	$(3, -3)$

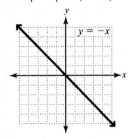

31. $y = -3x$

x	y	Ordered Pair
0	0	$(0, 0)$
−1	3	$(-1, 3)$
1	−3	$(1, -3)$

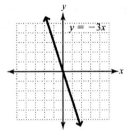

33. $y = \dfrac{1}{3}x$

x	y	Ordered Pair
0	0	$(0, 0)$
3	1	$(3, 1)$
−3	−1	$(-3, -1)$

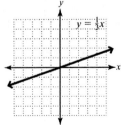

35. $y = -\dfrac{2}{3}x$

x	y	Ordered Pair
0	0	$(0, 0)$
−3	2	$(-3, 2)$
3	−2	$(3, -2)$

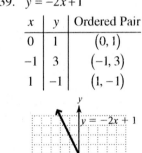

37. $y = 3x + 4$

x	y	Ordered Pair
0	4	$(0, 4)$
−1	1	$(-1, 1)$
−2	−2	$(-2, -2)$

39. $y = -2x + 1$

x	y	Ordered Pair
0	1	$(0, 1)$
−1	3	$(-1, 3)$
1	−1	$(1, -1)$

41. $y = \dfrac{3}{4}x + 2$

x	y	Ordered Pair
0	2	$(0, 2)$
4	5	$(4, 5)$
-4	-1	$(-4, -1)$

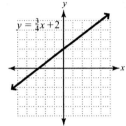

43. $y = -\dfrac{1}{3}x - 1$

x	y	Ordered Pair
0	-1	$(0, -1)$
3	-2	$(3, -2)$
-3	0	$(-3, 0)$

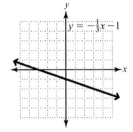

45. $y = 4$

x	y	Ordered Pair
0	4	$(0, 4)$
2	4	$(2, 4)$
-2	4	$(-2, 4)$

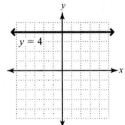

47. $x = 2$

x	y	Ordered Pair
2	0	$(2, 0)$
2	4	$(2, 4)$
2	-1	$(2, -1)$

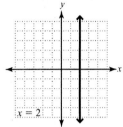

49. $2x + y = 4$

x-intercept	y-intercept
$y = 0: 2x + y = 4$	$x = 0: \quad 2x + y = 4$
$2x + 0 = 4$	$2(0) + y = 4$
$2x = 4$	$0 + y = 4$
$x = 2$	$y = 4$
$(2, 0)$	$(0, 4)$

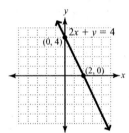

51. $3x + 2y = 6$

x-intercept	y-intercept
$y = 0: \quad 3x + 2y = 6$	$x = 0: \quad 3x + 2y = 6$
$3x + 2(0) = 6$	$3(0) + 2y = 6$
$3x = 6$	$2y = 6$
$x = 2$	$y = 3$
$(2, 0)$	$(0, 3)$

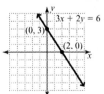

53. $2x - 3y = -6$

<u>x-intercept</u>   <u>y-intercept</u>

$y = 0: \quad 2x - 3y = -6 \qquad x = 0: \quad 2x - 3y = -6$

$\qquad 2x - 3(0) = -6 \qquad\qquad 2(0) - 3y = -6$

$\qquad\qquad 2x = -6 \qquad\qquad\qquad -3y = -6$

$\qquad\qquad\quad x = -3 \qquad\qquad\qquad\quad y = 2$

$\qquad\qquad (-3, 0) \qquad\qquad\qquad\qquad (0, 2)$

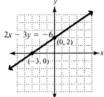

55. $y = x - 3$

<u>x-intercept</u>   <u>y-intercept</u>

$y = 0: \; y = x - 3 \qquad x = 0: \; y = x - 3$

$\qquad\quad 0 = x - 3 \qquad\qquad\quad y = 0 - 3$

$\qquad\quad 3 = x \qquad\qquad\qquad\quad y = -3$

$\qquad\quad (3, 0) \qquad\qquad\qquad (0, -3)$

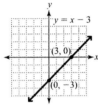

57. $y = -3x$

<u>x-intercept</u>   <u>y-intercept</u>

$y = 0: \; y = -3x \qquad x = 0: \; y = -3x$

$\qquad\quad 0 = -3x \qquad\qquad\quad y = -3(0)$

$\qquad\quad 0 = x \qquad\qquad\qquad\quad y = 0$

$\qquad\quad (0, 0) \qquad\qquad\qquad (0, 0)$

Extra points:

$x = -1: \; y = -3x \qquad x = 1: \; y = -3x$

$\qquad\qquad y = -3(-1) \qquad\qquad y = -3(1)$

$\qquad\qquad y = 3 \qquad\qquad\qquad\quad y = -3$

$\qquad\qquad (-1, 3) \qquad\qquad\qquad (1, -3)$

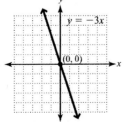

59. $y = -\dfrac{3}{4}x + 1$

<u>x-intercept</u>   <u>y-intercept</u>

$y = 0: \qquad\qquad\qquad\quad x = 0:$

$\qquad y = -\dfrac{3}{4}x + 1 \qquad\quad y = -\dfrac{3}{4}x + 1$

$\qquad 0 = -\dfrac{3}{4}x + 1 \qquad\quad y = -\dfrac{3}{4}(0) + 1$

$\qquad \dfrac{4}{3} \cdot \dfrac{3}{4}x = \dfrac{4}{3} \cdot 1 \qquad\qquad y = 1$

$\qquad\qquad x = \dfrac{4}{3} \qquad\qquad\qquad (0, 1)$

$\qquad\qquad \left(\dfrac{4}{3}, 0 \right)$

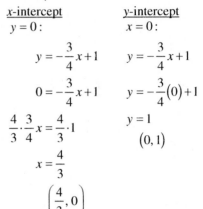

61. The graph gets steeper.

63. Adding b (where $b > 0$) shifts the graph up the y-axis.

65. a) $(-3, 1), (0, 1), (-2, -3), (1, -3)$

     b) $(1, 2,), (4, 2), (2, -2), (5, -2)$

     c) $\left[(x + 4), (y + 1) \right]$

67. a) $c = 5n + 60$ b) $c = 5n + 60$

         $c = 5(7) + 60$ $90 = 5n + 60$

         $c = 35 + 60$ $30 = 5n$

         $c = \$95$ $6 \text{ hr.} = n$

     c) The graph is shown below.

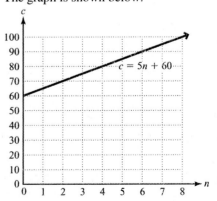

69. a) The number of minutes used over 450 was
    $600 - 450 = 150$.

    $c = 0.05n + 40$

    $c = 0.05(150) + 40$

    $c = 7.5 + 40$

    $c = \$47.50$

    b) $c = 0.05n + 40$

    $41.75 = 0.05n + 40$

    $1.75 = 0.05n$

    $35 \text{ min.} = n$

    c) The graph is shown below.

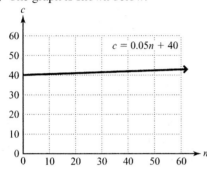

71. a) The number of copies beyond 200 was
    $300 - 200 = 100$.

    $c = 0.03n + 5$

    $c = 0.03(100) + 5$

    $c = 3 + 5$

    $c = \$8$

    b) $c = 0.03n + 5$

    $11 = 0.03n + 5$

    $6 = 0.03n$

    $200 = n$

    Since the 200 copies is then added to the "first 200 copies," the total number of copies is 400.

    c) The graph is shown below.

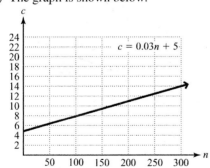

Review Exercises

1. Commutative property of addition

2. $\dfrac{3(5-8)+21}{8-3(6)} = \dfrac{3(-3)+21}{8-18} = \dfrac{-9+21}{-10} = \dfrac{12}{-10} = -\dfrac{6}{5}$

3. $\dfrac{y_2 - y_1}{x_2 - x_1} = \dfrac{-2-(-8)}{1-4} = \dfrac{-2+8}{-3} = \dfrac{6}{-3} = -2$

4. $\dfrac{y_2 - y_1}{x_2 - x_1} = \dfrac{5-13}{2-(-10)} = \dfrac{-8}{2+10} = \dfrac{-8}{12} = -\dfrac{2}{3}$

5. $12 - 3(n-2) = 6n - (2n+1)$

   $12 - 3n + 6 = 6n - 2n - 1$

   $18 - 3n = 4n - 1$

   $19 = 7n$

   $\dfrac{19}{7} = n$

6. $7x - 5 > 4x + 13 \qquad \{x | x > 6\} \qquad (6, \infty)$

   $3x - 5 > 13$

   $3x > 18$

   $x > 6$

Exercise Set 3.2

1. The graph of $y = 3x + 1$ is steeper than the graph of $y = 2x - 7$ because a slope of 3 is greater than a slope of 2.

3. The graph of $y = x - 4$ is steeper than the graph of $y = 0.2x + 3$ because a slope of 1 is greater than a slope of 0.2.

5. The graph of $y = 3x + 2$ has an upward incline from left to right because the slope is positive ($m = 3$).

7. The graph of $y = -\dfrac{4}{5}x - 3$ has a downward incline from left to right because the slope is negative $\left(m = -\dfrac{4}{5}\right)$.

9. From the graph, we can see that the line rises vertically 2 units and then runs horizontally 3 units for a slope of $m = \dfrac{2}{3}$. The y-intercept is $b = (0,1)$.

11. From the graph, we can see that the line is vertical, so it has an undefined slope. It does not cross the y-axis, so it has no y-intercept.

13. e   15. g

17. h   19. d

21. $y = \dfrac{2}{3}x + 5$

    $m = \dfrac{2}{3}$; y-intercept: $(0, 5)$

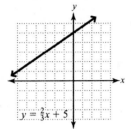

23. $y = -\dfrac{3}{4}x - 2$

    $m = -\dfrac{3}{4}$; y-intercept: $(0, -2)$

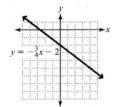

25. $y = x + 4$

    $m = 1$; y-intercept: $(0, 4)$

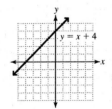

27. $y = -5x + \dfrac{2}{3}$

    $m = -5$; y-intercept: $\left(0, \dfrac{2}{3}\right)$

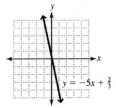

29. $2x + 3y = 6$

    $3y = -2x + 6$

    $y = -\dfrac{2}{3}x + 2$

    $m = -\dfrac{2}{3}$; y-intercept: $(0, 2)$

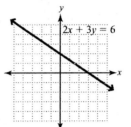

31. $x - 2y = -7$

    $-2y = -x - 7$

    $y = \dfrac{1}{2}x + \dfrac{7}{2}$

    $m = \dfrac{1}{2}$; y-intercept: $\left(0, \dfrac{7}{2}\right)$

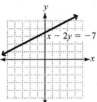

33. $2x - 7y = 8$

    $-7y = -2x + 8$

    $y = \dfrac{2}{7}x - \dfrac{8}{7}$

    $m = \dfrac{2}{7}$; y-intercept: $\left(0, -\dfrac{8}{7}\right)$

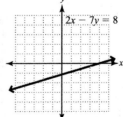

35. $-x + y = 0$

$$y = x$$

$$m = 1; \ y - \text{intercept}: (0, 0)$$

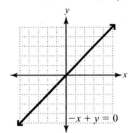

37. $(4, 2), (2, 6)$

$$m = \frac{y_2 - y_1}{x_2 - x_1} = \frac{6 - 2}{2 - 4} = -\frac{4}{2} = -2$$

39. $(-1, 3), (4, -6)$

$$m = \frac{y_2 - y_1}{x_2 - x_1} = \frac{-6 - 3}{4 - (-1)} = \frac{-9}{4 + 1} = -\frac{9}{5}$$

41. $(1, 4), (-3, 10)$

$$m = \frac{y_2 - y_1}{x_2 - x_1} = \frac{10 - 4}{-3 - 1} = \frac{6}{-4} = -\frac{3}{2}$$

43. $(8, 2), (8, -5)$

$$m = \frac{y_2 - y_1}{x_2 - x_1} = \frac{-5 - 2}{8 - 8} = -\frac{7}{0} \text{ is undefined.}$$

45. $(5, 12), (-1, 12)$

$$m = \frac{y_2 - y_1}{x_2 - x_1} = \frac{12 - 12}{-1 - 5} = \frac{0}{-6} = 0$$

47. $(0, 0), (10, -8)$

$$m = \frac{y_2 - y_1}{x_2 - x_1} = \frac{-8 - 0}{10 - 0} = \frac{-8}{10} = -\frac{4}{5}$$

49. a) The parallelogram is shown below.

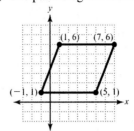

   b) Right: $m = \dfrac{6 - 1}{7 - 5} = \dfrac{5}{2}$

   Left: $m = \dfrac{6 - 1}{1 - (-1)} = \dfrac{5}{1 + 1} = \dfrac{5}{2}$

   Top: $m = \dfrac{6 - 6}{7 - 1} = \dfrac{0}{6} = 0$

   Bottom: $m = \dfrac{1 - 1}{5 - (-1)} = \dfrac{0}{5 + 1} = \dfrac{0}{6} = 0$

   c) Slopes of parallel sides are equal.

51. $m = \dfrac{29}{348} = \dfrac{1}{12}$

53. $m = \dfrac{215 - 210}{20 - 0} = \dfrac{5}{20} = 0.25$

55. a) $C(50, 15), D(100, 80)$

   b) $m = \dfrac{y_2 - y_1}{x_2 - x_1} = \dfrac{80 - 15}{100 - 50} = \dfrac{65}{50} = \dfrac{13}{10}$

   c) The coaster climbs 13 ft. vertically for every 10 ft. that it moves horizontally.

57. a)

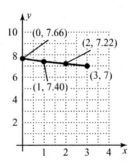

   b) $m = \dfrac{y_2 - y_1}{x_2 - x_1} = \dfrac{7 - 7.66}{3 - 0} = -\dfrac{0.66}{3} = -0.22$

   c) $7 - 0.22 = 6.78 = \$6.78$

59. a)

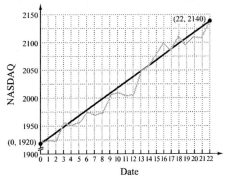

b) $m = \dfrac{y_2 - y_1}{x_2 - x_1} = \dfrac{2140 - 1920}{22 - 0} = \dfrac{220}{22} = 10$

c) $2140 + 10 = 2150$

61. a)

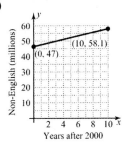

b) $m = \dfrac{y_2 - y_1}{x_2 - x_1} = \dfrac{58.1 - 47}{10 - 0} = \dfrac{11.1}{10} = 1.11$

63. a)

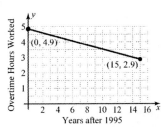

b) $m = \dfrac{y_2 - y_1}{x_2 - x_1} = \dfrac{2.9 - 4.9}{15 - 0} = -\dfrac{2}{15} = -0.1\overline{3}$

Review Exercises

1. $4n^2 - m(n + 2) = 4(-2)^2 - 7(-2 + 2)$
    $\qquad = 4(4) - 7(0)$
    $\qquad = 16 - 0$
    $\qquad = 16$

2. $4n^2 - m(n + 2) = 4(3)^2 - (-5)(3 + 2)$
    $\qquad = 4(9) + 5(5)$
    $\qquad = 36 + 25$
    $\qquad = 61$

3. $16x - 3y - 15 - x + 3y - 12$
    $= 16x - x - 3y + 3y - 15 - 12$
    $= 15x - 27$

4. $-5(3x - 2) = -5(3x) - 5(-2) = -15x + 10$

5. $V = lwh$
    $\dfrac{V}{lw} = \dfrac{lwh}{lw}$
    $\dfrac{V}{lw} = h$

6. $Ax + By = C$
    $By = C - Ax$
    $\dfrac{By}{B} = \dfrac{C - Ax}{B}$
    $y = \dfrac{C - Ax}{B}$

Exercise Set 3.3

1. $m = -4; (0, 3)$
    $y = mx + b$
    $y = -4x + 3$

3. $m = \dfrac{3}{5}; (0, -2)$
    $y = mx + b$
    $y = \dfrac{3}{5}x - 2$

5. $m = -0.2; (0, -1.5)$
    $y = mx + b$
    $y = -0.2x - 1.5$

7. $m = \dfrac{1}{2}; (0, 2)$
    $y = mx + b$
    $y = \dfrac{1}{2}x + 2$

9. $m = -2; (0, -4)$
    $y = mx + b$
    $y = -2x - 4$

11. $m = \dfrac{4 - 2}{3 - 0} = \dfrac{2}{3}; (0, 2)$
    $y = mx + b$
    $y = \dfrac{2}{3}x + 2$

13. $m = \dfrac{-3-(-6)}{-2-0} = -\dfrac{3}{2};\ (0,-6)$

$y = mx + b$

$y = -\dfrac{3}{2}x - 6$

15. $y - y_1 = m(x - x_1)$

$y - 1 = 2(x - 3)$

$y - 1 = 2x - 6$

$y = 2x - 5$

17. $y - y_1 = m(x - x_1)$

$y - 0 = -2(x - (-5))$

$y = -2(x + 5)$

$y = -2x - 10$

19. $y - y_1 = m(x - x_1)$

$y - 0 = -3(x - 0)$

$y = -3x$

21. $y - y_1 = m(x - x_1)$

$y - (-2) = \dfrac{2}{5}(x - 0)$

$y + 2 = \dfrac{2}{5}x$

$y = \dfrac{2}{5}x - 2$

23. $y - y_1 = m(x - x_1)$

$y - (-2) = \dfrac{4}{5}(x - (-2))$

$y + 2 = \dfrac{4}{5}(x + 2)$

$y + 2 = \dfrac{4}{5}x + \dfrac{8}{5}$

$y = \dfrac{4}{5}x - \dfrac{2}{5}$

25. $y - y_1 = m(x - x_1)$

$y - 3 = -\dfrac{3}{2}(x - (-1))$

$y - 3 = -\dfrac{3}{2}(x + 1)$

$y - 3 = -\dfrac{3}{2}x - \dfrac{3}{2}$

$y = -\dfrac{3}{2}x + \dfrac{3}{2}$

27. Find the slope: $m = \dfrac{3-(-1)}{2-4} = \dfrac{3+1}{-2} = \dfrac{4}{-2} = -2$

a) $y - 3 = -2(x - 2)$

$y - 3 = -2x + 4$

$y = -2x + 7$

b) $y = -2x + 7$

$2x + y = 7$

29. Find the slope: $m = \dfrac{-7-(-4)}{-2-1} = \dfrac{-7+4}{-3} = \dfrac{-3}{-3} = 1$

a) $y - y_1 = m(x - x_1)$

$y - (-4) = 1(x - 1)$

$y + 4 = x - 1$

$y = x - 5$

b) $y = x - 5$

$-x + y = -5$

$x - y = 5$

31. Find the slope: $m = \dfrac{0-(-6)}{3-(-6)} = \dfrac{0+6}{3+6} = \dfrac{6}{9} = \dfrac{2}{3}$

a) $y - y_1 = m(x - x_1)$

$y - 0 = \dfrac{2}{3}(x - 3)$

$y = \dfrac{2}{3}x - 2$

b) $y = \dfrac{2}{3}x - 2$

$3y = 2x - 6$

$-2x + 3y = -6$

$2x - 3y = 6$

33. Find the slope: $m = \dfrac{6-0}{0-(-5)} = \dfrac{6}{0+5} = \dfrac{6}{5}$

a) $y - y_1 = m(x - x_1)$

$y - 0 = \dfrac{6}{5}(x - (-5))$

$y = \dfrac{6}{5}(x + 5)$

$y = \dfrac{6}{5}x + 6$

b) $y = \dfrac{6}{5}x + 6$

$5y = 6x + 30$

$-6x + 5y = 30$

$6x - 5y = -30$

35. Find the slope: $m = \dfrac{3-2}{-9-4} = \dfrac{1}{-13} = -\dfrac{1}{13}$

a) $y - y_1 = m(x - x_1)$

$y - 2 = -\dfrac{1}{13}(x - 4)$

$y - 2 = -\dfrac{1}{13}x + \dfrac{4}{13}$

$y = -\dfrac{1}{13}x + \dfrac{30}{13}$

b) $y = -\dfrac{1}{13}x + \dfrac{30}{13}$

$13y = -x + 30$

$x + 13y = 30$

37. Find the slope: $m = \dfrac{9-2}{-5-(-2)} = \dfrac{7}{-5+2} = \dfrac{7}{-3}$

a) $y - y_1 = m(x - x_1)$

$y - 2 = -\dfrac{7}{3}(x - (-2))$

$y - 2 = -\dfrac{7}{3}(x + 2)$

$y - 2 = -\dfrac{7}{3}x - \dfrac{14}{3}$

$y = -\dfrac{7}{3}x - \dfrac{8}{3}$

b) $y = -\dfrac{7}{3}x - \dfrac{8}{3}$

$3y = -7x - 8$

$7x + 3y = -8$

39. Because these lines have equal slopes and different *y*-intercepts, they are parallel.

41. Because the slopes are $\dfrac{3}{4}$ and $-\dfrac{4}{3}$, these lines are perpendicular.

43. These slopes are neither equal nor of the form $\dfrac{a}{b}$ and $-\dfrac{b}{a}$. Therefore, these lines are neither parallel nor perpendicular.

45. $2x + 7y = 8$ 　　　　　　 $4x + 14y = -9$

$7y = -2x + 8$ 　　　　　 $14y = -4x - 9$

$y = -\dfrac{2}{7}x + \dfrac{8}{7}$ 　　　 $y = -\dfrac{4}{14}x - \dfrac{9}{14}$

　　　　　　　　　　　　 $y = -\dfrac{2}{7}x - \dfrac{9}{14}$

Because these lines have equal slopes and different *y*-intercepts, they are parallel.

47. $3x + 5y = 4$ 　　　　　 $5x - 3y = 2$

$5y = -3x + 4$ 　　　 $5x - 2 = 3y$

$y = -\dfrac{3}{5}x + \dfrac{4}{5}$ 　 $\dfrac{5}{3}x - \dfrac{2}{3} = y$

Because the slopes are $-\dfrac{3}{5}$ and $\dfrac{5}{3}$, these lines are perpendicular.

49. The first line is vertical and the second line is horizontal. Horizontal and vertical lines are perpendicular.

51. a) $m = -5$

$y - y_1 = m(x - x_1)$

$y - (-6) = -5(x - 2)$

$y + 6 = -5x + 10$

$y = -5x + 4$

b) $y = -5x + 4$

$5x + y = 4$

53. a) $m = 4$

$y - y_1 = m(x - x_1)$

$y - (-2) = 4(x - (-5))$

$y + 2 = 4(x + 5)$

$y + 2 = 4x + 20$

$y = 4x + 18$

b) $y = 4x + 18$

$4x - y = -18$

55. a) $m = \dfrac{2}{3}$

$y - y_1 = m(x - x_1)$

$y - (-4) = \dfrac{2}{3}(x - 3)$

$y + 4 = \dfrac{2}{3}x - 2$

$y = \dfrac{2}{3}x - 6$

b) $y = \dfrac{2}{3}x - 6$

$3y = 2x - 18$

$2x - 3y = 18$

57. $4x + 6y = 3$

$6y = -4x + 3$

$y = -\dfrac{2}{3}x + \dfrac{1}{2}$

a) $m = -\dfrac{2}{3}$

$y - y_1 = m(x - x_1)$

$y - (-3) = -\dfrac{2}{3}(x - 4)$

$y + 3 = -\dfrac{2}{3}x + \dfrac{8}{3}$

$y = -\dfrac{2}{3}x - \dfrac{1}{3}$

b) $y = -\dfrac{2}{3}x - \dfrac{1}{3}$

$3y = -2x - 1$

$2x + 3y = -1$

59. $2x + 5y - 30 = 0$

$5y = -2x + 30$

$y = -\dfrac{2}{5}x + 6$

a) $m = -\dfrac{2}{5}$

$y - y_1 = m(x - x_1)$

$y - (-7) = -\dfrac{2}{5}(x - (-3))$

$y + 7 = -\dfrac{2}{5}(x + 3)$

$y + 7 = -\dfrac{2}{5}x - \dfrac{6}{5}$

$y = -\dfrac{2}{5}x - \dfrac{41}{5}$

b) $y = -\dfrac{2}{5}x - \dfrac{41}{5}$

$5y = -2x - 41$

$2x + 5y = -41$

61. a) The slope of the given equation is $m = \dfrac{1}{3}$, so the slope of a perpendicular line is $m = -3$.

$y - y_1 = m(x - x_1)$

$y - (-1) = -3(x - 2)$

$y + 1 = -3x + 6$

$y = -3x + 5$

b) $y = -3x + 5$

$3x + y = 5$

63. a) The slope of the given equation is $m = -\dfrac{2}{5}$, so the slope of a perpendicular line is $m = \dfrac{5}{2}$.

$y - y_1 = m(x - x_1)$

$y - (-3) = \dfrac{5}{2}(x - 0)$

$y + 3 = \dfrac{5}{2}x$

$y = \dfrac{5}{2}x - 3$

b) $y = \dfrac{5}{2}x - 3$

$2y = 5x - 6$

$-5x + 2y = -6$

$5x - 2y = 6$

65. a) The slope of the given equation is $m = -3$, so the slope of a perpendicular line is $m = \dfrac{1}{3}$.

$$y - y_1 = m(x - x_1)$$
$$y - (-9) = \frac{1}{3}(x - (-2))$$
$$y + 9 = \frac{1}{3}(x + 2)$$
$$y + 9 = \frac{1}{3}x + \frac{2}{3}$$
$$y = \frac{1}{3}x - \frac{25}{3}$$

   b) $\qquad y = \dfrac{1}{3}x - \dfrac{25}{3}$
$$3y = x - 25$$
$$-x + 3y = -25$$
$$x - 3y = 25$$

67. $x + 4y = -10$
$$4y = -x - 10$$
$$y = -\frac{1}{4}x - \frac{5}{2}$$

   a) The slope of the given equation is $m = -\dfrac{1}{4}$, so the slope of a perpendicular line is $m = 4$.
$$y - y_1 = m(x - x_1)$$
$$y - (-3) = 4(x - (-2))$$
$$y + 3 = 4(x + 2)$$
$$y + 3 = 4x + 8$$
$$y = 4x + 5$$

   b) $\qquad y = 4x + 5$
$$-4x + y = 5$$
$$4x - y = -5$$

69. $2x - 3y = 15$
$$-3y = -2x + 15$$
$$y = \frac{2}{3}x - 5$$

   a) The slope of the given equation is $m = \dfrac{2}{3}$, so the slope of a perpendicular line is $m = -\dfrac{3}{2}$.
$$y - y_1 = m(x - x_1)$$
$$y - 4 = -\frac{3}{2}(x - 1)$$
$$y - 4 = -\frac{3}{2}x + \frac{3}{2}$$
$$y = -\frac{3}{2}x + \frac{11}{2}$$

   b) $\qquad y = -\dfrac{3}{2}x + \dfrac{11}{2}$
$$2y = -3x + 11$$
$$3x + 2y = 11$$

71. $y = -4$

73. $y = 0$

75. a) $m = \dfrac{313.9 - 282.4}{12 - 0} = \dfrac{31.5}{12} = 2.625$

   b) $p = 2.625t + 282.4$

   c) 2020 would be year 20.
$$p = 2.625(20) + 282.4$$
$$= 334.9$$
The model predicts 334,900,000 people in 2020.

77. a) $m = \dfrac{526 - 908}{8 - 0} = \dfrac{-382}{8} = -47.75$

   b) $b = -47.75t + 908$

   c) 2020 would be year 16.
$$b = -47.75(16) + 908$$
$$= 144$$
The predicted average number of barrels in 2020 is 144 thousand (or 144,000) barrels.

79. a)

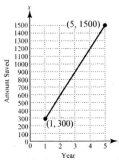

b) $m = \dfrac{1500 - 300}{5 - 1} = 300$

c) $s = 300t$

d) $s = 300(10)$

$s = \$3000$

81. a)

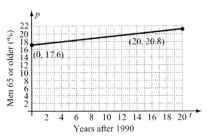

b) $m = \dfrac{20.8 - 17.6}{20 - 0} = \dfrac{3.2}{20} = 0.16$

c) $p = 0.16t + 17.6$

d) 2025 would be the year 35

$p = 0.16(35) + 17.6$

$= 23.2$

The percent is 23.2%.

Review Exercises

1. $-0.8 < -\dfrac{1}{8}$

2. $4(2x + 3) - 2(x - 6) = 7 - (3x + 1)$

$8x + 12 - 2x + 12 = 7 - 3x - 1$

$6x + 24 = 6 - 3x$

$9x = -18$

$x = -2$

3. $\dfrac{1}{4}x - \dfrac{5}{6} = \dfrac{3}{2}x - 1$

$\dfrac{1}{6} = \dfrac{5}{4}x$

$\dfrac{2}{15} = x$

4. $-4x - 7 \geq 13$ $\qquad \{x \mid x \leq -5\}$ $\qquad (-\infty, -5]$

$-4x \geq 20$

$x \leq -5$

$$\overset{\longleftarrow}{\underset{-8\,-7\,-6\,-5\,-4\,-3\,-2\,-1\ \ 0\ \ 1\ \ 2\ \ 3\ \ 4}{\mid\mid\mid\mid\mid\mid\mid\mid\mid\mid\mid\mid}}$$

5. $6x - 2(x - 1) \geq 9x - 8$

$6x - 2x + 2 \geq 9x - 8$

$4x + 2 \geq 9x - 8$

$-5x \geq -10$

$x \leq 2$

$\{x \mid x \leq 2\}$ $\qquad\qquad (-\infty, 2]$

$$\overset{\longleftarrow}{\underset{-2\ \ -1\ \ \ 0\ \ \ 1\ \ \ 2\ \ \ 3}{\mid\mid\mid\mid\mid\mid}}$$

6. $50w \geq 2000$

$w \geq 40$ ft.

Exercise Set 3.4

1. Test $(-1, 3)$.

$y \ \geq \ -x + 7$

$3 \ \overset{?}{\geq} \ -(-1) + 7$

$3 \ \overset{?}{\geq} \ 1 + 7$

$3 \ \geq \ 8$

This statement is false. No, $(-1, 3)$ is not a solution.

3. Test $(0, 0)$.

$2x + 3y \ \geq \ 0$

$2(0) + 3(0) \ \overset{?}{\geq} \ 0$

$0 + 0 \ \overset{?}{\geq} \ 0$

$0 \ \geq \ 0$

This statement is true. Yes, $(0, 0)$ is a solution.

5.   Test $(3, 4)$.

$$3y - x \;\; < \;\; 8$$
$$3(4) - 3 \;\overset{?}{<}\; 8$$
$$12 - 3 \;\overset{?}{<}\; 8$$
$$9 \;<\; 8$$

This statement is false. No, $(3, 4)$ is not a solution.

7.   Test $(5, 3)$.

$$y \;>\; \frac{3}{4}x - 5$$
$$3 \;\overset{?}{>}\; \frac{3}{4}(5) - 5$$
$$3 \;\overset{?}{>}\; \frac{15}{4} - 5$$
$$3 \;>\; -\frac{5}{4}$$

This statement is true. Yes, $(5, 3)$ is a solution.

9. $y \le x + 3$

Begin by graphing the related equation $y = x + 3$ with a solid line. Now choose $(0, 0)$ as a test point.

$$y \;\le\; x + 3$$
$$0 \;\overset{?}{\le}\; 0 + 3$$
$$0 \;\le\; 3$$

Because $(0, 0)$ satisfies the inequality, shade the region which includes $(0, 0)$.

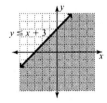

11. $y \ge -4x + 6$

Begin by graphing the related equation $y = -4x + 6$ with a solid line. Now choose $(0, 0)$ as a test point.

$$y \;\ge\; -4x + 6$$
$$0 \;\overset{?}{\ge}\; -4(0) + 6$$
$$0 \;\ge\; 6$$

Because $(0, 0)$ does not satisfy the inequality, shade the side of the line on the opposite side from $(0, 0)$.

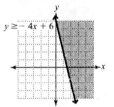

13. $y > x$

Begin by graphing the related equation $y = x$ with a dashed line. Now choose $(1, 0)$ as a test point.

$$y \;>\; x$$
$$0 \;\overset{?}{>}\; (1)$$
$$0 \;>\; 1$$

Because $(1, 0)$ does not satisfy the inequality, shade the side of the line on the opposite side from $(1, 0)$.

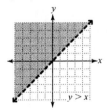

15. $y > -3x$

Begin by graphing the related equation $y = -3x$ with a dashed line. Now choose $(1, 0)$ as a test point.

$$y \;>\; -3x$$
$$0 \;\overset{?}{>}\; -3(1)$$
$$0 \;>\; -3$$

Because $(1, 0)$ satisfies the inequality, shade the region which includes $(1, 0)$.

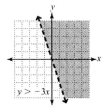

17. $y > \dfrac{2}{5}x$

Begin by graphing the related equation $y = \dfrac{2}{5}x$ with a dashed line. Now choose $(1, 0)$ as a test point.

$$y \;>\; \dfrac{2}{5}x$$
$$0 \;\overset{?}{>}\; \dfrac{2}{5}(1)$$
$$0 \;>\; \dfrac{2}{5}$$

Because $(1, 0)$ does not satisfy the inequality, shade the side of the line on the opposite side from $(1, 0)$.

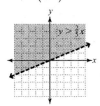

19. $x - y < 2$
$$-y < -x + 2$$
$$y > x - 2$$

Begin by graphing the related equation $y = x - 2$ with a dashed line. Now choose $(0, 0)$ as a test point.

$$x - y \;<\; 2$$
$$0 - 0 \;\overset{?}{<}\; 2$$
$$0 \;<\; 2$$

Because $(0, 0)$ satisfies the inequality, shade the region which includes $(0, 0)$.

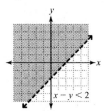

21. $x + 3y > -9$
$$3y > -x - 9$$
$$y > -\dfrac{1}{3}x - 3$$

Begin by graphing the related equation $y = -\dfrac{1}{3}x - 3$ with a dashed line. Now choose $(0, 0)$ as a test point.

$$x + 3y \;>\; -9$$
$$0 - 3(0) \;\overset{?}{>}\; -9$$
$$0 - 0 \;\overset{?}{>}\; -9$$
$$0 \;>\; -9$$

Because $(0, 0)$ satisfies the inequality, shade the region which includes $(0, 0)$.

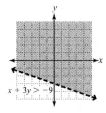

23. $x - 2y \geq -6$

$$-2y \geq -x - 6$$

$$y \leq \frac{1}{2}x + 3$$

Begin by graphing the related equation

$y = \frac{1}{2}x + 3$ with a solid line. Now choose $(0, 0)$

as a test point.

$$x - 2y \quad \geq \quad -6$$

$$0 - 2(0) \overset{?}{\geq} -6$$

$$0 - 0 \overset{?}{\geq} -6$$

$$0 \quad \geq \quad -6$$

Because $(0, 0)$ satisfies the inequality, shade the

region which includes $(0, 0)$.

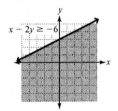

25. $3x - 2y > 6$

$$-2y > -3x + 6$$

$$y < \frac{3}{2}x - 3$$

Begin by graphing the related equation

$y = \frac{3}{2}x - 3$ with a dashed line. Now choose $(0, 0)$

as a test point.

$$3x - 2y \quad > \quad 6$$

$$3(0) - 2(0) \overset{?}{>} 6$$

$$0 - 0 \overset{?}{>} 6$$

$$0 \quad > \quad 6$$

Because $(0, 0)$ does not satisfy the inequality,

shade the side of the line on the opposite side

from $(0, 0)$.

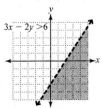

27. $5x - y \leq 0$

$$-y \leq -5x$$

$$y \geq 5x$$

Begin by graphing the related equation $y = 5x$

with a solid line. Now choose $(1, 0)$ as a test point.

$$5x - y \quad \leq \quad 0$$

$$5(1) - 0 \overset{?}{\leq} 0$$

$$5 - 0 \overset{?}{\leq} 0$$

$$5 \quad \leq \quad 0$$

Because $(1, 0)$ does not satisfy the inequality,

shade the side of the line on the opposite side from

$(1, 0)$.

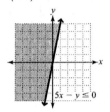

29. $4x + 2y \leq 3$

$$2y \leq -4x + 3$$

$$y \leq -2x + \frac{3}{2}$$

Begin by graphing the related equation

$y = -2x + \frac{3}{2}$ with a solid line. Now choose $(0, 0)$

as a test point.

$$4x + 2y \quad \leq \quad 3$$

$$4(0) + 2(0) \overset{?}{\leq} 3$$

$$0 + 0 \overset{?}{\leq} 3$$

$$0 \quad \leq \quad 3$$

Because $(0, 0)$ satisfies the inequality, shade the

region which includes $(0, 0)$.

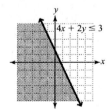

31. $x > 6$

Begin by graphing the related equation $x = 6$ with a dashed line. Now choose $(0, 0)$ as a test point.

$$x > 6$$
$$0 \overset{?}{>} 6$$

Because $(0, 0)$ does not satisfy the inequality, shade the side of the line on the opposite side from $(0, 0)$.

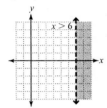

33. $y \le 7$

Begin by graphing the related equation $y = 7$ with a solid line. Now choose $(0, 0)$ as a test point.

$$y < 7$$
$$0 \overset{?}{<} 7$$

Because $(0, 0)$ satisfies the inequality, shade the region which includes $(0, 0)$.

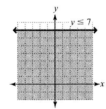

35. a) x represents the number of board games produced, and y represents the number of video games produced.

b) Because x and y represent nonnegative numbers, the graph should be in the first quadrant only $(x \ge 0 \text{ and } y \ge 0)$.

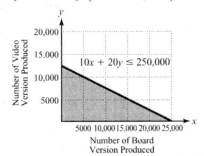

c) All combinations that cost exactly $250,000 to produce.

d) All combinations that cost less than $250,000 to produce.

e) Answers may vary. Two examples are (0, 12,500) and (25,000, 0).

f) Answers may vary. Two examples are (0, 500) and (10,000, 5000).

g) No, fractions of a game are not produced.

37. a) $2l + 2w \le 200$

b)

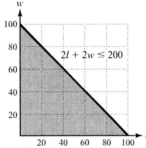

c) Combinations of length and width that make the perimeter exactly 200 ft.

d) All combinations of length and width that make the perimeter less than 200 ft.

e) Answers may vary. Two examples are (20, 80) and (60, 40).

f) Answers may vary. Two examples are (20, 40) and (80, 10).

39. a) $12x + 15y \ge 18,000$

b)

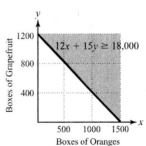

c) All sales combinations that raise exactly $18,000.

d) All sales combination that will raise more than $18,000.

e) Answers may vary. Two examples are (500, 800) and (1000, 400).

f) Answers may vary. Two examples are (500, 1000) and (800, 1200).

Review Exercises

1. $\{-2, 0, 3, 4\}$

2. $\{1, 2, 5, 7\}$

3. $x^2 - 4x + 1 = (-2)^2 - 4(-2) + 1$

$$= 4 + 8 + 1$$
$$= 13$$

4. $x - 3 = 0$

$$x = 3$$

The expression is undefined when the denominator equals 0, which occurs when $x = 3$.

5. $3x - 4(x + 8) = 5(x - 1) + 21$

$$3x - 4x - 32 = 5x - 5 + 21$$
$$-x - 32 = 5x + 16$$
$$-48 = 6x$$
$$-8 = x$$

6. x-intercept: $4x - 5(0) = 8$

$$4x = 8$$
$$x = 2$$
$$(2, 0)$$

y-intercept: $4(0) - 5y = 8$

$$-5y = 8$$
$$y = -\frac{8}{5}$$
$$\left(0, -\frac{8}{5}\right)$$

Exercise Set 3.5

1. Domain: {Sarah Lawrence College, New York University, Harvey Mudd College, Columbia College at Columbia University, Wesleyan University}
Range: {$61,236, $59,337, $58,913, $58,742, $58,502}
It is a function because every element in the domain is paired with exactly one element in the range.

3. Domain: {1, 2, 3, 4, 5, 6}
Range: {San Francisco 49ers, Dallas Cowboys, Pittsburgh Steelers, Green Bay Packers, New England Patriots, Oakland/LA Raiders, Washington Redskins, New York Giants, Miami Dolphins, Denver Broncos, Indianapolis/Baltimore Colts, Baltimore Ravens, Chicago Bears, New York Jets, Tampa Bay Buccaneers, Kansas City Chiefs, St. Louis/LA Rams, New Orleans Saints}
It is not a function because an element in the domain is paired with more than one element in the range. (More than one team has won each number of Superbowls.)

5. Domain: {1%, 2%, 5%, 6%, 8%, 35%, 42%}
Range: {Windows XP, Windows 7, Windows Vista, Mac OS X, iOS, Android, Linux, Other}
It is not a function because an element in the domain is paired with more than one element in the range.(1% is paired with both Android and Linux.)

7. Domain: {1, 2, 3, 4, 5}
Range: {*Thriller, Back in Black, Dark Side of the Moon, Bat out of Hell, Their Greatest Hits 1971–1975*}
It is a function because every element in the domain is paired with exactly one element in the range.

9. Domain: { −2, 6, 5, −1 }
Range: {4, −2, 1, −6}
It is a function because every element in the domain is paired with exactly one element in the range.

11. Domain: $\{-6, 2, -5\}$
Range: {2, −3, 7, 4}
It is not a function because an element in the domain is paired with more than one element in the range.

13. Domain: $\{2, -4, 5, 3\}$
Range: $\{-4, 3, -1\}$
It is a function because every element in the domain is paired with exactly one element in the range.

15. Domain: {2000, 2001, 2002, 2003, 2004, 2005, 2006, 2007, 2008 2009, 2010}
Range: {5920, 5915, 5534, 5574, 5764, 5734, 5840, 5657, 5214, 4551, 4690}
It is a function because every element in the domain is paired with exactly one element in the range.

17. Domain: $\mathbb{R}$ or $(-\infty, \infty)$

   Range: $\{y \mid y \leq 4\}$ or $(-\infty, 4]$

   It is a function because a vertical line at each x-value would intersect the graph at only one point.

19. Domain: $\{x \mid x \geq 0\}$ or $[0, \infty)$

   Range: $\mathbb{R}$ or $(-\infty, \infty)$

   It is not a function because there are values in the domain that correspond to two values in the range.

21. Domain: $\{x \mid -4 \leq x \leq 5\}$ or $[-4, 5]$

   Range: $\{-3, -2\}$

   It is a function because a vertical line at each x-value would intersect the graph at only one point.

23. Domain: $\{x \mid -4 \leq x \leq 0\}$ or $[-4, 0]$

   Range: $\{y \mid -3 \leq y \leq 1\}$ or $[-3, 1]$

   It is not a function because there are values in the domain that correspond to two values in the range.

25. $f(x) = -2x - 9$

   a) $f(0) = -2(0) - 9$
   $= 0 - 9$
   $= -9$

   b) $f(1) = -2(1) - 9$
   $= -2 - 9$
   $= -11$

   c) $f(-1) = -2(-1) - 9$
   $= 2 - 9$
   $= -7$

   d) $f(a+1) = -2(a+1) - 9$
   $= -2a - 2 - 9$
   $= -2a - 11$

27. $f(x) = 2x^2 - x + 7$

   a) $f(0) = 2(0)^2 - 0 + 7$
   $= 0 - 0 + 7$
   $= 7$

   b) $f(1) = 2(1)^2 - 1 + 7$
   $= 2(1) - 1 + 7$
   $= 2 - 1 + 7$
   $= 8$

   c) $f(-1) = 2(-1)^2 - (-1) + 7$
   $= 2(1) + 1 + 7$
   $= 2 + 1 + 7$
   $= 10$

   d) $f(a) = 2(a)^2 - a + 7$
   $= 2a^2 - a + 7$

29. $f(x) = \sqrt{3 - x}$

   a) $f(-1) = \sqrt{3 - (-1)}$
   $= \sqrt{3 + 1}$
   $= \sqrt{4}$
   $= 2$

   b) $f(12) = \sqrt{3 - 12}$
   $= \sqrt{-9}$
   not real

   c) $f(-2) = \sqrt{3 - (-2)}$
   $= \sqrt{3 + 2}$
   $= \sqrt{5}$

   d) $f(t) = \sqrt{3 - t}$

31. $f(x) = \dfrac{2}{5}x + 1$

   a) $f(0) = \dfrac{2}{5}(0) + 1$
   $= 0 + 1$
   $= 1$

   b) $f(-5) = \dfrac{2}{5}(-5) + 1$
   $= -2 + 1$
   $= -1$

   c) $f(-1) = \dfrac{2}{5}(-1) + 1$
   $= -\dfrac{2}{5} + 1$
   $= \dfrac{3}{5}$

   d) $f(r) = \dfrac{2}{5}r + 1$

33. $f(x) = x^2 - 2.1x - 3$

   a) $f(0) = (0)^2 - 2.1(0) - 3$

        $= 0 - 0 - 3$

        $= -3$

   b) $f(-2.2) = (-2.2)^2 - 2.1(-2.2) - 3$

        $= 4.84 + 4.62 - 3$

        $= 6.46$

   c) $f\left(\dfrac{2}{3}\right) = \left(\dfrac{2}{3}\right)^2 - 2.1\left(\dfrac{2}{3}\right) - 3$

        $= \dfrac{4}{9} - \dfrac{4.2}{3} - 3$

        ≈ -3.96

   d) $f(a) = (a)^2 - 2.1(a) - 3$

        $= a^2 - 2.1a - 3$

35. $f(x) = \sqrt{x^2 - 4x}$

   a) $f(4) = \sqrt{4^2 - 4(4)}$ b) $f(7) = \sqrt{7^2 - 4(7)}$

        $= \sqrt{16 - 16}$ $= \sqrt{49 - 28}$

        $= \sqrt{0}$ $= \sqrt{21}$

        $= 0$

   c) $f(2) = \sqrt{2^2 - 4(2)}$ d) $f(n) = \sqrt{n^2 - 4(n)}$

        $= \sqrt{4 - 8}$ $= \sqrt{n^2 - 4n}$

        $= \sqrt{-4}$

        not real

37. $f(x) = \left|3x^2 + 1\right|$

   a) $f(0) = \left|3(0)^2 + 1\right|$ b) $f\left(\dfrac{2}{3}\right) = \left|3\left(\dfrac{2}{3}\right)^2 + 1\right|$

        $= \left|3(0) + 1\right|$ $= \left|3\left(\dfrac{4}{9}\right) + 1\right|$

        $= \left|0 + 1\right|$ $= \left|\dfrac{4}{3} + 1\right|$

        $= \left|1\right|$

        $= 1$ $= \left|\dfrac{7}{3}\right| = \dfrac{7}{3}$

   c) $f(-1) = \left|3(-1)^2 + 1\right|$ d) $f(-2) = \left|3(-2)^2 + 1\right|$

        $= \left|3(1) + 1\right|$ $= \left|3(4) + 1\right|$

        $= \left|3 + 1\right|$ $= \left|12 + 1\right|$

        $= \left|4\right|$ $= \left|13\right|$

        $= 4$ $= 13$

39. $f(x) = \dfrac{3 - x}{x - 4}$

   a) $f(5) = \dfrac{3 - 5}{5 - 4}$ b) $f(4) = \dfrac{3 - 4}{4 - 4}$

        $= -2$ $= \dfrac{-1}{0}$

                             undefined

   c) $f(-2) = \dfrac{3 - (-2)}{-2 - 4}$ d) $f(3) = \dfrac{3 - 3}{3 - 4}$

        $= \dfrac{3 + 2}{-6}$ $= \dfrac{0}{-1}$

        $= -\dfrac{5}{6}$ $= 0$

41. $f(x) = \dfrac{x}{x^2 - 1}$

   a) $f(0) = \dfrac{0}{0^2 - 1} = \dfrac{0}{-1} = 0$

   b) $f(1) = \dfrac{1}{1^2 - 1} = \dfrac{1}{1 - 1} = \dfrac{1}{0}$ is undefined.

   c) $f(2) = \dfrac{2}{2^2 - 1} = \dfrac{2}{3}$

   d) $f(m) = \dfrac{m}{m^2 - 1}$

43. $f(x) = \dfrac{2x}{\sqrt{2 - x}}$

   a) $f(-2) = \dfrac{2(-2)}{\sqrt{2 - (-2)}}$ b) $f(1) = \dfrac{2(1)}{\sqrt{2 - 1}}$

        $= \dfrac{-4}{\sqrt{2 + 2}}$ $= \dfrac{2}{\sqrt{1}}$

        $= \dfrac{-4}{\sqrt{4}}$ $= \dfrac{2}{1}$

        $= \dfrac{-4}{2}$ $= 2$

        $= -2$

   c) $f(2) = \dfrac{2(2)}{\sqrt{2 - 2}}$ d) $f(6) = \dfrac{2(6)}{\sqrt{2 - 6}}$

        $= \dfrac{4}{\sqrt{0}}$ $= \dfrac{12}{\sqrt{-4}}$

        $= \dfrac{4}{0}$ not real

        undefined

45. a) $f(-4) = 2$

   b) $f(0) = 0$

   c) $f(2) = -1$

47. a) $f(-2) = -4$

   b) $f(0) = 0$

   c) $f(2) = -4$

49. a) $f(-4) = -3$

   b) $f(0) = 1$

   c) $f(3) = 2$

51. $f(x) = -3x + 2$

   Think of the function as the equation $y = -3x + 2$ and use the fact that the slope is $m = -3$ and the y-intercept is 2.

   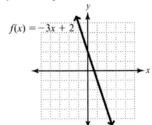

   Domain: $\mathbb{R}$ or $(-\infty, \infty)$

   Range: $\mathbb{R}$ or $(-\infty, \infty)$

53. $f(x) = \dfrac{2}{3}x - 1$

   Think of the function as the equation $y = \dfrac{2}{3}x - 1$ and use the fact that the slope is $m = \dfrac{2}{3}$ and the y-intercept is -1.

   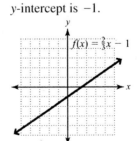

   Domain: $\mathbb{R}$ or $(-\infty, \infty)$

   Range: $\mathbb{R}$ or $(-\infty, \infty)$

55. $f(x) = -5x$

   Think of the function as the equation $y = -5x$ and use the fact that the slope is $m = -5$ and the y-intercept is 0.

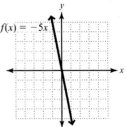

Domain: $\mathbb{R}$ or $(-\infty, \infty)$

Range: $\mathbb{R}$ or $(-\infty, \infty)$

57. $f(x) = |x| - 4$

x	$f(x)$
-2	-2
-4	0
0	-4
4	0
2	-2

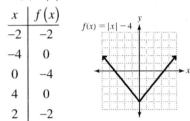

Domain: $(-\infty, \infty)$

Range: $\{y \mid -4 \le y < \infty\}$ or $[-4, \infty)$

59. $f(x) = |x + 3| - 2$

x	$f(x)$
-0	1
-1	0
-3	-2
-5	0
-6	1

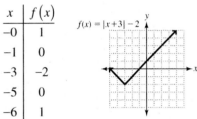

Domain: $(-\infty, \infty)$

Range: $\{y \mid -2 \le y < \infty\}$ or $[-2, \infty)$

61. $f(x) = 2|x - 1| + 3$

x	$f(x)$
-1	7
0	5
1	3
2	5
3	7

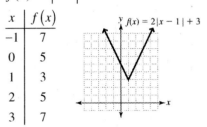

Domain: $\mathbb{R}$ or $(-\infty, \infty)$

Range: $\{y \mid 3 \le y < \infty\}$ or $[3, \infty)$

63. $f(x) = -|x| + 3$

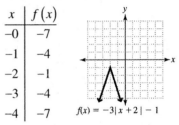

x	$f(x)$
-3	0
-2	1
0	3
2	1
3	0

Domain: $\mathbb{R}$ or $(-\infty, \infty)$

Range: $\{y \mid -\infty < y \le 3\}$ or $(-\infty, 3]$

65. $f(x) = -3|x + 2| - 1$

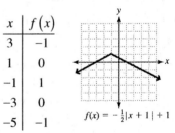

x	$f(x)$
-0	-7
-1	-4
-2	-1
-3	-4
-4	-7

Domain: $\mathbb{R}$ or $(-\infty, \infty)$

Range: $\{y \mid -\infty < y \le -1\}$ or $(-\infty, -1]$

67. $f(x) = -\dfrac{1}{2}|x + 1| + 1$

x	$f(x)$
3	-1
1	0
-1	1
-3	0
-5	-1

Domain: $\mathbb{R}$ or $(-\infty, \infty)$

Range: $\{y \mid -\infty < y \le 1\}$ or $(-\infty, 1]$

69. $f(x) = x^2 - 1$

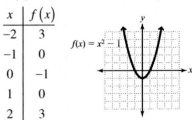

x	$f(x)$
-2	3
-1	0
0	-1
1	0
2	3

Domain: $\mathbb{R}$ or $(-\infty, \infty)$

Range: $\{y \mid -1 \le y < \infty\}$ or $[-1, \infty)$

71. $f(x) = -x^2 + 3$

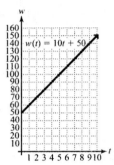

x	$f(x)$
-2	-1
-1	2
0	3
1	2
2	-1

Domain: $\mathbb{R}$ or $(-\infty, \infty)$

Range: $\{y \mid -\infty < y \le 3\}$ or $(-\infty, 3]$

73.   a)

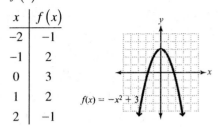

b) $w(t) = 10t + 50$

$$w(7.5) = 10(7.5) + 50$$
$$= 75 + 50$$
$$= \$125$$

75.   a) $C(V) = 1.225V + 6$

b)

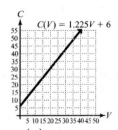

c) $C(V) = 1.225V + 6$

$$C(40) = 1.225(40) + 6$$
$$= 49 + 6$$
$$= \$55$$

77. a)

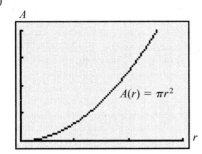

$A(r) = \pi r^2$

b) No, it is not a linear function.

c) The radius of the circle is 1.5 units.

d) $A(r) \approx \pi r^2$

$$A(1.5) \approx 3.14(1.5)^2$$
$$\approx 7.07 \text{ sq. units}$$

79. a)

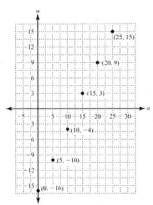

b) Yes.

$$m = \frac{15 - (-16)}{25 - 0} = \frac{15 + 16}{25} = \frac{31}{25} = 1.24$$

c) $w(a) = 1.24a - 16$

d) $w(a) = 1.24a - 16$

$$w(40) = 1.24(40) - 16$$
$$= 49.6 - 16$$
$$= 33.6$$

When the actual temperature is $40°$ F, the wind-chill temperature is $33.6°$ F.

Review Exercises

1. $x + 9y - 3x + 7 - 3y - 9$

$= x - 3x + 9y - 3y + 7 - 9$

$= -2x + 6y - 2$

2. $5(x+1) - x = 3x - (x+1)$

$5x + 5 - x = 3x - x - 1$

$4x + 5 = 2x - 1$

$2x + 5 = -1$

$2x = -6$

$x = -3$

3. $4x - (6x + 3) \geq 8x - 23$

$4x - 6x - 3 \geq 8x - 23$

$-2x - 3 \geq 8x - 23$

$-10x - 3 \geq -23$

$-10x \geq -20$

$x \leq 2$

$\{x \mid x \leq 2\} \quad (-\infty, 2]$

```
  ←─┼──┼──┼──┼──┼──┼─)─┼──┼──→
   −4 −3 −2 −1  0  1  2  3  4
```

4. Let the first integer be x and the second integer be $x + 2$.

$x + x + 2 = 74$

$2x + 2 = 74$

$2x = 72$

$x = 36$

The integers are 36 and 38.

5. If the width is x, then the length is $x + 6$.

$2(x + 6) + 2x = 108$

$2x + 12 + 2x = 108$

$12 + 4x = 108$

$4x = 96$

$x = 24$

The width is 24 in. and the length is $24 + 6 = 30$ in.

6. Create a table.

	how many	value	total
large	$40 - x$	2.25	$2.25(40 - x)$
small	x	1.50	$1.50x$

Translate to an equation and solve.

$2.25(40 - x) + 1.50x = 69$

$90 - 2.25x + 1.50x = 69$

$90 - 0.75x = 69$

$-0.75x = -21$

$28 = x$

28 small and $40 - 28 = 12$ large

Chapter 3 Review Exercises

1. Replace x with 2 and y with 1 and see if the equation is true.

    $$2x - y = -3$$
    $$2(2) - 1 \stackrel{?}{=} -3$$
    $$4 - 1 \stackrel{?}{=} -3$$
    $$3 \neq -3$$

    No, $(2, 1)$ is not a solution.

2. Replace x with $\frac{4}{5}$ and y with 2 and see if the equation is true.

    $$5x + y = 6$$
    $$5\left(\frac{4}{5}\right) + 2 \stackrel{?}{=} 6$$
    $$4 + 2 \stackrel{?}{=} 6$$
    $$6 = 6$$

    Yes, $\left(\frac{4}{5}, 2\right)$ is a solution.

3. Replace x with 0.4 and y with 1.2 and see if the equation is true.

    $$y + 2x = 4.1$$
    $$1.2 + 2(0.4) \stackrel{?}{=} 4.1$$
    $$1.2 + 0.8 \stackrel{?}{=} 4.1$$
    $$2 \neq 4.1$$

    No, $(0.4, 1.2)$ is not a solution.

4. Replace x with 5 and y with 2 and see if the equation is true.

    $$y = \frac{2}{5}x$$
    $$2 \stackrel{?}{=} \frac{2}{5}(5)$$
    $$2 = 2$$

    Yes, $(5, 2)$ is a solution.

5. $y = x - 4$

    x-intercept: $0 = x - 4$
    $$4 = x$$
    $$(4, 0)$$

y-intercept: $y = 0 - 4$
$$y = -4$$
$$(0, -4)$$

x	y	Ordered Pair
4	0	$(4, 0)$
0	-4	$(0, -4)$
2	-2	$(2, -2)$

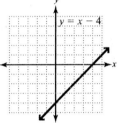

6. $y = 5x$

    x-intercept: $0 = 5x$
    $$0 = x$$
    $$(0, 0)$$

    y-intercept: $y = 5(0)$
    $$y = 0$$
    $$(0, 0)$$

x	y	Ordered Pair
0	0	$(0, 0)$
-1	-5	$(-1, -5)$
1	5	$(1, 5)$

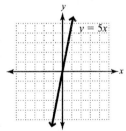

7. $y = \frac{2}{3}x - 3$

    x-intercept: $0 = \frac{2}{3}x - 3$
    $$3 = \frac{2}{3}x$$
    $$\frac{9}{2} = x$$
    $$\left(\frac{9}{2}, 0\right)$$

y-intercept: $y = \dfrac{2}{3}(0) - 3$

$\qquad\qquad\quad y = 0 - 3$

$\qquad\qquad\quad y = -3$

$\qquad\qquad\quad (0, -3)$

x	y	Ordered Pair
0	-3	$(0, -3)$
-3	-5	$(-3, -5)$
3	-1	$(3, -1)$

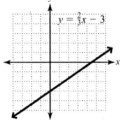

8. $y = -\dfrac{2}{7}x + 1$

x-intercept: $0 = -\dfrac{2}{7}x + 1$

$\qquad\qquad\quad -1 = -\dfrac{2}{7}x$

$\qquad\qquad\quad \dfrac{7}{2} = x$

$\qquad\qquad\quad \left(\dfrac{7}{2}, 0\right)$

y-intercept: $y = -\dfrac{2}{7}(0) + 1$

$\qquad\qquad\quad y = 0 + 1$

$\qquad\qquad\quad y = 1$

$\qquad\qquad\quad (0, 1)$

x	y	Ordered Pair
0	1	$(0, 1)$
7	-1	$(7, -1)$
-7	3	$(-7, 3)$

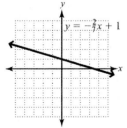

9. $2x - 3y = 6$

x-intercept: $2x - 3(0) = 6$

$\qquad\qquad\quad 2x - 0 = 6$

$\qquad\qquad\quad 2x = 6$

$\qquad\qquad\quad x = 3$

$\qquad\qquad\quad (3, 0)$

y-intercept: $2(0) - 3y = 6$

$\qquad\qquad\quad 0 - 3y = 6$

$\qquad\qquad\quad -3y = 6$

$\qquad\qquad\quad y = -2$

$\qquad\qquad\quad (0, -2)$

x	y	Ordered Pair
3	0	$(3, 0)$
0	-2	$(0, -2)$
-3	-4	$(-3, -4)$

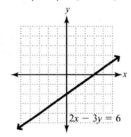

10. $3x + 4y = 28$

x-intercept: $3x + 4(0) = 28$

$\qquad\qquad\quad 3x = 28$

$\qquad\qquad\quad x = \dfrac{28}{3}$

$\qquad\qquad\quad \left(\dfrac{28}{3}, 0\right)$

y-intercept: $2(0) - 3y = 6$

$\qquad\qquad\quad 0 - 3y = 6$

$\qquad\qquad\quad -3y = 6$

$\qquad\qquad\quad y = -2$

$\qquad\qquad\quad (0, -2)$

x	y	Ordered Pair
0	7	$(0, 7)$
4	4	$(4, 4)$
8	1	$(8, 1)$

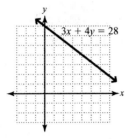

11. $x = 3$

x-intercept: $(3, 0)$

y-intercept: none

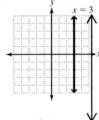

12. $y = -4$

x-intercept: none

y-intercept: $(0, -4)$

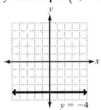

13. $y = -3x + 2$

$m = -3$; y-intercept: $(0, 2)$

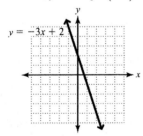

14. $y = \dfrac{5}{2}x + 3$

$m = \dfrac{5}{2}$; y-intercept: $(0, 3)$

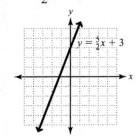

15. $2x + 5y = 15$

$$5y = -2x + 15$$

$$y = -\frac{2}{5}x + 3$$

$m = -\dfrac{2}{5}$; y-intercept: $(0, 3)$

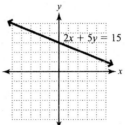

16. $y - 4x + 3 = 0$

$$y = 4x - 3$$

$m = 4$; y-intercept: $(0, -3)$

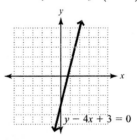

17. $(1, 8), (4, -1)$

$$m = \frac{y_2 - y_1}{x_2 - x_1} = \frac{-1 - 8}{4 - 1} = \frac{-9}{3} = -3$$

18. $(3, -1), (3, 2)$

$$m = \frac{y_2 - y_1}{x_2 - x_1} = \frac{2 - (-1)}{3 - 3} = \frac{2 + 1}{0} = \frac{3}{0} \text{ is undefined.}$$

19. $(7, -4), (2, -9)$

$$m = \frac{y_2 - y_1}{x_2 - x_1} = \frac{-9 - (-4)}{2 - 7} = \frac{-9 + 4}{-5} = \frac{-5}{-5} = 1$$

20. $(-1, -1), (3, -1)$

$$m = \frac{y_2 - y_1}{x_2 - x_1} = \frac{-1 - (-1)}{3 - (-1)} = \frac{-1 + 1}{3 + 1} = \frac{0}{4} = 0$$

21. $m = \dfrac{10}{12} = \dfrac{5}{6}$

22. a) $m = \dfrac{10.7 - 12.2}{4 - 0} = \dfrac{-1.5}{4} \approx -0.375$

   b) $y = -0.375x + 12.2$

   c) Year 2020 would be $x = 14$

$$y = -0.375x + 12.2$$
$$y = -0.375(14) + 12.2$$
$$= 6.95$$

The model predicts that only 6.95% of twelfth grade students will smoke daily in 2020.

23. $m = 2; (0, -4)$

$$y = mx + b$$
$$y = 2x - 4$$

24. $m = -\dfrac{2}{5}; (0, 4)$

$$y = mx + b$$
$$y = -\dfrac{2}{5}x + 4$$

25. $m = -0.3; (0, -1)$

$$y = mx + b$$
$$y = -0.3x - 1$$

26. $m = -3; (0, 0)$

$$y = mx + b$$
$$y = -3x$$

27. $m = \dfrac{2}{5}; (0, 2); y = \dfrac{2}{5}x + 2$

28. $m = -2; (0, -1); y = -2x - 1$

29. $m = -1$

$$y - y_1 = m(x - x_1)$$
$$y - 8 = -(x - 2)$$
$$y - 8 = -x + 2$$
$$y = -x + 10$$

30. $m = 6.2$

$$y - y_1 = m(x - x_1)$$
$$y - (-5) = 6.2(x - 3)$$
$$y + 5 = 6.2x - 18.6$$
$$y = 6.2x - 23.6$$

31. $m = \dfrac{2}{3}$

$$y - y_1 = m(x - x_1)$$
$$y - 0 = \dfrac{2}{3}(x - 4)$$
$$y = \dfrac{2}{3}x - \dfrac{8}{3}$$

32. $m = -3$

$$y - y_1 = m(x - x_1)$$
$$y - (-2) = -3(x - (-2))$$
$$y + 2 = -3(x + 2)$$
$$y + 2 = -3x - 6$$
$$y = -3x - 8$$

33. Find the slope: $m = \dfrac{9 - 3}{5 - 3} = \dfrac{6}{2} = 3$

   a) $y - y_1 = m(x - x_1)$

$$y - 3 = 3(x - 3)$$
$$y - 3 = 3x - 9$$
$$y = 3x - 6$$

   b) $\qquad y = 3x - 6$

$$-3x + y = -6$$
$$3x - y = 6$$

34. Find the slope: $m = \dfrac{-5 - 2}{-4 - (-2)} = \dfrac{-7}{-4 + 2} = \dfrac{-7}{-2} = \dfrac{7}{2}$

   a) $y - y_1 = m(x - x_1)$

$$y - 2 = \dfrac{7}{2}(x - (-2))$$
$$y - 2 = \dfrac{7}{2}(x + 2)$$
$$y - 2 = \dfrac{7}{2}x + 7$$
$$y = \dfrac{7}{2}x + 9$$

   b) $\qquad y = \dfrac{7}{2}x + 9$

$$2y = 7x + 18$$
$$-7x + 2y = 18$$
$$7x - 2y = -18$$

35. Find the slope: $m = \dfrac{-3-(-2)}{-2-4} = \dfrac{-3+2}{-6} = \dfrac{-1}{-6} = \dfrac{1}{6}$

    a) $y - y_1 = m(x - x_1)$

$$y - (-2) = \frac{1}{6}(x - 4)$$

$$y + 2 = \frac{1}{6}x - \frac{4}{6}$$

$$y + 2 = \frac{1}{6}x - \frac{2}{3}$$

$$y = \frac{1}{6}x - \frac{8}{3}$$

    b) $y = \dfrac{1}{6}x - \dfrac{8}{3}$

$$6y = x - 16$$

$$-x + 6y = -16$$

$$x - 6y = 16$$

36. Find the slope: $m = \dfrac{6-0}{0-6} = \dfrac{6}{-6} = -1$

    a) $y - y_1 = m(x - x_1)$

$$y - 6 = -(x - 0)$$

$$y - 6 = -x$$

$$y = -x + 6$$

    b) $y = -x + 6$

$$x + y = 6$$

37. Because these lines have equal slopes and different y-intercepts, they are parallel.

38. $7x + y = -9$ $7x - y = -6$

       $y = -7x - 9$ $-y = -7x - 6$

                           $y = 7x + 6$

These slopes are neither equal nor of the form $\dfrac{a}{b}$

and $-\dfrac{b}{a}$. Therefore, these lines are neither parallel nor perpendicular.

39. Because the slopes are $\dfrac{4}{3}$ and $-\dfrac{3}{4}$, these lines are perpendicular.

40. $x + y = -1$ $x - y = 5$

       $y = -x - 1$ $-y = -x + 5$

                          $y = x - 5$

Because the slopes are -1 and 1, these lines are perpendicular.

41. The slope of the given equation is $m = -\dfrac{3}{5}$, so the slope of a parallel line is $m = -\dfrac{3}{5}$. The y-intercept is $(0, -2)$.

$$y = mx + b$$

$$y = -\frac{3}{5}x - 2$$

42. Find m. $x - 5y = 10$

$$-5y = -x + 10$$

$$y = \frac{1}{5}x - 2$$

The slope of the given equation is $m = \dfrac{1}{5}$, so the slope of a parallel line is $m = \dfrac{1}{5}$.

$$y - y_1 = m(x - x_1)$$

$$y - 5 = \frac{1}{5}(x - 2)$$

$$y - 5 = \frac{1}{5}x - \frac{2}{5}$$

$$y = \frac{1}{5}x + \frac{23}{5}$$

43. The slope of the given equation is $m = -\dfrac{2}{3}$, so the slope of a perpendicular line is $m = \dfrac{3}{2}$.

$$y - y_1 = m(x - x_1)$$

$$y - (-2) = \frac{3}{2}(x - (-1))$$

$$y + 2 = \frac{3}{2}(x + 1)$$

$$y + 2 = \frac{3}{2}x + \frac{3}{2}$$

$$y = \frac{3}{2}x - \frac{1}{2}$$

44. The slope of the given equation is $m = 3$, so the slope of a perpendicular line is $m = -\frac{1}{3}$.

$$y - (-5) = -\frac{1}{3}(x - (-3))$$
$$y + 5 = -\frac{1}{3}(x + 3)$$
$$y + 5 = -\frac{1}{3}x - 1$$
$$y = -\frac{1}{3}x - 6$$

45. Test $(-3, -1)$.

$$\begin{array}{rcl} 3x + 5y & > & 1 \\ 3(-3) + 5(-1) & \overset{?}{>} & 1 \\ -9 - 5 & \overset{?}{>} & 1 \\ -14 & > & 1 \end{array}$$

This statement is false. No, $(-3, -1)$ is not a solution.

46. Test $(-4, 2)$.

$$\begin{array}{rcl} y & \geq & 3x + 1 \\ 2 & \overset{?}{\geq} & 3(-4) + 1 \\ 2 & \overset{?}{\geq} & -12 + 1 \\ 2 & \geq & -11 \end{array}$$

This statement is true. Yes, $(-4, 2)$ is a solution.

47. $y < -2x + 5$

Begin by graphing the related equation $y = -2x + 5$ with a dashed line. Now choose $(0, 0)$ as a test point.

$$\begin{array}{rcl} y & < & -2x + 5 \\ 0 & \overset{?}{<} & -2(0) + 5 \\ 0 & \overset{?}{<} & 0 + 5 \\ 0 & < & 5 \end{array}$$

Because $(0, 0)$ satisfies the inequality, shade the region which includes $(0, 0)$.

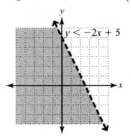

48. $3x - 4y > 12$

$$-4y > -3x + 12$$
$$y < \frac{3}{4}x - 3$$

Begin by graphing the related equation $y = \frac{3}{4}x - 3$ with a dashed line. Now choose $(0, 0)$ as a test point.

$$\begin{array}{rcl} 3x - 4y & > & 12 \\ 3(0) - 4(0) & \overset{?}{>} & 12 \\ 0 - 0 & \overset{?}{>} & 12 \\ 0 & > & 12 \end{array}$$

Because $(0, 0)$ does not satisfy the inequality, shade the side of the line on the opposite side from $(0, 0)$.

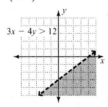

49. $-2x - 5y \leq -10$

$-5y \leq 2x - 10$

$y \geq -\dfrac{2}{5}x + 2$

Begin by graphing the related equation

$y = -\dfrac{2}{5}x + 2$ with a solid line. Now choose $(0, 0)$

as a test point.

$$-2x - 5y \;\; \leq \;\; -10$$
$$-2(0) - 5(0) \;\; \overset{?}{\leq} \;\; -10$$
$$0 - 0 \;\; \overset{?}{\leq} \;\; -10$$
$$0 \;\; \leq \;\; -10$$

Because $(0, 0)$ does not satisfy the inequality, shade the side of the line on the opposite side from $(0, 0)$.

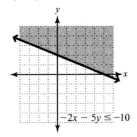

50. $y > \dfrac{4}{5}x$

Begin by graphing the related equation $y = \dfrac{4}{5}x$

with a dashed line. Now choose $(1, 0)$ as a test point.

$$y \;\; > \;\; \dfrac{4}{5}x$$
$$0 \;\; \overset{?}{>} \;\; \dfrac{4}{5}(1)$$
$$0 \;\; > \;\; \dfrac{4}{5}$$

Because $(1, 0)$ does not satisfy the inequality, shade the side of the line on the opposite side from $(1, 0)$.

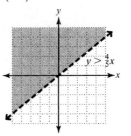

51. $x + y \geq 3$

$y \geq -x + 3$

Begin by graphing the related equation $y = -x + 3$

with a solid line. Now choose $(0, 0)$ as a test point.

$$x + y \;\; \geq \;\; 3$$
$$0 + 0 \;\; \overset{?}{\geq} \;\; 3$$
$$0 \;\; \geq \;\; 3$$

Because $(0, 0)$ does not satisfy the inequality, shade the side of the line on the opposite side from $(0, 0)$.

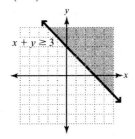

52. $y \geq -3$

Begin by graphing the related equation $y = -3$ with a solid line. Now choose $(0, 0)$ as a test point.

$$y \geq -3$$
$$0 \overset{?}{\geq} -3$$

Because $(0, 0)$ satisfies the inequality, shade the region which includes $(0, 0)$.

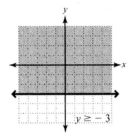

53. a) $35x + 50y \geq 70,000$

b)

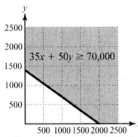

c) All combinations of chair sales that make the company break even ($70,000 revenue).

d) All combinations of chair sales that make the company a profit (more than $70,000 revenue).

e) Answers may vary. Some possible answers are (1000, 700) and (2000, 0).

f) Answers may vary. Two examples are (2000, 500) and (2000, 1000).

54. Domain: {McKinley, Logan, Pico de Orizaba, St. Elias, Popocatepetl}
Range: {20,320, 19,551, 18,555, 18,008, 17,930}
It is a function because every element in the domain is paired with exactly one element in the range.

55. Domain: {21, 23, 32, 35}
Range: {California, Indiana, New York, Ohio, Pennsylvania}
It is not a function because an element in the domain is paired with more than one element in the range.

56. Domain: $\{x \mid -4 \leq x \leq 5\}$ or $[-4, 5]$

Range: $\{-2, 2, 3\}$

It is a function because a vertical line at each x-value would intersect the graph at only one point.

57. Domain: $\{x \mid -3 \leq x \leq 3\}$ or $[-3, 3]$

Range: $\{y \mid 0 \leq y \leq 3\}$ or $[0, 3]$

It is a function because a vertical line at each x-value would intersect the graph at only one point.

58. Domain: $\mathbb{R}$ or $(-\infty, \infty)$

Range: $\mathbb{R}$ or $(-\infty, \infty)$

It is a function because a vertical line at each x-value would intersect the graph at only one point.

59. Domain: $\{x \mid x \leq 3\}$ or $(-\infty, 3]$

Range: $\mathbb{R}$ or $(-\infty, \infty)$

It is not a function because there are values in the domain that correspond to two values in the range.

60. $f(x) = x^2 - 4$

a) $f(2) = 2^2 - 4$
$\quad = 4 - 4$
$\quad = 0$

b) $f(0) = 0^2 - 4$
$\quad = 0 - 4$
$\quad = -4$

c) $f(-3) = (-3)^2 - 4$
$\quad = 9 - 4$
$\quad = 5$

d) $f(n) = n^2 - 4$

61. $g(x) = \dfrac{x - 3}{x + 5}$

a) $g(2) = \dfrac{2 - 3}{2 + 5}$
$\quad = -\dfrac{1}{7}$

b) $g(3) = \dfrac{3 - 3}{3 + 5}$
$\quad = 0$

c) $g(-5) = \dfrac{-5 - 3}{-5 + 5}$
$\quad = \dfrac{-8}{0}$
$\quad$ undefined

62. a) $f(-3) = -2$
c) $f(0) = 3$

b) $f(2) = 3$
d) $f(5) = 2$

63. a) $f(-3) = 0$
c) $f(3) = 0$

b) $f(0) = 3$
d) $f(4)$ undefined

64. a) $c(t) = 25t + 75$

    b) $c(1.5) = 25(1.5) + 75$
    $$= 37.5 + 75$$
    $$= \$112.50$$

    c) $150 = 25t + 75$
    $$75 = 25t$$
    $$3 \text{ hr} = t$$

    d)

    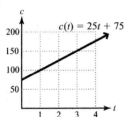

Chapter 3 Practice Test

1. $A(0, 5)$, because it is on the y-axis and is 5 units up from the origin.
   $B(3, 3)$, because it is 3 units to the right and 3 units up from the origin.
   $C(-2, -3)$, because it is 2 units to the left and 3 units down from the origin.
   $D(4, -2)$, because it is 4 units to the right and 2 units down from the origin.

2. Quadrant IV because the first coordinate is positive and the second coordinate is negative.

3. No $5 \overset{?}{=} -\dfrac{1}{4}(-3) + 3$

    $$5 \overset{?}{=} \dfrac{3}{4} + 3$$

    $$5 \neq 3\dfrac{3}{4}$$

4. $y = -\dfrac{4}{3}x + 5$

    $m = -\dfrac{4}{3}; b = (0, 5)$

    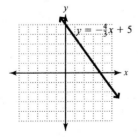

5. $x - 2y = -8$

    $$-2y = -x - 8$$

    $$y = \dfrac{1}{2}x + 4$$

    $$m = \dfrac{1}{2}; b = (0, 4)$$

    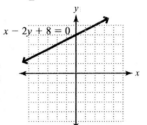

6. Let $(x_1, y_1) = (-1, -4)$ and $(x_2, y_2) = (-2, -4)$.

    $$m = \dfrac{y_2 - y_1}{x_2 - x_1} = \dfrac{-4 - (-4)}{-2 - (-1)} = \dfrac{-4 + 4}{-2 + 1} = \dfrac{0}{-1} = 0$$

7. Let $(x_1, y_1) = (-3, -9)$ and $(x_2, y_2) = (4, -1)$.

    $$m = \dfrac{y_2 - y_1}{x_2 - x_1} = \dfrac{-1 - (-9)}{4 - (-3)} = \dfrac{-1 + 9}{4 + 3} = \dfrac{8}{7}$$

8. $m = \dfrac{2}{7}$; y-intercept: $(0, b) = (0, 5)$

    $$y = mx + b$$

    $$y = \dfrac{2}{7}x + 5$$

9. Let $(x_1, y_1) = (4, 2)$ and $(x_2, y_2) = (-5, -1)$.

    $$m = \dfrac{y_2 - y_1}{x_2 - x_1} = \dfrac{-1 - 2}{-5 - 4} = \dfrac{-3}{-9} = \dfrac{1}{3}$$

    $$y - y_1 = m(x - x_1)$$

    $$y - 2 = \dfrac{1}{3}(x - 4)$$

    $$y - 2 = \dfrac{1}{3}x - \dfrac{4}{3}$$

    $$y = \dfrac{1}{3}x + \dfrac{2}{3}$$

10. Let $(x_1, y_1) = (1, 4)$ and $(x_2, y_2) = (-3, -1)$.

$$m = \frac{y_2 - y_1}{x_2 - x_1} = \frac{-1 - 4}{-3 - 1} = \frac{-5}{-4} = \frac{5}{4}$$

$$y - y_1 = m(x - x_1)$$

$$y - 4 = \frac{5}{4}(x - 1)$$

$$4(y - 4) = 4\left(\frac{5}{4}(x - 1)\right)$$

$$4(y - 4) = 5(x - 1)$$

$$4y - 16 = 5x - 5$$

$$-5x + 4y - 16 = -5$$

$$-5x + 4y = 11$$

$$-1(-5x + 4y) = -1(11)$$

$$5x - 4y = -11$$

11. The slope of the first line is $m = \frac{3}{4}$ and the slope of the second line is $m = \frac{4}{3}$. Because these slopes are neither equal nor negative reciprocals of one another, the graphs are neither parallel nor perpendicular.

12.

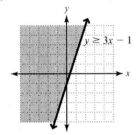

$y \geq 3x - 1$

13.

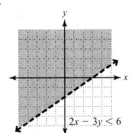

$2x - 3y \leq 6$

14. Domain: {Spanish, Chinese, Tagalog, Vietnamese, French}
Range: {37.6, 2.9, 1.6, 1.4, 1.3}
Because every value in the domain is paired with exactly one value in the range, it is a function.

15. Domain: $\{x | 0 \leq x \leq 3\}$, Range: $\{y | -3 \leq y \leq 3\}$

It is not a function because at least one element in the domain is paired with more than one element in the range.

16. $f(x) = 2x^2 - 7$

a) $f(-2) = 2(-2)^2 - 7$
$$= 8 - 7$$
$$= 1$$

b) $f(3) = 2(3)^2 - 7$
$$= 18 - 7$$
$$= 11$$

c) $f(t) = 2t^2 - 7$

17. $f(x) = \frac{4x}{x + 5}$

a) $f(-1) = \frac{4(-1)}{-1 + 5}$
$$= \frac{-4}{4}$$
$$= -1$$

b) $f(5) = \frac{4(5)}{5 + 5}$
$$= \frac{20}{10}$$
$$= 2$$

c) $f(-5) = \frac{4(-5)}{-5 + 5}$
$$= \frac{-20}{0}$$
undefined

18. a) Domain: $\{x | -3 \leq x \leq 3\}$ or $[-3, 3]$;
Range: $\{-2, 1, 3\}$

b) From the graph, we find that $f(1) = 3$.

19. a) $2l + 2w \leq 1000$

b)

$2l + 2w \leq 1000$

c) Answers may vary. Two possible answers are 300 ft. by 200 ft. or 200 ft. by 200 ft.

20. a) $c(w) = 0.45w + 3$

b)

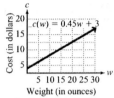

Cost (in dollars)

$c(w) = 0.45w + 3$

Weight (in ounces)

c) $c(32) = 0.45(32) + 3$
$= 14.4 + 3$
$= \$17.40$

Chapters 1–3 Cumulative Review

1. False. In interval notation, the inequality $x > 0$ is expressed as $(0, \infty)$.

2. True. For example, $(-2)^3 = (-2) \cdot (-2) \cdot (-2) = -8$.

3. True. For example,
$3(2 + 4) = 3(6) = 18$
$3(2 + 4) = 3(2) + 3(4) = 6 + 12 = 18$.

4. 10

5. principal or non negative

6. We perform operations in the following order:
   a) Within grouping symbols beginning with the innermost: parentheses ( ), brackets [ ], braces { }, absolute value| |, above and/or below fraction bars, and radicals $\sqrt{}$.
   b) Exponents/roots from left to right, in order as they occur.
   c) Multiplication/division from left to right, in order as they occur.
   d) Addition/subtraction from left to right in order as they occur.

7. $A \cap B = \varnothing$
$A \cup B = \{w, e, l, o, v, m, a, t, h\}$

8. $(-3)^2 = (-3)(-3) = 9$

9. $\sqrt[3]{27} = 3$

10. $-|4 + 3| - 8(1 - 5)^2 = -|7| - 8(-4)^2$
$= -7 - 8(16)$
$= -7 - 128$
$= -135$

11. $\dfrac{3x}{x - 5} = \dfrac{3(5)}{5 - 5} = \dfrac{15}{0}$ is undefined

12. a) $(-4n + 8) - 5n$ b) $-9n + 8$
    c) $-9(4) + 8 = -36 + 8 = -28$

13. Multiplicative inverse

14. Distributive property

15. $6(5 + y) - 3y = 2y + 1$
$30 + 6y - 3y = 2y + 1$
$30 + 3y = 2y + 1$
$30 + y = 1$
$y = -29$

16. $2x - 1 = x + 8$ and $2x - 1 = -(x + 8)$
$x = 9$ $2x - 1 = -x - 8$
                  $3x = -7$
                  $x = -\dfrac{7}{3}$

17. $3 + 2x \le 8 - 4x$
$3 + 6x \le 8$
$6x \le 5$
$x \le \dfrac{5}{6}$

a) $\left\{x \mid x \le \dfrac{5}{6}\right\}$

b) $\left(-\infty, \dfrac{5}{6}\right]$

c)

18. $3|x + 1| - 7 > 2$
$3|x + 1| > 9$
$|x + 1| > 3$
$x + 1 < -3$ or $x + 1 > 3$
$x < -4$ $x > 2$

a) $\{x \mid x < -4 \text{ or } x > 2\}$

b) $(-\infty, -4) \cup (2, \infty)$

c)

19. $d = rt$
$\dfrac{d}{r} = t$

20. $m = \dfrac{-1-(-1)}{-2-3} = \dfrac{-1+1}{-5} = \dfrac{0}{-5} = 0$

21. $2x + y = 6$ $\qquad$ $y = -\dfrac{1}{2}x + 1$

 $\qquad y = -2x + 6$

 These slopes are neither equal nor of the form $\dfrac{a}{b}$

 and $-\dfrac{b}{a}$. Therefore, these lines are neither parallel
 nor perpendicular.

22. $2x - y = 3$

 $\qquad -y = -2x + 3$

 $\qquad y = 2x - 3$

 The slope of the given equation is $m = 2,$ so the
 slope of a parallel line is $m = 2$.

 $y - y_1 = m(x - x_1)$

 $\qquad y - 8 = 2(x - 1)$

 $\qquad y - 8 = 2x - 2$

 $\qquad y = 2x + 6$

23. $2x - y \geq 6$

 $\qquad -y \geq -2x + 6$

 $\qquad y \leq 2x - 6$

 Begin by graphing the related equation $y = 2x - 6$
 with a solid line. Now choose $(0, 0)$ as a test
 point.

 $2x - y \overset{?}{\geq} 6$

 $2(0) - 0 \overset{?}{\geq} 6$

 $0 - 0 \overset{?}{\geq} 6$

 $0 \geq 6$

 Because $(0, 0)$ does not satisfy the inequality,
 shade the side of the line on the opposite side from
 $(0, 0)$.

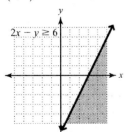

24. $f(x) = 3x^2 + 2$

 $f(-1) = 3(-1)^2 + 2$

 $\qquad = 3(1) + 2$

 $\qquad = 3 + 2$

 $\qquad = 5$

25. $f(x) = \dfrac{2}{5}x - 1$

 Think of the function as the equation $y = \dfrac{2}{5}x - 1$

 and use the fact that the slope is $m = \dfrac{2}{5}$ and the

 y-intercept is -1.

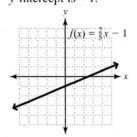

26. $\dfrac{3(4) + 3(4) + 4(3)}{10} = \dfrac{12 + 12 + 12}{10} = \dfrac{36}{10} = 3.6$

27. Let the first angle be x and the second angle be
 $180 - x$.

 $x = 2(180 - x)$

 $x = 360 - 2x$

 $3x = 360$

 $x = 120$

 The angles are 120 and 60 degrees.

28. Create a table.

	amount	value	total cost
standard	$2x$	35	$70x$
deluxe	x	75	$75x$

 Translate into an equation and solve.

 $70x + 75x = 580$

 $\qquad 145x = 580$

 $\qquad x = 4$

 He ordered 8 standard phones and 4 deluxe phones

29. Create a table.

Solutions	Concen-trate	Vol. of Solution	Vol. of HCl
10%	0.10	50	$0.10(50)$
25%	0.25	x	$0.25(x)$
15%	0.15	$x+50$	$0.15(x+50)$

Translate into an equation and solve.

$$0.10(50)+0.25x = 0.15(x+50)$$
$$5+0.25x = 0.15x+7.5$$
$$5+0.10x = 7.5$$
$$0.10x = 2.5$$
$$x = 25$$

Janet will need 25 ml of 25% solution.

30. $60 - $40 = 20

$$0.04t \le 20$$

$$t \le 500$$

Maximum: $500 + 500 = 1000$ minutes or less

Chapter 4

Systems of Linear Equations and Inequalities

Exercise Set 4.1

1. Is $(-1, 2)$ a solution of $\begin{cases} x - y = -3? \\ x + y = 1 \end{cases}$

   <u>Equation 1</u>   <u>Equation 2</u>

   $x - y = -3$ $x + y = 1$

   $-1 - 2 \overset{?}{=} -3$ $-1 + 2 \overset{?}{=} 1$

   $-3 = -3$ $1 = 1$

   Yes, $(-1, 2)$ is a solution for the system because it satisfies both equations.

3. Is $(-2, 3)$ a solution of $\begin{cases} 3x + 4y = 6? \\ x - 4y = 8 \end{cases}$

   <u>Equation 1</u>   <u>Equation 2</u>

   $3x + 4y = 6$ $x - 4y = 8$

   $3(-2) + 4(3) \overset{?}{=} 6$ $(-2) - 4(3) \overset{?}{=} 8$

   $-6 + 12 \overset{?}{=} 6$ $-2 - 12 \overset{?}{=} 8$

   $6 = 6$ $-14 \neq 8$

   No, $(-2, 3)$ is not a solution for the system because it does not satisfy both equations.

5. Is $\left(-\dfrac{3}{4}, -\dfrac{2}{3}\right)$ a solution of $\begin{cases} 4x + 3y = -5 \ ? \\ 12x - 6y = -5 \end{cases}$

   <u>Equation 1</u>   <u>Equation 2</u>

   $4x + 3y = -5$ $12x - 6y = -5$

   $4\left(-\dfrac{3}{4}\right) + 3\left(-\dfrac{2}{3}\right) \overset{?}{=} -5$ $12\left(-\dfrac{3}{4}\right) - 6\left(-\dfrac{2}{3}\right) \overset{?}{=} -5$

   $-3 - 2 \overset{?}{=} -5$ $-9 + 4 \overset{?}{=} -5$

   $-5 = -5$ $-5 = -5$

   Yes, $\left(-\dfrac{3}{4}, -\dfrac{2}{3}\right)$ is a solution for the system

   because it satisfies both equations.

7. Is $(3, -4)$ a solution of $\begin{cases} 3x + 2y = 1 \ ? \\ \dfrac{1}{3}x + \dfrac{1}{2}y = 3 \end{cases}$

   <u>Equation 1</u>   <u>Equation 2</u>

   $3x + 2y = 1$ $\dfrac{1}{3}x + \dfrac{1}{2}y = 3$

   $3(3) + 2(-4) \overset{?}{=} 1$ $\dfrac{1}{3}(3) + \dfrac{1}{2}(-4) \overset{?}{=} 3$

   $9 - 8 \overset{?}{=} 1$ $1 - 2 \overset{?}{=} 3$

   $1 = 1$ $-1 \neq 3$

   No, $(3, -4)$ is not a solution for the system because it does not satisfy both equations.

9. $\begin{cases} x + y = 5 \\ x - y = 3 \end{cases}$

11. $\begin{cases} 2l + 2w = 50 \\ w = l - 2 \end{cases}$

13. $\begin{cases} x + y = 5 \\ x - y = 3 \end{cases}$ Graph each equation.

    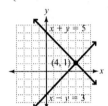

    The lines intersect at a single point, which appears to be $(4, 1)$.

15. $\begin{cases} y = 2x + 5 \\ y = -x - 4 \end{cases}$ Graph each equation.

    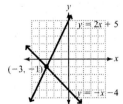

    The lines intersect at a single point, which appears to be $(-3, -1)$.

17. $\begin{cases} 2x - y = 3 \\ 2x - y = 8 \end{cases}$ Graph each equation.

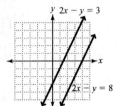

The lines appear to be parallel, so the system has no solution.

19. $\begin{cases} 3x + y = 4 \\ 6x + 2y = 8 \end{cases}$ Graph each equation.

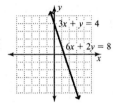

The lines appear to be identical, so the solution is the set of all ordered pairs that solve $3x + y = 4$.

21. $\begin{cases} x = 4 \\ y = -2 \end{cases}$ Graph each equation.

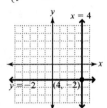

The lines intersect at a single point, which appears to be $(4, -2)$.

23. a) Because these lines intersect in a single point, this system is consistent with independent equations.
    b) This system has one solution.

25. a) Because the lines are parallel, this system is inconsistent.
    b) This system has no solution.

27. a) Because these lines coincide, this system is consistent with dependent equations.
    b) This system has an infinite number of solutions.

29. $\begin{cases} y = -x \\ y - x = 6 \end{cases} \rightarrow \begin{cases} y = -x \\ y = x + 6 \end{cases}$

The lines have different slopes. The graphs are different. This system is consistent with independent equations.

31. $\begin{cases} x + 3y = 1 \\ 2x + 6y = 2 \end{cases} \rightarrow \begin{cases} y = -\dfrac{1}{3}x + \dfrac{1}{3} \\ y = -\dfrac{1}{3}x + \dfrac{1}{3} \end{cases}$

The lines have the same slope and the same y-intercept. The graphs are identical. This system is consistent with dependent equations.

33. $\begin{cases} 3x + 2y = 12 \\ 6x + 4y = -12 \end{cases} \rightarrow \begin{cases} y = -\dfrac{3}{2}x + 6 \\ y = -\dfrac{3}{2}x - 3 \end{cases}$

The lines have the same slope but different y-intercepts. The graphs are parallel lines. This system is inconsistent.

35. $\begin{cases} 3x - y = 2 \\ y = 2x \end{cases}$

$$3x - (2x) = 2 \qquad\qquad y = 2(2)$$
$$x = 2 \qquad\qquad\qquad y = 4$$

Solution: $(2, 4)$

37. $\begin{cases} x + y = 1 \\ y = -2x - 1 \end{cases}$

$$x + y = 1 \qquad\qquad y = -2x - 1$$
$$x + (-2x - 1) = 1 \qquad y = -2(-2) - 1$$
$$-x - 1 = 1 \qquad\qquad y = 4 - 1$$
$$-x = 2 \qquad\qquad\quad y = 3$$
$$x = -2$$

Solution: $(-2, 3)$

39. $\begin{cases} y = -\dfrac{3}{4}x \\ x - 8y = -7 \end{cases}$

$$x - 8(y) = -7 \qquad\quad y = -\dfrac{3}{4}x$$
$$x - 8\left(-\dfrac{3}{4}x\right) = -7 \qquad y = -\dfrac{3}{4}(-1)$$
$$x + 6x = -7 \qquad\qquad y = \dfrac{3}{4}$$
$$7x = -7$$
$$x = -1$$

Solution: $\left(-1, \dfrac{3}{4}\right)$

41. $\begin{cases} 3x + 4y = 11 \\ x + 2y = 5 \end{cases}$

    Solve the second equation for x: $x = 5 - 2y$.

    $3(x) + 4y = 11$ $x = 5 - 2(y)$

    $3(5 - 2y) + 4y = 11$ $x = 5 - 2(2)$

    $15 - 6y + 4y = 11$ $x = 5 - 4$

    $15 - 2y = 11$ $x = 1$

    $-2y = -4$

    $y = 2$

    Solution: $(1, 2)$

43. $\begin{cases} 4x - y = -13 \\ 3x - 4y = -13 \end{cases}$

    Solve the first equation for y: $y = 4x + 13$.

    $3x - 4(4x + 13) = -13$ $y = 4(-3) + 13$

    $3x - 16x - 52 = -13$ $y = -12 + 13$

    $-13x - 52 = -13$ $y = 1$

    $-13x = 39$

    $x = -3$

    Solution: $(-3, 1)$

45. $\begin{cases} 4x + 3y = 2 \\ 3x + 2y = 2 \end{cases}$

    Solve the second equation for y.

    $3x + 2y = 2$

    $2y = -3x + 2$

    $y = -\dfrac{3}{2}x + 1$

    $4x + 3y = 2$ $y = -\dfrac{3}{2}x + 1$

    $4x + 3\left(-\dfrac{3}{2}x + 1\right) = 2$ $y = -\dfrac{3}{2}(2) + 1$

    $4x - \dfrac{9}{2}x + 3 = 2$ $y = -3 + 1$

    $-\dfrac{1}{2}x = -1$ $y = -2$

    $x = 2$

    Solution: $(2, -2)$

47. $\begin{cases} x - 3y = -4 \\ 5x - 15y = -6 \end{cases}$

    Solve the first equation for x: $x = 3y - 4$.

    $5(3y - 4) - 15y = -6$

    $15y - 20 - 15y = -6$

    $-20 = -6$ false

Because this last statement is false, the system is inconsistent and has no solution.

49. $\begin{cases} x - 2y = 6 \\ -2x + 4y = -12 \end{cases}$

    Solve the first equation for x: $x = 2y + 6$.

    $-2(2y + 6) + 4y = -12$

    $-4y - 12 + 4y = -12$

    $-12 = -12$ true

    Notice that $-12 = -12$ no longer has a variable and is true. The equations in the system are dependent; there are an infinite number of solutions that are all of the ordered pairs along $x - 2y = 6$.

51. Mistake: Did not distribute properly.

    Correct: $(3, 2)$

53. $\begin{cases} x + y = 1 \\ 2x - y = 2 \end{cases}$

    $\begin{array}{l} x + y = 1 \\ \underline{2x - y = 2} \\ 3x = 3 \\ x = 1 \end{array}$ $\begin{array}{l} 1 + y = 1 \\ y = 0 \end{array}$

    Solution: $(1, 0)$

55. $\begin{cases} 2x + 3y = 9 \\ 4x + 3y = 15 \end{cases}$

    $\begin{array}{l} -(2x + 3y = 9) \\ \underline{4x + 3y = 15} \\ -2x - 3y = -9 \\ \underline{4x + 3y = 15} \\ 2x = 6 \\ x = 3 \end{array}$ $\begin{array}{l} 2(3) + 3y = 9 \\ 6 + 3y = 9 \\ 3y = 3 \\ y = 1 \end{array}$

    Solution: $(3, 1)$

57. $\begin{cases} 4x + y = 8 \\ 5x + 3y = 3 \end{cases}$

    $\begin{array}{l} -3(4x + y = 8) \\ \underline{5x + 3y = 3} \\ -12x - 3y = -24 \\ \underline{5x + 3y = 3} \\ -7x = -21 \\ x = 3 \end{array}$ $\begin{array}{l} 4(3) + y = 8 \\ 12 + y = 8 \\ y = -4 \end{array}$

    Solution: $(3, -4)$

59. $\begin{cases} x+3y=-1 \\ 3x+6y=-1 \end{cases}$

$-3(x+3y=-1)$ $x+3\left(-\dfrac{2}{3}\right)=-1$

$\underline{3x+6y=-1}$ $x-2=-1$

$-3x-9y=3$ $x=1$

$\underline{3x+6y=-1}$

$-3y=2$

$y=-\dfrac{2}{3}$

Solution: $\left(1,-\dfrac{2}{3}\right)$

61. $\begin{cases} 3x=2y+7 \\ 4x-3y=10 \end{cases}$

$-3(3x-2y=7)$ $3(1)=2y+7$

$\underline{2(4x-3y=10)}$ $-4=2y$

$-9x+6y=-21$ $-2=y$

$\underline{8x-6y=20}$

$-x=-1$

$x=1$

Solution: $(1,-2)$

63. $\begin{cases} 3x-4y=2 \\ 4x+5y=6 \end{cases}$

$5(3x-4y=2)$ $3x-4y=2$

$\underline{4(4x+5y=6)}$ $3\left(\dfrac{34}{31}\right)-4y=2$

$15x-20y=10$

$\underline{16x+20y=24}$ $\dfrac{102}{31}-4y=2$

$31x=34$

$x=\dfrac{34}{31}$ $-4y=-\dfrac{40}{31}$

$y=\dfrac{10}{31}$

Solution: $\left(\dfrac{34}{31},\dfrac{10}{31}\right)$

65. $\begin{cases} \dfrac{1}{5}x+\dfrac{1}{2}y=\dfrac{1}{5} \\ \dfrac{1}{2}x+\dfrac{1}{3}y=-\dfrac{4}{3} \end{cases}$

$10\left(\dfrac{1}{5}x+\dfrac{1}{2}y=\dfrac{1}{5}\right)$ $\dfrac{1}{5}x+\dfrac{1}{2}(2)=\dfrac{1}{5}$

$\underline{6\left(\dfrac{1}{2}x+\dfrac{1}{3}y=-\dfrac{4}{3}\right)}$ $\dfrac{1}{5}x+1=\dfrac{1}{5}$

$3(2x+5y=2)$ $\dfrac{1}{5}x=-\dfrac{4}{5}$

$\underline{-2(3x+2y=-8)}$ $x=-4$

$6x+15y=6$

$\underline{-6x-4y=16}$

$11y=22$

$y=2$

Solution: $(-4,2)$

67. $\begin{cases} 0.4x-0.3y=1.3 \\ 0.3x+0.5y=1.7 \end{cases}$

$10(0.4x-0.3y=1.3)$ $0.4(4)-0.3y=1.3$

$\underline{10(0.3x+0.5y=1.7)}$ $1.6-0.3y=1.3$

$5(4x-3y=13)$ $-0.3y=-0.3$

$\underline{3(3x+5y=17)}$ $y=1$

$20x-15y=65$

$\underline{9x+15y=51}$

$29x=116$

$x=4$

Solution: $(4,1)$

69. $\begin{cases} 2x+3y=6 \\ 4x-18=-6y \end{cases}$

$-2(2x+3y=6)$

$\underline{4x+6y=18}$

$-4x-6y=-12$

$\underline{4x+6y=18}$

$0=6$ false

Both variables have been eliminated, and the resulting equation, $0=6$, is false. This system of equations is inconsistent. Therefore, there is no solution.

71. $\begin{cases} x+2y=4 \\ x-4=-2y \end{cases}$

$x+2y=4$

$\underline{-1(x+2y=4)}$

$x+2y=4$

$\underline{-x-2y=-4}$

$\qquad 0=0$

Both variables have been eliminated, and the resulting equation, $0=0$, is true. This means that the equations are dependent. There are an infinite number of solutions, which are all of the ordered pairs along the line $x+2y=4$.

73. Mistake: Subtracted 7 from 8 instead of adding.

Correct: $\left(\dfrac{15}{2}, \dfrac{1}{2} \right)$

75. a) 5000 b) \$4000 c) $n>5000$

77. a) $\begin{cases} c=45 \\ c=0.10n+30 \end{cases}$

b)

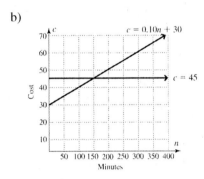

c) 150 min.
d) Plan 1
e) Plan 2

Review Exercises

1. $4(5x+7y-z)=20x+28y-4z$

2. $3x-4y+z-2x+y-z=x-3y$

3. $x+3y=6$

$3y=-x+6$

$y=-\dfrac{1}{3}x+2$

4. $\dfrac{1}{3}x-2y=10$

$x-6y=30$

$x=6y+30$

5. $\quad x+y+2z=7$

$\quad x+(-1)+2(3)=7$

$\quad x-1+6=7$

$\quad x=2$

6. $\quad x-3y+2z=6$

$\quad -2-3y+2(4)=6$

$\quad -2-3y+8=6$

$\quad 6-3y=6$

$\quad -3y=0$

$\quad y=0$

Exercise Set 4.2

1. Is $(3,-1,1)$ a solution of $\begin{cases} x+y+z=3 \\ 2x-2y-z=7 \\ 2x+y-2z=3 \end{cases}$?

Equation 1: $x+y+z=3$

$3+(-1)+1 \overset{?}{=} 3$

$3=3$

Equation 2: $2x-2y-z=7$

$2(3)-2(-1)-1 \overset{?}{=} 7$

$6+2-1 \overset{?}{=} 7$

$7=7$

Equation 3: $2x+y-2z=3$

$2(3)+(-1)-2(1) \overset{?}{=} 3$

$6-1-2 \overset{?}{=} 3$

$3=3$

Yes, $(3,-1,1)$ is a solution for the system because it satisfies all of the equations.

3.   Is $(1, 0, 2)$ a solution of $\begin{cases} 2x+3y-3z = -4 \\ -2x+4y-z = -4 \\ 3x-4y+2z = 5 \end{cases}$?

Equation 1: $2x+3y-3z = -4$

$$2(1)+3(0)-3(2) \overset{?}{=} -4$$

$$2+0-6 \overset{?}{=} -4$$

$$-4 = -4$$

Equation 2: $-2x+4y-z = -4$

$$-2(1)+4(0)-(2) \overset{?}{=} -4$$

$$-2+0-2 \overset{?}{=} -4$$

$$-4 = -4$$

Equation 3: $3x-4y+2z = 5$

$$3(1)-4(0)+2(2) \overset{?}{=} 5$$

$$3-0+4 \overset{?}{=} 5$$

$$7 \neq 5$$

No, $(1, 0, 2)$ is not a solution for the system because it does not satisfy all three equations.

5.   Is $(2, -2, 4)$ a solution of $\begin{cases} x+2y-z = -6 \\ 2x-3y+4z = 26 \\ -x+2y-3z = -18 \end{cases}$?

Equation 1: $x+2y-z = -6$

$$2+2(-2)-4 \overset{?}{=} -6$$

$$2-4-4 \overset{?}{=} -6$$

$$-6 = -6$$

Equation 2: $2x-3y+4z = 26$

$$2(2)-3(-2)+4(4) \overset{?}{=} 26$$

$$4+6+16 \overset{?}{=} 26$$

$$26 = 26$$

Equation 3: $-x+2y-3z = -18$

$$-(2)+2(-2)-3(4) \overset{?}{=} -18$$

$$-2-4-12 \overset{?}{=} -18$$

$$-18 = -18$$

Yes, $(2, -2, 4)$ is a solution for the system because it satisfies all of the equations.

7. $\begin{cases} x+y+z = 5 & \text{Eqtn. 1} \\ 2x+y-2z = -5 & \text{Eqtn. 2} \\ x-2y+z = 8 & \text{Eqtn. 3} \end{cases}$

Multiply equation 1 by -1 and add to equation 3. This makes equation 4.

$$\begin{aligned} -x-y-z &= -5 \\ \underline{x-2y+z} &= \underline{8} \\ -3y &= 3 \\ y &= -1 \quad \text{Eqtn. 4} \end{aligned}$$

Multiply equation 1 by -2 and add to equation 2. This makes equation 5.

$$\begin{aligned} -2x-2y-2z &= -10 \\ \underline{2x+y-2z} &= \underline{-5} \\ -y-4z &= -15 \quad \text{Eqtn. 5} \end{aligned}$$

Substitute equation 4 into equation 5 and solve for z.

$$\begin{aligned} -(-1)-4z &= -15 \\ 1-4z &= -15 \\ -4z &= -16 \\ z &= 4 \end{aligned}$$

Substitute the values for y and z into equation 1 to solve for x.

$$\begin{aligned} x-1+4 &= 5 \\ x+3 &= 5 \\ x &= 2 \end{aligned}$$

Solution: $(2, -1, 4)$

9. $\begin{cases} x + y + z = 2 & \text{Eqtn. 1} \\ 4x - 3y + 2z = 2 & \text{Eqtn. 2} \\ 2x + 3y - 2z = -8 & \text{Eqtn. 3} \end{cases}$

Multiply equation 1 by -2 and add to equation 3. This makes equation 4.

$\begin{aligned} -2x - 2y - 2z &= -4 \\ \underline{2x + 3y - 2z} &= \underline{-8} \\ y - 4z &= -12 \quad \text{Eqtn. 4} \end{aligned}$

Multiply equation 1 by -4 and add to equation 2. This makes equation 5.

$\begin{aligned} -4x - 4y - 4z &= -8 \\ \underline{4x - 3y + 2z} &= \underline{2} \\ -7y - 2z &= -6 \quad \text{Eqtn. 5} \end{aligned}$

Use equations 4 and 5 to make a system of equations in two variables. Solve for y.

$\begin{aligned} y - 4z &= -12 \\ \underline{-2(-7y - 2z} &= \underline{-6)} \\ y - 4z &= -12 \\ 14y + 4z &= 12 \\ 15y &= 0 \\ y &= 0 \end{aligned}$

Substitute the value for y into equation 4 and solve for z.

$\begin{aligned} 0 - 4z &= -12 \\ -4z &= -12 \\ z &= 3 \end{aligned}$

Substitute the values for y and z into equation 1 to solve for x.

$\begin{aligned} x + 0 + 3 &= 2 \\ x + 3 &= 2 \\ x &= -1 \end{aligned}$

Solution: $(-1, 0, 3)$

11. $\begin{cases} 2x + y + 2z = 5 & \text{Eqtn. 1} \\ 3x - 2y + 3z = 4 & \text{Eqtn. 2} \\ -2x + 3y + z = 8 & \text{Eqtn. 3} \end{cases}$

Multiply equation 1 by 2 and add to equation 2. This makes equation 4.

$\begin{aligned} 4x + 2y + 4z &= 10 \\ \underline{3x - 2y + 3z} &= \underline{4} \\ 7x + 7z &= 14 \quad \text{Eqtn. 4} \end{aligned}$

Multiply equation 1 by -3 and add to equation 3. This makes equation 5.

$\begin{aligned} -6x - 3y - 6z &= -15 \\ \underline{-2x + 3y + z} &= \underline{8} \\ -8x - 5z &= -7 \quad \text{Eqtn. 5} \end{aligned}$

Use equations 4 and 5 to make a system of equations in two variables. Solve.

$\begin{aligned} 8(7x + 7z &= 14) \\ 7(-8x - 5z &= -7) \\ 56x + 56z &= 112 \\ \underline{-56x - 35z} &= \underline{-49} \\ 21z &= 63 \\ z &= 3 \end{aligned}$

Substitute the value for z into equation 4 and solve for x.

$\begin{aligned} 7x + 7 \cdot 3 &= 14 \\ 7x + 21 &= 14 \\ 7x &= -7 \\ x &= -1 \end{aligned}$

Substitute the values for x and z into equation 1 to solve for y.

$\begin{aligned} 2(-1) + y + 2(3) &= 5 \\ -2 + y + 6 &= 5 \\ y + 4 &= 5 \\ y &= 1 \end{aligned}$

Solution: $(-1, 1, 3)$

13. $\begin{cases} x + 2y - z = 1 & \text{Eqtn. 1} \\ 2x + 4y - 2z = -8 & \text{Eqtn. 2} \\ 3x + y - 4z = 6 & \text{Eqtn. 3} \end{cases}$

Multiply equation 1 by -2 and add to equation 2. This makes equation 4.

$\begin{aligned} -2x - 4y + 2z &= -2 \\ \underline{2x + 4y - 2z} &= \underline{-8} \\ 0 &= -10 \end{aligned}$

Because this is a false statement, the system is inconsistent and there is no solution.

15. $\begin{cases} 4x - 2y + 3z = 6 & \text{Eqtn. 1} \\ 6x - 3y + 4.5z = 9 & \text{Eqtn. 2} \\ 12x - 6y + 9z = 18 & \text{Eqtn. 3} \end{cases}$

Multiply equation 2 by -2 and add to equation 3. This makes equation 4.

$\begin{array}{r} -12x + 6y - 9z = -18 \\ \underline{12x - 6y + 9z = 18} \\ 0 = 0 \quad \text{Eqtn. 4} \end{array}$

Multiply equation 1 by -3 and add to equation 3. This makes equation 5.

$\begin{array}{r} -12x + 6y - 9z = -18 \\ \underline{12x - 6y + 9z = 18} \\ 0 = 0 \quad \text{Eqtn. 5} \end{array}$

Because equation 4 and equation 5 are both true statements, the three equations are dependent. There are an infinite number of solutions.

17. $\begin{cases} x = 2y + z + 7 & \text{Eqtn. 1} \\ y = -3x + 2z + 1 & \text{Eqtn. 2} \\ 2x + y - z = 0 & \text{Eqtn. 3} \end{cases}$

Multiply equation 2 by -1 and add to equation 3. This makes equation 4.

$\begin{array}{r} -3x - y + 2z = -1 \\ \underline{2x + y - z = 0} \\ -x + z = -1 \quad \text{Eqtn. 4} \end{array}$

Multiply equation 2 by 2 and add to equation 1. This makes equation 5.

$\begin{array}{r} x - 2y - z = 7 \\ \underline{6x + 2y - 4z = 2} \\ 7x - 5z = 9 \quad \text{Eqtn. 5} \end{array}$

Use equations 4 and 5 to make a system of equations in two variables. Solve for x.

$-x + z = -1$ Multiply by 5.

$\begin{array}{r} 7x - 5z = 9 \\ -5x + 5z = -5 \\ \underline{7x - 5z = 9} \\ 2x = 4 \\ x = 2 \end{array}$

Substitute the value for x into equation 4 and solve for z.

$-x + z = -1$

$-2 + z = -1$

$z = 1$

Substitute the values for x and z into equation 2 to solve for y.

$y = -3x + 2z + 1$

$y = -3(2) + 2(1) + 1$

$y = -6 + 2 + 1$

$y = -3$

Solution: $(2, -3, 1)$

19. $\begin{cases} x - 4y + z = 1 & \text{Eqtn. 1} \\ 3x + 2y - z = -8 & \text{Eqtn. 2} \\ x + 6y + 2z = -3 & \text{Eqtn. 3} \end{cases}$

Add equations 1 and 2. This makes equation 4.

$\begin{array}{r} x - 4y + z = 1 \\ \underline{3x + 2y - z = -8} \\ 4x - 2y = -7 \quad \text{Eqtn. 4} \end{array}$

Multiply equation 2 by 2 and add to equation 3. This makes equation 5.

$\begin{array}{r} x + 6y + 2z = -3 \\ \underline{6x + 4y - 2z = -16} \\ 7x + 10y = -19 \quad \text{Eqtn. 5} \end{array}$

Use equations 4 and 5 to make a system of equations in two variables. Solve for x.

$4x - 2y = -7$ Multiply by 5.

$\begin{array}{r} 7x + 10y = -19 \\ 20x - 10y = -35 \\ \underline{7x + 10y = -19} \\ 27x = -54 \\ x = -2 \end{array}$

Substitute the value for x into equation 4 and solve for y.

$4(-2) - 2y = -7$

$-8 - 2y = -7$

$-2y = 1$

$y = -\dfrac{1}{2}$

Substitute the values for x and y into equation 1 to solve for z.

$-2 - 4\left(-\dfrac{1}{2}\right) + z = 1$

$-2 + 2 + z = 1$

$z = 1$

Solution: $\left(-2, -\dfrac{1}{2}, 1\right)$

21. $\begin{cases} 4x + 2y + 3z = 9 & \text{Eqtn. 1} \\ 2x - 4y - z = 7 & \text{Eqtn. 2} \\ 3x - 2z = 4 & \text{Eqtn. 3} \end{cases}$

Multiply equation 1 by 2 and add to equation 2. This makes equation 4.

$8x + 4y + 6z = 18$

$\underline{2x - 4y - z = 7}$

$10x + 5z = 25$ Eqtn. 4

Use equations 3 and 4 to make a system of equations in two variables. Solve for z.

$-10(3x - 2z = 4)$

$\underline{3(10x + 5z = 25)}$

$-30x + 20z = -40$

$\underline{30x + 15z = 75}$

$35z = 35$

$z = 1$

Substitute the value of z into equation 3 and solve for x.

$3x - 2 \cdot 1 = 4$

$3x - 2 = 4$

$3x = 6$

$x = 2$

Substitute the values for x and z into equation 1 to solve for y.

$4 \cdot 2 + 2y + 3 \cdot 1 = 9$

$8 + 2y + 3 = 9$

$2y = -2$

$y = -1$

Solution: $(2, -1, 1)$

23. $\begin{cases} 3x - 2z = -1 & \text{Eqtn. 1} \\ 4x + 5y = 23 & \text{Eqtn. 2} \\ y + 2z = -1 & \text{Eqtn. 3} \end{cases}$

Add equation 1 and equation 3. This makes equation 4.

$3x - 2z = -1$

$\underline{y + 2z = -1}$

$3x + y = -2$ Eqtn. 4

Use equation 2 and equation 4 to form a system. Multiply equation 4 by -5 and solve for x.

$4x + 5y = 23$

$\underline{3x + y = -2}$

$4x + 5y = 23$

$\underline{-15x - 5y = 10}$

$-11x = 33$

$x = -3$

Substitute the value for x in equation 1 and solve for z.

$3x - 2z = -1$

$3(-3) - 2z = -1$

$-9 - 2z = -1$

$-2z = 8$

$z = -4$

Substitute the value for z in equation 3 and solve for y.

$y + 2(-4) = -1$

$y - 8 = -1$

$y = 7$

Solution: $(-3, 7, -4)$

25. $\begin{cases} 3x + 2y = -2 & \text{Eqtn. 1} \\ 2x - 3z = 1 & \text{Eqtn. 2} \\ 0.4y - 0.5z = -2.1 & \text{Eqtn. 3} \end{cases}$

Multiply equation 1 by 2 and multiply equation 2 by -3. Add the equations to make equation 4.

$$6x + 4y \quad\;\; = -4$$
$$\underline{-6x \qquad +9z = -3}$$
$$4y + 9z = -7 \quad \text{Eqtn. 4}$$

Use equations 3 and 4 to make a system of equations in two variables. Solve for z.

$0.4y - 0.5z = -2.1$ \qquad Multiply by -10.

$$\underline{4y + 9z \quad = -7}$$
$$-4y + 5z \quad = 21$$
$$\underline{4y + 9z \quad = -7}$$
$$14z \quad = 14$$
$$z \quad = 1$$

Substitute the value of z into equation 4 and solve for y.

$$4y + 9 \cdot 1 = -7$$
$$4y + 9 = -7$$
$$4y = -16$$
$$y = -4$$

Substitute the value for y into equation 1 to solve for x.

$$3x + 2(-4) = -2$$
$$3x - 8 = -2$$
$$3x = 6$$
$$x = 2$$

Solution: $(2, -4, 1)$

27. $\begin{cases} \dfrac{3}{2}x + y - z = 0 & \text{Eqtn. 1} \\ 4y - 3z = -22 & \text{Eqtn. 2} \\ -0.2x + 0.3y = -2 & \text{Eqtn. 3} \end{cases}$

Multiply equation 1 by -3 and add to equation 2 to make equation 4.

$$-\frac{9}{2}x - 3y + 3z = 0$$
$$\underline{4y - 3z = -22}$$
$$-\frac{9}{2}x + y = -22 \quad \text{Eqtn. 4}$$

Use equations 3 and 4 to make a system of equations in two variables. Solve for x.

$-\dfrac{9}{2}x + y \quad = -22$ \qquad Multiply by -0.3.

$$\underline{-0.2x + 0.3y = -2}$$
$$1.35x - 0.3y = 6.6$$
$$\underline{-0.2x + 0.3y = -2}$$
$$1.15x \quad = 4.6$$
$$x = 4$$

Substitute the value of x into equation 4 and solve for y.

$$-\frac{9}{2} \cdot 4 + y = -22$$
$$-18 + y = -22$$
$$y = -4$$

Substitute the value for y into equation 2 to solve for z.

$$4(-4) - 3z = -22$$
$$-16 - 3z = -22$$
$$-3z = -6$$
$$z = 2$$

Solution: $(4, -4, 2)$

Review Exercises

1. Multiply each term of the expression inside the parenthesis by -0.10.

$-0.10(x - y + 200)$

$-0.10x + 0.10y + 20$

2. Use the distributive property to eliminate the parentheses.

$12x = 16\left(x - \dfrac{1}{6}\right)$

$12x = 16x - \dfrac{16}{6}$

$12x = 16x - \dfrac{8}{3}$

Isolate x by subtracting $16x$ from both sides of the equation, and then dividing both sides of the equation by -4.

$-4x = -\dfrac{8}{3}$

$x = \dfrac{2}{3}$

3. Use the distributive property to eliminate the parentheses.

$12.25x + 15.75(40 - x) = 539$

$12.25x + 630 - 15.75x = 539$

Combine the x terms.

$12.25x + 630 - 15.75x = 539$

$-3.5x + 630 = 539$

Isolate x by subtracting 630 from both sides of the equation, and then dividing both sides of the equation by -3.5

$-3.5x = -91$

$x = 26$

4. To solve for y, first multiply equation 1 by -6, and then add equation 1 to equation 2.

$(-6)\left(x = y + \dfrac{1}{30}\right)$

$\begin{aligned} -6x &= -6y - \dfrac{1}{5} \\ + \ 6x &= \ 10y \\ \hline 0 &= 4y - \dfrac{1}{5} \end{aligned}$

Simplify.

$4y = \dfrac{1}{5}$

$y = \dfrac{1}{20}$

Use the value of y in one of the equations to solve for x.

$6x = 10y$

$6x = 10\left(\dfrac{1}{20}\right)$

$6x = \dfrac{1}{2}$

$x = \dfrac{1}{12}$

Solution: $\left(\dfrac{1}{12}, \dfrac{1}{20}\right)$

Exercise Set 4.3

1. Let $x =$ the measure of the larger angle and $y =$ the measure of the smaller angle. The angles are complimentary.

$\begin{cases} x + y = 90 \\ x = y + 40 \end{cases}$

$\begin{array}{ll} x + y = 90 & x = y + 40 \\ (y + 40) + y = 90 & x = 25 + 40 \\ 2y = 50 & x = 65 \\ y = 25 & \end{array}$

The angles are $25°$ and $65°$.

3. Let $x =$ the measure of the larger angle and $y =$ the measure of the smaller angle. The angles are supplementary.

$\begin{cases} x + y = 180 \\ x = 4y \end{cases}$

$\begin{array}{ll} x + y = 180 & x = 4y \\ 4y + y = 180 & x = 4(36) \\ 5y = 180 & x = 144 \\ y = 36 & \end{array}$

The angles are $36°$ and $144°$.

5. The angle made be the incident ray and the glass boundary and I are complimentary.

$90 - I = 2I - 24$

$90 = 3I - 24$

$114 = 3I$

$38 = I$

The angle of incidence is $38°$.

7. Let x = the measure of the larger angle and y = the measure of the smaller angle. The angles are supplementary.
$$\begin{cases} x+y=180 \\ x=4y+10 \end{cases}$$

$$\begin{array}{ll} x+y=180 & x=4y+10 \\ (4y+10)+y=180 & x=4(34)+10 \\ 5y=170 & x=146 \\ y=34 \end{array}$$

The angles are $34°$ and $146°$.

9. Let w = the width of the base and l = the length of the base.
$$\begin{cases} 2l+2w=220.5 \\ w=l \end{cases}$$

$$2l+2w=220.5$$
$$2w+2w=220.5$$
$$4w=220.5$$
$$w=55.125$$

The width and length are 55.125 ft.

11. Let x = the number of smaller containers and y = the number of larger containers.
$$\begin{cases} x+y=42 \\ 6x+10y+308 \end{cases}$$

Multiply the first equation by 10, then subtract equation 2 from equation 1.

$$\begin{array}{ll} & x+y=42 \\ 10(x+y=42) & 28+y=42 \\ 10x+10y=420 & y=14 \end{array}$$

$$\underline{6x+10y=308}$$
$$4x=112$$
$$x=28$$

The vender sold 28 small containers and 14 large containers.

13. Let x = amount of money in million dollars that *The Twilight Saga; Breaking Dawn Part 1* grossed and y = the amount of money in million dollars that *The Twilight Saga; Breaking Dawn Part2* grossed.
$$\begin{cases} x+y=1542 \\ x+118=y \end{cases}$$

$$\begin{array}{ll} x+y=1542 & x+118=y \\ x+(x+118)=1542 & 712+118=y \\ 2x+118=1542 & 830=y \\ 2x=1424 \\ x=712 \end{array}$$

Breaking Dawn: Part 1 grossed \$712 million and *Breaking Dawn: Part 2* grossed \$830 million.

15. Let x = number of women in millions who are enrolled in a community college and y = number of men in millions who are enrolled in a community college.
$$\begin{cases} x+y=11.6 \\ y=x-1.8 \end{cases}$$

$$x+y=11.6$$
$$x+(x-1.8)=11.6$$
$$2x-1.8=11.6$$
$$2x=13.4$$
$$x=6.7$$

6.7 million women are enrolled in a community college.

17. Let x = salary in dollars of a math teacher and y = salary in dollars of a mathematician.
$$\begin{cases} x+y=161,430 \\ x=y-37,330 \end{cases}$$

$$\begin{array}{ll} x+y=161,430 & x=y-37,330 \\ (y-37,330)+y=161,430 & x=99,380-37,330 \\ 2y-37,430=161,430 & x=62,050 \\ 2y=198,760 \\ y=99,380 \end{array}$$

The average annual salary of a mathematician is \$99,380, and the average annual salary of a math teacher is \$62,050.

19. Let x = percent of shark attacks occurring in deep water and y = percent of shark attacks occurring in shallower waters.
$$\begin{cases} x + y = 100 \\ x = 6y \end{cases}$$

$$\begin{array}{ll} x + y = 100 & x + y = 100 \\ 6y + y = 100 & x + 14.3 = 100 \\ 7y = 100 & x \approx 85.7 \\ y \approx 14.3 \end{array}$$

About 85.7% of shark attacks occur in deep water.

21. Let x = number of 18-inch wreaths sold and y = number of 22-inch wreaths sold.

Categories	Selling Price	Number Sold	Revenue
18-inch	20	x	$20x$
22-inch	25	y	$25y$

$$\begin{cases} 20x + 25y = 1100 \\ 6 + 3x = y \end{cases}$$

$$\begin{array}{ll} 20x + 25y = 1100 & 6 + 3x = y \\ 20x + 25(6 + 3x) = 1100 & 6 + 3(10) = y \\ 20x + 150 + 75x = 1100 & 36 = y \\ 95x = 950 \\ x = 10 \end{array}$$

A total of 10 18-inch wreaths and 36 22-inch wreaths were sold.

23. Let u = number of Union soldiers and c = number of Confederate soldiers.
$$\begin{cases} u + c = 498,000 \\ u = 3c - 38,000 \end{cases}$$

$$\begin{array}{l} u + c = 498,000 \\ (3c - 38,000) + c = 498,000 \\ 4c - 38,000 = 498,000 \\ 4c = 536,000 \\ c = 134,000 \end{array}$$

$$u = 3c - 38,000$$
$$u = 3(134,000) - 38,000$$
$$u = 364,000$$

A total of 364,000 Union soldiers and 134,000 Confederate soldiers were lost.

25. Let j = Josh's time in hours and d = Dolph's time in hours.
$$\begin{cases} j = d + 0.05 \\ 12j = 15d \end{cases}$$

$$\begin{array}{l} 12j = 15d \\ 12(d + 0.05) = 15d \\ 12d + 0.6 = 15d \\ 0.6 = 3d \\ 0.2 = d \end{array}$$

It will take Dolph 0.2 hours to catch up to Josh.

27. Let p = Poloma's time in hours and f = Francis' time in hours.
$$\begin{cases} p = f + 0.5 \\ 65p = 70f \end{cases}$$

$$\begin{array}{l} 65p = 70f \\ 65(f + 0.5) = 70f \\ 65f + 32.5 = 70f \\ 32.5 = 5f \\ 6.5 = f \end{array}$$

Francis will catch up with Poloma in 6.5 hours at 10:00 P.M.

29. Let x = the boat's speed in mph and y = the speed of the current in mph.
$$\begin{cases} 3(x - y) = 36 \\ 3(x + y) = 48 \end{cases} \Rightarrow \begin{cases} x - y = 12 \\ x + y = 16 \end{cases}$$

$$\begin{array}{ll} x - y = 12 & x + y = 16 \\ \underline{x + y = 16} & 14 + y = 16 \\ 2x = 28 & y = 2 \\ x = 14 \end{array}$$

The boat's speed is 14 mph and the current's speed is 2 mph.

31. Let x = the speed of the plane in mph and y = the wind speed in mph.
$$\begin{cases} x + y = 650 \\ x - y = 550 \end{cases}$$

$$\begin{array}{ll} x + y = 650 & x + y = 650 \\ \underline{x - y = 550} & 600 + y = 650 \\ 2x = 1200 & y = 50 \\ x = 600 \end{array}$$

The plane's speed is 600 mph and the wind's speed is 50 mph.

33. Let x = the volume of 5% HCl mixture and y = the volume of 20% HCl.

Solution	Concentration	Volume of Solution	Volume of HCl
5%	0.05	x	$0.05x$
20%	0.20	y	$0.20y$
12.5%	0.125	10	$0.125(10)$

$\begin{cases} x+y=10 \\ 0.05x+0.20y=0.125(10) \end{cases}$

Solve the first equation for x: $x=10-y$

$$0.05x+0.20y=0.125(10)$$
$$0.05(10-y)+0.20y=1.25$$
$$0.5-0.05y+0.20y=1.25$$
$$0.15y=0.75$$
$$y=5$$

$x=10-y$
$x=10-5$
$x=5$

Combine 5 ml of 20% HCL mixture and 5 ml of 5% HCL mixture.

35. Let x = the volume of 10% antifreeze mixture and y = the volume of 16% antifreeze mixture.

Solution	Concentration	Volume of Solution	Volume of Antifreeze
10%	0.10	x	$0.10x$
20%	0.20	24	$0.20(24)$
16%	0.16	y	$0.16y$

$\begin{cases} y=x+24 \\ 0.16y=0.10x+0.20(24) \end{cases}$

Multiply the first equation by -0.16, then add the two equations.
$-0.16(y=x+24)$
$-0.16y=-0.16x-3.84$
$+\ 0.16y=0.10x+4.8$
$$0=-0.06x+0.96$$
$$0.06x=0.96$$
$$x=16$$

Add 16 oz. of the 10% mixture to the 20% mixture.

37. Let x = the volume of 40% pesticide mixture and y = the volume of 20% pesticide mixture.

Solution	Concentration	Volume of Solution	Volume of Pesdticide
40%	0.40	x	$0.40x$
10%	0.10	20	$0.10(20)$
20%	0.20	y	$0.20y$

$\begin{cases} y=x+20 \\ 0.20y=0.40x+0.10(20) \end{cases}$

Multiply the first equation by -0.20, then add the two equations.
$-0.20(y=x+20)$
$-0.20y=-0.20x-4$
$+\ 0.20y=0.40x+2$
$$0=0.20x-2$$
$$2=0.20x$$
$$10=x$$

Add 10 gallons of the 40% mixture to the 10% mixture.

39. Let x = the amount in the account at 9% and y = the amount in the account at 5%

Categories	Interest Rate	Amount in Account	Interest
9% account	0.09	x	$0.09x$
5% account	0.05	y	$0.05y$

$\begin{cases} x=4y \\ 0.09x+0.05y=1435 \end{cases}$

$0.09x+0.05y=1435$　　$x=4y$
$0.09(4y)+0.05y=1435$　　$x=4(3500)$
$0.36y+0.05y=1435$　　$x=14,000$
$0.41y=1435$
$y=3500$

He should invest \$3500 at 5% and \$14,000 at 9%.

41. Let x, y, and z represent the three numbers.
$$\begin{cases} x+y+z=16 & \text{Eqtn. 1} \\ x-2y=2 & \text{Eqtn. 2} \\ x-z=-2 & \text{Eqtn. 3} \end{cases}$$

Multiply equation 2 by -1 and add to equation 3 to make equation 4.

$$-x+2y=-2$$
$$\underline{x-z=-2}$$
$$2y-z=-4 \quad \text{Eqtn. 4}$$

Multiply equation 1 by -1 and add to equation 2. This makes equation 5.

$$-x-y-z=-16$$
$$\underline{x-2y=2}$$
$$-3y-z=-14 \quad \text{Eqtn. 5}$$

Use equations 4 and 5 to make a system of equations in two variables. Solve.

$$-1(2y-z=-4)$$
$$\underline{-3y-z=-14}$$
$$-2y+z=4$$
$$\underline{-3y-z=-14}$$
$$-5y \quad = -10$$
$$y=2$$

Substitute the value of y into equation 2 and solve for x.

$$x=2+2\cdot2$$
$$x=6$$

Substitute the value for x into equation 3 to solve for z.

$$6=z-2$$
$$8=z$$

The numbers are 6, 2, and 8.

43. Let x be the first integer, y be the second integer, and z be the third integer.
$$39=x+y+z$$
$$39=x+(x+1)+(x+2)$$
$$39=3x+3$$
$$36=3x$$
$$12=x$$

Use x to find y and z.
$$y=x+1 \quad z=x+2$$
$$y=12+1 \quad z=12+2$$
$$y=13 \qquad z=14$$

The consecutive integers are 12, 13, and 14.

45. Let x, y, and z represent the measures of the three angles of the triangle.
$$\begin{cases} x+y+z=180 & \text{Eqtn. 1} \\ x=3y & \text{Eqtn. 2} \\ y=z-5 & \text{Eqtn. 3} \end{cases}$$

Substitute equation 2 into equation 1. This becomes equation 4.
$$3y+y+z=180$$
$$4y+z=180 \quad \text{Eqtn. 4}$$

Use equations 3 and 4 to make a system of equations in two variables. Solve.

$$y-z=-5 \qquad \text{Eqtn. 3}$$
$$\underline{4y+z=180} \qquad \text{Eqtn. 4}$$
$$5y=175$$
$$y=35$$

Substitute the value of y into equation 2 and solve for x.
$$x=3\cdot35$$
$$x=105$$

Substitute the value for y into equation 3 and solve for z.
$$35=z-5$$
$$40=z$$

The angles are 105°, 35°, and 40°.

47. Let b = the cost of a burger, f = the cost of an order of fries, and d = the cost of a drink.
$$\begin{cases} b+f+d=5 & \text{Eqtn. 1} \\ 3b+2f+2d=12.50 & \text{Eqtn. 2} \\ 2b+4f+3d=14 & \text{Eqtn. 3} \end{cases}$$

Multiply equation 1 by -2 and add to equation 3 to make equation 4.

$$-2b-2f-2d=-10$$
$$\underline{2b+4f+3d=14}$$
$$2f+d=4 \qquad \text{Eqtn. 4}$$

Multiply equation 1 by -3 and add to equation 2 to make equation 5.

$$-3b-3f-3d=-15$$
$$\underline{3b+2f+2d=12.50}$$
$$-f-d=-2.5 \qquad \text{Eqtn. 5}$$

Use equations 4 and 5 to make a system of equations in two variables. Solve.

$$2f+d=4$$
$$\underline{-f-d=-2.5}$$
$$f=1.5$$

Substitute into equation 5 to solve for d.

$$-1.5 - d = -2.5$$
$$-d = -1$$
$$d = 1$$

Substitute into equation 1 to solve for b.
$$b + 1.5 + 1 = 5$$
$$b + 2.5 = 5$$
$$b = 2.5$$

A burger costs \$2.50, fries cost \$1.50, and a drink costs \$1.00.

49. Let c = the number of children's tickets sold, s = the number of student tickets sold, and a = the number of adult tickets sold.

$$\begin{cases} c + s + a = 500 & \text{Eqtn. 1} \\ 3c + 5s + 8a = 2500 & \text{Eqtn. 2} \\ a = s - 150 & \text{Eqtn. 3} \end{cases}$$

Multiply equation 1 by -3 and add to equation 2. This makes equation 4.
$$-3c - 3s - 3a = -1500$$
$$\underline{3c + 5s + 8a = 2500}$$
$$2s + 5a = 1000 \quad \text{Eqtn. 4}$$

Use equations 3 and 4 to make a systems of equations and solve.
$$2s + 5a = 1000$$
$$\underline{-s + a = -150} \quad \text{Multiply by 2.}$$
$$2s + 5a = 1000$$
$$\underline{-2s + 2a = -300}$$
$$7a = 700$$
$$a = 100$$

Substitute into equation 3 to solve for s.
$$-s + 100 = -150$$
$$-s = -250$$
$$s = 250$$

Substitute into equation 1 to solve for c.
$$c + 250 + 100 = 500$$
$$c = 150$$

150 child, 250 student, and 100 adult tickets were sold.

51. Let a = then number of 2-point shots, b = the number of 3-point shots, and let c = the number of 1-point shots.

$$\begin{cases} 2a + 3b + c = 14 & \text{Eqtn. 1} \\ a + b + c = 7 & \text{Eqtn. 2} \\ c = a + 2 & \text{Eqtn. 3} \end{cases}$$

Substitute equation 3 into equation 1. This makes equation 4.
$$2a + 3b + c = 14$$
$$2a + 3b + (a + 2) = 14$$
$$3a + 3b = 12 \quad \text{Eqtn. 4}$$

Substitute equation 3 into equation 2. This makes equation 5.
$$a + b + c = 7$$
$$a + b + a + 2 = 7$$
$$2a + b = 5 \quad \text{Eqtn. 5}$$

Set up a system using equation 4 and equation 5.

$$3a + 3b = 12$$
$$-3(2a + b = 5)$$
$$3a + 3b = 12$$
$$\underline{-6a - 3b = -15}$$
$$-3a = -3$$
$$a = 1$$

Substitute 1 for a in equation 5 and solve for b.
$$2a + b = 5$$
$$2(1) + b = 5$$
$$2 + b = 5$$
$$b = 3$$

Substitute 1 for a in equation 3 and solve for c.
$$c = a + 2$$
$$c = 1 + 2$$
$$c = 3$$

John made three 3-point field goals, one 2-point field goal, and three free throws.

53. Let z = the number of pounds of zinc, t = the number of pounds of tin, and c = the number of pounds of copper.

$$\begin{cases} z + t + c = 1000 & \text{Eqtn. 1} \\ t = 3z & \text{Eqtn. 2} \\ c = 20 + 15t & \text{Eqtn. 3} \end{cases}$$

Substitute equation 2 into equations 1 and 3 to make into a system of equations in two variables.

$z + 3z + c = 1000$ $\qquad t = 3 \cdot 20$

$\underline{c = 20 + 15(3z)}$ $\qquad t = 60$

$\quad 4z + c = 1000$

$\underline{45z - c = -20}$

$\qquad 49z = 980$

$\qquad\quad z = 20$

$c = 20 + 15 \cdot 60$

$c = 920$

20 lbs. of zinc, 60 lbs. of tin, and 920 lbs. of copper were used.

55. Let h = the number of pounds of Black Forest ham, t = the number of pounds of turkey breast, and r = the number of pounds of roast beef.

$$\begin{cases} h + t + r = 10 & \text{Eqtn. 1} \\ 11.96h + 8.76t + 9.16r = 10(9.80) & \text{Eqtn. 2} \\ t = h + r & \text{Eqtn. 3} \end{cases}$$

Substitute equation 3 into equations 1 and 2 to make a system of linear equations in two variables.

$h + h + r + r = 10$

$\underline{11.96h + 8.76(h + r) + 9.16r = 98}$

$2h + 2r = 10 \qquad$ Multiply by -8.96.

$\underline{20.72h + 17.92r = 98}$

$-17.92h - 17.92r = -89.6$

$\underline{20.72h + 17.92r = 98}$

$\qquad\quad 2.8h = 8.4$

$\qquad\qquad h = 3$

$2 \cdot 3 + 2r = 10 \qquad 3 + t + 2 = 10$

$\qquad 2r = 4 \qquad\qquad\quad t = 5$

$\qquad\quad r = 2$

3 lbs. of ham, 5 lbs. of turkey, and 2 lbs. of roast beef were used for the party tray.

57. Let x = the amount invested at 4%, y = the amount invested at 6%, and z = the amount invested at 7%.

$$\begin{cases} x + y + z = 8000 & \text{Eqtn. 1} \\ 0.04x + 0.06y + 0.07z = 475 & \text{Eqtn. 2} \\ z - x = 1500 & \text{Eqtn. 3} \end{cases}$$

Multiply equation 1 by -0.06 and add to equation 2. This makes equation 4.

$-0.06x - 0.06y - 0.06z = -480$

$\underline{0.04x + 0.06y + 0.07z = 475}$

$\quad -0.02x + 0.01z = -5 \qquad$ Eqtn. 4

Multiply equation 3 by -0.01 and add to equation 4.

$0.01x - 0.01z = -15$

$\underline{-0.02x + 0.01z = -5}$

$\quad -0.01x \qquad = -20$

$\qquad\quad x \qquad = 2000$

$z - x = 1500 \qquad\qquad x + y + z = 8000$

$z - 2000 = 1500 \qquad 2000 + y + 3500 = 8000$

$\qquad z = 3500 \qquad\qquad\qquad y = 2500$

A total of \$2000 is invested at 4%, \$2500 is invested at 6%, and \$3500 is invested at 7%.

59. Let b = the number of calories burned while bicycling, w = the number of calories burned while walking, and c = the number of calories burned while climbing stairs.

$$\begin{cases} b = 120 + w & \text{Eqtn. 1} \\ c = 180 + b & \text{Eqtn. 2} \\ c = 2w & \text{Eqtn. 3} \end{cases}$$

Substitute equation 3 into equation 2. Use this new equation with equation 1 to form a system of equations in two variables.

$b - w = 120 \qquad\qquad c = 2 \cdot 300$

$\underline{-b + 2w = 180} \qquad\quad c = 600$

$\qquad w = 300$

$b = 120 + 300$

$b = 420$

The number of calories burned while bicycling is 420, while walking is 300, and while stair climbing is 600.

61. $\begin{cases} a + v_0 + h_0 = 234 & \text{Eqtn. 1} \\ 9a + 3v_0 + h_0 = 306 & \text{Eqtn. 2} \\ 36a + 6v_0 + h_0 = 174 & \text{Eqtn. 3} \end{cases}$

Multiply equation 1 by -1 and add it to equation 2. This will make equation 4.

$-a - v_0 - h_0 = -234$

$\underline{9a + 3v_0 + h_0 = 306}$

$8a + 2v_0 \quad\quad = 72 \quad\quad$ Eqtn. 4

Multiply equation 1 by -1 and add it to equation 3. This will make equation 5.

$-a - v_0 - h_0 = -234$

$\underline{36a + 6v_0 + h_0 = 174}$

$35a + 5v_0 \quad\quad = -60 \quad\quad$ Eqtn. 5

Use equations 4 and 5 to make a system of linear equations in two variables. Solve.

$8a + 2v_0 = 72 \quad\quad$ Multiply by -5.

$\underline{35a + 5v_0 = -60} \quad\quad$ Multiply by 2.

$-40a - 10v_0 = -360$

$\underline{70a + 10v_0 = -120}$

$30a \quad\quad = -480$

$a \quad\quad = -16$

$8(-16) + 2v_0 = 72 \quad\quad -16 + 100 + h_0 = 234$

$-128 + 2v_0 = 72 \quad\quad 84 + h_0 = 234$

$2v_0 = 200 \quad\quad h_0 = 150$

$v_0 = 100$

$h = -16t^2 + 100t + 150$

Review Exercises

1. $-3(2x - 4y - z + 9) = -6x + 12y + 3z - 27$

2. $x - 0.5y = 8$

   $6 - 0.5y = 8$

   $-0.5y = 2$

   $y = -4$

3. $f(x) = -2x + 1$

   $f(1) = -2(1) + 1$

   $= -2 + 1$

   $= -1$

4. The y-intercept is. $(0, 1)$.

5. The slope is $m = -2$.

6.

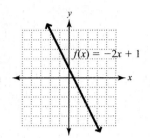

Exercise Set 4.4

1. $\begin{bmatrix} 14 & 7 & | & 6 \\ 7 & 6 & | & 8 \end{bmatrix}$

3. $\begin{bmatrix} 7 & -6 & | & 1 \\ 0 & -2 & | & 5 \end{bmatrix}$

5. $\begin{bmatrix} 1 & -3 & 1 & | & 4 \\ 2 & -4 & 2 & | & -4 \\ 6 & -2 & 5 & | & -4 \end{bmatrix}$

7. $\begin{bmatrix} 4 & 6 & -2 & | & -1 \\ 8 & 3 & 0 & | & -12 \\ 0 & -1 & 2 & | & 4 \end{bmatrix}$

9. $\begin{bmatrix} 1 & -3 & | & -2 \\ 0 & 1 & | & 2 \end{bmatrix}$ represents the system $\begin{cases} x - 3y = -2 \\ \quad\quad y = 2 \end{cases}$

   $x - 3y = -2$

   $x - 3(2) = -2$

   $x - 6 = -2$

   $x = 4$

   Solution: $(4, 2)$

11. $\begin{bmatrix} 1 & -4 & -8 & | & 6 \\ 0 & 1 & -2 & | & -7 \\ 0 & 0 & 1 & | & 1 \end{bmatrix}$ represents the system

    $\begin{cases} x - 4y - 8z = 6 \\ \quad\quad y - 2z = -7 \\ \quad\quad\quad\quad z = 1 \end{cases}$

    $y - 2z = -7 \quad\quad\quad\quad x - 4y - 8z = 6$

    $y - 2(1) = -7 \quad\quad\quad x - 4(-5) - 8(1) = 6$

    $y - 2 = -7 \quad\quad\quad\quad x + 20 - 8 = 6$

    $y = -5 \quad\quad\quad\quad\quad x = -6$

    Solution: $(-6, -5, 1)$

13. $\begin{bmatrix} 1 & 3 & | & -1 \\ -2 & 5 & | & 6 \end{bmatrix}$

    $2R_1 + R_2 \rightarrow \begin{bmatrix} 1 & 3 & | & -1 \\ 0 & 11 & | & 4 \end{bmatrix}$

15. $\begin{bmatrix} 1 & -2 & 4 & | & 6 \\ 0 & 2 & -1 & | & -5 \\ 0 & 8 & -6 & | & -3 \end{bmatrix}$

$-4R_2 + R_3 \rightarrow \begin{bmatrix} 1 & -2 & 4 & | & 6 \\ 0 & 2 & -1 & | & -5 \\ 0 & 0 & -2 & | & 17 \end{bmatrix}$

17. $\begin{bmatrix} 4 & 8 & | & -10 \\ -1 & 3 & | & 2 \end{bmatrix}$

$\frac{1}{4}R_1 \rightarrow \begin{bmatrix} 1 & 2 & | & -2.5 \\ -1 & 3 & | & 2 \end{bmatrix}$

19. Replace R_2 with $3R_1 + R_2$.

21. Replace R_3 with $-2R_2 + R_3$.

23. $\begin{cases} x - y = 5 \\ x + y = -1 \end{cases} \begin{bmatrix} 1 & -1 & | & 5 \\ 1 & 1 & | & -1 \end{bmatrix}$

$-1R_1 + R_2 \rightarrow \begin{bmatrix} 1 & -1 & | & 5 \\ 0 & 2 & | & -6 \end{bmatrix}$

$\frac{1}{2}R_2 \rightarrow \begin{bmatrix} 1 & -1 & | & 5 \\ 0 & 1 & | & -3 \end{bmatrix}$

$\begin{cases} x - y = 5 \\ y = -3 \end{cases}$

$x - y = 5$ Solution: $(2, -3)$
$x - (-3) = 5$
$x + 3 = 5$
$x = 2$

25. $\begin{cases} x + y = 3 \\ 3x - y = 1 \end{cases}$

$\begin{bmatrix} 1 & 1 & | & 3 \\ 3 & -1 & | & 1 \end{bmatrix}$

$-3R_1 + R_2 \rightarrow \begin{bmatrix} 1 & 1 & | & 3 \\ 0 & -4 & | & -8 \end{bmatrix}$

$-\frac{1}{4}R_2 \rightarrow \begin{bmatrix} 1 & 1 & | & 3 \\ 0 & 1 & | & 2 \end{bmatrix}$

$\begin{cases} x + y = 3 \\ y = 2 \end{cases}$

$x + y = 3$ Solution: $(1, 2)$
$x + 2 = 3$
$x = 1$

27. $\begin{cases} x + 2y = -7 \\ 2x - 4y = 2 \end{cases}$

$\begin{bmatrix} 1 & 2 & | & -7 \\ 2 & -4 & | & 2 \end{bmatrix}$

$-2R_1 + R_2 \rightarrow \begin{bmatrix} 1 & 2 & | & -7 \\ 0 & -8 & | & 16 \end{bmatrix}$

$-\frac{1}{8}R_2 \rightarrow \begin{bmatrix} 1 & 2 & | & -7 \\ 0 & 1 & | & -2 \end{bmatrix}$

$\begin{cases} x + 2y = -7 \\ y = -2 \end{cases}$

$x + 2y = -7$ Solution: $(-3, -2)$
$x + 2(-2) = -7$
$x - 4 = -7$
$x = -3$

29. $\begin{cases} -2x + 5y = 4 \\ x - 2y = -2 \end{cases}$

$$\begin{bmatrix} -2 & 5 & | & 4 \\ 1 & -2 & | & -2 \end{bmatrix}$$

$R_1 + 2R_2 \rightarrow \begin{bmatrix} -2 & 5 & | & 4 \\ 0 & 1 & | & 0 \end{bmatrix}$

$-\dfrac{1}{2}R_1 \rightarrow \begin{bmatrix} 1 & -\dfrac{5}{2} & | & -2 \\ 0 & 1 & | & 0 \end{bmatrix}$

$\begin{cases} x - \dfrac{5}{2}y = -2 \\ \qquad y = 0 \end{cases}$

$x - \dfrac{5}{2}y = -2$ Solution: $(-2, 0)$

$x - \dfrac{5}{2}(0) = -2$

$x - 0 = -2$

$x = -2$

31. $\begin{cases} 4x - 3y = -2 \\ 2x - 3y = -10 \end{cases}$

$$\begin{bmatrix} 4 & -3 & | & -2 \\ 2 & -3 & | & -10 \end{bmatrix}$$

$R_1 - 2R_2 \rightarrow \begin{bmatrix} 4 & -3 & | & -2 \\ 0 & 3 & | & 18 \end{bmatrix}$

$\dfrac{1}{4}R_1 \rightarrow \begin{bmatrix} 1 & -\dfrac{3}{4} & | & -\dfrac{1}{2} \\ 0 & 3 & | & 18 \end{bmatrix}$

$\dfrac{1}{3}R_2 \rightarrow \begin{bmatrix} 1 & -\dfrac{3}{4} & | & -\dfrac{1}{2} \\ 0 & 1 & | & 6 \end{bmatrix}$

$\begin{cases} x - \dfrac{3}{4}y = -\dfrac{1}{2} \\ \qquad y = 6 \end{cases}$

$x - \dfrac{3}{4}y = -\dfrac{1}{2}$ Solution: $(4, 6)$

$x - \dfrac{3}{4}(6) = -\dfrac{1}{2}$

$x - \dfrac{9}{2} = -\dfrac{1}{2}$

$x = 4$

33. $\begin{cases} 5x + 2y = 12 \\ 2x + 3y = -4 \end{cases}$

$$\begin{bmatrix} 5 & 2 & | & 12 \\ 2 & 3 & | & -4 \end{bmatrix}$$

$-2R_1 + 5R_2 \rightarrow \begin{bmatrix} 5 & 2 & | & 12 \\ 0 & 11 & | & -44 \end{bmatrix}$

$\dfrac{1}{5}R_1 \rightarrow \begin{bmatrix} 1 & \dfrac{2}{5} & | & \dfrac{12}{5} \\ 0 & 11 & | & -44 \end{bmatrix}$

$\dfrac{1}{11}R_2 \rightarrow \begin{bmatrix} 1 & \dfrac{2}{5} & | & \dfrac{12}{5} \\ 0 & 1 & | & -4 \end{bmatrix}$

$\begin{cases} x + \dfrac{2}{5}y = \dfrac{12}{5} \\ \qquad y = -4 \end{cases}$

$x + \dfrac{2}{5}y = \dfrac{12}{5}$ Solution: $(4, -4)$

$x + \dfrac{2}{5}(-4) = \dfrac{12}{5}$

$x - \dfrac{8}{5} = \dfrac{12}{5}$

$x = 4$

35. $\begin{cases} x + y + z = 3 \\ 2x + y - 3z = -10 \\ 2x + 2y + z = 3 \end{cases}$

$$\begin{bmatrix} 1 & 1 & 1 & | & 3 \\ 2 & 1 & -3 & | & -10 \\ 2 & 2 & 1 & | & 3 \end{bmatrix}$$

$-2R_1 + R_2 \rightarrow \begin{bmatrix} 1 & 1 & 1 & | & 3 \\ 0 & -1 & -5 & | & -16 \\ 2 & 2 & 1 & | & 3 \end{bmatrix}$

$-2R_1 + R_3 \rightarrow \begin{bmatrix} 1 & 1 & 1 & | & 3 \\ 0 & -1 & -5 & | & -16 \\ 0 & 0 & -1 & | & -3 \end{bmatrix}$

$\begin{matrix} -1R_2 \rightarrow \\ -1R_3 \rightarrow \end{matrix} \begin{bmatrix} 1 & 1 & 1 & | & 3 \\ 0 & 1 & 5 & | & 16 \\ 0 & 0 & 1 & | & 3 \end{bmatrix}$

$\begin{cases} x + y + z = 3 \\ \quad y + 5z = 16 \\ \qquad z = 3 \end{cases}$

$$y + 5z = 16$$
$$y + 5(3) = 16$$
$$y + 15 = 16$$
$$y = 1$$

$$x + y + z = 3$$
$$x + 1 + 3 = 3$$
$$x + 4 = 3$$
$$x = -1$$

Solution: $(-1, 1, 3)$

37. $\begin{cases} x - y + z = -1 \\ 2x - 2y + z = 0 \\ x + 3y + 2z = 1 \end{cases}$

$$\begin{bmatrix} 1 & -1 & 1 & | & -1 \\ 2 & -2 & 1 & | & 0 \\ 1 & 3 & 2 & | & 1 \end{bmatrix}$$

$$-2R_1 + R_2 \rightarrow \begin{bmatrix} 1 & -1 & 1 & | & -1 \\ 0 & 0 & -1 & | & 2 \\ 1 & 3 & 2 & | & 1 \end{bmatrix}$$

$$\begin{matrix} R_3 \rightarrow \\ R_2 \rightarrow \end{matrix} \begin{bmatrix} 1 & -1 & 1 & | & -1 \\ 1 & 3 & 2 & | & 1 \\ 0 & 0 & -1 & | & 2 \end{bmatrix}$$

$$-R_1 + R_2 \begin{bmatrix} 1 & -1 & 1 & | & -1 \\ 0 & 4 & 1 & | & 2 \\ 0 & 0 & -1 & | & 2 \end{bmatrix}$$

$$\begin{matrix} \frac{1}{4}R_2 \rightarrow \\ -1R_3 \rightarrow \end{matrix} \begin{bmatrix} 1 & -1 & 1 & | & -1 \\ 0 & 1 & \frac{1}{4} & | & \frac{1}{2} \\ 0 & 0 & 1 & | & -2 \end{bmatrix}$$

$$\begin{cases} x - y + z = -1 \\ y + \frac{1}{4}z = \frac{1}{2} \\ z = -2 \end{cases}$$

$$y + \frac{1}{4}z = \frac{1}{2}$$
$$y + \frac{1}{4}(-2) = \frac{1}{2}$$
$$y - \frac{1}{2} = \frac{1}{2}$$
$$y = 1$$

$$x - y + z = -1$$
$$x - 1 - 2 = -1$$
$$x - 3 = -1$$
$$x = 2$$

Solution: $(2, 1, -2)$

39. $\begin{cases} 2x - y + z = 8 \\ x - 2y + 3z = 11 \\ 2x + 3y - z = -6 \end{cases}$

$$\begin{bmatrix} 2 & -1 & 1 & | & 8 \\ 1 & -2 & 3 & | & 11 \\ 2 & 3 & -1 & | & -6 \end{bmatrix}$$

$$\begin{matrix} R_2 \rightarrow \\ R_1 \rightarrow \end{matrix} \begin{bmatrix} 1 & -2 & 3 & | & 11 \\ 2 & -1 & 1 & | & 8 \\ 2 & 3 & -1 & | & -6 \end{bmatrix}$$

$$-2R_1 + R_2 \rightarrow \begin{bmatrix} 1 & -2 & 3 & | & 11 \\ 0 & 3 & -5 & | & -14 \\ 2 & 3 & -1 & | & -6 \end{bmatrix}$$

$$-2R_1 + R_3 \rightarrow \begin{bmatrix} 1 & -2 & 3 & | & 11 \\ 0 & 3 & -5 & | & -14 \\ 0 & 7 & -7 & | & -28 \end{bmatrix}$$

$$\frac{1}{7}R_3 \rightarrow \begin{bmatrix} 1 & -2 & 3 & | & 11 \\ 0 & 3 & -5 & | & -14 \\ 0 & 1 & -1 & | & -4 \end{bmatrix}$$

$$\begin{matrix} R_3 \rightarrow \\ R_2 \rightarrow \end{matrix} \begin{bmatrix} 1 & -2 & 3 & | & 11 \\ 0 & 1 & -1 & | & -4 \\ 0 & 3 & -5 & | & -14 \end{bmatrix}$$

$$-3R_2 + R_3 \rightarrow \begin{bmatrix} 1 & -2 & 3 & | & 11 \\ 0 & 1 & -1 & | & -4 \\ 0 & 0 & -2 & | & -2 \end{bmatrix}$$

$$-\frac{1}{2}R_3 \rightarrow \begin{bmatrix} 1 & -2 & 3 & | & 11 \\ 0 & 1 & -1 & | & -4 \\ 0 & 0 & 1 & | & 1 \end{bmatrix}$$

$$\begin{cases} x - 2y + 3z = 11 \\ y - z = -4 \\ z = 1 \end{cases}$$

$$y - z = -4$$
$$y - 1 = -4$$
$$y = -3$$

$$x - 2y + 3z = 11$$
$$x - 2(-3) + 3(1) = 11$$
$$x + 6 + 3 = 11$$
$$x = 2$$

Solution: $(2, -3, 1)$

41. $\begin{cases} 3x+2y-3z=1 \\ -2x+3y-4z=7 \\ 5x-2y+z=-5 \end{cases}$

$$\begin{bmatrix} 3 & 2 & -3 & | & 1 \\ -2 & 3 & -4 & | & 7 \\ 5 & -2 & 1 & | & -5 \end{bmatrix}$$

$$R_2+R_1 \to \begin{bmatrix} 1 & 5 & -7 & | & 8 \\ -2 & 3 & -4 & | & 7 \\ 5 & -2 & 1 & | & -5 \end{bmatrix}$$

$$2R_1+R_2 \to \begin{bmatrix} 1 & 5 & -7 & | & 8 \\ 0 & 13 & -18 & | & 23 \\ 5 & -2 & 1 & | & -5 \end{bmatrix}$$

$$-5R_1+R_3 \to \begin{bmatrix} 1 & 5 & -7 & | & 8 \\ 0 & 13 & -18 & | & 23 \\ 0 & -27 & 36 & | & -45 \end{bmatrix}$$

$$\begin{matrix} -R_3-2R_2 \to \\ \frac{1}{9}R_3 \to \end{matrix} \begin{bmatrix} 1 & 5 & -7 & | & 8 \\ 0 & 1 & 0 & | & -1 \\ 0 & -3 & 4 & | & -5 \end{bmatrix}$$

$$3R_2+R_3 \to \begin{bmatrix} 1 & 5 & -7 & | & 8 \\ 0 & 1 & 0 & | & -1 \\ 0 & 0 & 4 & | & -8 \end{bmatrix}$$

$$\frac{1}{4}R_3 \to \begin{bmatrix} 1 & 5 & -7 & | & 8 \\ 0 & 1 & 0 & | & -1 \\ 0 & 0 & 1 & | & -2 \end{bmatrix}$$

$$\begin{cases} x+5y-7z=8 \\ y=-1 \\ z=-2 \end{cases}$$

$$x+5y-7z=8$$
$$x+5(-1)-7(-2)=8$$
$$x-5+14=8$$
$$x=-1$$

Solution: $(-1,-1,-2)$

43. $\begin{cases} 3x-6y+z=-10 \\ 7y-z=2 \\ 2x+4z=14 \end{cases}$

$$\begin{bmatrix} 3 & -6 & 1 & | & -10 \\ 0 & 7 & -1 & | & 2 \\ 2 & 0 & 4 & | & 14 \end{bmatrix}$$

$$\begin{matrix} R_3 \to \\ \\ R_1 \to \end{matrix} \begin{bmatrix} 2 & 0 & 4 & | & 14 \\ 0 & 7 & -1 & | & 2 \\ 3 & -6 & 1 & | & -10 \end{bmatrix}$$

$$\frac{1}{2}R_1 \to \begin{bmatrix} 1 & 0 & 2 & | & 7 \\ 0 & 7 & -1 & | & 2 \\ 3 & -6 & 1 & | & -10 \end{bmatrix}$$

$$-3R_1+R_3 \to \begin{bmatrix} 1 & 0 & 2 & | & 7 \\ 0 & 7 & -1 & | & 2 \\ 0 & -6 & -5 & | & -31 \end{bmatrix}$$

$$6R_2+7R_3 \to \begin{bmatrix} 1 & 0 & 2 & | & 7 \\ 0 & 7 & -1 & | & 2 \\ 0 & 0 & -41 & | & -205 \end{bmatrix}$$

$$\begin{matrix} \frac{1}{7}R_2 \to \\ -\frac{1}{41}R_3 \to \end{matrix} \begin{bmatrix} 1 & 0 & 2 & | & 7 \\ 0 & 1 & -\frac{1}{7} & | & \frac{2}{7} \\ 0 & 0 & 1 & | & 5 \end{bmatrix}$$

$$\begin{cases} x+2z=7 \\ y-\frac{1}{7}z=\frac{2}{7} \\ z=5 \end{cases}$$

$$y-\frac{1}{7}z=\frac{2}{7} \qquad\qquad x+2z=7$$
$$y-\frac{1}{7}(5)=\frac{2}{7} \qquad\qquad x+2(5)=7$$
$$y-\frac{5}{7}=\frac{2}{7} \qquad\qquad x+10=7$$
$$y=1 \qquad\qquad\qquad x=-3$$

Solution: $(-3,1,5)$

45. $\begin{cases} 2x+5y-z=10 \\ x-y-z=14 \\ x-6y=20 \end{cases}$

$$\begin{bmatrix} 2 & 5 & -1 & | & 10 \\ 1 & -1 & -1 & | & 14 \\ 1 & -6 & 0 & | & 20 \end{bmatrix}$$

$$\begin{matrix} R_2 \to \\ R_1 \to \\ \end{matrix} \begin{bmatrix} 1 & -1 & -1 & | & 14 \\ 2 & 5 & -1 & | & 10 \\ 1 & -6 & 0 & | & 20 \end{bmatrix}$$

$$\begin{matrix} \\ -2R_1+R_2 \to \\ -R_1+R_3 \to \end{matrix} \begin{bmatrix} 1 & -1 & -1 & | & 14 \\ 0 & 7 & 1 & | & -18 \\ 0 & -5 & 1 & | & 6 \end{bmatrix}$$

$$\tfrac{1}{7}R_2 \to \begin{bmatrix} 1 & -1 & -1 & | & 14 \\ 0 & 1 & \frac{1}{7} & | & -\frac{18}{7} \\ 0 & -5 & 1 & | & 6 \end{bmatrix}$$

$$5R_2+R_3 \to \begin{bmatrix} 1 & -1 & -1 & | & 14 \\ 0 & 1 & \frac{1}{7} & | & -\frac{18}{7} \\ 0 & 0 & \frac{12}{7} & | & -\frac{48}{7} \end{bmatrix}$$

$$\tfrac{7}{12}R_3 \to \begin{bmatrix} 1 & -1 & -1 & | & 14 \\ 0 & 1 & \frac{1}{7} & | & -\frac{18}{7} \\ 0 & 0 & 1 & | & -4 \end{bmatrix}$$

$$\begin{cases} x-y-z=14 \\ y+\frac{1}{7}z=-\frac{18}{7} \\ z=-4 \end{cases}$$

$y+\dfrac{1}{7}z=-\dfrac{18}{7}$ $x-y-z=14$

$y+\dfrac{1}{7}(-4)=-\dfrac{18}{7}$ $x-(-2)-(-4)=14$

$y-\dfrac{4}{7}=-\dfrac{18}{7}$ $x+2+4=14$

$y=-2$ $x=8$

Solution: $(8,-2,-4)$

47. $(3,-4)$ 49. $(-6,8)$

51. $(4,-3,5)$ 53. $(7,-6,5)$

55. Mistake: In the second step, $4R_1+R_2$ is calculated incorrectly.

Correct: The correct calculation is $\begin{bmatrix} 1 & 3 & | & 13 \\ 0 & 11 & | & 26 \end{bmatrix}$.

The solution is $\left(\dfrac{65}{11}, \dfrac{26}{11} \right)$.

57. Let x = the cost of a grilled chicken sandwich and y = the cost of a drink.

$$\begin{cases} 3x+2y=12.90 \\ 7x+4y=29.30 \end{cases}$$

$$\begin{bmatrix} 3 & 2 & | & 12.9 \\ 7 & 4 & | & 29.3 \end{bmatrix}$$

$$-7R_1+3R_2 \to \begin{bmatrix} 3 & 2 & | & 12.9 \\ 0 & -2 & | & -2.4 \end{bmatrix}$$

$$\begin{matrix} \tfrac{1}{3}R_1 \to \\ -\tfrac{1}{2}R_2 \to \end{matrix} \begin{bmatrix} 1 & \frac{2}{3} & | & 4.3 \\ 0 & 1 & | & 1.2 \end{bmatrix}$$

$$\begin{cases} x+\frac{2}{3}y=4.3 \\ \phantom{x+\frac{2}{3}}y=1.2 \end{cases}$$

$x+\dfrac{2}{3}y=4.3$

$x+\dfrac{2}{3}(1.2)=4.3$

$x+0.8=4.3$

$x=3.5$

The chicken sandwich costs \$3.50 and the drink costs \$1.20.

59. Let n = the length of the Nile and a = the length of the Amazon.

$$\begin{cases} n + a = 8050 \\ n = 250 + a \end{cases}$$

$$\begin{bmatrix} 1 & 1 & | & 8050 \\ 1 & -1 & | & 250 \end{bmatrix}$$

$$-R_1 + R_2 \rightarrow \begin{bmatrix} 1 & 1 & | & 8050 \\ 0 & -2 & | & -7800 \end{bmatrix}$$

$$-\tfrac{1}{2}R_2 \rightarrow \begin{bmatrix} 1 & 1 & | & 8050 \\ 0 & 1 & | & 3900 \end{bmatrix}$$

$$\begin{cases} n + a = 8050 \\ \quad\ \ a = 3900 \end{cases}$$

$$n + a = 8050$$

$$n + 3900 = 8050$$

$$n = 4150$$

The Amazon River is 3900 miles long and the Nile River is 4150 miles long.

61. Let c = the amount invested in the CD at 5% interest and m = the amount invested in the money market account at 6% interest.

$$\begin{cases} c + m = 10,000 \\ 0.05c + 0.06m = 536 \end{cases}$$

$$\begin{bmatrix} 1 & 1 & | & 10,000 \\ 0.05 & 0.06 & | & 536 \end{bmatrix}$$

$$-0.05R_1 + R_2 \rightarrow \begin{bmatrix} 1 & 1 & | & 10,000 \\ 0 & 0.01 & | & 36 \end{bmatrix}$$

$$100R_2 \rightarrow \begin{bmatrix} 1 & 1 & | & 10,000 \\ 0 & 1 & | & 3600 \end{bmatrix}$$

$$\begin{cases} c + m = 10,000 \\ \quad\ \ m = 3600 \end{cases}$$

$$c + m = 10,000$$

$$c + 3600 = 10,000$$

$$c = 6400$$

$6400 was invested in the CD and $3600 was invested in the money market account.

63. Let c = the cost of a CD, b = the cost of a book, and d = the cost of a DVD.

$$\begin{cases} 2c + 4b + 3d = 164 \\ 5c + 2b + 2d = 160 \\ c + b = d \end{cases}$$

$$\begin{bmatrix} 2 & 4 & 3 & | & 164 \\ 5 & 2 & 2 & | & 160 \\ 1 & 1 & -1 & | & 0 \end{bmatrix}$$

$$\begin{matrix} R_3 \rightarrow \\ \\ R_1 \rightarrow \end{matrix} \begin{bmatrix} 1 & 1 & -1 & | & 0 \\ 5 & 2 & 2 & | & 160 \\ 2 & 4 & 3 & | & 164 \end{bmatrix}$$

$$\begin{matrix} -5R_1 + R_2 \rightarrow \\ -2R_1 + R_3 \rightarrow \end{matrix} \begin{bmatrix} 1 & 1 & -1 & | & 0 \\ 0 & -3 & 7 & | & 160 \\ 0 & 2 & 5 & | & 164 \end{bmatrix}$$

$$\begin{matrix} -\tfrac{1}{3}R_2 \rightarrow \\ 2R_2 + 3R_3 \rightarrow \end{matrix} \begin{bmatrix} 1 & 1 & -1 & | & 0 \\ 0 & 1 & -\tfrac{7}{3} & | & -\tfrac{160}{3} \\ 0 & 0 & 29 & | & 812 \end{bmatrix}$$

$$\tfrac{1}{29}R_3 \rightarrow \begin{bmatrix} 1 & 1 & -1 & | & 0 \\ 0 & 1 & -\tfrac{7}{3} & | & -\tfrac{160}{3} \\ 0 & 0 & 1 & | & 28 \end{bmatrix}$$

$$\begin{cases} c + b - d = 0 \\ b - \tfrac{7}{3}d = -\tfrac{160}{3} \\ \quad\quad\ d = 28 \end{cases}$$

$$b - \frac{7}{3}d = -\frac{160}{3} \qquad\qquad c + b - d = 0$$

$$b - \frac{7}{3}(28) = -\frac{160}{3} \qquad\qquad c + 12 - 28 = 0$$

$$b - \frac{196}{3} = -\frac{160}{3} \qquad\qquad c - 16 = 0$$

$$b = 12 \qquad\qquad c = 16$$

The CDs cost $16 each, the books cost $12 each, and the DVDs cost $28 each.

65. Let t = the number of touchdowns, e = the number of extra points, and f = the number of field goals.

$$\begin{cases} 6t + e + 3f = 31 \\ t = e \\ t = f + 3 \end{cases} = \begin{cases} 6t + e + 3f = 31 \\ t - e = 0 \\ t - f = 3 \end{cases}$$

$$\begin{bmatrix} 6 & 1 & 3 & | & 31 \\ 1 & -1 & 0 & | & 0 \\ 1 & 0 & -1 & | & 3 \end{bmatrix}$$

$$\begin{matrix} R_3 \to \\ \\ R_1 \to \end{matrix} \begin{bmatrix} 1 & 0 & -1 & | & 3 \\ 1 & -1 & 0 & | & 0 \\ 6 & 1 & 3 & | & 31 \end{bmatrix}$$

$$\begin{matrix} \\ -R_1 + R_2 \to \\ -6R_1 + R_3 \to \end{matrix} \begin{bmatrix} 1 & 0 & -1 & | & 3 \\ 0 & -1 & 1 & | & -3 \\ 0 & 1 & 9 & | & 13 \end{bmatrix}$$

$$\begin{matrix} \\ -1R_2 \to \\ R_2 + R_3 \to \end{matrix} \begin{bmatrix} 1 & 0 & -1 & | & 3 \\ 0 & 1 & -1 & | & 3 \\ 0 & 0 & 10 & | & 10 \end{bmatrix}$$

$$\begin{matrix} \\ \\ \frac{1}{10}R_3 \to \end{matrix} \begin{bmatrix} 1 & 0 & -1 & | & 3 \\ 0 & 1 & -1 & | & 3 \\ 0 & 0 & 1 & | & 1 \end{bmatrix}$$

$$\begin{cases} t - f = 3 \\ e - f = 3 \\ f = 1 \end{cases}$$

$$\begin{matrix} e - f = 3 & \qquad t - f = 3 \\ e - 1 = 3 & \qquad t - 1 = 3 \\ e = 4 & \qquad t = 4 \end{matrix}$$

Green Bay scored 4 touchdowns, 4 extra points, and 1 field goal.

67.
$$\begin{cases} 4v_1 - v_2 = 30 \\ -2v_1 + 5v_2 - v_3 = 10 \\ -v_2 + 5v_3 = 4 \end{cases}$$

$$\begin{bmatrix} 4 & -1 & 0 & | & 30 \\ -2 & 5 & -1 & | & 10 \\ 0 & -1 & 5 & | & 4 \end{bmatrix}$$

$$\begin{matrix} \\ R_3 \to \\ R_2 \to \end{matrix} \begin{bmatrix} 4 & -1 & 0 & | & 30 \\ 0 & -1 & 5 & | & 4 \\ -2 & 5 & -1 & | & 10 \end{bmatrix}$$

$$\begin{matrix} \\ -1R_2 \to \\ R_1 + 2R_3 \to \end{matrix} \begin{bmatrix} 4 & -1 & 0 & | & 30 \\ 0 & 1 & -5 & | & -4 \\ 0 & 9 & -2 & | & 50 \end{bmatrix}$$

$$\begin{matrix} \\ \\ -9R_2 + R_3 \to \end{matrix} \begin{bmatrix} 4 & -1 & 0 & | & 30 \\ 0 & 1 & -5 & | & -4 \\ 0 & 0 & 43 & | & 86 \end{bmatrix}$$

$$\begin{matrix} \frac{1}{4}R_1 \to \\ \\ \frac{1}{43}R_3 \to \end{matrix} \begin{bmatrix} 1 & -\frac{1}{4} & 0 & | & \frac{15}{2} \\ 0 & 1 & -5 & | & -4 \\ 0 & 0 & 1 & | & 2 \end{bmatrix}$$

$$\begin{cases} v_1 - \frac{1}{4}v_2 = \frac{15}{2} \\ v_2 - 5v_3 = -4 \\ v_3 = 2 \end{cases}$$

$$\begin{matrix} v_2 - 5v_3 = -4 & \qquad v_1 - \frac{1}{4}v_2 = \frac{15}{2} \\ v_2 - 5(2) = -4 & \qquad v_1 - \frac{1}{4}(6) = \frac{15}{2} \\ v_2 - 10 = -4 & \qquad v_1 - \frac{3}{2} = \frac{15}{2} \\ v_2 = 6 & \qquad v_1 = 9 \end{matrix}$$

$$v_1 = 9, \quad v_2 = 6, \quad v_3 = 2$$

Review Exercises

1. $(-3)(4) - (-4)(5) = -12 + 20 = 8$

2. $2(2-8) + 3\left[-1 - (-6)\right] - 4(4-6)$
$$= 2(-6) + 3(5) - 4(-2)$$
$$= -12 + 15 + 8$$
$$= 11$$

3. $\dfrac{(-5)(-1) - (9)(3)}{(2)(-1) - (3)(3)} = \dfrac{5 - 27}{-2 - 9} = \dfrac{-22}{-11} = 2$

4. $\dfrac{2\left(\dfrac{5}{4}\right)-8\left(\dfrac{1}{4}\right)}{\left(\dfrac{1}{2}\right)\left(\dfrac{5}{4}\right)-\left(\dfrac{3}{2}\right)\left(\dfrac{1}{4}\right)}=\dfrac{\dfrac{5}{2}-2}{\dfrac{5}{8}-\dfrac{3}{8}}=\dfrac{\dfrac{1}{2}}{\dfrac{2}{8}}=2$

5. $\dfrac{(1.1)(-0.2)-(1.7)(-0.5)}{(0.2)(-0.2)-(0.5)(-0.5)}=\dfrac{-0.22+0.85}{-0.04+0.25}$

$=\dfrac{0.63}{0.21}$

$=3$

6. $\dfrac{1\left[-45-(-9)\right]+7(-9-2)-2(-27-30)}{1(6-3)-2(-9-2)-2\left[9-(-4)\right]}$

$=\dfrac{1\left[-45+9\right]+7(-11)-2(-57)}{1(3)-2(-11)-2\left[9+4\right]}$

$=\dfrac{1\left[-36\right]-77+114}{3+22-2\cdot13}$

$=\dfrac{-36-77+114}{3+22-26}$

$=\dfrac{1}{-1}$

$=-1$

Exercise Set 4.5

1. $\begin{cases} x+y>4 \\ x-y<6 \end{cases}$

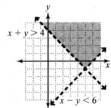

3. $\begin{cases} x+y<-2 \\ x-y>6 \end{cases}$

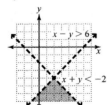

5. $\begin{cases} 2x+y\geq-1 \\ x-y\leq 5 \end{cases}$

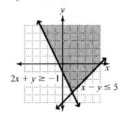

7. $\begin{cases} x+y<3 \\ x-2y\geq 2 \end{cases}$

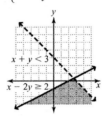

9. $\begin{cases} 2x+y>9 \\ y>\dfrac{1}{4}x \end{cases}$

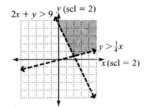

11. $\begin{cases} y<x \\ y>-x+1 \end{cases}$

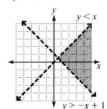

13. $\begin{cases} 3x>-4y \\ 3x+4y\leq-8 \end{cases}$

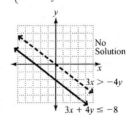

15. $\begin{cases} y < 3x+1 \\ 3x-y \le 4 \end{cases}$

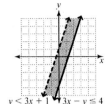

17. $\begin{cases} x > 2 \\ 3x-2y > 6 \end{cases}$

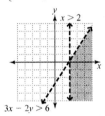

19. $\begin{cases} x \ge -3y \\ y \ge 2x \end{cases}$

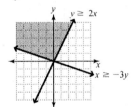

21. $\begin{cases} y > 2 \\ x < -1 \end{cases}$

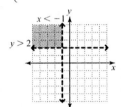

23. $\begin{cases} 2x+y \ge 1 \\ x-y > -1 \\ x > 2 \end{cases}$

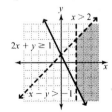

25. $x+y=3$ should be a dashed line.

27. The wrong area is shaded. It should be the region containing the point $(1, 5)$.

29. a) $V + M \ge 1100$

$M \ge 500$

b) $V \le 800$

$M \le 800$

c)

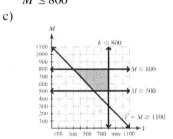

d) Answers will vary. Some possible (V, M) pairs are $(500, 700)$, $(600, 500)$, and $(700, 600)$.

31. a) $\begin{cases} R + D \ge 100 \\ 15.95R + 20.95D \ge 1700 \end{cases}$

b)

c) Answers will vary. Some possible (R, D) pairs are $(30, 90)$, $(40, 80)$, and $(50, 60)$.

Review Exercises

1. $3^2 = 9$

2. $-10^2 = -10 \cdot 10 = -100$

3. $\sqrt{25} = 5$

4. $\sqrt{\dfrac{16}{49}} = \dfrac{\sqrt{16}}{\sqrt{49}} = \dfrac{4}{7}$

5. $f(x) = 3x - 5$

   $f(3) = 3(3) - 5$

   $\quad = 9 - 5$

   $\quad = 4$

6. $f(x) = 3x^2 - 2x + 5$

   $f(-2) = 3(-2)^2 - 2(-2) + 5$

   $\quad\quad = 3(4) + 4 + 5$

   $\quad\quad = 12 + 4 + 5$

   $\quad\quad = 21$

Chapter 4 Review Exercises

1.   Is $(-2, 7)$ a solution of $\begin{cases} x + y = 5 \\ 3x - y = -13 \end{cases}$?

   Equation 1   Equation 2

   $x + y = 5$ $3x - y = -13$

   $\overset{?}{-2 + 7 = 5}$ $\overset{?}{3(-2) - 7 = -13}$

   $5 = 5$ $-13 = -13$

   Yes, $(-2, 7)$ is a solution for the system because it satisfies both equations.

2.   Is $(3, 2)$ a solution of $\begin{cases} 2x + y = 8? \\ x - y = -1 \end{cases}$

   Equation 1   Equation 2

   $2x + y = 8$ $x - y = -1$

   $\overset{?}{2(3) + 2 = 8}$ $\overset{?}{3 - 2 = -1}$

   $\overset{?}{6 + 2 = 8}$ $1 \neq -1$

   $8 = 8$

   No, $(3, 2)$ is not a solution for the system because it does not satisfy the second equation.

3. $\begin{cases} y = 2x + 1 \\ x + y = -8 \end{cases}$ Graph each equation.

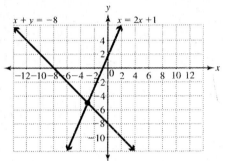

   The lines intersect at a single point, which appears to be $(-3, -5)$.

4. $\begin{cases} x - y = 6 \\ 2x + y = 6 \end{cases}$ Graph each equation.

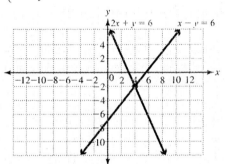

   The lines intersect at a single point, which appears to be $(4, -2)$.

5. $\begin{cases} y = -3x - 4 \\ 9x + 3y = 12 \end{cases}$ Graph each equation.

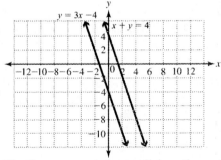

   The lines appear to be parallel, so the system has no solution.

6. $\begin{cases} 5x - 2y = 10 \\ y = \dfrac{5}{2}x - 5 \end{cases}$ Graph each equation.

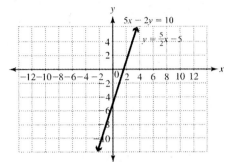

The lines appear to be identical, so the solution is the set of all ordered pairs that solve $5x - 2y = 10$.

7. a) Because these lines intersect in a single point, this system is consistent with independent equations.
    b) This system has one solution.

8. a) Because the lines are parallel, this system is inconsistent.
    b) This system has no solution.

9. $\begin{cases} x - 3y = 10 \\ 2x + 3y = 5 \end{cases} \rightarrow \begin{cases} y = \dfrac{1}{3}x - \dfrac{10}{3} \\ y = -\dfrac{2}{3}x + \dfrac{5}{3} \end{cases}$

   a) Because these lines have different slopes, the graphs are different and the lines intersect in a single point. This system is consistent with independent equations.
   b) This system has one solution.

10. $\begin{cases} x - 3y = 15 \\ 2x - 6y = 30 \end{cases} \rightarrow \begin{cases} y = \dfrac{1}{3}x - 5 \\ y = \dfrac{1}{3}x - 5 \end{cases}$

    a) The lines have the same slope and the same y-intercept, so the graphs are identical. Because these lines coincide, this system is consistent with dependent equations.
    b) This system has an infinite number of solutions.

11. $\begin{cases} y = -3x + 5 \\ x + 2y = -10 \end{cases}$

    Solve the second equation for x: $x = -2y - 10$.

    $y = -3x + 5$ $\qquad\qquad$ $x = -2(-7) - 10$
    $y = -3(-2y - 10) + 5$ $\qquad$ $x = 14 - 10$
    $y = 6y + 30 + 5$ $\qquad\qquad$ $x = 4$
    $-5y = 35$
    $y = -7$

    Solution: $(4, -7)$

12. $\begin{cases} 6x + 3y = 11 \\ x - 6y = 4 \end{cases}$

    Solve the second equation for x: $x = 6y + 4$.

    $6x + 3y = 11$ $\qquad$ $x = 6\left(-\dfrac{1}{3}\right) + 4$
    $6(6y + 4) + 3y = 11$ $\qquad$ $x = -2 + 4$
    $36y + 24 + 3y = 11$ $\qquad$ $x = 2$
    $39y = -13$
    $y = -\dfrac{1}{3}$

    Solution: $\left(2, -\dfrac{1}{3}\right)$

13. $\begin{cases} 2x + 3y = 9 \\ y = -\dfrac{2}{3}x + 3 \end{cases}$

    Solve the first equation for y: $y = -\dfrac{2}{3}x + 3$.

    Notice that the first equation is the same as the second equation. The equations in the system are dependent; there are an infinite number of solutions that are all of the ordered pairs that solve $2x + 3y = 9$.

14. $\begin{cases} 12x - 3y = -6 \\ y - 4x = -2 \end{cases}$

    Solve the second equation for y: $y = 4x - 2$.

    $12x - 3y = -6$
    $12x - 3(4x - 2) = -6$
    $12x - 12x + 6 = -6$
    $6 \neq -6$

    Because this last statement is false, the system is inconsistent and has no solution.

15. $\begin{cases} x+y=7 \\ x-y=-5 \end{cases}$

$\begin{aligned} x+y &= 7 \\ x-y &= -5 \\ \hline 2x &= 2 \end{aligned}$ $\qquad \begin{aligned} x+y &= 7 \\ 1+y &= 7 \\ y &= 6 \end{aligned}$

$x=1$

Solution: $(1, 6)$

16. $\begin{cases} 3x+5y=2 \\ 2x-3y=14 \end{cases}$

$\begin{aligned} 3(3x+5y=2) \\ 5(2x-3y=14) \\ \hline 9x+15y=6 \\ 10x-15y=70 \\ \hline 19x=76 \end{aligned}$ $\qquad \begin{aligned} 3x+5y &= 2 \\ 3(4)+5y &= 2 \\ 12+5y &= 2 \\ 5y &= -10 \\ y &= -2 \end{aligned}$

$x=4$

Solution: $(4, -2)$

17. $\begin{cases} 0.45x+0.2y=-2.6 \\ 0.6x-0.5y=-7.3 \end{cases}$

$\begin{aligned} 5(0.45x+0.2y=-2.6) \\ 2(0.6x-0.5y=-7.3) \\ \hline 2.25x+y=-13 \\ 1.2x-y=-14.6 \\ \hline 3.45x=-27.6 \\ x=-8 \end{aligned}$

$\begin{aligned} 0.45(-8)+0.2y &= -2.6 \\ -3.6+0.2y &= -2.6 \\ 0.2y &= 1 \\ y &= 5 \end{aligned}$

Solution: $(-8, 5)$

18. $\begin{cases} \dfrac{2}{3}x-\dfrac{3}{5}y=\dfrac{8}{15} \\ x+\dfrac{1}{4}y=-\dfrac{3}{2} \end{cases}$

$-\dfrac{2}{3}\left(x+\dfrac{1}{4}y=-\dfrac{3}{2}\right)$ $\qquad x+\dfrac{1}{4}y=-\dfrac{3}{2}$

$-\dfrac{2}{3}x-\dfrac{2}{12}y=\dfrac{6}{6}$ $\qquad x+\dfrac{1}{4}(-2)=-\dfrac{3}{2}$

$\begin{aligned} \dfrac{2}{3}x-\dfrac{3}{5}y &= \dfrac{8}{15} \\ \hline -\dfrac{46}{60}y &= \dfrac{23}{15} \\ y &= -2 \end{aligned}$ $\qquad x=-1$

Solution: $(-1, -2)$

19. $\begin{cases} x+y+z=-2 & \text{Eqtn. 1} \\ 2x+3y+4z=-10 & \text{Eqtn. 2} \\ 3x-2y-3z=12 & \text{Eqtn. 3} \end{cases}$

Multiply equation 1 by –2 and add to equation 2. This makes equation 4.

$\begin{aligned} -2x-2y-2z &= 4 \\ 2x+3y+4z &= -10 \\ \hline y+2z &= -6 \quad \text{Eqtn. 4} \end{aligned}$

Multiply equation 1 by –3 and add to equation 3. This makes equation 5.

$\begin{aligned} -3x-3y-3z &= 6 \\ 3x-2y-3z &= 12 \\ \hline -5y-6z &= 18 \quad \text{Eqtn. 5} \end{aligned}$

Use equations 4 and 5 to make a system of equations in two variables. Solve.

$\begin{aligned} y+2z &= -6 \qquad \text{Multiply by 5.} \\ -5y-6z &= 18 \\ \hline 5y+10z &= -30 \\ -5y-6z &= 18 \\ \hline 4z &= -12 \\ z &= -3 \end{aligned}$

$\begin{aligned} y+2(-3) &= -6 \\ y-6 &= -6 \\ y &= 0 \end{aligned}$ $\qquad \begin{aligned} x+0-3 &= -2 \\ x-3 &= -2 \\ x &= 1 \end{aligned}$

Solution: $(1, 0, -3)$

20. $\begin{cases} x + y + z = 0 & \text{Eqtn. 1} \\ 2x - 4y + 3z = -12 & \text{Eqtn. 2} \\ 3x - 3y + 4z = 2 & \text{Eqtn. 3} \end{cases}$

Multiply equation 1 by –2 and add to equation 2. This makes equation 4.

$-2x - 2y - 2z = 0$

$\underline{2x - 4y + 3z = -12}$

$\qquad -6y + z = -12 \quad$ Eqtn 4

Multiply equation 1 by –3 and add to equation 3. This makes equation 5.

$-3x - 3y - 3z = 0$

$\underline{3x - 3y + 4z = 2}$

$\qquad -6y + z = 2 \quad$ Eqtn. 5

Use equations 4 and 5 to form a system of equations in two variables. Solve.

$-6y + z = -12 \qquad$ Multiply by -1.

$\underline{-6y + z = 2}$

$6y - z = 12$

$\underline{-6y + z = 2}$

$\qquad 0 \neq 14$

Because this last statement is false, the system is inconsistent and has no solution.

21. $\begin{cases} 3x + 4y = -3 & \text{Eqtn. 1} \\ -2y + 3z = 12 & \text{Eqtn. 2} \\ 4x - 3z = 6 & \text{Eqtn. 3} \end{cases}$

Multiply equation 2 by 2 and add to equation 1. This makes equation 4.

$3x + 4y = -3$

$\underline{-4y + 6z = 24}$

$\quad 3x + 6z = 21 \quad$ Eqtn. 4

Use equations 3 and 4 to make a system of equations in two variables. Solve.

$4x - 3z = 6 \qquad$ Multiply by 2.

$\underline{3x + 6z = 21}$

$8x - 6z = 12$

$\underline{3x + 6z = 21}$

$\quad 11x = 33$

$\qquad x = 3$

$4 \cdot 3 - 3z = 6 \qquad\qquad 3 \cdot 3 + 4y = -3$

$12 - 3z = 6 \qquad\qquad 9 + 4y = -3$

$-3z = -6 \qquad\qquad 4y = -12$

$z = 2 \qquad\qquad\quad y = -3$

Solution: $(3, -3, 2)$

22. $\begin{cases} 2x + 2y - 3z = 5 & \text{Eqtn. 1} \\ x + 3y - 4z = 0 & \text{Eqtn. 2} \\ x + 2y + z = 5 & \text{Eqtn. 3} \end{cases}$

Multiply equation 2 by –2 and add to equation 1. This makes equation 4.

$2x + 2y - 3z = 5$

$\underline{-2x - 6y + 8z = 0}$

$\quad -4y + 5z = 5 \quad$ Eqtn. 4

Multiply equation 2 by –1 and add to equation 3. This makes equation 5.

$-x - 3y + 4z = 0$

$\underline{x + 2y + z = 5}$

$\quad -y + 5z = 5 \quad$ Eqtn. 5

Use equations 4 and 5 to make a system of equations in two variables. Solve.

$-4y + 5z = 5 \qquad$ Multiply by -1.

$\underline{-y + 5z = 5}$

$4y - 5z = -5$

$\underline{-y + 5z = 5}$

$\quad 3y = 0$

$\quad y = 0$

$-y + 5z = 5 \qquad\qquad x + 2y + z = 5$

$0 + 5z = 5 \qquad\qquad x + 2(0) + 1 = 5$

$5z = 5 \qquad\qquad\qquad x + 1 = 5$

$z = 1 \qquad\qquad\qquad\quad x = 4$

Solution: $(4, 0, 1)$

23. Let h = the number of hotdogs sold and b = the number of hamburgers sold.

    $$\begin{cases} h + b = 226 & \text{Eqtn. 1} \\ 4h + 6b = 1048 & \text{Eqtn. 2} \end{cases}$$

    Multiply equation 1 by -4 and add equations 1 and 2.

    $-4(h + b = 226)$

    $\begin{array}{ll} -4h - 4b = -904 & h + b = 226 \\ \underline{4h + 6b = 1048} & h + 72 = 226 \\ 2b = 144 & h = 154 \\ b = 72 & \end{array}$

    The vender sold 154 hotdogs and 72 hamburgers.

24. Let x be the greater angle and y be the lesser angle. The sum of the two angles is 180.

    $$\begin{cases} x + y = 180 & \text{Eqtn.1} \\ x = 4y - 10 & \text{Eqtn.2} \end{cases}$$

    Add the equations.

    $\begin{array}{ll} x + y = 180 & \\ \underline{x - 4y = -10} & x + y = 180 \\ 5y = 190 & x + 38 = 180 \\ y = 38 & x = 142 \end{array}$

    The measure of the greater angle is 142º and the measure of the lesser angle is 38º.

25. Complete a table.

Categories	Rate	Time	Distance
Linea	4	x	$4x$
Tanya	12	y	$12y$

    Translate to a system of equations.

    $$\begin{cases} 4x = 12y \\ x = y + \dfrac{1}{2} \end{cases}$$

    $4x = 12y$

    $4\left(y + \dfrac{1}{2}\right) = 12y$

    $4y + 2 = 12y$

    $2 = 8y$

    $\dfrac{1}{4} = y$

    Tanya will catch up with Linea 0.25 hours after 8:45, at 9:00 A.M.

26. Let x be the speed of the kayaker in still water and y be the speed of the current.
    Complete a table.

Categories	Rate	Time	Distance
against	$x - y$	0.75	$0.75(x - y)$
with	$x + y$	0.25	$0.25(x + y)$

    Translate to a system of equations.

    $$\begin{cases} 0.25(x + y) = 1 \\ 0.75(x - y) = 1 \end{cases} \rightarrow \begin{cases} x + y = 4 \\ x - y = \dfrac{4}{3} \end{cases}$$

    $x + y = 4$

    $\begin{array}{l} x + y = 4 \\ \underline{x - y = \dfrac{4}{3}} \\ 2x = 4\dfrac{4}{3} \\ x = 2\dfrac{2}{3} \end{array}$

    $\begin{array}{l} 2\dfrac{2}{3} + y = 4 \\ y = 1\dfrac{1}{3} \end{array}$

    The speed in still water is $2\dfrac{2}{3}$ mph. The current changes the speed by $1\dfrac{1}{3}$ hr.

27. Create a table.

Categories	APR	principal	interest
stocks	0.08	x	$0.08x$
savings	0.02	y	$0.02y$

    Translate to a system of equations.

    $$\begin{cases} x = 3y \\ 0.08x + 0.02y = 624 \end{cases}$$

    Use substitution.

    $\begin{array}{ll} 0.08x + 0.02y = 624 & x = 3y \\ 0.08(3y) + 0.02y = 624 & x = 3(2400) \\ 0.24y + 0.02y = 624 & x = 7200 \\ 0.26y = 624 & \\ y = 2400 & \end{array}$

    He invested $7200 in stocks and $2400 in savings.

28. Let x = the volume of 5% HCl mixture and y = the volume of 20% HCl.

Solution	Concentration	Volume of Solution	Volume of HCl
20%	0.20	x	$0.20x$
50%	0.50	y	$0.50y$
30%	0.30	300	$0.30(300)$

$$\begin{cases} x + y = 300 \\ 0.02x + 0.50y = 0.30(300) \end{cases}$$

Solve the first equation for x: $x = 300 - y$.

$$0.20x + 0.50y = 0.30(300)$$
$$0.20(300 - y) + 0.50y = 0.30(300)$$
$$60 - 0.20y + 0.50y = 90$$
$$0.30y = 30$$
$$y = 100$$

$x = 300 - y$

$x = 300 - 100$

$x = 200$

Combine 200 ml of 20% HCL mixture and 100 ml of 50% HCL mixture.

29. Let v = the number of vitamin supplements, f = the number rolls of film, and c = the number of bags of candy.

$$\begin{cases} 8v + 4f + 3c = 32 & \text{Eqtn. 1} \\ v + f + c = 8 & \text{Eqtn. 2} \\ f = c - 1 & \text{Eqtn. 3} \end{cases}$$

Multiply the second equation by -8, then add equations 1 and 2.

$8v + 4f + 3c = 32$

$\underline{-8(v + f + c = 8)}$

$-4f - 5c = -32$

Substitute the value of f.

$v + f + c = 8$

$-4f - 5c = -32 \qquad f = c - 1 \qquad v + 3 + 4 = 8$

$-4(c-1) - 5c = -32 \qquad f = 4 - 1 \qquad v = 1$

$-9c = -36 \qquad f = 3$

$c = 4$

John bought 1 of the vitamin supplements, 3 rolls of films, and 4 bags of candy.

30. Let f = the number of first-place finishes, s = the number of second-place finishes, and t = the number of third-place finishes.

$$\begin{cases} 5f + 3s + t = 38 & \text{Eqtn. 1} \\ f = 1 + s & \text{Eqtn. 2} \\ t = 3s & \text{Eqtn. 3} \end{cases}$$

Substitute equations 2 and 3 into equation 1 and solve for s.

$5f + 3s + t = 38 \qquad t = 3s \qquad f = 1 + s$

$5(1+s) + 3s + 3s = 38 \qquad t = 3(3) \qquad f = 1 + 3$

$5 + 5s + 3s + 3s = 38 \qquad t = 9 \qquad f = 4$

$11s = 33$

$s = 3$

There were 4 first-place finishes, 3 second-place finishes, and 9 third-place finishes.

31. $\begin{cases} x - 4y = 8 \\ x + 2y = 2 \end{cases}$

$$\begin{bmatrix} 1 & -4 & | & 8 \\ 1 & 2 & | & 2 \end{bmatrix}$$

$$-1R_1 + R_2 \rightarrow \begin{bmatrix} 1 & -4 & | & 8 \\ 0 & 6 & | & -6 \end{bmatrix}$$

$$\frac{1}{6}R_2 \rightarrow \begin{bmatrix} 1 & -4 & | & 8 \\ 0 & 1 & | & -1 \end{bmatrix}$$

$$\begin{cases} x - 4y = 8 \\ y = -1 \end{cases}$$

$x - 4y = 8 \qquad$ Solution: $(4, -1)$

$x - 4(-1) = 8$

$x + 4 = 8$

$x = 4$

32. $\begin{cases} x+y+z=2 \\ 2x+y+2z=1 \\ 3x+2y+z=1 \end{cases}$

$$\begin{bmatrix} 1 & 1 & 1 & | & 2 \\ 2 & 1 & 2 & | & 1 \\ 3 & 2 & 1 & | & 1 \end{bmatrix}$$

$$\begin{matrix} \\ -2R_1+R_2 \to \\ -3R_1+R_3 \to \end{matrix} \begin{bmatrix} 1 & 1 & 1 & | & 2 \\ 0 & -1 & 0 & | & -3 \\ 0 & -1 & -2 & | & -5 \end{bmatrix}$$

$$\begin{matrix} \\ -1R_2 \to \\ \\ \end{matrix} \begin{bmatrix} 1 & 1 & 1 & | & 2 \\ 0 & 1 & 0 & | & 3 \\ 0 & -1 & -2 & | & -5 \end{bmatrix}$$

$$\begin{matrix} \\ \\ R_2+R_3 \to \end{matrix} \begin{bmatrix} 1 & 1 & 1 & | & 2 \\ 0 & 1 & 0 & | & 3 \\ 0 & 0 & -2 & | & -2 \end{bmatrix}$$

$$\begin{matrix} \\ \\ -\frac{1}{2}R_3 \to \end{matrix} \begin{bmatrix} 1 & 1 & 1 & | & 2 \\ 0 & 1 & 0 & | & 3 \\ 0 & 0 & 1 & | & 1 \end{bmatrix}$$

$$\begin{cases} x+y+z=2 \\ y=3 \\ z=1 \end{cases}$$

$x+y+z=2$ Solution: $(-2,3,1)$

$x+3+1=2$

$x+4=2$

$x=-2$

33. $\begin{cases} 2x+y \ge -4 \\ -3x+y \ge -1 \end{cases}$

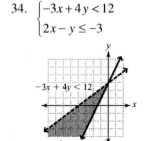

34. $\begin{cases} -3x+4y < 12 \\ 2x-y \le -3 \end{cases}$

35. a) $\begin{cases} x+y \le 500 \\ 40x+50y \ge 16,000 \\ x \ge 0 \\ y \ge 0 \end{cases}$

b)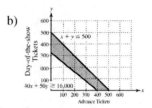

c) Answers may vary. One example is $(450,50)$

Chapter 4 Practice Test

1. Is $(-1,3)$ a solution of $\begin{cases} 3x+2y=3 \\ 4x-y=-7 \end{cases}$?

Equation 1	Equation 2
$3x+2y=3$	$4x-y=-7$
$3(-1)+2(3) \overset{?}{=} 3$	$4(-1)-3 \overset{?}{=} -7$
$-3+6 \overset{?}{=} 3$	$-4-3 \overset{?}{=} -7$
$3=3$	$-7=-7$

Yes, $(-1,3)$ is a solution for the system because it satisfies both equations.

2. Is $(2,2,3)$ a solution of $\begin{cases} 2x-3y+z=1 \\ -2x+y-3z=11 \\ 3x+y+3z=14 \end{cases}$?

Equation 1: $2x-3y+z=1$

$$2(2)-3(2)+(3) \overset{?}{=} 1$$

$$4-6+3 \overset{?}{=} 1$$

$$1=1$$

Equation 2: $-2x+y-3z=11$

$$-2(2)+(2)-3(3) \overset{?}{=} 11$$

$$-4+2-9 \overset{?}{=} 11$$

$$-11 \ne 11$$

Equation 3: $\qquad 3x + y + 3z = 14$

$$3(2) + (2) + 3(3) \overset{?}{=} 14$$

$$6 + 2 + 9 \overset{?}{=} 14$$

$$17 \neq 14$$

No, $(2,2,3)$ is not a solution for the system because it does not satisfy all three equations.

3. $\begin{cases} x + 3y = 1 \\ -2x + y = 5 \end{cases}$ $\qquad$ Graph each equation.

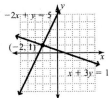

The lines intersect at a single point, which appears to be $(-2, 1)$.

4. $\begin{cases} 2x + y = 15 \\ y = 7 - x \end{cases}$

$$2x + y = 15 \qquad\qquad y = 7 - x$$
$$2x + (7 - x) = 15 \qquad y = 7 - 8$$
$$x + 7 = 15 \qquad\qquad y = -1$$
$$x = 8$$

Solution: $(8, -1)$

5. $\begin{cases} x - 2y = 1 \\ 3x - 5y = 4 \end{cases}$

Solve the first equation for x. $x = 2y + 1$

$$3x - 5y = 4 \qquad\qquad x = 2y + 1$$
$$3(2y + 1) - 5y = 4 \qquad x = 2 \cdot 1 + 1$$
$$6y + 3 - 5y = 4 \qquad x = 3$$
$$y = 1$$

Solution: $(3, 1)$

6. $\begin{cases} 3x - 2y = -8 \\ 2x + 3y = -14 \end{cases}$

$3x - 2y = -8 \qquad$ Multiply by 3.

$\underline{2x + 3y = -14} \qquad$ Multiply by 2.

$9x - 6y = -24$

$\underline{4x + 6y = -28}$

$13x \qquad = -52$

$x \qquad = -4$

$$3x - 2y = -8$$
$$3(-4) - 2y = -8$$
$$-12 - 2y = -8$$
$$-2y = 4$$
$$y = -2$$

Solution: $(-4, -2)$

7. $\begin{cases} 4x + 6y = 2 \\ 6x + 9y = 3 \end{cases}$

$4x + 6y = 2 \qquad$ Multiply by -3.

$\underline{6x + 9y = 3} \qquad$ Multiply by 2.

$-12x - 18y = -6$

$\underline{12x + 18y = 6}$

$0 = 0$

Notice that $0 = 0$ no longer has a variable and is true. The equations in the system are dependent; there are an infinite number of solutions that are all ordered pairs that solve $4x + 6y = 2$.

8. $\begin{cases} x + 2y + z = 2 & \text{Eqtn. 1} \\ x + 4y - z = 12 & \text{Eqtn. 2} \\ 3x - 3y - 2z = -11 & \text{Eqtn. 3} \end{cases}$

Multiply equation 1 by -3 and add to equation 3.
This makes equation 4.

$-3x - 6y - 3z = -6$

$\underline{3x - 3y - 2z = -11}$

$-9y - 5z = -1 \qquad \text{Eqtn. 4}$

Multiply equation 1 by -1 and add to equation 2.
This makes equation 5.

$-x - 2y - z = -2$

$\underline{x + 4y - z = 12}$

$2y - 2z = 10 \qquad \text{Eqtn. 5}$

Use equations 4 and 5 to make a system of
equations in two variables. Solve.

$-9y - 5z = -17 \qquad \text{Multiply by 2.}$

$\underline{2y - 2z = 10} \qquad \text{Multipy by 9.}$

$-18y - 10z = -34$

$\underline{18y - 18z = 90}$

$-28z = 56$

$z = -2$

$2y - 2(-2) = 10 \qquad \text{Eqtn. 5}$

$2y + 4 = 10$

$2y = 6$

$y = 3$

$x + 2 \cdot 3 - 2 = 2 \qquad \text{Eqtn. 1}$

$x + 6 - 2 = 2$

$x + 4 = 2$

$x = -2$

Solution: $(-2, 3, -2)$

9. $\begin{cases} x + 2y - 3z = -9 & \text{Eqtn. 1} \\ 3x - y + 2z = -8 & \text{Eqtn. 2} \\ 4x - 3y + 3z = -13 & \text{Eqtn. 3} \end{cases}$

Multiply equation 1 by -3 and add to equation 2.
This makes equation 4.

$-3x - 6y + 9z = 27$

$\underline{3x - y + 2z = -8}$

$-7y + 11z = 19 \qquad \text{Eqtn. 4}$

Multiply equation 1 by -4 and add to equation 3.
This makes equation 5.

$-4x - 8y + 12z = 36$

$\underline{4x - 3y + 3z = -13}$

$-11y + 15z = 23 \qquad \text{Eqtn. 5}$

Use equations 4 and 5 to make a system of
equations in two variables. Solve.

$-7y + 11z = 19 \qquad \text{Multiply by } -11.$

$\underline{-11y + 15z = 23} \qquad \text{Multipy by 7.}$

$77y - 121z = -209$

$\underline{-77y + 105z = 161}$

$-16z = -48$

$z = 3$

$-7y + 11 \cdot 3 = 19 \qquad \text{Eqtn. 4}$

$-7y + 33 = 19$

$-7y = -14$

$y = 2$

$x + 2 \cdot 2 - 3 \cdot 3 = -9 \qquad \text{Eqtn. 1}$

$x + 4 - 9 = -9$

$x - 5 = -9$

$x = -4$

Solution: $(-4, 2, 3)$

10. $\begin{cases} x + 2y = -6 \\ 3x + 4y = -10 \end{cases}$

$$\begin{bmatrix} 1 & 2 & | & -6 \\ 3 & 4 & | & -10 \end{bmatrix}$$

$-3R_1 + R_2 \quad \begin{bmatrix} 1 & 2 & | & -6 \\ 0 & -2 & | & 8 \end{bmatrix}$

$-\dfrac{1}{2}R_2 \quad \begin{bmatrix} 1 & 2 & | & -6 \\ 0 & 1 & | & -4 \end{bmatrix}$

$\begin{cases} x + 2y = -6 \\ \quad\quad y = -4 \end{cases}$

$x + 2(-4) = -6$

$x - 8 = -6$

$x = 2$

Solution: $(2, -4)$

11. $\begin{cases} x + y + z = 6 & \text{Eqtn. 1} \\ 3x - 2y + 3z = -7 & \text{Eqtn. 2} \\ 4x - 2y + z = -12 & \text{Eqtn. 3} \end{cases}$

$$\begin{bmatrix} 1 & 1 & 1 & | & 6 \\ 3 & -2 & 3 & | & -7 \\ 4 & -2 & 1 & | & -12 \end{bmatrix}$$

$-3R_1 + R_2, -4R_1 + R_2 \quad \begin{bmatrix} 1 & 1 & 1 & | & 6 \\ 0 & -5 & 0 & | & -25 \\ 0 & -6 & -3 & | & -36 \end{bmatrix}$

$-\dfrac{1}{5}R_2 \quad \begin{bmatrix} 1 & 1 & 1 & | & 6 \\ 0 & 1 & 0 & | & 5 \\ 0 & -6 & -3 & | & -36 \end{bmatrix}$

$6R_2 + R_3 \quad \begin{bmatrix} 1 & 1 & 1 & | & 6 \\ 0 & 1 & 0 & | & 5 \\ 0 & 0 & -3 & | & -6 \end{bmatrix}$

$-\dfrac{1}{3}R_3 \quad \begin{bmatrix} 1 & 1 & 1 & | & 6 \\ 0 & 1 & 0 & | & 5 \\ 0 & 0 & 1 & | & 2 \end{bmatrix}$

$\begin{cases} x + y + z = -6 \\ \quad\quad y = 5 \\ \quad\quad\quad z = 2 \end{cases}$

$x + 5 + 2 = 6$

$x = -1$

$(-1, 5, 2)$

12. $D = \begin{vmatrix} 3 & 4 \\ 2 & -3 \end{vmatrix} = 3(-3) - 2 \cdot 4 = -17$

$D_x = \begin{vmatrix} 14 & 4 \\ -19 & -3 \end{vmatrix} = 14(-3) - (-19)(4) = 34$

$D_y = \begin{vmatrix} 3 & 14 \\ 2 & -19 \end{vmatrix} = 3(-19) - 2(14) = -85$

$x = \dfrac{D_x}{D} = \dfrac{34}{-17} = -2 \qquad y = \dfrac{D_y}{D} = \dfrac{-85}{-17} = 5$

Solution: $(-2, 5)$

13. $D = \begin{vmatrix} 1 & 1 & 1 \\ 3 & -2 & 4 \\ 2 & 5 & -1 \end{vmatrix} = 1\begin{vmatrix} -2 & 4 \\ 5 & -1 \end{vmatrix} - 3\begin{vmatrix} 1 & 1 \\ 5 & -1 \end{vmatrix}$

$\quad + 2\begin{vmatrix} 1 & 1 \\ -2 & 4 \end{vmatrix}$

$= 1(2 - 20) - 3(-1 - 5) + 2(4 + 2)$

$= -18 + 18 + 12$

$= 12$

$D_x = \begin{vmatrix} -1 & 1 & 1 \\ 0 & -2 & 4 \\ -11 & 5 & -1 \end{vmatrix} = -1\begin{vmatrix} -2 & 4 \\ 5 & -1 \end{vmatrix} + 0 - 11\begin{vmatrix} 1 & 1 \\ -2 & 4 \end{vmatrix}$

$= -1(2 - 20) + 0 - 11(4 + 2)$

$= 18 + 0 - 66$

$= -48$

$D_y = \begin{vmatrix} 1 & -1 & 1 \\ 3 & 0 & 4 \\ 2 & -11 & -1 \end{vmatrix} = 1\begin{vmatrix} 0 & 4 \\ -11 & -1 \end{vmatrix} - 3\begin{vmatrix} -1 & 1 \\ -11 & -1 \end{vmatrix}$

$\quad + 2\begin{vmatrix} -1 & 1 \\ 0 & 4 \end{vmatrix}$

$= 1(0 + 44) - 3(1 + 11) + 2(-4 - 0)$

$= 44 - 36 - 8$

$= 0$

$D_z = \begin{vmatrix} 1 & 1 & -1 \\ 3 & -2 & 0 \\ 2 & 5 & -11 \end{vmatrix} = 1\begin{vmatrix} -2 & 0 \\ 5 & -11 \end{vmatrix} - 3\begin{vmatrix} 1 & -1 \\ 5 & -11 \end{vmatrix}$

$\quad + 2\begin{vmatrix} 1 & -1 \\ -2 & 0 \end{vmatrix}$

$= 1(22 - 0) - 3(-11 + 5) + 2(0 - 2)$

$= 22 + 18 - 4$

$= 36$

$x = \dfrac{D_x}{D} = \dfrac{-48}{12} = -4 \qquad y = \dfrac{D_y}{D} = \dfrac{0}{12} = 0$

$z = \dfrac{D_z}{D} = \dfrac{36}{12} = 3$

Solution: $(-4, 0, 3)$

14. $\begin{cases} 2x - 3y < 1 \\ x + 2y \le -2 \end{cases}$

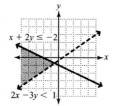

15. Let a be the number of accountants and let w be the number of waiters/waitresses. We are given that the combined number of accountants and waiters/waitresses who got headaches is 163. We are also given that 9 more accountants than waiters/waitresses got headaches. Translate to a system of equations.

$\begin{cases} a + w = 163 \\ 9 + w = a \end{cases}$

Substitute $9 + w$ for a in the first equation.

$\begin{array}{ll} a + w = 163 & 9 + w = a \\ 9 + w + w = 163 & 9 + 77 = a \\ 2w = 154 & 86 = a \\ w = 77 & \end{array}$

Therefore, 77 waiters/waitresses and 86 accountants got headaches.

16. Let l be the number of people considering the internet to be a library and let h be the number of people considering the internet to be a highway. We are given that a total of 240 people were polled, and that 3 times as many people responded "library" as opposed to "highway." Translate to a system of equations.

$\begin{cases} l + h = 240 \\ 3h = l \end{cases}$

Substitute $3h$ for l in the first equation.

$\begin{array}{ll} l + h = 240 & 3h = l \\ 3h + h = 240 & 3(60) = l \\ 4h = 240 & 180 = l \\ h = 60 & \end{array}$

So, 180 people considered the internet to be a library.

17. Let x represent the rate of the boat in still water and let y represent the amount that the current increases or decreases the boat's rate. We are told that with the current, the boat took 3 hours to go 30 miles. We are also given that against the current, the boat took 3 hours to go 12 miles. Create a table.

Categories	Rate	Time	Distance
With	$x + y$	3	30
Against	$x - y$	3	12

Translate to a system of equations. $\begin{cases} 3(x + y) = 30 \\ 3(x - y) = 12 \end{cases}$

Divide each equation by 3 to get the following system.

$\begin{cases} x + y = 10 \\ x - y = 4 \end{cases}$

Using elimination, solve the system.

$\begin{array}{ll} x + y = 10 & x + y = 10 \\ \underline{x - y = 4} & 7 + y = 10 \\ 2x = 14 & y = 3 \\ x = 7 & \end{array}$

The speed of the boat in still water is 7 mph.

18. Let x represent the amount invested at 6% interest (fund 1) and let y represent the amount invested at 8% interest (fund 2). We are given that a total of $12,000 was invested in the two funds. We are also given that the interest earned after one year was $880. Create a table.

Categories	APR	Principal	Interest
fund 1	0.06	x	$0.06x$
fund 2	0.08	y	$0.08y$

Translate to a system of equations.

$\begin{cases} x + y = 12,000 \\ 0.06x + 0.08y = 880 \end{cases}$

Solve the first equation for x. $x = 12,000 - y$, and then substitute into the second equation.

$0.06x + 0.08y = 880$

$0.06(12,000 - y) + 0.08y = 880$

$720 - 0.06y + 0.08y = 880$

$0.02y = 160$

$y = 8000$

$x = 12,000 - y$

$x = 12,000 - 8,000$

$x = 4000$

So, $4000 was invested in fund 1 at 6% and $8000 was invested in fund 2 at 8%.

19. Let a represent the number of adult tickets, c the number of child tickets, and s the number of student tickets. Translate to a system of equations.

$$\begin{cases} c+s+a=800 & \text{Eqtn. 1} \\ 3c+5s+8a=4750 & \text{Eqtn. 2} \\ a=50+2c & \text{Eqtn. 3} \end{cases}$$

Multiply equation 1 by -5 and add to equation 2. This makes equation 4.

$$\begin{array}{r} -5c-5s-5a=-4000 \\ \underline{3c+5s+8a=4750} \\ -2c+3a=750 \quad \text{Eqtn. 4} \end{array}$$

Use equations 3 and 4 to form a system of equations in two variables. Solve.

$$\begin{array}{r} -1(-2c+a=50) \\ \underline{-2c+3a=750} \\ 2c-a=-50 \\ \underline{-2c+3a=750} \\ 2a=700 \\ a=350 \end{array}$$

$350 = 50 + 2c \qquad$ Eqtn. 3

$300 = 2c$

$150 = c$

$150 + s + 350 = 800 \qquad$ Eqtn. 1

$\qquad s + 500 = 800$

$\qquad s = 300$

Therefore, there were 150 child tickets, 300 student tickets, and 350 adult tickets sold.

20. a) Let l represent the length of the garden and let w represent the width. Translate to a system of equations.

$$\begin{cases} 2l+2w \le 200 \\ l \ge w+10 \\ l > 0 \\ w > 0 \end{cases}$$

b)

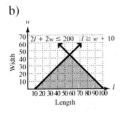

c) Answers may vary. One example is (60, 20).

Chapters 1–4 Cumulative Review

1. False. The slope is -3.

2. False; $-(-4)^3 = 64$.

3. True.

$$\begin{array}{ll} x=y & 9y-13=-4x \\ 1=1 & 9(1)-13=-4(1) \\ & -4=-4 \end{array}$$

4. $4(2x+3)=8x+12$

5. $x > 5$ or $x < -5$

6. The slope of the second line is the negative inverse of the slope of the first line. Thus, the lines are perpendicular.

7. $-\left(6-2^2\right)^2 + 15$

$= -(6-4)^2 + 15$

$= -2^2 + 15$

$= 11$

8. $-|2-8| + 3(2)$

$= -|-6| + 6$

$= -6 + 6$

$= 0$

9. $\dfrac{3}{8}x + 4y + 12 - \dfrac{1}{2}x - \dfrac{7}{10}y - 7$

$= \dfrac{3}{8}x + \dfrac{40}{10}y + 5 - \dfrac{4}{8}x - \dfrac{7}{10}y$

$= -\dfrac{1}{8}x + \dfrac{33}{10}y + 5$

10. $3.6 - 4(0.7p - 15) = 2.2p - 1.4$

$3.6 - 2.8p + 60 = 2.2p - 1.4$

$65 = 5p$

$p = 13$

11. $2(x-3) + 4x = 6(x-2)$

$= 2x - 6 + 4x = 6x - 12$

$= 6x - 6 = 6x - 12$

$= 6x \ne 6x - 6$

This is a contradiction. Thus, there is not solution.

12. $4z - 32 - 5z < 6z + 13 + 2z$

$-45 < 9z$

$z > -5$

13. $|4x+12|+2 < 10$

$4x+12+2 < 10$ or $-(4x+12)+2 < 10$

$4x+14 < 10$ $-4x-12+2 < 10$

$4x < -4$ $-4x < 20$

$x < -1$ $x > -5$

The solution set is $\{x \mid -5 < x < -1\}$ or $(-5, -1)$.

14. $A = P + Prt$

$A - P = Prt$

$t = \dfrac{A-P}{Pr}$

15. You cannot divide by 0, so $\dfrac{x}{x+2}$ is undefined when $x = -2$.

16. The formula for the area of a trapezoid is

$\dfrac{1}{2}(b_1 + b_2)h.$

$\dfrac{1}{2}(140+100)130$

$= 120 \cdot 130$

$= 15,600$

Subtract the area of the house.

$15,600 - (40 \cdot 50)$

$= 15,600 - 2000$

$= 13,600$

The area to be sodded is $13,600 \text{ ft}^2$.

17. $y = 3x + 1$

x	y	Ordered Pair
0	1	$(0, 1)$
$-\dfrac{1}{3}$	0	$\left(\dfrac{1}{3}, 0\right)$
1	4	$(1, 4)$

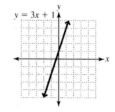

18. $4x + 5y = 10$

x	y	Ordered Pair
0	2	$(0, 2)$
5	-2	$(5, -2)$
$\dfrac{5}{2}$	0	$\left(\dfrac{5}{2}, 0\right)$

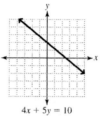

$4x + 5y = 10$

19. $x = -4$

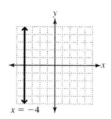

$x = -4$

20. $2x - 7y < 14$

$-7y < -2x + 14$

$y > \dfrac{2}{7}x - 2$

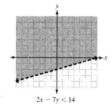

$2x - 7y < 14$

21. $(y+1) = -\dfrac{1}{5}(x+2)$

$y + 1 = -\dfrac{1}{5}x - \dfrac{2}{5}$

$y = -\dfrac{1}{5}x - \dfrac{7}{5}$

22. $(4, 7), (-2, -1)$

$m = \dfrac{y_2 - y_1}{x_2 - x_1} = \dfrac{-1+7}{-2-4} = \dfrac{6}{-6} = -1$

$(y+1) = -1(x+2)$

$y + 1 = -x - 2$

$y = -x - 3$

23. The slope of $3x + 4y = 4$ is $-\dfrac{3}{4}$.

$$(y - 2) = -\frac{3}{4}(x + 4)$$

$$y - 2 = -\frac{3}{4}x - 3$$

$$y + \frac{3}{4}x = -1$$

$$4y + 3x = -4$$

24. $\begin{cases} 4x - 3y = -2 \\ 6x - 7y = 7 \end{cases}$

$\begin{array}{r} 3(4x - 3y = -2) \\ -2(6x - 7y = 7) \\ \hline 5y = -20 \\ y = -4 \end{array}$
$\quad$
$\begin{array}{r} 4x - 3y = -2 \\ 4x - 3(-4) = -2 \\ 4x = -14 \\ x = \dfrac{-14}{4} = -3.5 \end{array}$

The solution is $(-3.5, -4)$.

25. $\begin{cases} x + y + z = 5 & \text{Eqtn.1} \\ 2x + y - 2z = -5 & \text{Eqtn.2} \\ x - 2y + z = 8 & \text{Eqtn.3} \end{cases}$

Subtract equation 3 from equation 1.
$$\begin{array}{r} x + y + z = 5 \\ x - 2y + z = 8 \\ \hline 3y = -3 \\ y = -1 \end{array}$$

Multiply equation 1 by -2, then add it to equation 2 and solve for z.
$$\begin{array}{r} -2(x + y + z = 5) \\ 2x + y - 2z = -5 \\ \hline -y - 4z = -15 \\ -(-1) - 4z = -15 \\ -4z = -16 \\ z = 4 \end{array}$$

Use equation 1 to solve for x.
$$\begin{array}{r} x + y + z = 5 \\ x + (-1) + 4 = 5 \\ x = 2 \end{array}$$
The solution is $(2, -1, 4)$.

26. $\begin{cases} 4x - 3y < 12 \\ 3x + y > 6 \end{cases}$

Find the y-intercepts.

$$4x - 3y < 12$$
$$-3y < -4x + 12$$
$$y > \frac{4}{3}x - 4$$

$$3x + y > 6$$
$$y > -3x + 6$$

The y-intercepts are $(0, -4)$ and $(0, 6)$.

Solve the system of equations.

$$\begin{array}{r} 4x - 3y = 12 \\ 3(3x + y = 6) \\ \hline 13x = 30 \\ x = \dfrac{30}{13} \end{array}$$
$\quad$
$$\begin{array}{r} 4x - 3y = 12 \\ 4\left(\dfrac{30}{13}\right) - 3y = 12 \\ -3y = \dfrac{39}{13} \\ y = -1 \end{array}$$

The solution is $\left(\dfrac{30}{13}, -1\right)$.

Graph the inequalities.

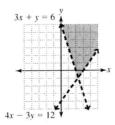

27. Let l = the length and w = the width.

$$\begin{cases} 2l + 2w = 420 \\ l = w + 50 \end{cases}$$

Multiply the second equation by -2, then add it to the first equation.
$$\begin{array}{r} 2l + 2w = 420 \\ -2(l = w + 50) \\ \hline 4w = 320 \\ w = 80 \end{array}$$
$\quad$
$$\begin{array}{l} l = w + 50 \\ l = 80 + 50 \\ l = 130 \end{array}$$

The length is 130 ft and the width is 80 ft.

28. The sum of the angles is 180°. Let x = the greater angle and y = the lesser angle.

$$\begin{cases} x + y = 180 \\ x = 2y - 15 \end{cases}$$

Solve the system of equations.
$$\begin{array}{r} x + y = 180 \\ -(x = 2y - 15) \\ \hline 3y = 195 \\ y = 65 \end{array}$$
$\quad$
$$\begin{array}{l} x + y = 180 \\ x + 65 = 180 \\ x = 115 \end{array}$$

The greater angle is 115° and the lesser angle is 65°.

29. Let x represent the amount invested at 3% interest and let y represent the amount invested at 2.5% interest. We are given that a total of $50,000 was invested. We are also given that the interest earned after one year was $1425. Create a table.

Categories	APR	Principal	Interest
investment 1	0.03	x	$0.03x$
investment 2	0.025	y	$0.025y$

Translate to a system of equations.

$$\begin{cases} x + y = 50,000 \\ 0.03x + 0.025y = 1425 \end{cases}$$

Solve the first equation for x. $x = 50,000 - y$, and then substitute into the second equation.

$$0.03x + 0.025y = 1425$$

$$0.03(50,000 - y) + 0.025y = 1425$$

$$1500 - 0.03y + 0.025y = 1425$$

$$-0.005y = -75$$

$$y = 15,000$$

$x = 50,000 - y$

$x = 50,000 - 15,000$

$x = 35,000$

So, $35,000 was invested at 3% and $15,000 was invested at 2.5%.

30. Let x = the number of gallons inside paint and y = the number of gallons of outside paint.

$$\begin{cases} x + y = 24 \\ 22x + 28y = 588 \end{cases}$$

Solve the first equation for x. $x = -y + 24$, and then substitute into the second equation.

$$22(-y + 24) + 28y = 588$$

$$-22y + 528 + 28y = 588$$

$$6y = 60$$

$$y = 10$$

$x = -y + 24$

$x = -10 + 24$

$x = 14$

The contractor bought 14 gallons of inside paint and 10 gallons of outside paint.

Chapter 5

Exponents, Polynomials, and Polynomial Functions

Exercise Set 5.1

1. $mn^2 \cdot m^3 n^3 = m^{1+3} n^{2+3} = m^4 n^5$

3. $3^{10} \cdot 3^2 = 3^{10+2} = 3^{12}$

5. $(-3)^5 (-3)^3 = (-3)^{5+3} = (-3)^8$ or 3^8

7. $\left(-4p^4 q^3\right)\left(3p^2 q^2\right) = -12 p^{4+2} q^{3+2} = -12 p^6 q^5$

9. $\left(6r^3 st^4\right)\left(-3r^4 s^5 t^8\right) = 6(-3) r^{3+4} s^{1+5} t^{4+8}$
$\qquad = -18 r^7 s^6 t^{12}$

11. $\left(1.2u^3 t^9\right)\left(3.1u^4 t^2\right) = (1.2)(3.1) u^{3+4} t^{9+2} = 3.72 u^7 t^{11}$

13. $15^0 = 1$

15. $4x^0 = 4(1) = 4$

17. $\left(3xy^2\right)^0 = 1$

19. $-5^2 = -(25) = -25$

21. Mistake: Assigned the minus sign to the base.
Correct: It should be $-(2 \cdot 2 \cdot 2 \cdot 2) = -16$.

23. $\dfrac{h^5}{h^2} = h^{5-2} = h^3$

25. $\dfrac{a^2}{a^7} = a^{2-7} = a^{-5} = \dfrac{1}{a^5}$

27. $\dfrac{6x^2 y^6}{3xy^3} = 2x^{2-1} y^{6-3} = 2xy^3$

29. $\dfrac{18r^3 s^7 t}{-12r^6 s^4 t} = -\dfrac{3}{2} r^{3-6} s^{7-4} t^{1-1} = -\dfrac{3}{2} r^{-3} s^3 t^0 = -\dfrac{3s^3}{2r^3}$

31. $\dfrac{15u^{-8} v^3 w^4}{-21u^3 v^3 w^{-2}} = -\dfrac{5}{7} u^{-8-3} v^{3-3} w^{4-(-2)}$
$\qquad = -\dfrac{5}{7} u^{-11} v^0 w^6$
$\qquad = -\dfrac{5w^6}{7u^{11}}$

33. $\dfrac{8a^{-2} b^3 c^8}{2a^3 b^{-4} c^5} = 4a^{-2-3} b^{3-(-4)} c^{8-5} = 4a^{-5} b^7 c^3 = \dfrac{4b^7 c^3}{a^5}$

35. $2^{-3} = \dfrac{1}{2^3} = \dfrac{1}{8}$

37. $-3^{-4} = -\dfrac{1}{3^4} = -\dfrac{1}{81}$

39. $\dfrac{1}{5^{-3}} = 5^3 = 125$

41. $\dfrac{4}{x^{-3}} = 4x^3$

43. $\dfrac{1}{4a^{-6}} = \dfrac{a^6}{4}$

45. Mistake: Multiplied 4 by -1.
Correct: It should be $\dfrac{1}{4}$.

47. $\left(x^3\right)^4 = x^{3 \cdot 4} = x^{12}$

49. $\left(2x^5\right)^4 = 2^4 x^{5 \cdot 4} = 16x^{20}$

51. $\left(-5x^3 y\right)^2 = (-5)^2 x^{3 \cdot 2} y^{1 \cdot 2} = 25x^6 y^2$

53. $\left(\dfrac{3}{4} a^2 b^4\right)^3 = \left(\dfrac{3}{4}\right)^3 a^{2 \cdot 3} b^{4 \cdot 3} = \dfrac{27}{64} a^6 b^{12}$

55. $\left(-0.3r^2 t^4 u\right)^3 = (-0.3)^3 r^{2 \cdot 3} t^{4 \cdot 3} u^{1 \cdot 3} = -0.027 r^6 t^{12} u^3$

57. $\left(\dfrac{c}{d}\right)^4 = \dfrac{c^4}{d^4}$

59. $\left(\dfrac{3}{x}\right)^3 = \dfrac{3^3}{x^3} = \dfrac{27}{x^3}$

61. $\left(-\dfrac{4}{x^3}\right)^2 = \dfrac{(-4)^2}{\left(x^3\right)^2} = \dfrac{16}{x^6}$

63. $\left(\dfrac{a}{b}\right)^{-6} = \left(\dfrac{b}{a}\right)^6 = \dfrac{b^6}{a^6}$

65. $\left(\dfrac{x^2}{y^3}\right)^{-4} = \left(\dfrac{y^3}{x^2}\right)^4 = \dfrac{y^{12}}{x^8}$

67. $\left(\dfrac{4}{x^{-2}}\right)^{-3} = \left(\dfrac{x^{-2}}{4}\right)^{3} = \dfrac{x^{-6}}{4^3} = \dfrac{1}{64x^6}$

69. $\left(\dfrac{3x^2}{y^3}\right)^{-4} = \left(\dfrac{y^3}{3x^2}\right)^{4} = \dfrac{\left(y^3\right)^4}{(3)^4\left(x^2\right)^4} = \dfrac{y^{12}}{81x^8}$

71. $\left(4x^2y^3\right)^2\left(-2x^3y^4\right)^3$

$= 4^2\left(x^2\right)^2\left(y^3\right)^2 \cdot (-2)^3\left(x^3\right)^3\left(y^4\right)^3$

$= 16x^4y^6 \cdot (-8)x^9y^{12}$

$= -128x^{13}y^{18}$

73. $\dfrac{\left(3q^4p^2\right)^2}{\left(5q^2p^2\right)^3} = \dfrac{3^2\left(q^4\right)^2\left(p^2\right)^2}{5^3\left(q^2\right)^3\left(p^2\right)^3} = \dfrac{9q^8p^4}{125q^6p^6} = \dfrac{9q^2}{125p^2}$

75. $\left(3h^3t^5\right)^{-2}\left(9h^2t^4\right)^2$

$= \dfrac{1}{(3)^2\left(h^3\right)^2\left(t^5\right)^2} \cdot (9)^2\left(h^2\right)^2\left(t^4\right)^2$

$= \dfrac{1}{9h^6t^{10}} \cdot 81h^4t^8$

$= \dfrac{81h^4t^8}{9h^6t^{10}}$

$= \dfrac{9}{h^2t^2}$

77. $\dfrac{\left(9u^3v^2\right)^{-1}}{\left(2u^4v^2\right)^4} = \dfrac{1}{9u^3v^2} \cdot \dfrac{1}{2^4\left(u^4\right)^4\left(v^2\right)^4}$

$= \dfrac{1}{9u^3v^2} \cdot \dfrac{1}{16u^{16}v^8}$

$= \dfrac{1}{144u^{19}v^{10}}$

79. $\dfrac{\left(2u^2v^3\right)^{-3}\left(4u^{-2}v^3\right)}{\left(3u^2v^{-3}\right)^{-2}} = \dfrac{\left(4u^{-2}v^3\right)\left(3u^2v^{-3}\right)^2}{\left(2u^2v^3\right)^3}$

$= \dfrac{\left(4u^{-2}v^3\right)(3)^2\left(u^2\right)^2\left(v^{-3}\right)^2}{(2)^3\left(u^2\right)^3\left(v^3\right)^3}$

$= \dfrac{\left(4u^{-2}v^3\right)\cdot 9u^4v^{-6}}{8u^6v^9}$

$= \dfrac{36u^2v^{-3}}{8u^6v^9}$

$= \dfrac{9}{2u^4v^{12}}$

81. $\dfrac{\left(-4a^{-2}b^3c^2\right)^{-1}\left(2abc\right)^{-2}}{\left(3a^5b^2\right)^3\left(2a^{-1}b^{-3}c\right)^{-3}}$

$= \dfrac{\left(2a^{-1}b^{-3}c\right)^3}{\left(3a^5b^2\right)^3\left(-4a^{-2}b^3c^2\right)\left(2abc\right)^2}$

$= \dfrac{8a^{-3}b^{-9}c^3}{\left(27a^{15}b^6\right)\left(-4a^{-2}b^3c^2\right)\left(4a^2b^2c^2\right)}$

$= \dfrac{8a^{-3}b^{-9}c^3}{-432a^{15}b^{11}c^4}$

$= -\dfrac{1}{54a^{18}b^{20}c}$

83. $2.9 \times 10^6 = 2{,}900{,}000$

85. $2 \times 10^{11} = 200{,}000{,}000{,}000$

87. $7.3 \times 10^{10} = 73{,}000{,}000{,}000$

89. $1.7 \times 10^{-7} = 0.00000017$

91. 1.675×10^{-24}

$= 0.000000000000000000000001675$

93. 6.645×10^{-27}

$= 0.000000000000000000000000006645$

95. $50{,}150{,}000 = 5.015 \times 10^7$

97. $16{,}780{,}000{,}000{,}000 = 1.678 \times 10^{13}$

99. $9{,}309{,}000{,}000 = 9.309 \times 10^9$

101. $0.00000055 = 5.5 \times 10^{-7}$

103. $0.0000001 = 1 \times 10^{-7}$

105. $0.0000000000000005 = 5 \times 10^{-17}$

107. $6 \times 10^5, 7.4 \times 10^6, 8.3 \times 10^6, 1.2 \times 10^7, 2.4 \times 10^8$

109. $(2100)(30,000) = (2.1 \times 10^3)(3.0 \times 10^4)$
$$= 2.1 \times 3.0 \times 10^3 \times 10^4$$
$$= 6.3 \times 10^{3+4}$$
$$= 6.3 \times 10^7$$

111. $(-32,000)(410,000) = (-3.2 \times 10^4)(4.1 \times 10^5)$
$$= -3.2 \times 4.1 \times 10^4 \times 10^5$$
$$= -13.12 \times 10^{4+5}$$
$$= -13.12 \times 10^9$$
$$= -1.312 \times 10^{10}$$

113. $(0.00081)(220,000) = (8.1 \times 10^{-4})(2.2 \times 10^5)$
$$= 8.1 \times 2.2 \times 10^{-4} \times 10^5$$
$$= 17.82 \times 10^{-4+5}$$
$$= 17.82 \times 10^1$$
$$= 1.782 \times 10^2$$

115. $\dfrac{8,400,000}{210} = \dfrac{8.4 \times 10^6}{2.1 \times 10^2}$
$$= \dfrac{8.4}{2.1} \times \dfrac{10^6}{10^2}$$
$$= 4 \times 10^{6-2}$$
$$= 4 \times 10^4$$

117. $\dfrac{930,000,000,000,000}{-3,000,000} = \dfrac{9.3 \times 10^{14}}{-3.0 \times 10^6}$
$$= \dfrac{9.3}{-3.0} \times \dfrac{10^{14}}{10^6}$$
$$= -3.1 \times 10^{14-6}$$
$$= -3.1 \times 10^8$$

119. $\dfrac{0.0057}{0.0000095} = \dfrac{5.7 \times 10^{-3}}{9.5 \times 10^{-6}}$
$$= \dfrac{5.7}{9.5} \times \dfrac{10^{-3}}{10^{-6}}$$
$$= 0.6 \times 10^{-3-(-6)}$$
$$= 0.6 \times 10^3$$
$$= 6 \times 10^2$$

121. $30,000^3 = (3 \times 10^4)^3$
$$= 3^3 \times (10^4)^3$$
$$= 27 \times 10^{12}$$
$$= 2.7 \times 10^{13}$$

123. $20,000,000^3 = (2 \times 10^7)^3 = 2^3 \times (10^7)^3 = 8 \times 10^{21}$

125. $0.0005^2 = (5 \times 10^{-4})^2$
$$= 5^2 \times (10^{-4})^2$$
$$= 25 \times 10^{-8}$$
$$= 2.5 \times 10^{-7}$$

127. $(-0.000004)^{-2} = (-4 \times 10^{-6})^{-2}$
$$= (-4)^{-2} \times (10^{-6})^{-2}$$
$$= \dfrac{1}{16} \times 10^{12}$$
$$= 0.0625 \times 10^{12}$$
$$= 6.25 \times 10^{10}$$

129. $E = hf$
$$E = (6.626 \times 10^{-34})(4.2 \times 10^{14})$$
$$\approx 27.8 \times 10^{-20}$$
$$= 2.78 \times 10^{-19} \text{ joules}$$

131. $$E = hf$$
$$1.4 \times 10^{-19} = (6.626 \times 10^{-34})f$$
$$\dfrac{1.4 \times 10^{-19}}{6.626 \times 10^{-34}} = f$$
$$0.211 \times 10^{15} \approx f$$
$$2.11 \times 10^{14} \text{ Hz} = f$$

133. $E = mc^2$
$$E = (4.2 \times 10^{-12})(3 \times 10^8)^2$$
$$= (4.2 \times 10^{-12})(9 \times 10^{16})$$
$$= 37.8 \times 10^4$$
$$= 3.78 \times 10^5 \text{ joules}$$

Review Exercises

1. Change the operation from subtraction to addition and the subtrahend, $-6,$ to its additive inverse, 6.

2. Associative property of addition

3. $-8(n-4) = -8 \cdot n - 8 \cdot (-4) = -8n + 32$

4. $7x + 8y - 4x - 12 + 13y - 5$
   $= 7x - 4x + 8y + 13y - 12 - 5$
   $= 3x + 21y - 17$

5. $10x^2 - x + 9 - 12x + 3x^2$
   $= 10x^2 + 3x^2 - x - 12x + 9$
   $= 13x^2 - 13x + 9$

6. $2(u+8) - (5u-3) - 7$
   $= 2u + 16 - 5u + 3 - 7$
   $= 2u - 5u + 16 + 3 - 7$
   $= -3u + 12$

Exercise Set 5.2

1. $-7uv^3$
   $d = 1 + 3 = 4;$ monomial

3. $25 - x^2$
   $d = 2;$ binomial

5. $5p^3 + 2p^2 - 1$
   $d = 3;$ trinomial

7. $4g^2 - 5g - 11g^3 - 8g^4 + 7$
   $d = 4;$ this polynomial has more than three terms, so it has no special polynomial name.

9. $7x + 5x^3 - 19$
   $d = 3;$ trinomial

11. -7
    $d = 0;$ monomial

13. $\dfrac{1}{3}pq - 3p^2q$
    $d = 2 + 1 = 3;$ binomial

15. $\left(5x^2 - 3x + 1\right) + \left(2x^2 + 7x - 3\right) = 7x^2 + 4x - 2$

17. $\left(p^4 - 3p^3 + 4p - 1\right) + \left(p^4 - 2p^3 - 4p + 7\right)$
    $= 2p^4 - 5p^3 + 6$

19. $\left(4u^3 - 6u^2 + u + 11\right) + \left(-5u^3 - 3u^2 + u - 5\right)$
    $= -u^3 - 9u^2 + 2u + 6$

21. $\left(\dfrac{2}{3}u^4 + \dfrac{3}{4}u^3 - u^2 - u + 3\right)$
    $\quad + \left(\dfrac{2}{3}u^4 - \dfrac{1}{4}u^3 + 3u^2 + 4u - 1\right)$
    $= \dfrac{4}{3}u^4 + \dfrac{1}{2}u^3 + 2u^2 + 3u + 2$

23. $\left(3.1t^4 - 2.1t^3 + 7t^2 + 5.8t + 4\right)$
    $\quad + \left(4.2t^4 + 3.6t^3 - 8t^2 - 3.1t + 3\right)$
    $= 7.3t^4 + 1.5t^3 - t^2 + 2.7t + 7$

25. $\left(7a^3 - a^2 - a\right) - \left(8a^3 - 3a + 1\right)$
    $= \left(7a^3 - a^2 - a\right) + \left(-8a^3 + 3a - 1\right)$
    $= -a^3 - a^2 + 2a - 1$

27. $\left(3m^4 + 2m^3 - m^2 + 1\right) - \left(6m^3 - 2m^2 - m + 3\right)$
    $= \left(3m^4 + 2m^3 - m^2 + 1\right) + \left(-6m^3 + 2m^2 + m - 3\right)$
    $= 3m^4 - 4m^3 + m^2 + m - 2$

29. $\left(-7r^3 + 3r^2 - 7r - 4\right) - \left(-5r^3 + 2r^2 - 6r - 1\right)$
    $= \left(-7r^3 + 3r^2 - 7r - 4\right) + \left(5r^3 - 2r^2 + 6r + 1\right)$
    $= -2r^3 + r^2 - r - 3$

31. $\left(\dfrac{4}{5}y^3 - \dfrac{1}{3}y^2 + 7y - \dfrac{2}{7}\right) - \left(\dfrac{1}{5}y^3 + \dfrac{2}{3}y^2 + \dfrac{3}{4}y - \dfrac{1}{7}\right)$
    $= \left(\dfrac{4}{5}y^3 - \dfrac{1}{3}y^2 + 7y - \dfrac{2}{7}\right)$
    $\quad + \left(-\dfrac{1}{5}y^3 - \dfrac{2}{3}y^2 - \dfrac{3}{4}y + \dfrac{1}{7}\right)$
    $= \dfrac{3}{5}y^3 - y^2 + \dfrac{25}{4}y - \dfrac{1}{7}$

33. $\left(-4.5w^3 - 5.1w^2 + 2.7w + 4.1\right)$
    $\quad - \left(3.8w^3 - 1.4w^2 + 3.4w - 2.6\right)$
    $= \left(-4.5w^3 - 5.1w^2 + 2.7w + 4.1\right)$
    $\quad + \left(-3.8w^3 + 1.4w^2 - 3.4w + 2.6\right)$
    $= -8.3w^3 - 3.7w^2 - 0.7w + 6.7$

35. $\left(-5w^4 - 3w^3 - 8w^2 + w - 14\right)$

   $+\left(-3w^4 + 6w^3 + w^2 + 12w + 5\right)$

   $= -8w^4 + 3w^3 - 7w^2 + 13w - 9$

37. $\left(6x^4 + 4x^3 - 3x + 5\right) - \left(-2x^3 - 5x^2 + 9x - 8\right)$

   $= \left(6x^4 + 4x^3 - 3x + 5\right) + \left(2x^3 + 5x^2 - 9x + 8\right)$

   $= 6x^4 + 6x^3 + 5x^2 - 12x + 13$

39. $\left(-\dfrac{7}{3}g^4 + \dfrac{4}{5}g^3 + \dfrac{7}{5}\right) + \left(\dfrac{4}{3}g^4 - \dfrac{2}{3}g^2 + \dfrac{2}{5}\right)$

   $= -g^4 + \dfrac{4}{5}g^3 - \dfrac{2}{3}g^2 + \dfrac{9}{5}$

41. $\left(4y^2 - 8y + 1\right) + \left(5y^2 + 3y + 2\right) - \left(6y^2 - 9y + 2\right)$

   $= \left(4y^2 - 8y + 1\right) + \left(5y^2 + 3y + 2\right) + \left(-6y^2 + 9y - 2\right)$

   $= 3y^2 + 4y + 1$

43. $\left(-6a^3 - 5a^2 + 10\right) - \left(9a^3 + 5a^2 - 3a - 1\right)$

   $\quad + \left(-4a^3 - 2a^2 + 6a - 2\right)$

   $= \left(-6a^3 - 5a^2 + 10\right) + \left(-9a^3 - 5a^2 + 3a + 1\right)$

   $\quad + \left(-4a^3 - 2a^2 + 6a - 2\right)$

   $= -19a^3 - 12a^2 + 9a + 9$

45. $\left(8a^2 + 5ab - 2b^2\right) - \left(10ab - 6a^2 - 8b^2\right)$

   $= \left(8a^2 + 5ab - 2b^2\right) + \left(-10ab + 6a^2 + 8b^2\right)$

   $= 14a^2 - 5ab + 6b^2$

47. $\left(7x^3y^4 - 2x^2y^3 - xy + 4\right)$

   $\quad + \left(-5x^3y^4 + 7x^2y^3 + 8xy - 12\right)$

   $= 2x^3y^4 + 5x^2y^3 + 7xy - 8$

49. $\left(x^2y^2 + 8xy^2 - 12x^2y - 4xy + 7y^2 - 9\right)$

   $\quad - \left(-3x^2y^2 + 2xy^2 + 4x^2y - xy + 6y^2 + 4\right)$

   $= \left(x^2y^2 + 8xy^2 - 12x^2y - 4xy + 7y^2 - 9\right)$

   $\quad + \left(3x^2y^2 - 2xy^2 - 4x^2y + xy - 6y^2 - 4\right)$

   $= 4x^2y^2 + 6xy^2 - 16x^2y - 3xy + y^2 - 13$

51. $(x + 10) + (x - 4) + (2x - 1) = 4x + 5$

53. $2(a - 10) + 2(3a + 12) = 2a - 20 + 6a + 24$

   $\qquad\qquad\qquad\qquad = 8a + 4$

55. Because the degree of this polynomial is 0, the function is a constant function.

57. Because the degree of this polynomial is 2, the function is a quadratic function.

59. Because the degree of this polynomial is 1, the function is a linear function.

61. Because the degree of this polynomial is 3, the function is a cubic function.

63. $h(x) = -2$

x	$h(x)$
-2	-2
0	-2
2	-2

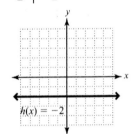

Domain: $\{x \mid -\infty < x < \infty\}$ or $(-\infty, \infty)$

Range: $\{y \mid y = -2\}$ or $[-2, -2]$

65. $c(x) = -5x + 1$

x	$c(x)$
-1	6
0	1
1	-4

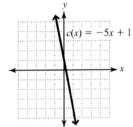

Domain: $\{x \mid -\infty < x < \infty\}$ or $(-\infty, \infty)$

Range: $\{y \mid -\infty < y < \infty\}$ or $(-\infty, \infty)$

67. $g(x) = x^2 + 2x - 3$

x	$g(x)$
-3	0
-2	-3
-1	-4
0	-3
1	0

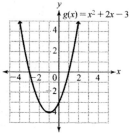

Domain: $\{x \mid -\infty < x < \infty\}$ or $(-\infty, \infty)$

Range: $\{y \mid y \geq -4\}$ or $[-4, \infty)$

69. $f(x) = x^3 - 2$

x	$f(x)$
-2	-10
-1	-3
0	-2
1	-1
2	6

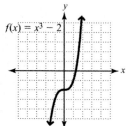

Domain: $\{x \mid -\infty < x < \infty\}$ or $(-\infty, \infty)$

Range: $\{y \mid -\infty < y < \infty\}$ or $(-\infty, \infty)$

71. $f(x) = -4;\ g(x) = 3$

a) $(f + g)(x) = -4 + 3 = -1$

b) $(f - g)(x) = -4 - 3 = -7$

73. $f(x) = 4x + 3;\ g(x) = -x - 1$

a) $(f + g)(x) = (4x + 3) + (-x - 1)$

$= 3x + 2$

b) $(f - g)(x) = (4x + 3) - (-x - 1)$

$= (4x + 3) + (x + 1)$

$= 5x + 4$

75. $f(x) = x^2 - 1;\ g(x) = 3x^2 + 2$

a) $(f + g)(x) = (x^2 - 1) + (3x^2 + 2)$

$= 4x^2 + 1$

b) $(f - g)(x) = (x^2 - 1) - (3x^2 + 2)$

$= (x^2 - 1) + (-3x^2 - 2)$

$= -2x^2 - 3$

77. $f(x) = x^3 + 2x + 5;\ g(x) = -2x^3 + 2$

a) $(f + g)(x) = (x^3 + 2x + 5) + (-2x^3 + 2)$

$= -x^3 + 2x + 7$

b) $(f - g)(x) = (x^3 + 2x + 5) - (-2x^3 + 2)$

$= (x^3 + 2x + 5) + (2x^3 - 2)$

$= 3x^3 + 2x + 3$

79. a) $P(x) = R(x) - C(x)$

$= (0.2x^2 + x + 3000) - (18x + 2000)$

$= (0.2x^2 + x + 3000) + (-18x - 2000)$

$= 0.2x^2 - 17x + 1000$

b) $P(1000) = 0.2x^2 - 17x + 1000$

$= 0.2(1000)^2 - 17(1000) + 1000$

$= 184,000$

The profit is $184,000.

81. a) $R(x) = P(x) + C(x)$

$= (190x + 260) + (230x + 580)$

$= 420x + 840$

b) $R(x) = 420x + 840$

$R(300) = 420(300) + 840$

$= 126,000 + 840$

$= 126,840$

The revenue is $126,840.

83. $h(t) = -16t^2 + 200$

$h(0.5) = -16(0.5)^2 + 200$

$= -4 + 200$

$= 196$

After 0.5 second, the height is 196 feet.

85. $V(r) = 1.5r^2 - 0.8r + 9.5$

$V(6) = 1.5(6)^2 - 0.8(6) + 9.5$

$= 58.7$

The voltage is 58.7 V.

87. a) The completed table is shown below.

Year	Years since 2002	Fatalities
2002	0	34.3
2004	2	29.0
2006	4	25.54
2008	6	23.92
2010	8	24.14

b)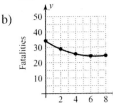

Years since 2002

c) $F(x) = 0.23x^2 - 3.11x + 34.3$

$F(18) = 0.23(18)^2 - 3.11(18) + 34.3$

$= 74.52 - 55.98 + 34.3$

$= 52.84$

The model predicts that the fatality rate will be 52.84 per 100 million vehicle miles in 2020.

Review Exercises

1. $8(3x - 9) = 24x - 72$

2. $-6(2y + 7) = -12y - 42$

3. $5x - 3(x + 2) = 7x - 8$

$5x - 3x - 6 = 7x - 8$

$2x - 6 = 7x - 8$

$2 = 5x$

$\dfrac{2}{5} = x$

4. $12x + 9 < 4(2x - 5)$

$12x + 9 < 8x - 20$

$4x < -29$

$x < -\dfrac{29}{4}$

5. $-11 < 2x - 3 \le 9$

$-8 < 2x \le 12$

$-4 < x \le 6$

a) $\{x \mid -4 < x \le 6\}$ b) $(-4, 6]$

c) [number line from −6 to 6, open interval graphed from −4 to 6]

6. $2x - 7 \le -5$ or $2x - 7 \ge 5$

$2x \le 2$ $\qquad\qquad$ $2x \ge 12$

$x \le 1$ $\qquad\qquad$ $x \ge 6$

a) $\{x \mid x \le 1 \text{ or } x \ge 6\}$

b) $(-\infty, 1] \cup [6, \infty)$

c) [number line from −5 to 7]

Exercise Set 5.3

1. $5x^3(x^2 + 3x - 2) = 5x^3 \cdot x^2 + 5x^3 \cdot 3x + 5x^3 \cdot (-2)$

$= 5x^5 + 15x^4 - 10x^3$

3. $-6x^3(2x^2 + 4x - 3)$

$= -6x^3 \cdot 2x^2 - 6x^3 \cdot 4x - 6x^3(-3)$

$= -12x^5 - 24x^4 + 18x^3$

5. $4n^4(n^3 + 7n^2 - 2n - 3)$

$= 4n^4 \cdot n^3 + 4n^4 \cdot 7n^2 + 4n^4 \cdot (-2n) + 4n^4 \cdot (-3)$

$= 4n^7 + 28n^6 - 8n^5 - 12n^4$

7. $9a^2b^4(3a^4b^2 - 2ab^8)$

$= 9a^2b^4 \cdot 3a^4b^2 + 9a^2b^4 \cdot (-2ab^8)$

$= 27a^6b^6 - 18a^3b^{12}$

9. $\dfrac{1}{4}m^2np^3(2mn^5 - 5m^2n^2 + 8m^2n^3p^2)$

$= \dfrac{1}{4}m^2np^3 \cdot 2mn^5 + \dfrac{1}{4}m^2np^3 \cdot (-5m^2n^2)$

$\qquad + \dfrac{1}{4}m^2np^3 \cdot 8m^2n^3p^2$

$= \dfrac{1}{2}m^3n^6p^3 - \dfrac{5}{4}m^4n^3p^3 + 2m^4n^4p^5$

11. $-0.3p^2q^2(8p^3q^7 - 2q + 7p^6)$

$= -0.3p^2q^2 \cdot 8p^3q^7 - 0.3p^2q^2 \cdot (-2q)$

$\qquad - 0.3p^2q^2 \cdot 7p^6$

$= -2.4p^5q^9 + 0.6p^2q^3 - 2.1p^8q^2$

13. $-5x^5 y^2 \left(4x^2 y + 2xy^4 - 5x^8 y^5\right)$

$= -5x^5 y^2 \cdot 4x^2 y - 5x^5 y^2 \cdot 2xy^4$

$\qquad -5x^5 y^2 \cdot \left(-5x^8 y^5\right)$

$= -20x^7 y^3 - 10x^6 y^6 + 25x^{13} y^7$

15. $-3r^2 s \left(2r^4 s^3 - 6r^2 s^2 - 3rs + 3\right)$

$= -3r^2 s \cdot 2r^4 s^3 - 3r^2 s \cdot \left(-6r^2 s^2\right) - 3r^2 s \cdot \left(-3rs\right)$

$\qquad -3r^2 s \cdot 3$

$= -6r^6 s^4 + 18r^4 s^3 + 9r^3 s^2 - 9r^2 s$

17. $-0.2a^2 b^2 \left(2.1a - 6ab + 3a^2 b - a^2 b^2\right)$

$= -0.2a^2 b^2 \cdot 2.1a - 0.2a^2 b^2 \cdot \left(-6ab\right)$

$\qquad -0.2a^2 b^2 \cdot 3a^2 b - 0.2a^2 b^2 \cdot \left(-a^2 b^2\right)$

$= -0.42a^3 b^2 + 1.2a^3 b^3 - 0.6a^4 b^3 + 0.2a^4 b^4$

19. $\dfrac{1}{3} abc^6 \left(9a^2 b^2 c^2 - 4a^2 bc + 12a\right)$

$= \dfrac{1}{3} abc^6 \cdot 9a^2 b^2 c^2 + \dfrac{1}{3} abc^6 \cdot \left(-4a^2 bc\right)$

$\qquad + \dfrac{1}{3} abc^6 \cdot 12a$

$= 3a^3 b^3 c^8 - \dfrac{4}{3} a^3 b^2 c^7 + 4a^2 bc^6$

21. a) $x + 5$
    b) $2x$
    c) $(x + 5)(2x)$
    d) $x^2 + 5x + x^2 + 5x = 2x^2 + 10x$
    e) They are equivalent because they both describe the area of the figure.

23. a) $x + 3$
    b) $x + 2$
    c) $(x + 3)(x + 2)$
    d) $x^2 + 3x + 2x + 6 = x^2 + 5x + 6$
    e) They are equivalent because they both describe the area of the figure.

25. $(2x + 3)(3x + 4) = 2x \cdot 3x + 2x \cdot 4 + 3 \cdot 3x + 3 \cdot 4$

$= 6x^2 + 8x + 9x + 12$

$= 6x^2 + 17x + 12$

27. $(3x - 1)(5x + 2) = 3x \cdot 5x + 3x \cdot 2 + (-1) \cdot 5x + (-1) \cdot 2$

$= 15x^2 + 6x - 5x - 2$

$= 15x^2 + x - 2$

29. $(2m - 5n)(3m - 2n)$

$= 2m \cdot 3m + 2m \cdot (-2n) + (-5n) \cdot 3m + (-5n) \cdot (-2n)$

$= 6m^2 - 4mn - 15mn + 10n^2$

$= 6m^2 - 19mn + 10n^2$

31. $(5m - 3n)(3m + 4n)$

$= 5m \cdot 3m + 5m \cdot 4n + (-3n) \cdot 3m + (-3n) \cdot 4n$

$= 15m^2 + 20mn - 9mn - 12n^2$

$= 15m^2 + 11mn - 12n^2$

33. $\left(t^2 - 5\right)\left(t^2 - 2\right)$

$= t^2 \cdot t^2 + t^2 \cdot (-2) + (-5) \cdot t^2 + (-5)(-2)$

$= t^4 - 2t^2 - 5t^2 + 10$

$= t^4 - 7t^2 + 10$

35. $\left(a^2 + 6b^2\right)\left(a^2 - b^2\right)$

$= a^2 \cdot a^2 + a^2 \cdot \left(-b^2\right) + 6b^2 \cdot a^2 + 6b^2 \cdot \left(-b^2\right)$

$= a^4 - a^2 b^2 + 6a^2 b^2 - 6b^4$

$= a^4 + 5a^2 b^2 - 6b^4$

37. $(3x - 1)\left(4x^2 - 2x + 1\right)$

$= 3x \cdot 4x^2 + 3x \cdot (-2x) + 3x \cdot 1$

$\qquad + (-1) \cdot 4x^2 + (-1) \cdot (-2x) + (-1) \cdot 1$

$= 12x^3 - 6x^2 + 3x - 4x^2 + 2x - 1$

$= 12x^3 - 10x^2 + 5x - 1$

39. $(7a + 1)\left(3a^2 - 2a - 2\right)$

$= 7a \cdot 3a^2 + 7a \cdot (-2a) + 7a \cdot (-2)$

$\qquad + 1 \cdot 3a^2 + 1 \cdot (-2a) + 1 \cdot (-2)$

$= 21a^3 - 14a^2 - 14a + 3a^2 - 2a - 2$

$= 21a^3 - 11a^2 - 16a - 2$

41. $(7c + 2)\left(2c^2 - 4c - 3\right)$

$= 7c \cdot 2c^2 + 7c \cdot (-4c) + 7c \cdot (-3)$

$\qquad + 2 \cdot 2c^2 + 2 \cdot (-4c) + 2 \cdot (-3)$

$= 14c^3 - 28c^2 - 21c + 4c^2 - 8c - 6$

$= 14c^3 - 24c^2 - 29c - 6$

43. $(4p+10q)(3p^2+2pq+q^2)$

$= 4p \cdot 3p^2 + 4p \cdot 2pq + 4p \cdot q^2$
$\quad + 10q \cdot 3p^2 + 10q \cdot 2pq + 10q \cdot q^2$
$= 12p^3 + 8p^2q + 4pq^2 + 30p^2q + 20pq^2 + 10q^3$
$= 12p^3 + 38p^2q + 24pq^2 + 10q^3$

45. $(2y-3z)(6y^2-2yz+4z^2)$

$= 2y \cdot 6y^2 + 2y \cdot (-2yz) + 2y \cdot 4z^2$
$\quad + (-3z) \cdot 6y^2 + (-3z) \cdot (-2yz) + (-3z) \cdot 4z^2$
$= 12y^3 - 4y^2z + 8yz^2 - 18y^2z + 6yz^2 - 12z^3$
$= 12y^3 - 22y^2z + 14yz^2 - 12z^3$

47. $(3u^2-2u-1)(3u^2+2u+1)$

$= 3u^2 \cdot 3u^2 + 3u^2 \cdot 2u + 3u^2 \cdot 1$
$\quad + (-2u) \cdot 3u^2 + (-2u) \cdot 2u + (-2u) \cdot 1$
$\quad + (-1) \cdot 3u^2 + (-1) \cdot 2u + (-1) \cdot 1$
$= 9u^4 + 6u^3 + 3u^2 - 6u^3 - 4u^2 - 2u - 3u^2 - 2u - 1$
$= 9u^4 - 4u^2 - 4u - 1$

49. $(x+y)^2 = x^2 + 2xy + y^2$

51. $(4w+3)^2 = (4w)^2 + 2(4w)(3) + 3^2$
$\quad\quad = 16w^2 + 24w + 9$

53. $(4t-3w)^2 = (4t)^2 - 2(4t)(3w) + (3w)^2$
$\quad\quad = 16t^2 - 24tw + 9w^2$

55. $(9-5y^2)^2 = 9^2 - 2(9)(5y^2) + (5y^2)^2$
$\quad\quad = 81 - 90y^2 + 25y^4$
$\quad\quad = 25y^4 - 90y^2 + 81$

57. $[(x+1)-y]^2$

$= (x+1)^2 - 2y(x+1) + y^2$
$= x^2 + 2x + 1 - 2xy - 2y + y^2$
$= x^2 + y^2 - 2xy + 2x - 2y + 1$

59. $[p-(q+5)]^2$

$= p^2 - 2p(q+5) + (q+5)^2$
$= p^2 - 2pq - 10p + q^2 + 10q + 25$
$= p^2 + q^2 - 2pq - 10p + 10q + 25$

61. $x-8$

63. $3m-2n$

65. m^2-n^2

67. $-2j+5k$

69. $(2x-7)(2x+7) = (2x)^2 - 7^2 = 4x^2 - 49$

71. $(2q-5)(2q+5) = (2q)^2 - 5^2 = 4q^2 - 25$

73. $(x+2y)(x-2y) = x^2 - (2y)^2 = x^4 - 4y^2$ *not good*

75. $[(s+1)+t][(s+1)-t] = (s+1)^2 - t^2$
$\quad\quad = s^2 + 2s + 1 - t^2$
$\quad\quad = s^2 - t^2 + 2s + 1$

77. $[3b+(c+2)][3b-(c+2)] = (3b)^2 - (c+2)^2$
$\quad\quad = 9b^2 - (c^2 + 4c + 4)$
$\quad\quad = 9b^2 - c^2 - 4c - 4$

79. $A = yz(3x+y)$
$\quad = 3xyz + y^2z$

81. $A = \frac{1}{2}h[(h+1)+(h+6)]$
$\quad = \frac{1}{2}h(2h+7)$
$\quad = h^2 + \frac{7}{2}h$

83. Mistake: Combined like terms incorrectly.
Correct: $6x^2 - 37x + 45$

85. Mistake: Multiplied $2x$ by $2x$ incorrectly.
Correct: $4x^2 - 49$

87. $-4wv(-w^4v^3 - 4x^2wv^7 + 5xw^2v)$
$\quad = 4v^4w^5 + 16v^8w^2x^2 - 20v^2w^3x$

89. $(8r^2-3s)(8r^2+3s) = (8r^2)^2 - (3s)^2$
$\quad\quad = 64r^4 - 9s^2$

91. $(2u^2+3v^2)^2 = (2u^2)^2 + 2(2u^2)(3v^2) + (3v^2)^2$
$\quad\quad = 4u^4 + 12u^2v^2 + 9v^4$

93. $\left(2a^2 - 5b^2\right)\left(a^2 - 4b^2\right)$

$= 2a^4 - 8a^2b^2 - 5a^2b^2 + 20b^4$

$= 2a^4 - 13a^2b^2 + 20b^4$

95. $\left(5q^2 - 3t\right)\left(3q^2 + 4t\right)$

$= 15q^4 + 20tq^2 - 9tq^2 - 12t^2$

$= 15q^4 + 11tq^2 - 12t^2$

97. $3r^2s^3\left(r^4 - \dfrac{1}{9}r^3s - \dfrac{1}{6}r^2s^2 + \dfrac{1}{3}rs - 1\right)$

$= 3r^6s^3 - \dfrac{1}{3}r^5s^4 - \dfrac{1}{2}r^4s^5 + r^3s^4 - 3r^2s^3$

99. $-0.1t^2r^3\left(3.5t^2r^3 - 8t^3r^2 + 2.2t^3r - 2tr\right)$

$= -0.35t^4r^6 + 0.8t^5r^5 - 0.22t^5r^4 + 0.2t^3r^4$

101. $\left[(3m-4)+n\right]\left[(3m-4)-n\right]$

$= (3m-4)^2 - n^2$

$= 9m^2 - 24m + 16 - n^2$

$= 9m^2 - n^2 - 24m + 16$

103. $\left(x^2 + 3xy + y^2\right)\left(4x^2 + 2xy - y^2\right)$

$= 4x^4 + 2x^3y - x^2y^2 + 12x^3y + 6x^2y^2$

$\quad - 3xy^3 + 4x^2y^2 + 2xy^3 - y^4$

$= 4x^4 + 14x^3y + 9x^2y^2 - xy^3 - y^4$

105. $\left[\left(x^2 - 4\right) + 3\right]^2$

$= \left(x^2 - 4\right)^2 + 6\left(x^2 - 4\right) + 9$

$= x^4 - 8x^2 + 16 + 6x^2 - 24 + 9$

$= x^4 - 2x^2 + 1$

107. $f(x) = 3x - 2; \quad g(x) = 2x - 1$

a) $(f \cdot g)(x) = (3x-2)(2x-1)$

$\qquad = 6x^2 - 3x - 4x + 2$

$\qquad = 6x^2 - 7x + 2$

b) $(f \cdot g)(-4) = 6(-4)^2 - 7(-4) + 2$

$\qquad = 96 + 28 + 2$

$\qquad = 126$

c) $(f \cdot g)(2) = 6(2)^2 - 7(2) + 2$

$\qquad = 24 - 14 + 2$

$\qquad = 12$

109. $f(x) = 3x + 4; \quad g(x) = 4x^2 - 3x + 2$

a) $(f \cdot g)(x) = (3x+4)\left(4x^2 - 3x + 2\right)$

$\qquad = 12x^3 - 9x^2 + 6x + 16x^2 - 12x + 8$

$\qquad = 12x^3 + 7x^2 - 6x + 8$

b) $(f \cdot g)(5) = 12(5)^3 + 7(5)^2 - 6(5) + 8$

$\qquad = 12(125) + 7(25) - 30 + 8$

$\qquad = 1500 + 175 - 30 + 8$

$\qquad = 1653$

c) $(f \cdot g)(-3) = 12(-3)^3 + 7(-3)^2 - 6(-3) + 8$

$\qquad = 12(-27) + 7(9) + 18 + 8$

$\qquad = -324 + 63 + 18 + 8$

$\qquad = -235$

111. $f(x) = x^2 + 3x - 1$

$f(n+1) = (n+1)^2 + 3(n+1) - 1$

$\qquad = n^2 + 2n + 1 + 3n + 3 - 1$

$\qquad = n^2 + 5n + 3$

113. $f(x) = 2x^2 - 1$

$f(t-6) = 2(t-6)^2 - 1$

$\qquad = 2\left(t^2 - 12t + 36\right) - 1$

$\qquad = 2t^2 - 24t + 72 - 1$

$\qquad = 2t^2 - 24t + 71$

115. $\qquad f(x) = 2x - 3$

$f(x+h) - f(x) = 2(x+h) - 3 - (2x-3)$

$\qquad = 2x + 2h - 3 - 2x + 3$

$\qquad = 2h$

117. $f(x) = 2x^2 + 3x - 4$

$f(x+h) - f(x)$

$= 2(x+h)^2 + 3(x+h) - 4 - \left(2x^2 + 3x - 4\right)$

$= 2\left(x^2 + 2xh + h^2\right) + 3x + 3h - 4 - 2x^2 - 3x + 4$

$= 2x^2 + 4xh + 2h^2 + 3x + 3h - 4 - 2x^2 - 3x + 4$

$= 4xh + 2h^2 + 3h$

119. Let l = the length of the room and $l - 2$ = the width of the room.

Area $= l(l-2) = l^2 - 2l$

The area of the room is $\left(l^2 - 2l\right)$ square feet.

121. Volume $= 4x(x+1)(x+5)$
$= 4x(x^2 + 6x + 5)$
$= 4x^3 + 24x^2 + 20x$

The volume is $(4x^3 + 24x^2 + 20x)$ cubic centimeters.

123. The width is w. The length is $3w$. The height is $(3w + 5)$.
$V = w(3w)(3w+5) = 3w^2(3w+5)$
$= 9w^3 + 15w^2$

The volume is $(9w^3 + 15w^2)$ cubic feet.

125. The radius is r. The height is $(r+3)$.
$V = \pi r^2(r+3) = \pi r^3 + 3\pi r^2$

The volume is $(\pi r^3 + 3\pi r^2)$ cubic inches.

Review Exercises

1. $(-4.26 \times 10^6)(2.1 \times 10^5) = (-4.26 \cdot 2.1) \times (10^{6+5})$
$= -8.946 \times 10^{11}$

2. $\dfrac{1.5 \times 10^5}{2.4 \times 10^{-4}} = \dfrac{1.5}{2.4} \times \dfrac{10^5}{10^{-4}}$
$= 0.625 \times 10^{5-(-4)}$
$= 0.625 \times 10^9$
$= 6.25 \times 10^8$

3. $\dfrac{5}{6}x - \dfrac{1}{8} = \dfrac{3}{4}x + \dfrac{1}{2}$
$24\left(\dfrac{5}{6}x - \dfrac{1}{8}\right) = 24\left(\dfrac{3}{4}x + \dfrac{1}{2}\right)$
$20x - 3 = 18x + 12$
$2x - 3 = 12$
$2x = 15$
$x = \dfrac{15}{2}$

4. $d = 30t - n$
$d + n = 30t$
$\dfrac{d+n}{30} = t$

5. Use $m = -\dfrac{1}{5}$.
$y - 5 = -\dfrac{1}{5}(x - (-3))$
$y - 5 = -\dfrac{1}{5}(x + 3)$
$y - 5 = -\dfrac{1}{5}x - \dfrac{3}{5}$
$y = -\dfrac{1}{5}x + \dfrac{22}{5}$

6. Let s = the number of smaller boxes sold and l = the number of larger boxes sold.
$\begin{cases} s + l = 84 \\ 3.5s + 4.8l = 349.9 \end{cases}$

Solve the first equation for s: $s = 84 - l$.
$3.5(84 - l) + 4.8l = 349.9 \qquad s = 84 - 43$
$294 - 3.5l + 4.8l = 349.9 \qquad\quad = 41$
$1.3l = 55.9$
$l = 43$

Therefore, 43 large boxes and 41 small boxes were sold.

Exercise Set 5.4

1. $\dfrac{28a^4 - 7a^3 + 21a^2}{7a^2} = \dfrac{28a^4}{7a^2} - \dfrac{7a^3}{7a^2} + \dfrac{21a^2}{7a^2}$
$= 4a^2 - a + 3$

3. $\dfrac{12u^5 - 6u^4 - 15u^3 + 3u^2}{3u^2}$
$= \dfrac{12u^5}{3u^2} - \dfrac{6u^4}{3u^2} - \dfrac{15u^3}{3u^2} + \dfrac{3u^2}{3u^2}$
$= 4u^3 - 2u^2 - 5u + 1$

5. $\dfrac{24a^5b^4 - 8a^4b^2 + 16a^2b}{8ab}$
$= \dfrac{24a^5b^4}{8ab} - \dfrac{8a^4b^2}{8ab} + \dfrac{16a^2b}{8ab}$
$= 3a^4b^3 - a^3b + 2a$

7. $\dfrac{36u^3v^4 + 12uv^5 - 15u^2v^2}{3uv^9}$
$= \dfrac{36u^3v^4}{3uv^9} + \dfrac{12uv^5}{3uv^9} - \dfrac{15u^2v^2}{3uv^9}$
$= \dfrac{12u^2}{v^5} + \dfrac{4}{v^4} - \dfrac{5u}{v^7}$

9. $\dfrac{30x^3y^2 - 45xy^3 - xy}{-5xy}$

$= \dfrac{30x^3y^2}{-5xy} - \dfrac{45xy^3}{-5xy} - \dfrac{xy}{-5xy}$

$= -6x^2y + 9y^2 + \dfrac{1}{5}$

11. $\dfrac{12t^6u^5v - 2t^4u^4 - 16t^3u^2 + 3tu^2}{-4t^3u^2}$

$= \dfrac{12t^6u^5v}{-4t^3u^2} - \dfrac{2t^4u^4}{-4t^3u^2} - \dfrac{16t^3u^2}{-4t^3u^2} + \dfrac{3tu^2}{-4t^3u^2}$

$= -3t^3u^3v + \dfrac{1}{2}tu^2 + 4 - \dfrac{3}{4t^2}$

13. $\dfrac{18a^3b^2c^2 + 12a^2b^3c - 3a^2b}{9a^2bc}$

$= \dfrac{18a^3b^2c^2}{9a^2bc} + \dfrac{12a^2b^3c}{9a^2bc} - \dfrac{3a^2b}{9a^2bc}$

$= 2abc + \dfrac{4}{3}b^2 - \dfrac{1}{3c}$

15. $\dfrac{25xy^2}{10x} = \dfrac{5y^2}{2}$

17. a) $\dfrac{6x^2}{3x} - \dfrac{12x}{3x} + \dfrac{18}{3x} = 2x - 4 + \dfrac{6}{x}$

b) Area $= 6(5)^2 - 12(5) + 18$ length $= 3(5) = 15$

$= 150 - 60 + 18$

$= 108$

width $= 2(5) - 4 + \dfrac{6}{5} = 7.2$

19. $x + 6 \overline{)\,x^2 + 10x + 24}$ quotient $x + 4$

$\underline{x^2 + 6x}$

$4x + 24$

$\underline{4x + 24}$

0

21. $p + 1 \overline{)\,3p^2 + 4p + 1}$ quotient $3p + 1$

$\underline{3p^2 + 3p}$

$p + 1$

$\underline{p + 1}$

0

23. $n + 1 \overline{)\,5n^2 - 3n - 8}$ quotient $5n - 8$

$\underline{5n^2 + 5n}$

$-8n - 8$

$\underline{-8n - 8}$

0

25. $5x + 4 \overline{)\,15x^2 + 2x - 8}$ quotient $3x - 2$

$\underline{15x^2 + 12x}$

$-10x - 8$

$\underline{-10x - 8}$

0

27. $3y + 4 \overline{)\,6y^2 + 5y - 4}$ quotient $2y - 1$

$\underline{6y^2 + 8y}$

$-3y - 4$

$\underline{-3y - 4}$

0

29. $2x - 3 \overline{)\,2x^3 - 13x^2 + 27x - 18}$ quotient $x^2 - 5x + 6$

$\underline{2x^3 - 3x^2}$

$-10x^2 + 27x$

$\underline{-10x^2 + 15x}$

$12x - 18$

$\underline{12x - 18}$

0

31. $x + 2 \overline{)\,x^3 + 0x^2 + 0x + 8}$ quotient $x^2 - 2x + 4$

$\underline{x^3 + 2x^2}$

$-2x^2 + 0x$

$\underline{-2x^2 - 4x}$

$4x + 8$

$\underline{4x + 8}$

0

33.
$$
\begin{array}{r}
y^3 + 2y^2 + 4y + 8 \\
y - 2 \overline{)\, y^4 + 0y^3 + 0y^2 + 0y - 16} \\
\underline{y^4 - 2y^3} \\
2y^3 + 0y^2 \\
\underline{2y^3 - 4y^2} \\
4y^2 + 0y \\
\underline{4y^2 - 8y} \\
8y - 16 \\
\underline{8y - 16} \\
0
\end{array}
$$

35.
$$
\begin{array}{r}
4z^2 - 10z + 25 \\
2z + 5 \overline{)\, 8z^3 + 0z^2 + 0z + 125} \\
\underline{8z^3 + 20z^2} \\
-20z^2 + 0z \\
\underline{-20z^2 - 50z} \\
50z + 125 \\
\underline{50z + 125} \\
0
\end{array}
$$

37.
$$
\begin{array}{r}
v^2 + 3v + 2 \\
2v + 1 \overline{)\, 2v^3 + 7v^2 + 7v + 2} \\
\underline{2v^3 + v^2} \\
6v^2 + 7v \\
\underline{6v^2 + 3v} \\
4v + 2 \\
\underline{4v + 2} \\
0
\end{array}
$$

39.
$$
\begin{array}{r}
7a^2 - 2a + 3 \\
3a - 2 \overline{)\, 21a^3 - 20a^2 + 13a - 6} \\
\underline{21a^3 - 14a^2} \\
-6a^2 + 13a \\
\underline{-6a^2 + 4a} \\
9a - 6 \\
\underline{9a - 6} \\
0
\end{array}
$$

41.
$$
\begin{array}{r}
6q^2 + 5q + 9 \\
2q - 1 \overline{)\, 12q^3 + 4q^2 + 13q + 9} \\
\underline{12q^3 - 6q^2} \\
10q^2 + 13q \\
\underline{10q^2 - 5q} \\
18q + 9 \\
\underline{18q - 9} \\
18
\end{array}
$$

Answer: $6q^2 + 5q + 9 + \dfrac{18}{2q - 1}$

43.
$$
\begin{array}{r}
3x^2 - 6x + 5 \\
x^2 + 0x + 4 \overline{)\, 3x^4 - 6x^3 + 17x^2 - 24x + 20} \\
\underline{3x^4 + 0x^3 + 12x^2} \\
-6x^3 + 5x^2 - 24x \\
\underline{-6x^3 - 0x^2 - 24x} \\
5x^2 + 0x + 20 \\
\underline{5x^2 + 0x + 20} \\
0
\end{array}
$$

45.
$$
\begin{array}{r}
c^3 - 2c^2 + c - 4 \\
3c + 2 \overline{)\, 3c^4 - 4c^3 - c^2 - 10c - 8} \\
\underline{3c^4 + 2c^3} \\
-6c^3 - c^2 \\
\underline{-6c^3 - 4c^2} \\
3c^2 - 10c \\
\underline{3c^2 + 2c} \\
-12c - 8 \\
\underline{-12c - 8} \\
0
\end{array}
$$

47.
$$
\begin{array}{r}
5u^2 - 11u + 32 \\
u + 2 \overline{)\, 5u^3 - u^2 + 10u + 2} \\
\underline{5u^3 + 10u^2} \\
-11u^2 + 10u \\
\underline{-11u^2 - 22u} \\
32u + 2 \\
\underline{32u + 64} \\
-62
\end{array}
$$

Answer: $5u^2 - 11u + 32 - \dfrac{62}{u + 2}$

49. The width is 6, the height is $t+2$, and the volume is $6t^3 + 30t^2 + 12t - 48$. We are looking for the length.

$$whl = V$$

$$6(t+2)l = 6t^3 + 30t^2 + 12t - 48$$

$$l = \left(6t^3 + 30t^2 + 12t - 48\right) \div \left(6t + 12\right)$$

a)
$$
\begin{array}{r}
t^2 + 3t - 4 \\
6t+12 \overline{)6t^3 + 30t^2 + 12t - 48} \\
\underline{6t^3 + 12t^2} \\
18t^2 + 12t \\
\underline{18t^2 + 36t} \\
-24t - 48 \\
\underline{-24t - 48}
\end{array}
$$

b) height: $2 + 2 = 4$ ft.;

    length: $2^2 + 3(2) - 4 = 4 + 6 - 4 = 6$ ft. ;

    volume: $6(2)^3 + 30(2)^2 + 12(2) - 48 = 144$ ft.^{3}

51. a) $(f/g)(x) = \dfrac{24x^4 - 42x^3 - 30x^2}{6x^2}$

$$= \dfrac{24x^4}{6x^2} - \dfrac{42x^3}{6x^2} - \dfrac{30x^2}{6x^2}$$

$$= 4x^2 - 7x - 5$$

b) $(f/g)(4) = 4(4)^2 - 7(4) - 5 = 64 - 28 - 5 = 31$

c) $(f/g)(-2) = 4(-2)^2 - 7(-2) - 5$

$$= 16 + 14 - 5$$

$$= 25$$

53. a) $(h/k)(x) = \dfrac{6x^3 + 7x^2 - 9x + 2}{2x - 1}$

$$
\begin{array}{r}
3x^2 + 5x - 2 \\
2x-1 \overline{)6x^3 + 7x^2 - 9x + 2} \\
\underline{6x^3 - 3x^2} \\
10x^2 - 9x \\
\underline{10x^2 - 5x} \\
-4x + 2 \\
\underline{-4x + 2} \\
0
\end{array}
$$

$$(h/k)(x) = 3x^2 + 5x - 2$$

b) $(h/k)(3) = 3(3)^2 + 5(3) - 2$

$$= 27 + 15 - 2$$

$$= 40$$

c) $(h/k)(-1) = 3(-1)^2 + 5(-1) - 2$

$$= 3 - 5 - 2$$

$$= -4$$

55. a) $(n/p)(x) = \dfrac{x^3 - x^2 - 7x + 4}{x - 3}$

$$
\begin{array}{r}
x^2 + 2x - 1 \\
x-3 \overline{)x^3 - x^2 - 7x + 4} \\
\underline{x^3 - 3x^2} \\
2x^2 - 7x \\
\underline{2x^2 - 6x} \\
-x + 4 \\
\underline{-x + 3} \\
1
\end{array}
$$

$$(n/p)(x) = x^2 + 2x - 1 + \dfrac{1}{x - 3}$$

b) $(n/p)(4) = 4^2 + 2(4) - 1 + \dfrac{1}{4 - 3}$

$$= 16 + 8 - 1 + 1$$

$$= 24$$

c) $(n/p)(x) = (-2)^2 + 2(-2) - 1 + \dfrac{1}{-2 - 3}$

$$= 4 - 4 - 1 - \dfrac{1}{5}$$

$$= -\dfrac{6}{5}$$

Review Exercises

1. $9x - (6x + 2) = 5x - 2\left[4x - (6 - x)\right]$

$$9x - 6x - 2 = 5x - 2(4x - 6 + x)$$

$$3x - 2 = 5x - 8x + 12 - 2x$$

$$3x - 2 = -5x + 12$$

$$8x = 14$$

$$x = \dfrac{7}{4}$$

2. $7x - 5 \geq 8 - (2x + 1)$

$$7x - 5 \geq 8 - 2x - 1$$

$$7x - 5 \geq 7 - 2x$$

$$9x \geq 12$$

$$x \geq \dfrac{4}{3}$$

3. $5x - 2y = 10$

$-2y = -5x + 10$

$y = \frac{5}{2}x - 5$

Use $m = -\frac{2}{5}$.

$y - (-1) = -\frac{2}{5}(x - 3)$

$y + 1 = -\frac{2}{5}x + \frac{6}{5}$

$y = -\frac{2}{5}x + \frac{1}{5}$ or $2x + 5y = 1$

4. Yes, because every value in the domain corresponds to only one value in the range. (It passes the vertical line test.)

5. $\begin{cases} 2x + 3y - z = 7 & \text{Eqtn. 1} \\ x + y - 5z = -14 & \text{Eqtn. 2} \\ -3x - 5y + z = -12 & \text{Eqtn. 3} \end{cases}$

Multiply equation 2 by -2 and add to equation 1. This makes equation 4.

$2x + 3y - z = 7$

$\underline{-2x - 2y + 10z = 28}$

$\qquad y + 9z = 35 \quad$ Eqtn. 4

Multiply equation 2 by 3 and add to equation 3. This makes equation 5.

$3x + 3y - 15z = -42$

$\underline{-3x - 5y + z = -12}$

$\quad -2y - 14z = -54 \quad$ Eqtn. 5

Multiply equation 4 by 2 and add to equation 5.

$2y + 18z = 70$

$\underline{-2y - 14z = -54}$

$\qquad\quad 4z = 16$

$\qquad\qquad z = 4$

$y + 9z = 35 \qquad\qquad x + y - 5z = -14$

$y + 9(4) = 35 \qquad\quad x + (-1) - 5(4) = -14$

$y + 36 = 35 \qquad\qquad x - 1 - 20 = -14$

$\qquad y = -1 \qquad\qquad\qquad\quad x = 7$

Solution: $(7, -1, 4)$

6. $\begin{cases} x - y > 3 \\ 2x + y \leq 4 \end{cases}$

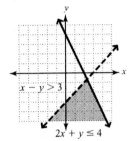

Exercise Set 5.5

1. a) $x + 3$ b) $x^2 + 7x + 12$ c) $x + 4$

3. a) $x - 3$ b) $x^3 + 4x^2 - 25x + 7$

   c) $x^2 + 7x - 4 - \dfrac{5}{x - 3}$

5. a) $x - 4$ b) $2x^4 - 8x^3 - 5x^2 + 17x + 10$

   c) $2x^3 - 5x - 3 - \dfrac{2}{x - 4}$

7. $\underline{2|}\ \ 1 \quad -5 \quad 6$

$\qquad\quad\ \underline{\quad 2 \quad -6}$

$\qquad\ \ 1 \quad -3 \quad 0$

Answer: $x - 3$

9. $\underline{4|}\ \ 2 \quad -1 \quad -5$

$\qquad\quad\ \underline{\quad 8 \quad 28}$

$\qquad\ \ 2 \quad 7 \quad 23$

Answer: $2x + 7 + \dfrac{23}{x - 4}$

11. $\underline{2|}\ \ 1 \quad -1 \quad -5 \quad 6$

$\qquad\quad\ \underline{\quad 2 \quad 2 \quad -6}$

$\qquad\ \ 1 \quad 1 \quad -3 \quad 0$

Answer: $x^2 + x - 3$

13. $\underline{-2|}\ \ 3 \quad 8 \quad 3 \quad -2$

$\qquad\qquad\ \underline{\ -6 \ -4 \quad 2}$

$\qquad\quad\ \ 3 \quad 2 \quad -1 \quad 0$

Answer: $3x^2 + 2x - 1$

15. $\underline{-3|}\ \ 3 \quad -1 \quad -22 \quad 24$

$\qquad\qquad\ \underline{\ -9 \quad 30 \quad -24}$

$\qquad\quad\ \ 3 \quad -10 \quad 8 \quad 0$

Answer: $3x^2 - 10x + 8$

17. $\underline{-2|}$ 2 -5 -1 6
           -4 18 -34
     _____
              2   -9 17 -28

Answer: $2x^2 - 9x + 17 - \dfrac{28}{x+2}$

19. $\underline{2|}$ 1 -2 1 -3
           2   0   2
     _____
              1   0   1 -1

Answer: $x^2 + 1 - \dfrac{1}{x-2}$

21. $\underline{-3|}$ 2   1   0 -4
           -6 15 -45
     _____
              2   -5 15 -49

Answer: $2x^2 - 5x + 15 - \dfrac{49}{x+3}$

23. $\underline{3|}$ 3 -4 0   2
           9   15   45
     _____
              3   5   15   47

Answer: $3x^2 + 5x + 15 + \dfrac{47}{x-3}$

25. $\underline{2|}$ 1   0 -7 6
           2   4 -6
     _____
              1   2 -3 0

Answer: $x^2 + 2x - 3$

27. $\underline{3|}$ 1   0   0   27
           3   9   27
     _____
              1   3   9   54

Answer: $x^2 + 3x + 9 + \dfrac{54}{x-3}$

29. $\underline{-2|}$ 1   0   0   0   16
           -2 4 -8 16
     _____
              1   -2 4 -8 32

Answer: $x^3 - 2x^2 + 4x - 8 + \dfrac{32}{x+2}$

31. $-\dfrac{2}{3}\bigg|$ 6   1 -11 -6
           -4 2   6
     _____
              6   -3 -9 0

Answer: $6x^2 - 3x - 9$

33. $\underline{3|}$ 3 -8 8 -3 $P(3) = 30$
           9   3   33
     _____
              3   1   11   30

35. $\underline{2|}$ 6 -13 1   2 $P(2) = 0$
           12 -2 -2
     _____
              6   -1 -1 0

37. $\underline{-3|}$ 3   0 -11 -7 $P(-3) = -55$
           -9 27 -48
     _____
              3   -9 16 -55

39. $\underline{1|}$ 1 -1 -2 1 -2 $P(1) = -3$
           1   0 -2 -1
     _____
              1   0 -2 -1 -3

41. $\underline{-1|}$ 2   0 -6 -5 4 $P(-1) = 5$
           -2 2   4   1
     _____
              2   -2 -4 -1 5

43. $\underline{-2|}$ 1 -1 0   3 -3 -3 $P(-2) = -33$
           -2 6 -12 18 -30
     _____
              1   -3 6 -9 15 -33

45. a) $\underline{2|}$ 9 -10 -16 $9x + 8$
           18   16
     _____
              9   8   0

    b) Base: $9(12) + 8 = 108 + 8 = 116$ in.
        Height: $12 - 2 = 10$ in.
        Area: $116(10) = 1160$ in.^{2}

47. a) $\underline{-4|}$ 2 -7 -38 88
           -8 60 -88
     _____
              2   -15 22   0

        $\underline{2|}$ 2 -15 22
           4 -22
     _____
              2   -11 0

        The height of the room is $2y - 11$.

    b) height $= 2(10) - 11 = 9$ ft.
        length $= 10 + 4 = 14$ ft.
        width $= 10 - 2 = 8$ ft.
        Volume $= 9(14)(8) = 1008$ ft.^{3}

Review Exercises

1. $\{2, 3, 5, 7, 11\}$

2. $m = \dfrac{5-(-5)}{-2-3} = \dfrac{10}{-5} = -2$

3. $2x - 3 < 1$ or $2x - 3 > -5$

     $2x < 4$ $2x > -2$

     $x < 2$ $x > -1$

     a) $\{x \mid x \text{ is a real number}\}$ b) $(-\infty, \infty)$

     c)

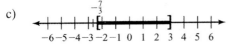

4. $-8 \le 3x - 1 \le 8$

     $-7 \le 3x \le 9$

     $-\dfrac{7}{3} \le x \le 3$

     a) $\left\{x \mid -\dfrac{7}{3} \le x \le 3\right\}$ b) $\left[-\dfrac{7}{3}, 3\right]$

     c)

5. $y > 2$

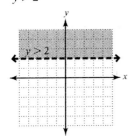

6. $\begin{cases} l = 2w \\ 100 = 2l + 2w \end{cases}$

     $100 = 2(2w) + 2w$ $l = 2\left(16\dfrac{2}{3}\right)$

     $100 = 4w + 2w$

     $100 = 6w$ $= 33\dfrac{1}{3}$ ft.

     $16\dfrac{2}{3}$ ft. $= w$

Chapter 5 Review Exercises

1. $\left(\dfrac{2}{3}\right)^{-3} = \left(\dfrac{3}{2}\right)^{3} = \dfrac{27}{8}$

2. $-4^2 = -(4 \cdot 4) = -16$

3. $2^{-5} = \dfrac{1}{2^5} = \dfrac{1}{32}$

4. $14x^0 = 14 \cdot 1 = 14$

5. $\left(4x^2\right)\left(5xy^3\right) = 20x^{2+1}y^3 = 20x^3 y^3$

6. $\left(-\dfrac{2}{3}m^2 n^4\right)\left(3mn^5\right) = -2m^{2+1}n^{4+5} = -2m^3 n^9$

7. $\dfrac{6x^4}{2x^9} = 3x^{4-9} = 3x^{-5} = \dfrac{3}{x^5}$

8. $-\dfrac{7a^3 bc^9}{3abc^{15}} = -\dfrac{7}{3}a^{3-1}b^{1-1}c^{9-15}$

                $= -\dfrac{7}{3}a^2 b^0 c^{-6}$

                $= -\dfrac{7a^2}{3c^6}$

9. $\left(3m^5 n\right)^2 = 3^2 m^{5 \cdot 2} n^2 = 9m^{10} n^2$

10. $\dfrac{\left(6m^2 n^3 p\right)^{-1}}{\left(2m^6 n^2\right)^3} = \dfrac{1}{\left(6m^2 n^3 p\right)\left(2m^6 n^2\right)^3}$

           $= \dfrac{1}{6m^2 n^3 p \cdot (2)^3 \left(m^6\right)^3 \left(n^2\right)^3}$

           $= \dfrac{1}{6m^2 n^3 p \cdot 8m^{18} n^6}$

           $= \dfrac{1}{48m^{20} n^9 p}$

11. $\left(3j^2 k^5\right)^2 \left(0.1jk^3\right)^3$

     $= 3^2 \left(j^2\right)^2 \left(k^5\right)^2 \cdot (0.1)^3 \left(j\right)^3 \left(k^3\right)^3$

     $= 9j^4 k^{10} \cdot 0.001 j^3 k^9$

     $= 0.009 j^7 k^{19}$

12. $\dfrac{\left(2s^3 t\right)^{-1}}{\left(3st\right)^2 \left(s^4 t\right)^8} = \dfrac{1}{\left(2s^3 t\right)\left(3st\right)^2 \left(s^4 t\right)^8}$

       $= \dfrac{1}{\left(2s^3 t\right)\left(9s^2 t^2\right)\left(s^{32} t^8\right)}$

       $= \dfrac{1}{18s^{37} t^{11}}$

13. $\dfrac{\left(5m^7 n^3\right)^{-2}}{\left(3m\right)^{-1}\left(3m\right)^2} = \dfrac{\left(5m^7 n^3\right)^{-2}}{\left(3m\right)^{-1+2}}$

$= \dfrac{1}{\left(3m\right)^1 \left(5m^7 n^3\right)^2}$

$= \dfrac{1}{3m\left(25m^{14} n^6\right)}$

$= \dfrac{1}{75m^{15} n^6}$

14. $\left(\dfrac{2a}{b^4}\right)^{-3} = \left(\dfrac{b^4}{2a}\right)^3 = \dfrac{b^{12}}{8a^3}$

15. 1.6736×10^{-24}

$= 0.0000000000000000000000016737$

16. $1.65 \times 10^{10} = 16,500,000,000$

17. $0.000000000753 = 7.53 \times 10^{-10}$

18. $300,000,000 = 3 \times 10^8$

19. $\left(5.1 \times 10^4\right)\left(-2 \times 10^6\right) = 5.1 \times (-2) \times \left(10^{4+6}\right)$

$= -10.2 \times 10^{10}$

$= -1.02 \times 10^{11}$

20. $\dfrac{8.12 \times 10^{-8}}{2 \times 10^{-5}} = \dfrac{8.12}{2} \times \dfrac{10^{-8}}{10^{-5}}$

$= 4.06 \times 10^{-8-(-5)}$

$= 4.06 \times 10^{-3}$

21. $E = hf$

$= \left(6.626 \times 10^{-34}\right)\left(6.1 \times 10^{14}\right)$

$= 40.4186 \times 10^{-20}$

$= 4.04186 \times 10^{-19}$

The energy is 4.04186×10^{-19} joules.

22. $-\dfrac{1}{4} + 2c^2 + 8.7c^3 + \dfrac{2}{5}c$

$d = 3$; this polynomial has more than three terms, so it has no special polynomial name.

23. $7m^2 + 1$

$d = 2$; binomial

24. $-8xy^4$

$d = 1 + 4 = 5$; monomial

25. $9h^5 + 4h^3 - h^2 + 1$

$d = 5$; this polynomial has more than three terms, so it has no special polynomial name.

26. $\left(3c^3 + 2c^2 - c - 1\right) + \left(8c^2 + c - 10\right)$

$= 3c^3 + 10c^2 - 11$

27. $\left(y^2 + 3y + 6\right) - \left(-5y^2 + 3y - 8\right)$

$= \left(y^2 + 3y + 6\right) + \left(5y^2 - 3y + 8\right)$

$= 6y^2 + 14$

28. $\left(x^2 y^3 - xy^2 + 2x^2 y - 4xy + 7y^2 - 9\right)$

$+ \left(-3x^2 y^2 + 2xy^2 + 4x^2 y - xy + 6y^2 + 4\right)$

$= x^2 y^3 - 3x^2 y^2 + xy^2 + 6x^2 y - 5xy + 13y^2 - 5$

29. $\left(4hk - 8k^3\right) - \left(5kh + 3k - 2k^3\right)$

$= \left(4hk - 8k^3\right) + \left(-5hk - 3k + 2k^3\right)$

$= -hk - 3k - 6k^3$

30. a) $a + (5a - b) + b + (6a + b) = 12a + b$

b) $12a + b = 12(12) + 9$

$= 144 + 9$

$= 153$

The perimeter is 153 cm.

31. linear

32. quadratic

33. cubic

34. constant

35. $f(x) = x^2 - 3$

x	$f(x)$
-2	1
-1	-2
0	-3
1	-2
2	1

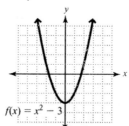

$f(x) = x^2 - 3$

Domain: $\{x \mid -\infty < x < \infty\}$ or $(-\infty, \infty)$

Range: $\{y \mid y \geq -3\}$ or $[-3, \infty)$

36. $g(x) = -\dfrac{1}{3}x + 2$

x	$g(x)$
-3	3
0	2
3	1

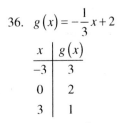

Domain: $\{x \mid -\infty < x < \infty\}$ or $(-\infty, \infty)$

Range: $\{y \mid -\infty < y < \infty\}$ or $(-\infty, \infty)$

37. $h(x) = x^3 - 1$

x	$h(x)$
-2	-9
-1	-2
0	-1
1	0
2	7

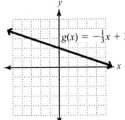

Domain: $\{x \mid -\infty < x < \infty\}$ or $(-\infty, \infty)$

Range: $\{y \mid -\infty < y < \infty\}$ or $(-\infty, \infty)$

38. $w(x) = 7$

x	$w(x)$
-2	7
0	7
2	7

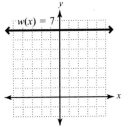

Domain: $\{x \mid -\infty < x < \infty\}$ or $(-\infty, \infty)$

Range: $\{y \mid y = 7\}$ or $[7, 7]$

39. $f(x) = x^3 + 2x + 5; \quad g(x) = -2x^3 + 2$

   a) $(f+g)(x) = (x^3 + 2x + 5) + (-2x^3 + 2)$
   $$= -x^3 + 2x + 7$$

   b) $(f+g)(2) = -2^3 + 2(2) + 7 = -8 + 4 + 7 = 3$

   c) $(f-g)(x) = (x^3 + 2x + 5) - (-2x^3 + 2)$
   $$= (x^3 + 2x + 5) + (2x^3 - 2)$$
   $$= 3x^3 + 2x + 3$$

   d) $(f-g)(-4) = 3(-4)^3 + 2(-4) + 3$
   $$= 3(-64) - 8 + 3$$
   $$= -192 - 8 + 3$$
   $$= -197$$

40. $h(t) = -16(t)^2 + 200$
   $$h(3) = -16(3)^2 + 200$$
   $$= 56$$
   The height after 3 seconds is 56 feet.

41. a) $P(x) = R(x) - C(x)$
   $$P(x) = (0.2x^2 + x + 1000) - (10x + 2000)$$
   $$= (0.2x^2 + x + 1000) + (-10x - 2000)$$
   $$= 0.2x^2 - 9x - 1000$$

   b) $P(200) = 0.2(200)^2 - 9(200) - 1000$
   $$= 5200$$
   The profit is $5200.

42. $3x^2y^3\left(2x^2-3x+1\right)$

$\quad = 3x^2y^3\cdot 2x^2+3x^2y^3\cdot\left(-3x\right)+3x^2y^3\cdot 1$

$\quad = 6x^4y^3-9x^3y^3+3x^2y^3$

43. $-pq\left(2p^2-3pq+3q^2\right)$

$\quad = -pq\cdot 2p^2-pq\cdot\left(-3pq\right)-pq\cdot 3q^2$

$\quad = -2p^3q+3p^2q^2-3pq^3$

44. $(x-3)(2x+8)=2x^2+8x-6x-24$

$\qquad\qquad\qquad\quad = 2x^2+2x-24$

45. $(9w-1)(3w+2)=27w^2+18w-3w-2$

$\qquad\qquad\qquad\quad = 27w^2+15w-2$

46. $(3r-s)(6r+7s)=18r^2+21rs-6rs-7s^2$

$\qquad\qquad\qquad\quad = 18r^2+15rs-7s^2$

47. $\left(x^2-1\right)\left(x^2+2\right)=x^4+2x^2-x^2-2$

$\qquad\qquad\qquad\quad = x^4+x^2-2$

48. $(x-1)\left(2x^3-3x^2+4x+7\right)$

$\quad = 2x^4-3x^3+4x^2+7x-2x^3+3x^2-4x-7$

$\quad = 2x^4-5x^3+7x^2+3x-7$

49. $\left(3t^2-t+1\right)\left(4t^2+t-1\right)$

$\quad = 12t^4+3t^3-3t^2-4t^3-t^2+t+4t^2+t-1$

$\quad = 12t^4-t^3+2t-1$

50. $(3a-5)(3a+5)=\left(3a\right)^2-5^2=9a^2-25$

51. $(2p-1)^2=\left(2p\right)^2-2\left(2p\right)\left(1\right)+1^2=4p^2-4p+1$

52. $(8k+3)^2=\left(8k\right)^2+2\left(8k\right)\left(3\right)+3^2=64k^2+48k+9$

53. $\left(4h^2+7\right)\left(4h^2-7\right)=\left(4h^2\right)^2-7^2=16h^4-49$

54. $5m+2$

55. The width is w. The length is $w + 10$.
    The height is $w + 5$.

$\quad w(w+10)(w+5)=w\left(w^2+15w+50\right)$

$\qquad\qquad\qquad\qquad = w^3+15w^2+50w$

The volume of the tank is $\left(w^3+15w^2+50w\right)$

cubic inches.

56. $f(x)=2x+7;\quad g(x)=x^2+4x-1$

a) $(f\cdot g)(x)=(2x+7)\left(x^2+4x-1\right)$

$\qquad\qquad\quad = 2x^3+8x^2-2x+7x^2+28x-7$

$\qquad\qquad\quad = 2x^3+15x^2+26x-7$

b) $(f\cdot g)(-1)=2(-1)^3+15(-1)^2+26(-1)-7$

$\qquad\qquad\quad = 2(-1)+15(1)-26-7$

$\qquad\qquad\quad = -2+15-26-7$

$\qquad\qquad\quad = -20$

57. $f(x)=4x^2-x$

$\quad f(x-4)=4(x-4)^2-(x-4)$

$\qquad\qquad = 4\left(x^2-8x+16\right)-x+4$

$\qquad\qquad = 4x^2-32x+64-x+4$

$\qquad\qquad = 4x^2-33x+68$

58. $\dfrac{20m^5-5m^4+15m^3-5m^2}{5m^2}$

$\quad = \dfrac{20m^5}{5m^2}-\dfrac{5m^4}{5m^2}+\dfrac{15m^3}{5m^2}-\dfrac{5m^2}{5m^2}$

$\quad = 4m^3-m^2+3m-1$

59. $\dfrac{8x^2-2x+3}{2x}=\dfrac{8x^2}{2x}-\dfrac{2x}{2x}+\dfrac{3}{2x}=4x-1+\dfrac{3}{2x}$

60.
$$\begin{array}{r}
x+3 \\
x+2\overline{\smash)x^2+5x+7} \\
\underline{x^2+2x} \\
3x+7 \\
\underline{3x+6} \\
1
\end{array}$$

Answer: $x+3+\dfrac{1}{x+2}$

61.
$$\begin{array}{r}
x-2 \\
x-2\overline{\smash)x^2-4x+4} \\
\underline{x^2-2x} \\
-2x+4 \\
\underline{-2x+4} \\
0
\end{array}$$

62.
$$2x-1 \overline{\smash{\big)}\, 4x^2+4x-3}$$
quotient: $2x+3$

$\underline{4x^2-2x}$

$\quad 6x-3$

$\quad \underline{6x-3}$

$\qquad 0$

63. $(f/g)(x) = \dfrac{9x^4-7x^2-10x-5}{3x-1}$

$$3x-1 \overline{\smash{\big)}\, 9x^4+0x^3-7x^2-10x-5}$$
quotient: $3x^3+x^2-2x-4$

$\underline{9x^4-3x^3}$

$\quad 3x^3-7x^2$

$\quad \underline{3x^3-x^2}$

$\qquad -6x^2-10x$

$\qquad \underline{-6x^2+2x}$

$\qquad\quad -12x-5$

$\qquad\quad \underline{-12x+4}$

$\qquad\qquad -9$

$(f/g)(x) = 3x^3+x^2-2x-4-\dfrac{9}{3x-1}$

64. a)
$$3y+5 \overline{\smash{\big)}\, 3y^4+11y^3+22y^2+23y+5}$$
quotient: y^3+2y^2+4y+1

$\underline{3y^4+5y^3}$

$\quad 6y^3+22y^2$

$\quad \underline{6y^3+10y^2}$

$\qquad 12y^2+23y$

$\qquad \underline{12y^2+20y}$

$\qquad\quad 3y+5$

$\qquad\quad \underline{3y+5}$

$\qquad\qquad 0$

The current is $\left(y^3+2y^2+4y+1\right)$ amps.

b) $y^3+2y^2+4y+1 = 3^3+2(3)^2+4(3)+1 = 58$

The current is 58 amps if $y=3$.

65.
$$\begin{array}{r|rrr} 1 & 5 & -3 & 2 \\ & & 5 & 2 \\ \hline & 5 & 2 & 4 \end{array}$$

$5x+2+\dfrac{4}{x-1}$

66.
$$\begin{array}{r|rrrr} 3 & 2 & -6 & -5 & 15 \\ & & 6 & 0 & -15 \\ \hline & 2 & 0 & -5 & 0 \end{array}$$

$2x^2-5$

67.
$$\begin{array}{r|rrrr} 2 & 4 & 0 & -2 & 5 \\ & & 8 & 16 & 28 \\ \hline & 4 & 8 & 14 & 33 \end{array}$$

$4x^2+8x+14+\dfrac{33}{x-2}$

68.
$$\begin{array}{r|rrrr} 2 & 1 & 0 & 0 & -8 \\ & & 2 & 4 & 8 \\ \hline & 1 & 2 & 4 & 0 \end{array}$$

x^2+2x+4

69.
$$\begin{array}{r|rrrr} 1 & 5 & -1 & 8 & 3 \\ & & 5 & 4 & 12 \\ \hline & 5 & 4 & 12 & 15 \end{array}$$
$\qquad P(1)=15$

70.
$$\begin{array}{r|rrrr} -2 & 7 & 0 & -2 & 1 \\ & & -14 & 28 & -52 \\ \hline & 7 & -14 & 26 & -51 \end{array}$$
$\qquad P(-2)=-51$

Chapter 5 Practice Test

1. $7.2\times 10^{-3} = 0.0072$

2. $0.00357 = 3.57\times 10^{-3}$

3. $5x^3\left(3x^2y\right) = 15x^{3+2}y = 15x^5y$

4. $\left(3x^4y\right)^{-2} = \dfrac{1}{\left(3x^4y\right)^2} = \dfrac{1}{9x^8y^2}$

5. $\dfrac{8u^7v}{2u^4v^5} = 4u^{7-4}v^{1-5} = 4u^3v^{-4} = \dfrac{4u^3}{v^4}$

6. $\left(-\dfrac{2}{3}t^5u^2v\right)^4 = \left(-\dfrac{2}{3}\right)^4\left(t^5\right)^4\left(u^2\right)^4\left(v\right)^4$

$\qquad = \dfrac{16}{81}t^{20}u^8v^4$

7. $\left(6\times 10^5\right)\left(2.1\times 10^4\right) = \left(6\cdot 2.1\right)\times 10^{5+4}$

$\qquad\qquad = 12.6\times 10^9$

$\qquad\qquad = 1.26\times 10^{10}$

8. $\dfrac{8.4 \times 10^{10}}{4.2 \times 10^3} = \dfrac{8.4}{4.2} \times \dfrac{10^{10}}{10^3}$

$= 2 \times 10^{10-3}$

$= 2 \times 10^7$

9. The graph of a constant function is a horizontal line.

10. The graph of a cubic function resembles an S-shaped curve.

11. Domain: $\{x \mid -\infty < x < \infty\}$ or $(-\infty, \infty)$;

Range: $\{y \mid y \geq -1\}$ or $[-1, \infty)$

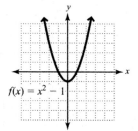

$f(x) = x^2 - 1$

12. $(a^2 - 4ab + b^2) - (a^2 + 4ab + b^2)$

$= (a^2 - 4ab + b^2) + (-a^2 - 4ab - b^2)$

$= -8ab$

13. $(6r^5 - 9r^4 - 2r^2 + 8)$

$\qquad + (5r^5 + 2r^4 + 7r^3 - 8r^2 - r - 9)$

$= 11r^5 - 7r^4 + 7r^3 - 10r^2 - r - 1$

14. $(3x - 4y)(5x + 2y)$

$= 3x \cdot 5 + 3x \cdot 2y + (-4y) \cdot 5x + (-4y) \cdot 2y$

$= 15x^2 + 6xy - 20xy - 8y^2$

$= 15x^2 - 14xy - 8y^2$

15. $4m^3 n^9 (3m^2 - 2mn + 5n^2)$

$= 4m^3 n^9 \cdot 3m^2 + 4m^3 n^9 \cdot (-2mn) + 4m^3 n^9 \cdot 5n^2$

$= 12m^5 n^9 - 8m^4 n^{10} + 20m^3 n^{11}$

16. $(7k - 2j)^2 = (7k)^2 - 2(7k)(2j) + (2j)^2$

$= 49k^2 - 28jk + 4j^2$

17. $(2x - 3)(7x^2 + 3x - 5)$

$= 2x \cdot 7x^2 + 2x \cdot 3x + 2x \cdot (-5) + (-3) \cdot 7x^2$

$\quad + (-3) \cdot 3x + (-3) \cdot (-5)$

$= 14x^3 + 6x^2 - 10x - 21x^2 - 9x + 15$

$= 14x^3 - 15x^2 - 19x + 15$

18. $(4h - 3)(4h + 3) = (4h)^2 - (3)^2$

$= 16h^2 - 9$

19. $(f \cdot g)(x) = (5x - 8)(3x + 2)$

$= 15x^2 + 10x - 24x - 16$

$= 15x^2 - 14x - 16$

20. $(f \cdot g)(x) = (5x - 8)(3x + 2)$

$= 15x^2 + 10x - 24x - 16$

$= 15x^2 - 14x - 16$

$(f \cdot g)(2) = 15(2)^2 - 14(2) - 16$

$= 60 - 28 - 16$

$= 16$

21. $\dfrac{8k^3 - 4k^2 + 2k}{2k} = \dfrac{8k^3}{2k} - \dfrac{4k^2}{2k} + \dfrac{2k}{2k}$

$= 4k^2 - 2k + 1$

22.
$$
\begin{array}{r}
m^2 + 3m + 2 \\
3m+1{\overline{\smash{\big)}\,3m^3 + 10m^2 + 9m + 2}} \\
\underline{3m^3 + m^2} \\
9m^2 + 9m \\
\underline{9m^2 + 3m} \\
6m + 2 \\
\underline{6m + 2} \\
0
\end{array}
$$

23. $\begin{array}{r|rrrr} -3 & 2 & 5 & 0 & -8 \\ & & -6 & 3 & -9 \\ \hline & 2 & -1 & 3 & -17 \end{array}$

$2x^2 - x + 3 - \dfrac{17}{x + 3}$

24. $\begin{array}{r|rrrr} -3 & 1 & 0 & -5 & 4 \\ & & -3 & 9 & -12 \\ \hline & 1 & -3 & 4 & -8 \end{array}$ $P(-3) = -8$

25. a) $P(x) = R(x) - C(x)$

$\qquad = \left(0.1x^2 + x + 5000\right) - \left(15x + 3000\right)$

$\qquad = \left(0.1x^2 + x + 5000\right) + \left(-15x - 3000\right)$

$\qquad = 0.1x^2 - 14x + 2000$

   b) $P(1000) = 0.1(1000)^2 - 14(1000) + 2000$

$\qquad\qquad = 100{,}000 - 14{,}000 + 2000$

$\qquad\qquad = \$88{,}000$

Chapters 1–5 Cumulative Review

1. False; the identity for multiplication is 1.

2. True

3. False; the boundary line of the graph of
    $3x + 4y < 15$ is a dashed line.

4. True

5. identity

6. $3x + 7 = 5; \; 3x + 7 = -5$

7. inconsistent

8. $x - c$

9. Let $x_2 = 6$, $x_1 = -2$, $y_2 = -3$, and $y_1 = 3$.

$\sqrt{(x_2 - x_1)^2 + (y_2 - y_1)^2} = \sqrt{(6 - (-2))^2 + (-3 - 3)^2}$

$\qquad\qquad = \sqrt{(6 + 2)^2 + (-6)^2}$

$\qquad\qquad = \sqrt{(8)^2 + 36}$

$\qquad\qquad = \sqrt{64 + 36}$

$\qquad\qquad = \sqrt{100}$

$\qquad\qquad = 10$

10. The expression is undefined if the denominator
    equals 0.
    $x + 5 = 0$
    $\quad x = -5$
    The expression is undefined if $x = -5$.

11. $\begin{vmatrix} -4 & 5 \\ -3 & 2 \end{vmatrix} = (-4)(2) - (-3)(5)$

$\qquad = -8 - (-15)$

$\qquad = -8 + 15$

$\qquad = 7$

12. $\left(-2a^4 b^{-3}\right)\left(4a^{-2} b^{-1}\right) = -8a^{4+(-2)} b^{-3+(-1)}$

$\qquad\qquad = -8a^2 b^{-4}$

$\qquad\qquad = -\dfrac{8a^2}{b^4}$

13. $\dfrac{8x^{-2} y^3 z^8}{2x^3 y^{-4} z^4} = 4x^{-2-3} y^{3-(-4)} z^{8-4}$

$\qquad\qquad = 4x^{-5} y^7 z^4$

$\qquad\qquad = \dfrac{4y^7 z^4}{x^5}$

14. $\left(-3x^4 y^{-2}\right)^3 = (-3)^3 \left(x^4\right)^3 \left(y^{-2}\right)^3$

$\qquad\qquad = -27x^{4 \cdot 3} y^{-2 \cdot 3}$

$\qquad\qquad = -27x^{12} y^{-6}$

$\qquad\qquad = -\dfrac{27x^{12}}{y^6}$

15. $\left(5x^3 - 7x^2 + 2x - 3\right) + \left(-8x^3 + 7x^2 + 3x - 5\right)$

$\qquad = -3x^3 + 5x - 8$

16. $\left(3x^2 + 6x - 2\right) - \left(-4x^2 - 7x + 1\right)$

$\qquad = \left(3x^2 + 6x - 2\right) + \left(4x^2 + 7x - 1\right)$

$\qquad = 7x^2 + 13x - 3$

17. $(5x + 2y)(3x - 2y) = 15x^2 - 10xy + 6xy - 4y^2$

$\qquad\qquad = 15x^2 - 4xy - 4y^2$

18. $\quad 3x + 10 \overline{\smash{\big)}\, 6x^3 + 14x^2 - 17x + 10}$

$\qquad\qquad\quad \dfrac{2x^2 - 2x + 1}{}$

$\qquad\qquad \underline{6x^3 + 20x^2}$

$\qquad\qquad\qquad -6x^2 - 17x$

$\qquad\qquad\qquad \underline{-6x^2 - 20x}$

$\qquad\qquad\qquad\qquad 3x + 10$

$\qquad\qquad\qquad\qquad \underline{3x + 10}$

$\qquad\qquad\qquad\qquad\qquad 0$

19. $\qquad P = 2\pi r + 2d$

$\qquad P - 2\pi r = 2d$

$\qquad \dfrac{P - 2\pi r}{2} = \dfrac{2d}{2}$

$\qquad \dfrac{P - 2\pi r}{2} = d$

20. $-6 \le -4x - 2 \le 6$

$-4 \le -4x \le 8$

$1 \ge x \ge -2$

The solution set is $\{x \mid -2 \le x \le 1\}$ or $[-2, 1]$.

21. $3|2x - 5| > 9$

$|2x - 5| > 3$

$2x - 5 < -3$ or $2x - 5 > 3$

$\quad 2x < 2 \qquad\qquad 2x > 8$

$\quad\ x < 1 \qquad\qquad\ x > 4$

The solution set is $\{x \mid x < 1 \text{ or } x > 4\}$ or

$(-\infty, 1) \cup (4, \infty)$.

22. $\begin{cases} 5x - 3y = 2 \\ 3x - y = -2 \end{cases}$

Solve the second equation for y.

$3x - y = -2$

$3x + 2 = y$

Substitute for y in the first equation.

$\quad\quad 5x - 3y = 2 \qquad\qquad 3x + 2 = y$

$5x - 3(3x + 2) = 2 \qquad 3(-2) + 2 = y$

$\quad 5x - 9x - 6 = 2 \qquad\quad -6 + 2 = y$

$\quad\quad -4x - 6 = 2 \qquad\qquad\quad -4 = y$

$\quad\quad\quad\ -4x = 8$

$\quad\quad\quad\quad\ x = -2$

The solution is $(-2, -4)$.

23. $y = \dfrac{3}{4}x - 2$

x	y	Ordered Pair
0	-2	$(0, -2)$
$\dfrac{8}{3}$	0	$\left(\dfrac{8}{3}, 0\right)$
4	1	$(4, 1)$

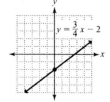

24. $2x - 3y > -6$

$-3y > -2x - 6$

$y < \dfrac{2}{3}x + 2$

Begin by graphing the related equation

$y = \dfrac{2}{3}x + 2$ with a dashed line. Now choose $(0, 0)$

as a test point.

$2x - 3y \ \ > \ \ -6$

$2(0) - 3(0) \ \overset{?}{>} \ -6$

$0 - 0 \ \overset{?}{>} \ -6$

$0 \ \ > \ \ -6$

Because $(0, 0)$ satisfies the inequality, shade the

region which includes $(0, 0)$.

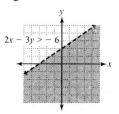

25. $5x - 3y = 8$

$-3y = -5x + 8$

$y = \dfrac{5}{3}x - \dfrac{8}{3}$

The slope of the given equation is $m = \dfrac{5}{3}$, so the

slope of a parallel line is $m = \dfrac{5}{3}$.

$y - y_1 = m(x - x_1)$

$y - 2 = \dfrac{5}{3}(x - (-6))$

$y - 2 = \dfrac{5}{3}(x + 6)$

$y - 2 = \dfrac{5}{3}x + 10$

$y = \dfrac{5}{3}x + 12$

$3y = 5x + 36$

$-5x + 3y = 36$

$5x - 3y = -36$

26. No, it is not a function because there are values in the domain that correspond to two values in the range. (It fails the vertical line test.)

27. The consecutive integers are x, $x + 1$, and $x + 2$.

$$x + (x + 1) + (x + 2) = 162$$
$$3x + 3 = 162$$
$$3x = 159$$
$$x = 53$$

The seat numbers are 53, 53 + 1 = 54, and 53 + 2 = 55.

28. Create a table.

	rate	time	distance
Franz	50	x	$50x$
Hanz	60	$x - 1$	$60(x - 1)$

Translate the information in the table to an equation and solve.

$$50x = 60(x - 1)$$
$$50x = 60x - 60$$
$$-10x = -60$$
$$x = 6$$

Hanz will catch up in $6 + 1 = 5$ hours.

29. Create a table.

Categories	How Many	Value	Total Amount
azaleas	x	6	$6x$
rosebushes	$2x + 7$	10	$10(2x + 7)$

Translate the information in the table to an equation and solve.

$$6x + 10(2x + 7) = 798$$
$$6x + 20x + 70 = 798$$
$$26x + 70 = 798$$
$$26x = 728$$
$$x = 28$$

A total of 28 azaleas and 2(28) + 7 = 63 rosebushes were sold.

30. Let w represent the width of the rectangle and $3w + 2$ represent the length.

$$P = 2w + 2l$$
$$84 = 2w + 2(3w + 2)$$
$$84 = 2w + 6w + 4$$
$$80 = 8w$$
$$10 = w$$

The width is 10 feet and the length is $3(10) + 2 = 32$ feet.

Chapter 6
Factoring

1. $4x^3y^2 = 2^2 \cdot x^3 \cdot y^2$
   $24x^2y = 2^3 \cdot 3 \cdot x^2 \cdot y$
   $\text{GCF} = 2^2 \cdot x^2 \cdot y = 4x^2y$

3. $12u^3v^6 = 2^2 \cdot 3 \cdot u^3 \cdot v^6$
   $28u^8v^8 = 2^2 \cdot 7 \cdot u^8 \cdot v^8$
   $\text{GCF} = 2^2 \cdot u^3 \cdot v^6 = 4u^3v^6$

5. $10a^3b^2c^8 = 2 \cdot 5 \cdot a^3 \cdot b^2 \cdot c^8$
   $14abc^8 = 2 \cdot 7 \cdot a \cdot b \cdot c^8$
   $20a^7b^3c^4 = 2^2 \cdot 5 \cdot a^7 \cdot b^3 \cdot c^4$
   $\text{GCF} = 2 \cdot a \cdot b \cdot c^4 = 2abc^4$

7. $5(a+b) = 5 \cdot (a+b)$
   $7(a+b) = 7 \cdot (a+b)$
   $\text{GCF} = a+b$

9. $15c^4d - 20c^2 = 5c^2\left(\dfrac{15c^4d - 20c^2}{5c^2}\right)$
   $= 5c^2\left(\dfrac{15c^4d}{5c^2} - \dfrac{20c^2}{5c^2}\right)$
   $= 5c^2\left(3c^2d - 4\right)$

11. $x^5 - x^3 + x^2 = x^2\left(\dfrac{x^5 - x^3 + x^2}{x^2}\right)$
    $= x^2\left(\dfrac{x^5}{x^2} - \dfrac{x^3}{x^2} + \dfrac{x^2}{x^2}\right)$
    $= x^2\left(x^3 - x + 1\right)$

13. $25xy - 50xz + 100x^2$
    $= 25x\left(\dfrac{25xy - 50xz + 100x^2}{25x}\right)$
    $= 25x\left(\dfrac{25xy}{25x} - \dfrac{50xz}{25x} + \dfrac{100x^2}{25x}\right)$
    $= 25x\left(y - 2z + 4x\right)$

15. $-14u^2v^2 - 7uv^2 + 7uv$
    $= -7uv\left(\dfrac{-14u^2v^2 - 7uv^2 + 7uv}{-7uv}\right)$
    $= -7uv\left(\dfrac{-14u^2v^2}{-7uv} - \dfrac{7uv^2}{-7uv} + \dfrac{7uv}{-7uv}\right)$
    $= -7uv\left(2uv + v - 1\right)$

17. $9a^7b^3 + 3a^4b^2 - 6a^2b$
    $= 3a^2b\left(\dfrac{9a^7b^3 + 3a^4b^2 - 6a^2b}{3a^2b}\right)$
    $= 3a^2b\left(\dfrac{9a^7b^3}{3a^2b} + \dfrac{3a^4b^2}{3a^2b} - \dfrac{6a^2b}{3a^2b}\right)$
    $= 3a^2b\left(3a^5b^2 + a^2b - 2\right)$

19. $3w^3v^4 + 39w^2v + 18wv^2$
    $= 3wv\left(\dfrac{3w^3v^4 + 39w^2v + 18wv^2}{3wv}\right)$
    $= 3wv\left(\dfrac{3w^3v^4}{3wv} + \dfrac{39w^2v}{3wv} + \dfrac{18wv^2}{3wv}\right)$
    $= 3wv\left(w^2v^3 + 13w + 6v\right)$

21. $18ab^3c - 36a^2b^2c + 24a^5b^2c^8$
    $= 6ab^2c\left(\dfrac{18ab^3c - 36a^2b^2c + 24a^5b^2c^8}{6ab^2c}\right)$
    $= 6ab^2c\left(\dfrac{18ab^3c}{6ab^2c} - \dfrac{36a^2b^2c}{6ab^2c} + \dfrac{24a^5b^2c^8}{6ab^2c}\right)$
    $= 6ab^2c\left(3b - 6a + 4a^4c^7\right)$

23. $-8x^2y + 16xy^2 - 12xy$
    $= -4xy\left(\dfrac{-8x^2y + 16xy^2 - 12xy}{-4xy}\right)$
    $= -4xy\left(\dfrac{-8x^2y}{-4xy} + \dfrac{16xy^2}{-4xy} - \dfrac{12xy}{-4xy}\right)$
    $= -4xy\left(2x - 4y + 3\right)$

25. $m(n-3) + 4(n-3)$
    $= (n-3)\left(\dfrac{m(n-3) + 4(n-3)}{(n-3)}\right)$
    $= (n-3)(m+4)$

27. $6(b+2c)-a(b+2c)$

$= (b+2c)\left(\dfrac{6(b+2c)-a(b+2c)}{(b+2c)}\right)$

$= (b+2c)(6-a)$

29. $ax+ay+bx+by = (ax+ay)+(bx+by)$

$= a(x+y)+b(x+y)$

$= (x+y)(a+b)$

31. $u^3+3u^2+3u+9 = (u^3+3u^2)+(3u+9)$

$= u^2(u+3)+3(u+3)$

$= (u+3)(u^2+3)$

33. $mn+np-3m-3p = (mn+np)+(-3m-3p)$

$= n(m+p)-3(m+p)$

$= (m+p)(n-3)$

35. $cd+d+c+1 = (cd+d)+(c+1)$

$= d(c+1)+1(c+1)$

$= (c+1)(d+1)$

37. $2a^2-a+2a-1 = (2a^2-a)+(2a-1)$

$= a(2a-1)+1(2a-1)$

$= (2a-1)(a+1)$

39. $3ax+6ay+8by+4bx = (3ax+6ay)+(8by+4bx)$

$= 3a(x+2y)+4b(2y+x)$

$= (x+2y)(3a+4b)$

41. $h^2+8h-hk-8k = (h^2+8h)+(-hk-8k)$

$= h(h+8)-k(h+8)$

$= (h+8)(h-k)$

43. $3x^2+3y^2-ax^2-ay^2$

$= (3x^2+3y^2)+(-ax^2-ay^2)$

$= 3(x^2+y^2)-a(x^2+y^2)$

$= (x^2+y^2)(3-a)$

45. $3p^3-6p^2q+2pq^2-4q^3$

$= (3p^3-6p^2q)+(2pq^2-4q^3)$

$= 3p^2(p-2q)+2q^2(p-2q)$

$= (p-2q)(3p^2+2q^2)$

47. $2x^3-8x^2y-3xy^2+12y^3$

$= (2x^3-8x^2y)-(3xy^2-12y^3)$

$= 2x^2(x-4y)-3y^2(x-4y)$

$= (x-4y)(2x^2-3y^2)$

49. $2ab+2bx-2ac-2cx$

$= 2(ab+bx-ac-cx)$

$= 2[(ab+bx)+(-ac-cx)]$

$= 2[b(a+x)-c(a+x)]$

$= 2(b-c)(a+x)$

51. $12xy+4y+30x+10$

$= 2(6xy+2y+15x+5)$

$= 2[(6xy+2y)+(15x+5)]$

$= 2[2y(3x+1)+5(3x+1)]$

$= 2(3x+1)(2y+5)$

53. $3a^2y-12a^2+9ay-36a$

$= 3a(ay-4a+3y-12)$

$= 3a[(ay-4a)+(3y-12)]$

$= 3a[a(y-4)+3(y-4)]$

$= 3a(y-4)(a+3)$

55. $3m^3+6m^2n-10m^2-20mn$

$= m(3m^2+6mn-10m-20n)$

$= m[(3m^2+6mn)-(10m+20n)]$

$= m[3m(m+2n)-10(m+2n)]$

$= m(m+2n)(3m-10)$

57. $15st^2-5t^2-30st+10t$

$= 5t(3st-t-6s+2)$

$= 5t[(3st-t)-(6s-2)]$

$= 5t[t(3s-1)-2(3s-1)]$

$= 5t(3s-1)(t-2)$

59. $5x^3y - 20x^2y + 5xy - 20y$

$\quad = 5y\left(x^3 - 4x^2 + x - 4\right)$

$\quad = 5y\left[\left(x^3 - 4x^2\right) + (x - 4)\right]$

$\quad = 5y\left[x^2(x - 4) + 1(x - 4)\right]$

$\quad = 5y(x - 4)\left(x^2 + 1\right)$

61. Mistake: Did not factor out the GCF, which is $12x^2y$.

Correct: $12x^2y\left(2y^2 + 3x\right)$

63. Mistake: Incorrect power of b in the parentheses.

Correct: $9a^2b\left(b^2c - 2a^2\right)$

65. It is correct.

67. It is not correct.

Correct form: $2x\left(2x^2 + 7x + 4\right)$

69. $10x(x + 4) - 4x(x) = 10x^2 + 40x - 4x^2$

$\quad\quad\quad\quad\quad\quad\quad = 6x^2 + 40x$

$\quad\quad\quad\quad\quad\quad\quad = 2x(3x + 20)$

71. Divide the figure into two rectangles by drawing a horizontal line at the bottom of the side measuring x. Add the area of each rectangle and then subtract the area of the hearth.

Area of top rectangle : $A = 3x \cdot x = 3x^2$

Area of bottom rectangle:

$A = (2x + 4)(6x + 1) = 12x^2 + 26x + 4$

Area of hearth: $1 \cdot 4 = 4$

a) $3x^2 + \left(12x^2 + 26x + 4\right) - 4 = 15x^2 + 26x$

b) $15x^2 + 26x = x(15x + 26)$

c) $A = 3(15 \cdot 3 + 26) = 3 \cdot 71 = 213$ ft.^{2}

73. Add the volume of the cylinder and the sphere. R is the radius of the sphere and $R = 3r$.

a) $\dfrac{4}{3}\pi R^3 + \pi r^2 h = \dfrac{4}{3}\pi(3r)^3 + \pi r^2(18)$

$\quad\quad\quad\quad\quad\quad = \dfrac{4}{3}\pi\left(27r^3\right) + 18\pi r^2$

$\quad\quad\quad\quad\quad\quad = 36\pi r^3 + 18\pi r^2$

b) $18\pi r^2(2r + 1)$

c) $18\pi(5)^2(2 \cdot 5 + 1) = 18\pi(25)(11)$

$\quad\quad\quad\quad\quad\quad\quad \approx 15,550.9$ ft.^{3}

75. a) $p - pr$ $\quad\quad\quad\quad$ b) $p(1 - r)$

c) $54.95(1 - 0.4) = \$32.97$

77. a) $p + prt$ $\quad\quad\quad\quad$ b) $p(1 + rt)$

c) $850(1 + (0.03)(0.5)) = \$862.75$

79. a) $2\pi r^2 + 2\pi rh$ $\quad\quad$ b) $2\pi r(r + h)$

c) $2\pi(15)(15 + 5) \approx 1884.96$ in.^{2}

Review Exercises

1. $-2.45 \times 10^7 = -24,500,000$

2. $0.000092 = 9.2 \times 10^{-5}$

3. $(x + 3)(x + 5) = x^2 + 5x + 3x + 15$

$\quad\quad\quad\quad\quad\quad = x^2 + 8x + 15$

4. $(x - 6)(x - 4) = x^2 - 4x - 6x + 24$

$\quad\quad\quad\quad\quad\quad = x^2 - 10x + 24$

5. $(2x + 7)(3x - 1) = 6x^2 - 2x + 21x - 7$

$\quad\quad\quad\quad\quad\quad = 6x^2 + 19x - 7$

6. $2x(4x - 5)(x + 3) = 2x\left(4x^2 + 12x - 5x - 15\right)$

$\quad\quad\quad\quad\quad\quad = 2x\left(4x^2 + 7x - 15\right)$

$\quad\quad\quad\quad\quad\quad = 8x^3 + 14x^2 - 30x$

Exercise Set 6.2

1. $r^2 + 8r + 7 = (r + 7)(r + 1)$

3. $w^2 - 2w - 3 = (w + 1)(w - 3)$

5. $a^2 + 9a + 18 = (a + 6)(a + 3)$

7. $y^2 - 13y + 36 = (y - 9)(y - 4)$

9. $m^2 + 2n - 8 = (m + 4)(m - 2)$

11. $b^2 - 6b - 40 = (b - 10)(b + 4)$

13. $x^2 - 5x - 18$ is prime.

15. $3st^2 + 24st + 21s = 3s\left(t^2 + 8t + 7\right)$

$\quad\quad\quad\quad\quad\quad = 3s(t + 7)(t + 1)$

17. $5y^3 - 65y^2 + 60y = 5y\left(y^2 - 13y + 12\right)$
    $$= 5y(y-12)(y-1)$$

19. $6au^3 + 6au^2 - 36au = 6au\left(u^2 + u - 6\right)$
    $$= 6au(u+3)(u-2)$$

21. $3x^2y^3 - 12x^2y^2 + 30x^2y = 3x^2y\left(y^2 - 4y + 10\right)$

23. $p^2 + 11pq + 18q^2 = (p+9q)(p+2q)$

25. $u^2 - 13uv + 42v^2 = (u-7v)(u-6v)$

27. $x^2 - 5xy - 14y^2 = (x-7y)(x+2y)$

29. $a^2 - ab - 42b^2 = (a+6b)(a-7b)$

31. $3a^2 + 10a + 7 = (3a+7)(a+1)$

33. $2w^2 - 3w - 2 = (2w+1)(w-2)$

35. $4r^2 - r + 2$ is prime.

37. $4q^2 - 9q + 2 = (4q-1)(q-2)$

39. $6b^2 + 7b - 3 = (3b-1)(2b+3)$

41. $16m^2 + 24m + 9 = (4m+3)(4m+3) = (4m+3)^2$

43. $4x^2 + 5x - 6 = (4x-3)(x+2)$

45. $2w^2 + 15wv + 7v^2 = (2w+v)(w+7v)$

47. $5x^2 - 16xy + 3y^2 = (5x-y)(x-3y)$

49. $16x^2 - 10xy + y^2 = (2x-y)(8x-y)$

51. $6a^2 - 13ab - 10b^2$ is prime.

53. $3t^2 + 19tu - 14u^2 = (3t-2u)(t+7u)$

55. $3m^2 - 10mn - 8n^2 = (3m+2n)(m-4n)$

57. $12a^2 - 17ab + 6b^2 = (3a-2b)(4a-3b)$

59. $22m^3 + 200m^2 + 18m = 2m\left(11m^2 + 100m + 9\right)$
    $$= 2m(11m+1)(m+9)$$

61. $4u^2v + 2uv^2 - 30v^3 = 2v\left(2u^2 + uv - 15v^2\right)$
    $$= 2v(2u-5v)(u+3v)$$

63. $3y^2 + 16y + 5 = 3y^2 + 15y + y + 5$
    $$= \left(3y^2 + 15y\right) + (y+5)$$
    $$= 3y(y+5) + 1(y+5)$$
    $$= (y+5)(3y+1)$$

65. $6c^2 + 11c + 6$ is prime.

67. $3t^2 - 17t + 10 = 3t^2 - 15t - 2t + 10$
    $$= \left(3t^2 - 15t\right) - (2t - 10)$$
    $$= 3t(t-5) - 2(t-5)$$
    $$= (t-5)(3t-2)$$

69. $6x^2 - x - 15 = 6x^2 - 10x + 9x - 15$
    $$= \left(6x^2 - 10x\right) + (9x - 15)$$
    $$= 2x(3x-5) + 3(3x-5)$$
    $$= (3x-5)(2x+3)$$

71. $32x^2y + 24xy - 36y$
    $$= 4y\left(8x^2 + 6x - 9\right)$$
    $$= 4y\left[\left(8x^2 + 12x\right) + (-6x - 9)\right]$$
    $$= 4y\left[4x(2x+3) - 3(2x+3)\right]$$
    $$= 4y(2x+3)(4x-3)$$

73. $15a^2b - 25ab^2 - 10b^3$
    $$= 5b\left(3a^2 - 5ab - 2b^2\right)$$
    $$= 5b\left[\left(3a^2 - 6ab\right) + \left(ab - 2b^2\right)\right]$$
    $$= 5b\left[3a(a-2b) + b(a-2b)\right]$$
    $$= 5b(a-2b)(3a+b)$$

75. $x^4 - x^2 - 2$
    Let $u = x^2$.
    $u^2 - u - 2 = (u+1)(u-2)$
    $$= \left(x^2 + 1\right)\left(x^2 - 2\right)$$

77. $8r^4 + 2r^2 - 3$
    Let $u = r^2$.
    $8u^2 + 2u - 3 = (4u+3)(2u-1)$
    $$= \left(4r^2 + 3\right)\left(2r^2 - 1\right)$$

79. $15x^4 - 11x^2 + 2$

    Let $u = x^2$.

    $15u^2 - 11u + 2 = (5u - 2)(3u - 1)$
    $$= (5x^2 - 2)(3x^2 - 1)$$

81. $y^6 - 16y^3 + 48$

    Let $u = y^3$.

    $u^2 - 16u + 48 = (u - 12)(u - 4)$
    $$= (y^3 - 12)(y^3 - 4)$$

83. $7(x+1)^2 + 8(x+1) + 1$

    Let $u = x + 1$.

    $7u^2 + 8u + 1 = (7u + 1)(u + 1)$
    $$= \left[7(x+1) + 1\right]\left[(x+1) + 1\right]$$
    $$= (7x + 7 + 1)(x + 2)$$
    $$= (7x + 8)(x + 2)$$

85. $3(a+2)^2 - 10(a+2) - 8$

    Let $u = a + 2$.

    $3u^2 - 10u - 8 = (3u + 2)(u - 4)$
    $$= \left[3(a+2) + 2\right]\left[(a+2) - 4\right]$$
    $$= (3a + 6 + 2)(a - 2)$$
    $$= (3a + 8)(a - 2)$$

87. Mistake: The signs are incorrect.

    Correct: $(x + 2)(x - 3)$

89. Mistake: The GCF monomial was not factored out.

    Correct: $4(x + 2)^2$

91. $x^2 + bx + 16$

    To make the trinomial factorable, the value of b must be the sum of factor pairs of 16. Because b must be a natural number, both factors of 16 must be positive. Factor pairs of 16, and their resulting sums, are the following:

    $1 + 16 = 17$

    $2 + 8 = 10$

    $4 + 4 = 8$.

    The natural number values of b that make the trinomial factorable are 8, 10, and 17.

93. $x^2 + bx - 63$

    To make the trinomial factorable, the value of b must be the sum of factor pairs of -63. Because we wish b to be a natural number, we can disregard all negative values of b. Factor pairs of -63, and their resulting sums, are the following:

    $-1 + 63 = 62$

    $1 + (-63) = -62$

    $-3 + 21 = 18$

    $3 + (-21) = -18$

    $-7 + 9 = 2$

    $7 + (-9) = -2$.

    The natural number values of b that make the trinomial factorable are 2, 18, and 62.

95. $x^2 + 9x + c$

    To make the trinomial factorable, the value of c must be the product of pairs of numbers whose sum is 9. Because c must be a natural number, the value of c must be positive, which means both numbers must be positive. Possibilities are the following:

    $1(8) = 8$

    $2(7) = 14$

    $3(6) = 18$

    $4(5) = 20$.

    Possible natural number values of c that make the trinomial factorable are 8, 14, 18, or 20. Answers may vary.

97. $x^2 + x - c$

    To make the trinomial factorable, the value of c must be the product of pairs of numbers whose sum is 1. The value of $-c$ is negative, which means that in the number pairs, one number must be positive and one must be negative. Possibilities are the following:

    $-1(2) = -2$

    $-2(3) = -6$

    $-3(4) = -12$

    $-4(5) = -20$.

    A few possible natural number values of c that make the trinomial factorable are 2, 6, 12, and 20. Answers may vary.

Review Exercises

1. $(2x + 3)(2x + 3) = 4x^2 + 12x + 9$

2. $(4y - 1)(4y - 1) = 16y^2 - 1$

3. $(n-2)(n^2+2n+4)$

$= n^3 + 2n^2 + 4n - 2n^2 - 4n - 8$

$= n^3 - 8$

4.
$$
\begin{array}{r}
6x^2 + x + 4 \\
x-1 \overline{)6x^3 - 5x^2 + 3x - 2} \\
\underline{6x^3 - 6x^2} \\
x^2 + 3x \\
\underline{x^2 - x} \\
4x - 2 \\
\underline{4x - 4} \\
2
\end{array}
$$

Answer: $6x^2 + x + 4 + \dfrac{2}{x-1}$

5. $5(x-6) + 2x \ge 3x - 4$

$5x - 30 + 2x \ge 3x - 4$

$7x - 30 \ge 3x - 4$

$4x \ge 26$

$x \ge \dfrac{13}{2}$

a) $\left\{ x \,\middle|\, x \ge \dfrac{13}{2} \right\}$ b) $\left[\dfrac{13}{2}, \infty \right)$

c) ![number line from -2 to 10 with bracket at 13/2]
$$-2\;-1\;\;0\;\;1\;\;2\;\;3\;\;4\;\;5\;\;6\;\;7\;\;8\;\;9\;\;10$$

6. $-5 < 2x + 7 < 5$

$-12 < 2x < -2$

$-6 < x < -1$

a) $\{ x \,|\, -6 < x < -1 \}$ b) $(-6, -1)$

c) ![number line from -9 to 3 with open interval from -6 to -1]
$$-9\;-8\;-7\;-6\;-5\;-4\;-3\;-2\;-1\;\;0\;\;1\;\;2\;\;3$$

Exercise Set 6.3

1. $x^2 + 10x + 25 = (x+5)^2$

3. $b^2 - 4b + 4 = (b-2)^2$

5. $25u^2 - 30u + 9 = (5u-3)^2$

7. $n^2 + 24mn + 144m^2 = (n+12m)^2$

9. $9q^2 - 30pq + 25p^2 = (3q-5p)^2$

11. $4p^2 - 28pq + 49q^2 = (2p-7q)^2$

13. $a^2 - y^2 = (a+y)(a-y)$

15. $25x^2 - 4 = (5x+2)(5x-2)$

17. $100u^2 - 49v^2 = (10u+7v)(10u-7v)$

19. $9x^2 - 36 = 9(x^2-4) = 9(x+2)(x-2)$

21. $x^4 - 16 = (x^2+4)(x^2-4) = (x^2+4)(x+2)(x-2)$

23. $9(x-3)^2 - 16$

$= \left[3(x-3)+4 \right]\left[3(x-3)-4 \right]$

$= [3x-9+4][3x-9-4]$

$= (3x-5)(3x-13)$

25. $16z^2 - 9(x-y)^2$

$= \left[4z + 3(x-y) \right]\left[4z - 3(x-y) \right]$

$= (4z+3x-3y)(4z-3x+3y)$

27. $m^3 - 27 = (m-3)(m^2+3m+9)$

29. $125x^3 + 27 = (5x+3)(25x^2-15x+9)$

31. $27x^3 - 8 = (3x-2)(9x^2+6x+4)$

33. $u^3 + 125v^3 = (u+5v)(u^2-5uv+25v^2)$

35. $27m^3 - 125m^6n^3$

$= m^3(27 - 125m^3n^3)$

$= m^3(3-5mn)(9+15mn+25m^2n^2)$

37. $(u+3)^3 + 8$

$= \left[(u+3)+2 \right]\left[(u+3)^2 - 2(u+3) + 4 \right]$

$= (u+5)\left[(u^2+6u+9) - 2u - 6 + 4 \right]$

$= (u+5)(u^2+4u+7)$

39. $27 - (a+b)^3$

$= \left[3-(a+b) \right]\left[9 + 3(a+b) + (a+b)^2 \right]$

$= (3-a-b)(9+3a+3b+a^2+2ab+b^2)$

41. $64x^3 + 27(y+z)^3$

   $= \left[4x + 3(y+z)\right]$

   $\quad \cdot \left[16x^2 - 12x(y+z) + 9(y+z)^2\right]$

   $= (4x + 3y + 3z)$

   $\quad \cdot \left(16x^2 - 12xy - 12xz + 9\left(y^2 + 2yz + z^2\right)\right)$

   $= (4x + 3y + 3z)$

   $\quad \cdot \left(16x^2 - 12xy - 12xz + 9y^2 + 18yz + 9z^2\right)$

43. $64d^3 - 27(x+y)^3$

   $= \left[4d - 3(x+y)\right]$

   $\quad \cdot \left[16d^2 + 12d(x+y) + 9(x+y)^2\right]$

   $= (4d - 3x - 3y)$

   $\quad \cdot \left[16d^2 + 12dx + 12dy + 9\left(x^2 + 2xy + y^2\right)\right]$

   $= (4d - 3x - 3y)$

   $\quad \cdot \left(16d^2 + 12dx + 12dy + 9x^2 + 18xy + 9y^2\right)$

45. $16x^2 + bx + 25$ will be a perfect square trinomial if
   $b = 2\sqrt{a}\sqrt{c} = 2\sqrt{16}\sqrt{25} = 2(4)(5) = 40.$

47. $4x^2 - bx + 81$ will be a perfect square trinomial if
   $b = 2\sqrt{a}\sqrt{c} = 2\sqrt{4}\sqrt{81} = 2(2)(9) = 36.$

49. $\qquad b = 2\sqrt{a}\sqrt{c}$

   $\dfrac{b}{2\sqrt{a}} = \sqrt{c}$

   $\left(\dfrac{b}{2\sqrt{a}}\right)^2 = c$

   $x^2 + 8x + c$ will be a perfect square trinomial if
   $c = \left(\dfrac{b}{2\sqrt{a}}\right)^2 = \left(\dfrac{8}{2\sqrt{1}}\right)^2 = \left(\dfrac{8}{2}\right)^2 = 4^2 = 16.$

51. $\qquad b = 2\sqrt{a}\sqrt{c}$

   $\dfrac{b}{2\sqrt{a}} = \sqrt{c}$

   $\left(\dfrac{b}{2\sqrt{a}}\right)^2 = c$

   $9x^2 - 24x + c$ will be a perfect square trinomial if
   $c = \left(\dfrac{b}{2\sqrt{a}}\right)^2 = \left(\dfrac{24}{2\sqrt{9}}\right)^2 = \left(\dfrac{24}{2\cdot 3}\right)^2$

   $\quad = \left(\dfrac{24}{6}\right)^2 = 4^2 = 16.$

53. $12m^3n^2 + 20m^2n^4 = 4m^2n^2\left(3m + 5n^2\right)$

55. $x^2 + 8x + 15 = (x+5)(x+3)$

57. $\left(x^2 - 16\right) = (x+4)(x-4)$

59. $12c^2 - 8c - 15 = (2c-3)(6c+5)$

61. $ax - xy - ay + y^2 = (ax - xy) - \left(ay - y^2\right)$

   $\qquad = x(a-y) - y(a-y)$

   $\qquad = (a-y)(x-y)$

63. $4x^2 - 28x + 49 = (2x-7)^2$

65. $25x^2 + 36y^2$ is prime.

67. $12a^3b^2c + 3a^2b^2c^2 + 9abc^3$

   $= 3abc\left(4a^2b + abc + 3c^2\right)$

69. $b^3 + 125 = (b+5)\left(b^2 - 5b + 25\right)$

71. $15x^2 - 12x - 2$ is prime.

73. $6b^2 + b - 2 = (3b+2)(2b-1)$

75. $ab - 36ab^3 = ab\left(1 - 36b^2\right) = ab(1+6b)(1-6b)$

77. $16c^2 - 24c + 9 = (4c-3)^2$

79. $18x^3 - 3x^2 - 36x = 3x\left(6x^2 - x - 12\right)$

   $\qquad = 3x(2x-3)(3x+4)$

81. $x^4 - 16 = \left(x^2+4\right)\left(x^2-4\right)$

   $\qquad = \left(x^2+4\right)(x+2)(x-2)$

83. $20a^2 - 9a + 20$ is prime.

85. $12u^2 - 84uv + 147v^2 = 3\left(4u^2 - 28uv + 49v^2\right)$
$$= 3\left(2u - 7v\right)^2$$

87. $36a^4b^3 - 39a^3b^4 - 12a^2b^5$
$$= 3a^2b^3\left(12a^2 - 13ab - 4b^2\right)$$
$$= 3a^2b^3\left(4a + b\right)\left(3a - 4b\right)$$

89. $6x^5y + 5x^4y^2 - 12x^3y^3 = x^3y\left(6x^2 + 5xy - 12y^2\right)$

91. $x^2y + 3x - xy^2 - 3y = \left(x^2y + 3x\right) - \left(xy^2 + 3y\right)$
$$= x\left(xy + 3\right) - y\left(xy + 3\right)$$
$$= \left(xy + 3\right)\left(x - y\right)$$

93. $27 + \left(x - 1\right)^3$
$$= \left[3 + \left(x - 1\right)\right]\left[9 - 3\left(x - 1\right) + \left(x - 1\right)^2\right]$$
$$= \left(3 + x - 1\right)\left(9 - 3x + 3 + x^2 - 2x + 1\right)$$
$$= \left(x + 2\right)\left(x^2 - 5x + 13\right)$$

95. $5m^5 + 10m^3n^2 + 5mn^4 = 5m\left(m^4 + 2m^2n^2 + n^4\right)$
$$= 5m\left(m^2 + n^2\right)^2$$

97. $\left(n - 3\right)^2 + 6\left(n - 3\right) + 9$
Let $u = n - 3$.
$$u^2 + 6u + 9 = \left(u + 3\right)^2$$
$$= \left(n - 3 + 3\right)^2$$
$$= n^2$$

99. $64t^6 - t^3u^3 = t^3\left(64t^3 - u^3\right)$
$$= t^3\left(4t - u\right)\left(16t^2 + 4tu + u^2\right)$$

101. $2x^3 + 3x^2 - 2xy^2 - 3y^2$
$$= \left(2x^3 + 3x^2\right) + \left(-2xy^2 - 3y^2\right)$$
$$= x^2\left(2x + 3\right) - y^2\left(2x + 3\right)$$
$$= \left(2x + 3\right)\left(x^2 - y^2\right)$$
$$= \left(2x + 3\right)\left(x + y\right)\left(x - y\right)$$

103. $36y^{2\cdot} - \left(x^2 - 8x + 16\right)$
$$= 36y^2 - \left(x - 4\right)^2$$
$$= \left(6y + \left(x - 4\right)\right)\left(6y - \left(x - 4\right)\right)$$
$$= \left(6y + x - 4\right)\left(6y - x + 4\right)$$

105. $\left(x \cdot 12x\right) - \left(8 \cdot 5\right) = 12x^2 - 40 = 4\left(3x^2 - 10\right)$

107. $\left(6x \cdot 3x \cdot x\right) - \left(9 \cdot 3 \cdot 1\right) = 18x^3 - 27 = 9\left(2x^3 - 3\right)$

109. $6x^2 - 11x + 3 = \left(3x - 1\right)\left(2x - 3\right)$
Length: $3x - 1$; Width: $2x - 3$

111. $15x^3 + 55x^2 + 30x = 5x\left(3x^2 + 11x + 6\right)$
$$= 5x\left(3x + 2\right)\left(x + 3\right)$$
Length: $3x + 2$; Width: $x + 3$; Height: $5x$

Review Exercises

1. $9x - 5\left(3x + 2\right) = 12x - \left(4x - 3\right)$
$$9x - 15x - 10 = 12x - 4x + 3$$
$$-6x - 10 = 8x + 3$$
$$-13 = 14x$$
$$-\frac{13}{14} = x$$

2. $2x - 5 = -7$ or $2x - 5 = 7$
$$2x = -2 \qquad\qquad 2x = 12$$
$$x = -1 \qquad\qquad x = 6$$

3. $\begin{cases} x + y + z = 6000 & \text{Eqtn. 1} \\ x + y - 2z = 0 & \text{Eqtn. 2} \\ 0.04x + 0.06y + 0.08z = 370 & \text{Eqtn. 3} \end{cases}$
Multiply equation 2 by -1 and add to equation 1.
Multiply equation 3 by -25 and add to equation 1.
$$\begin{array}{ll} -x - y + 2z = 0 & -x - 1.5y - 2z = -9250 \\ \underline{x + y + z = 6000} & \underline{x + y + z = 6000} \\ \qquad\quad 3z = 6000 & \quad -0.5y - z = -3250 \end{array}$$
$$\begin{cases} x + y + z = 6000 \\ 3z = 6000 \\ -0.5y - z = -3250 \end{cases}$$
$$\begin{array}{ll} 3z = 6000 & -0.5y - 2000 = -3250 \\ z = 2000 & -0.5y = -1250 \\ & y = 2500 \end{array}$$
$$x + 2500 + 2000 = 6000$$
$$x = 1500$$
$1500 at 4%, $2500 at 6%, $2000 at 8%

4. $f(x) = -\dfrac{3}{4}x - 2$

Think of the function as the equation

$y = -\dfrac{3}{4}x - 2$, and use the fact that the slope is

$m = -\dfrac{3}{4}$ and the y-intercept is -2.

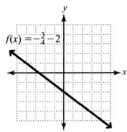

$f(x) = -\frac{3}{4} - 2$

5. $y > 2x - 3$

Begin by graphing the related equation $y = 2x - 3$ with a dashed line. Now choose $(0, 0)$ as a test point.

$\begin{array}{ccl} y & > & 2x - 3 \\ & \overset{?}{} & \\ 0 & > & 2(0) - 3 \\ & \overset{?}{} & \\ 0 & > & 0 - 3 \\ \\ 0 & > & -3 \end{array}$

Because $(0, 0)$ satisfies the inequality, shade the region which includes $(0, 0)$.

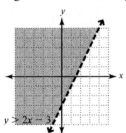

$y > 2x - 3$

6.   Find the slope: $m = \dfrac{1 - (-2)}{-4 - 3} = -\dfrac{3}{7}$.

$y - 1 = -\dfrac{3}{7}\left(x - (-4)\right)$

$y - 1 = -\dfrac{3}{7}(x + 4)$

$y - 1 = -\dfrac{3}{7}x - \dfrac{12}{7}$

$y = -\dfrac{3}{7}x - \dfrac{5}{7}$

$3x + 7y = -5$

Exercise Set 6.4

1. $x(x + 4) = 0$

$x = 0$ or $x + 4 = 0$

$x = -4$

3. $(x - 3)(x + 2) = 0$

$x - 3 = 0$ or $x + 2 = 0$

$x = 3$ $x = -2$

5. $(3b - 2)(2b + 5) = 0$

$3b - 2 = 0$ or $2b + 5 = 0$

$3b = 2$ $2b = -5$

$b = \dfrac{2}{3}$ $b = -\dfrac{5}{2}$

7. $x(x - 3)(2x + 5) = 0$

$x = 0$ or $x - 3 = 0$ or $2x + 5 = 0$

$x = 3$ $x = -\dfrac{5}{2}$

9. $(x + 2)(x - 5)(x + 7) = 0$

$x + 2 = 0$ or $x - 5 = 0$ or $x + 7 = 0$

$x = -2$ $x = 5$ $x = -7$

11. $(2x + 3)(3x - 1)(4x + 7) = 0$

$2x + 3 = 0$ or $3x - 1 = 0$ or $4x + 7 = 0$

$x = -\dfrac{3}{2}$ $x = \dfrac{1}{3}$ $x = -\dfrac{7}{4}$

13. $d^2 - 4d = 0$

$d(d - 4) = 0$

$d = 0$ or $d - 4 = 0$

$d = 4$

15. $x^2 - 9 = 0$

$(x + 3)(x - 3) = 0$

$x + 3 = 0$ or $x - 3 = 0$

$x = -3$ $x = 3$

17. $m^2 + 14m + 45 = 0$

$(m + 5)(m + 9) = 0$

$m + 5 = 0$ or $m + 9 = 0$

$m = -5$ $m = -9$

19. $2x^2 + 3x - 2 = 0$

$(x+2)(2x-1) = 0$

$x+2 = 0 \quad \text{or} \quad 2x-1 = 0$

$x = -2 \qquad\qquad x = \dfrac{1}{2}$

21. $6a^2 + 13a - 5 = 0$

$(3a-1)(2a+5) = 0$

$3a-1 = 0 \quad \text{or} \quad 2a+5 = 0$

$a = \dfrac{1}{3} \qquad\qquad a = -\dfrac{5}{2}$

23. $x^2 + 6x + 9 = 0$

$(x+3)^2 = 0$

$x+3 = 0$

$x = -3$

25. $6y^2 = 3y$

$6y^2 - 3y = 0$

$3y(2y-1) = 0$

$3y = 0 \quad \text{or} \quad 2y-1 = 0$

$y = 0 \qquad\qquad y = \dfrac{1}{2}$

27. $x^2 = 25$

$x^2 - 25 = 0$

$(x+5)(x-5) = 0$

$x+5 = 0 \quad \text{or} \quad x-5 = 0$

$x = -5 \qquad\qquad x = 5$

29. $n^2 + 6n = 27$

$n^2 + 6n - 27 = 0$

$(n+9)(n-3) = 0$

$n+9 = 0 \quad \text{or} \quad n-3 = 0$

$n = -9 \qquad\qquad n = 3$

31. $p^2 = 3p - 2$

$p^2 - 3p + 2 = 0$

$(p-2)(p-1) = 0$

$p-2 = 0 \quad \text{or} \quad p-1 = 0$

$p = 2 \qquad\qquad p = 1$

33. $2v^2 + 5 = -7v$

$2v^2 + 7v + 5 = 0$

$(2v+5)(v+1) = 0$

$2v+5 = 0 \quad \text{or} \quad v+1 = 0$

$v = -\dfrac{5}{2} \qquad\qquad v = -1$

35. $a(a-5) = 14$

$a^2 - 5a = 14$

$a^2 - 5a - 14 = 0$

$(a-7)(a+2) = 0$

$a-7 = 0 \quad \text{or} \quad a+2 = 0$

$a = 7 \qquad\qquad a = -2$

37. $4x(x+7) = -49$

$4x^2 + 28x = -49$

$4x^2 + 28x + 49 = 0$

$(2x+7)^2 = 0$

$2x+7 = 0$

$2x = -7$

$x = -\dfrac{7}{2}$

39. $(x+1)(x+2) = 20$

$x^2 + 3x + 2 = 20$

$x^2 + 3x - 18 = 0$

$(x+6)(x-3) = 0$

$x+6 = 0 \quad \text{or} \quad x-3 = 0$

$x = -6 \qquad\qquad x = 3$

41. $x^3 + x^2 - 6x = 0$

$x(x^2 + x - 6) = 0$

$x(x+3)(x-2) = 0$

$x = 0 \quad \text{or} \quad x+3 = 0 \quad \text{or} \quad x-2 = 0$

$x = -3 \qquad\qquad x = 2$

43. $12x(x+1)+3=5(2x+1)+2$

$12x^2+12x+3=10x+5+2$

$12x^2+2x-4=0$

$2(6x^2+x-2)=0$

$2(3x+2)(2x-1)=0$

$3x+2=0$ or $2x-1=0$

$x=-\dfrac{2}{3}$ $x=\dfrac{1}{2}$

45. $(2v+1)(v-2)=-v(v+2)$

$2v^2-3v-2=-v^2-2v$

$3v^2-v-2=0$

$(3v+2)(v-1)=0$

$3v+2=0$ or $v-1=0$

$v=-\dfrac{2}{3}$ $v=1$

47. $9x(x^2+x)=3x(x-2)+5x$

$9x^3+9x^2=3x^2-6x+5x$

$9x^3+6x^2+x=0$

$x(9x^2+6x+1)=0$

$x(3x+1)^2=0$

$x=0$ or $3x+1=0$

$3x=-1$

$x=-\dfrac{1}{3}$

49. $x^3+2x^2-15=13x-(4x-3)$

$x^3+2x^2-15=13x-4x+3$

$x^3+2x^2-9x-18=0$

$(x^3+2x^2)-(9x+18)=0$

$x^2(x+2)-9(x+2)=0$

$(x+2)(x^2-9)=0$

$(x+2)(x+3)(x-3)=0$

$x+2=0$ or $x+3=0$ or $x-3=0$

$x=-2$ $x=-3$ $x=3$

51. $x=-3$ $x=2$

$x+3=0$ $x-2=0$

$(x+3)(x-2)=0$

$x^2+x-6=0$

53. $x=-\dfrac{2}{3}$ $x=4$

$3x=-2$ $x-4=0$

$3x+2=0$

$(3x+2)(x-4)=0$

$3x^2-10x-8=0$

55. $x=-1$ $x=0$ $x=3$

$x+1=0$ $x-0=0$ $x-3=0$

$(x+1)(x-0)(x-3)=0$

$(x^2+x)(x-3)=0$

$x^3-3x^2+x^2-3x=0$

$x^3-2x^2-3x=0$

57. b   59. a   61. c

63. $f(x)=x^2-25$

$x^2-25=0$

$(x+5)(x-5)=0$

$x+5=0$ or $x-5=0$

$x=-5$ $x=5$

$(-5,0)(5,0)$

x	$f(x)$
0	-25
-3	-16
3	-16

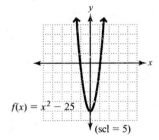

$f(x)=x^2-25$

(scl = 5)

65. $f(x) = x^2 - 6x + 5$

$x^2 - 6x + 5 = 0$

$(x - 5)(x - 1) = 0$

$x - 5 = 0$ or $x - 1 = 0$

$x = 5$ $x = 1$

$(5, 0), (1, 0)$

x	$f(x)$
2	-3
3	-4
4	-3

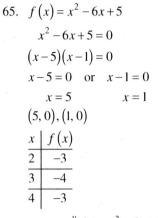

67. $f(x) = x(x + 4)(x - 1)$

$x(x + 4)(x - 1) = 0$

$x = 0$ or $x + 4 = 0$ or $x - 1 = 0$

$x = -4$ $x = 1$

$(0, 0), (-4, 0), (1, 0)$

x	$f(x)$
-3	12
-2	12
-1	6
2	12

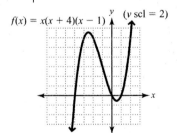

69. $f(x) = x^3 - x^2 - 6x$

$x^3 - x^2 - 6x = 0$

$x(x^2 - x - 6) = 0$

$x(x - 3)(x + 2) = 0$

$x = 0$ or $x - 3 = 0$ or $x + 2 = 0$

$x = 3$ $x = -2$

$(0, 0), (3, 0), (-2, 0)$

x	$f(x)$
-1	4
1	-6
2	-8

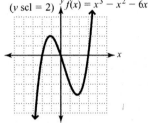

71. $(3, 0), \ (6, 0)$

73. $\left(-\dfrac{5}{3}, 0\right), \left(\dfrac{1}{2}, 0\right)$

75. $(-1, 0), (0, 0), (2, 0)$

77. $-5, 8$

79. $-0.25, 2.5$

81. $-4, 0, 3$

83. $x^2 + (x + 2)^2 = (x + 4)^2$

$x^2 + x^2 + 4x + 4 = x^2 + 8x + 16$

$x^2 - 4x - 12 = 0$

$(x - 6)(x + 2) = 0$

$x - 6 = 0$ or $x + 2 = 0$

$x = 6$ $x = -2$

             x cannot be negative

The lengths of the sides are 6, $6 + 2 = 8$, and $6 + 4 = 10$.

85. $$n^2 + (3n+3)^2 = (4n-3)^2$$
$$n^2 + 9n^2 + 18n + 9 = 16n^2 - 24n + 9$$
$$-6n^2 + 42n = 0$$
$$-6n(n-7) = 0$$
$$-6n = 0 \quad \text{or} \quad n-7 = 0$$
$$n = 0 \qquad\qquad n = 7$$

n cannot be zero
The lengths of the sides are 7, $3(7)+3 = 24$, and $4(7)-3 = 25$.

87. Let x represent the number.
$$x^2 + 55 = 16x$$
$$x^2 - 16x + 55 = 0$$
$$(x-5)(x-11) = 0$$
$$x-5 = 0 \quad \text{or} \quad x-11 = 0$$
$$x = 5 \qquad\qquad x = 11$$

There are two numbers: 5 and 11.

89. Let x represent the first consecutive positive odd integer. Then let $x+2$ and $x+4$ represent the other two consecutive positive odd integers.
$$x^2 + (x+2)^2 + (x+4)^2 = 155$$
$$x^2 + x^2 + 4x + 4 + x^2 + 8x + 16 = 155$$
$$3x^2 + 12x + 20 = 155$$
$$3x^2 + 12x - 135 = 0$$
$$3(x^2 + 4x - 45) = 0$$
$$3(x+9)(x-5) = 0$$
$$x+9 = 0 \quad \text{or} \quad x-5 = 0$$
$$x = -9 \qquad\qquad x = 5$$

x cannot be negative
The numbers are 5, $5+2 = 7$, and $5+4 = 9$.

91. Let x be the width and $x+4$ be the length.
$$x(x+4) = 320$$
$$x^2 + 4x - 320 = 0$$
$$(x+20)(x-16) = 0$$
$$x+20 = 0 \quad \text{or} \quad x-16 = 0$$
$$x = -20 \qquad\qquad x = 16$$

x cannot be negative
The dimensions are 16m by 20m.

93. The area is $(22 \text{ ft}) \cdot (28 \text{ ft}) = 616 \text{ ft}^2$.
$$\frac{22}{7}r^2 = 616$$
$$r^2 = 196$$
$$r^2 - 196 = 0$$
$$(r+14)(r-14) = 0$$
$$r+14 = 0 \quad \text{or} \quad r-14 = 0$$
$$r = -14 \qquad\qquad r = 14$$

r cannot be negative
The radius is 14 ft.

95. $$\frac{1}{2}h(10+h) = 85.5$$
$$5h + 0.5h^2 = 85.5$$
$$0.5h^2 + 5h - 85.5 = 0$$
$$0.5(h^2 + 10h - 171) = 0$$
$$0.5(h+19)(h-9) = 0$$
$$h+19 = 0 \quad \text{or} \quad h-9 = 0$$
$$h = -19 \qquad\qquad h = 9$$

h cannot be negative
The height is 9 in.

97. $$h^2 + 10^2 = (h+2)^2$$
$$h^2 + 100 = h^2 + 4h + 4$$
$$96 = 4h$$
$$24 = h$$

Height: 24 ft. Wire length: 26 ft.

99. $$9 = -16t^2 + 4t + 29$$
$$0 = -16t^2 + 4t + 20$$
$$0 = -4(4t^2 - t - 5)$$
$$0 = -4(4t-5)(t+1)$$
$$4t-5 = 0 \quad \text{or} \quad t+1 = 0$$
$$t = 1.25 \qquad\qquad t = \cancel{-1}$$

Time would be 1.25 sec.

101. $4840 = 4000\left(1 + \dfrac{r}{1}\right)^{2 \cdot 1}$

$1.21 = (1+r)^2$

$1.21 = 1 + 2r + r^2$

$0 = r^2 + 2r - 0.21$

$0 = 100r^2 + 200r - 21$

$0 = (10r + 21)(10r - 1)$

$10r + 21 = 0 \qquad \text{or} \qquad 10r - 1 = 0$

$r = -2.1 \qquad\qquad r = 0.1$

r is not negative

$r = 0.1 = 10\%$

Review Exercises

1. $0.0004203 = 4.203 \times 10^{-4}$

2. $\begin{array}{r} x^2 - 3x + 2 \\ x+4\overline{\smash{\big)}\,x^3 + x^2 - 10x + 11} \\ \underline{x^3 + 4x^2} \\ -3x^2 - 10x \\ \underline{-3x^2 - 12x} \\ 2x + 11 \\ \underline{2x + 8} \\ 3 \end{array}$

   Answer: $x^2 - 3x + 2 + \dfrac{3}{x+4}$

3. $2x - 5 \le -1 \quad \text{or} \quad 2x - 5 \ge 1$

   $2x \le 4 \qquad\qquad 2x \ge 6$

   $x \le 2 \qquad\qquad x \ge 3$

   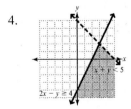

4.

5. No

6. $(f+g)(x) = (5x^3 + 6x - 19) + (3x^2 - 6x - 8)$

   $= 5x^3 + 3x^2 - 27$

Chapter 6 Review Exercises

1. $16x^4y^2 = 2^4 \cdot x^4 \cdot y^2$

   $24x^3y^9 = 2^3 \cdot 3 \cdot x^3 \cdot y^9$

   $\text{GCF} = 2^3 \cdot x^3 \cdot y^2 = 8x^3y^2$

2. $20mn^3 = 2^2 \cdot 5 \cdot m \cdot n^3$

   $15m^4n^2 = 3 \cdot 5 \cdot m^4 \cdot n^2$

   $\text{GCF} = 5 \cdot m \cdot n^2 = 5mn^2$

3. $2u^4v^3 = 2 \cdot u^4 \cdot v^3$

   $3xy^9 = 3 \cdot x \cdot y^9$

   $\text{GCF} = 1$

4. $4(x+1) = 2^2 \cdot (x+1)$

   $6(x+1) = 2 \cdot 3 \cdot (x+1)$

   $\text{GCF} = 2 \cdot (x+1) = 2(x+1)$

5. $u^6 - u^4 - u^2 = u^2\left(\dfrac{u^6 - u^4 - u^2}{u^2}\right)$

   $= u^2\left(u^4 - u^2 - 1\right)$

6. $13d^2 - 26de = 13d\left(\dfrac{13d^2 - 26de}{13d}\right)$

   $= 13d\left(d - 2e\right)$

7. $4h^2k^6 - 2h^5k = 2h^2k\left(\dfrac{4h^2k^6 - 2h^5k}{2h^2k}\right)$

   $= 2h^2k\left(2k^5 - h^3\right)$

8. $12cd - 4c^2d^2 + 10c^4d^4$

   $= 2cd\left(\dfrac{12cd - 4c^2d^2 + 10c^4d^4}{2cd}\right)$

   $= 2cd\left(6 - 2cd + 5c^3d^3\right)$

9. $16p^6q^3 - 12p^8q^5 + 13p^7q^2$

   $= p^6q^2\left(\dfrac{16p^6q^3 - 12p^8q^5 + 13p^7q^2}{p^6q^2}\right)$

   $= p^6q^2\left(16q - 12p^2q^3 + 13p\right)$

10. $9w^6v^8 + 6wv^6 - 12w^3v^3$

$= 3wv^3 \left(\dfrac{9w^6v^8 + 6wv^6 - 12w^3v^3}{3wv^3} \right)$

$= 3wv^3 \left(3w^5v^5 + 2v^3 - 4w^2 \right)$

11. $17(w+3) - m(w+3)$

$= (w+3) \left(\dfrac{17(w+3) - m(w+3)}{(w+3)} \right)$

$= (w+3)(17-m)$

12. $2y(x+5) + (x+5)$

$= (x+5) \left(\dfrac{2y(x+5) + (x+5)}{(x+5)} \right)$

$= (x+5)(2y+1)$

13. $3m + mn + 6 + 2n = (3m + mn) + (6 + 2n)$

$= m(3+n) + 2(3+n)$

$= (3+n)(m+2)$

$= (n+3)(m+2)$

14. $a^3 + 3a^2 + 3a + 9 = (a^3 + 3a^2) + (3a + 9)$

$= a^2(a+3) + 3(a+3)$

$= (a+3)(a^2+3)$

15. $2x + 2y - ax - ay = (2x + 2y) + (-ax - ay)$

$= 2(x+y) - a(x+y)$

$= (x+y)(2-a)$

16. $bc^2 + bd^2 - 5c^2 - 5d^2$

$= (bc^2 + bd^2) + (-5c^2 - 5d^2)$

$= b(c^2 + d^2) - 5(c^2 + d^2)$

$= (c^2 + d^2)(b-5)$

17. $x^2y - x^2s - ry + rs = (x^2y - x^2s) + (-ry + rs)$

$= x^2(y-s) - r(y-s)$

$= (y-s)(x^2-r)$

18. $4k^2 - k + 4k - 1 = (4k^2 - k) + (4k - 1)$

$= k(4k-1) + 1(4k-1)$

$= (4k-1)(k+1)$

19. $8uv^2 - 4v^2 + 20uv - 10v$

$= 2v(4uv - 2v + 10u - 5)$

$= 2v\left[(4uv - 2v) + (10u - 5) \right]$

$= 2v\left[2v(2u-1) + 5(2u-1) \right]$

$= 2v(2u-1)(2v+5)$

20. $5c^2d^2 - 5d^3 + 20c^2d - 20d^2$

$= 5d\left(c^2d - d^2 + 4c^2 - 4d \right)$

$= 5d\left[(c^2d - d^2) + (4c^2 - 4d) \right]$

$= 5d\left[d(c^2 - d) + 4(c^2 - d) \right]$

$= 5d(c^2 - d)(d+4)$

21. $a^2 - 10a + 9 = (a-9)(a-1)$

22. $m^2 + 20m + 51 = (m+3)(m+17)$

23. $y^2 + 2y - 48 = (y+8)(y-6)$

24. $x^2 - 7x - 30 = (x-10)(x+3)$

25. $3x^2 - x - 14 = (3x-7)(x+2)$

26. $16h^2 + 10h + 1 = (8h+1)(2h+1)$

27. $4u^2 - 2u + 3$ is prime.

28. $6t^2 + t - 15 = (3t+5)(2t-3)$

29. $s^2 - 11st + 10t^2 = (s-10t)(s-t)$

30. $6u^2 + 13uv + 6v^2 = (3u+2v)(2u+3v)$

31. $5m^2 - 16mn + 3n^2 = (5m-n)(m-3n)$

32. $4x^2 + 5xy - 8y^2$ is prime.

33. $b^4 - 7b^3 - 18b^2 = b^2\left(b^2 - 7b - 18 \right)$

$= b^2(b-9)(b+2)$

34. $2x^3 - 16x^2 - 40x = 2x\left(x^2 - 8x - 20 \right)$

$= 2x(x-10)(x+2)$

35. $2u^2 + 9u + 10 = (2u^2 + 5u) + (4u + 10)$

$= u(2u+5) + 2(2u+5)$

$= (2u+5)(u+2)$

36. $3m^2 - 8m + 4 = (3m^2 - 2m) + (-6m + 4)$
$$= m(3m - 2) - 2(3m - 2)$$
$$= (3m - 2)(m - 2)$$

37. $10u^2 + 7u - 3 = (10u^2 + 10u) + (-3u - 3)$
$$= 10u(u + 1) - 3(u + 1)$$
$$= (u + 1)(10u - 3)$$

38. $6y^2 - 13y - 5 = 6y^2 + 2y - 15y - 5$
$$= (6y^2 + 2y) + (-15y - 5)$$
$$= 2y(3y + 1) - 5(3y + 1)$$
$$= (3y + 1)(2y - 5)$$

39. $x^4 + 5x^2 + 4$
Let $u = x^2$.
$$u^2 + 5u + 4 = (u + 4)(u + 1)$$
$$= (x^2 + 4)(x^2 + 1)$$

40. $3c^4 + 13c^2 + 4$
Let $u = c^2$.
$$3u^2 + 13u + 4 = (3u + 1)(u + 4)$$
$$= (3c^2 + 1)(c^2 + 4)$$

41. $2h^6 + 9h^3 + 4$
Let $u = h^3$.
$$2u^2 + 9u + 4 = (2u + 1)(u + 4)$$
$$= (2h^3 + 1)(h^3 + 4)$$

42. $3(k + 1)^2 - 2(k + 1) - 5$
Let $u = k + 1$.
$$3u^2 - 2u - 5 = (3u - 5)(u + 1)$$
$$= [3(k + 1) - 5][(k + 1) + 1]$$
$$= (3k + 3 - 5)(k + 2)$$
$$= (3k - 2)(k + 2)$$

43. $x^2 + 6x + 9 = (x + 3)^2$

44. $y^2 + 12y + 36 = (y + 6)^2$

45. $m^2 - 4m + 4 = (m - 2)^2$

46. $w^2 - 14w + 49 = (w - 7)^2$

47. $9d^2 + 30d + 25 = (3d + 5)^2$

48. $4c^2 - 28c + 49 = (2c - 7)^2$

49. $h^2 - 9 = (h + 3)(h - 3)$

50. $p^2 - 64 = (p + 8)(p - 8)$

51. $9d^2 - 4 = (3d + 2)(3d - 2)$

52. $81k^2 - 100 = (9k + 10)(9k - 10)$

53. $2w^2 - 50 = 2(w^2 - 25)$
$$= 2(w + 5)(w - 5)$$

54. $4q^2 - 36 = 4(q^2 - 9)$
$$= 4(q + 3)(q - 3)$$

55. $25y^2 - 9z^2 = (5y + 3z)(5y - 3z)$

56. $x^4 - 81 = (x^2 + 9)(x^2 - 9)$
$$= (x^2 + 9)(x + 3)(x - 3)$$

57. $c^3 - 27 = (c - 3)(c^2 + 3c + 9)$

58. $m^3 + 64 = (m + 4)(m^2 - 4m + 16)$

59. $27b^3 + 8a^3 = (3b + 2a)(9b^2 - 6ab + 4a^2)$

60. $64d^3 - 8c^3 = 8(8d^3 - c^3)$
$$= 8(2d - c)(4d^2 + 2cd + c^2)$$

61. $9d^2 - 6d + 1 = (3d - 1)^2$

62. $3m^2 - 3n^2 = 3(m^2 - n^2) = 3(m + n)(m - n)$

63. $2a^2 - 5a - 12 = (2a + 3)(a - 4)$

64. $x^2 + 9$ is prime.

65. $3p^4 + 3p^3 - 90p^2 = 3p^2(p^2 + p - 30)$
$$= 3p^2(p + 6)(p - 5)$$

66. $x^4 - 16 = (x^2 + 4)(x^2 - 4)$
$$= (x^2 + 4)(x + 2)(x - 2)$$

67. $15a^3b^2c^7 + 3a^2b^4c^2 + 5a^9bc^3$
    $= a^2bc^2\left(15abc^5 + 3b^3 + 5a^7c\right)$

68. $w^3 - (2+y)^3$
    $= \left[w - (2+y)\right]\left[w^2 + w(2+y) + (2+y)^2\right]$
    $= (w - 2 - y)\left[w^2 + 2w + wy + 4 + 4y + y^2\right]$

69. $(x+4)(x-1) = 0$
    $x+4 = 0$ or $x-1 = 0$
        $x = -4$ $x = 1$

70. $x^2 - 64 = 0$
    $(x+8)(x-8) = 0$
    $x+8 = 0$ or $x-8 = 0$
        $x = -8$ $x = 8$

71. $w^2 + 2w - 3 = 0$
    $(w+3)(w-1) = 0$
    $w+3 = 0$ or $w-1 = 0$
        $w = -3$ $w = 1$

72. $6y^2 = 7y - 1$
       $6y^2 - 7y + 1 = 0$
    $(6y-1)(y-1) = 0$
    $6y-1 = 0$ or $y-1 = 0$
         $y = \dfrac{1}{6}$ $y = 1$

73. $4x(x+7) + 9 = -40$
    $4x^2 + 28x + 49 = 0$
        $(2x+7)^2 = 0$
          $2x+7 = 0$
               $x = -\dfrac{7}{2}$

74. $b^3 - 2b^2 + 20 = 2b^2 - (16 - 9b)$
           $b^3 - 2b^2 + 20 = 2b^2 - 16 + 9b$
        $b^3 - 4b^2 - 9b + 36 = 0$
    $\left(b^3 - 4b^2\right) + (-9b + 36) = 0$
        $b^2(b-4) - 9(b-4) = 0$
          $(b-4)\left(b^2 - 9\right) = 0$
         $(b-4)(b+3)(b-3) = 0$
    $b-4 = 0$ or $b+3 = 0$ or $b-3 = 0$
       $b = 4$ $b = -3$ $b = 3$

75. $f(x) = x^2 - 4$
         $x^2 - 4 = 0$
    $(x+2)(x-2) = 0$
    $x+2 = 0$ or $x-2 = 0$
       $x = -2$ $x = 2$
    $(-2, 0), (2, 0)$

x	$f(x)$
0	-4
-1	-3
1	-3

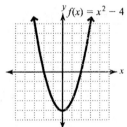

76. $f(x) = x(x+3)(x-2)$

$x(x+3)(x-2) = 0$

$x = 0$ or $x+3 = 0$ or $x-2 = 0$

$x = -3$ $x = 2$

$(0,0), (-3,0), (2,0)$

x	$f(x)$
-2	8
-1	6
1	-4

$f(x) = x(x - 2)(x + 3)$

(y scl = 2)

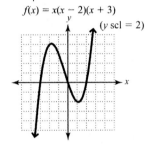

77. Let x be the first natural number and let $x + 10$ be the second natural number.

$x(x+10) = 56$

$x^2 + 10x - 56 = 0$

$(x+14)(x-4) = 0$

$x+14 = 0$ or $x-4 = 0$

$x = -14$ $x = 4$

x cannot be negative

The two numbers are 4 and $4 + 10 = 14$.

78. Let x be the first consecutive positive integer and let $x+1$ be the second consecutive positive integer.

$x(x+1) = 110$

$x^2 + x - 110 = 0$

$(x+11)(x-10) = 0$

$x+11 = 0$ or $x-10 = 0$

$x = -11$ $x = 10$

x cannot be negative

The two numbers are 10 and $10 + 1 = 11$

79. Let x be the first consecutive positive even integer and let $x + 2$ be the second consecutive positive even integer.

$x(x+2) = 288$

$x^2 + 2x - 288 = 0$

$(x+18)(x-16) = 0$

$x+18 = 0$ or $x-16 = 0$

$x = -18$ $x = 16$

x cannot be negative

The two numbers are 16 and $16 + 2 = 18.$.

80. Let x be the width of the room and let $x+3$ be the length.

$x(x+3) = 88$

$x^2 + 3x - 88 = 0$

$(x+11)(x-8) = 0$

$x+11 = 0$ or $x-8 = 0$

$x = -11$ $x = 8$

x cannot be negative

The dimensions of the room are 8 ft. by $8+3 = 11$ ft.

81. $\dfrac{1}{2}h(3h+36) = 672$

$1.5h^2 + 18h - 672 = 0$

$15h^2 + 180h - 6720 = 0$

$15(h^2 + 12h - 448) = 0$

$15(h+28)(h-16) = 0$

$h+28 = 0$ or $h-16 = 0$

$h = -28$ $h = 16$

h cannot be negative

The height is 16 ft. and the base is 48 ft.

82. $x^2 + (x+7)^2 = (2x+3)^2$

$x^2 + x^2 + 14x + 49 = 4x^2 + 12x + 9$

$2x^2 - 2x - 40 = 0$

$2(x^2 - x - 20) = 0$

$2(x-5)(x+4) = 0$

$x-5 = 0$ or $x+4 = 0$

$x = 5$ $x = -4$

x cannot be negative

The sides of the triangle are 5, 12, and 13.

83. The figure shows a rectangle with a base of 35 ft. and a height of 5 ft. with a triangle on top of it. The triangle has a height of $h-5$.

$$35^2 + (h-5)^2 = (2h+3)^2$$

$$1225 + h^2 - 10h + 25 = 4h^2 + 12h + 9$$

$$3h^2 + 22h - 1241 = 0$$

$$(h-17)(3h+73) = 0$$

$$h - 17 = 0 \quad \text{or} \quad 3h + 73 = 0$$

$$h = 17 \qquad\qquad h = -\frac{73}{3}$$

h cannot be negative

The height is 17 ft. The distance from the ear is $2h + 3 = 2(17) + 3 = 37$ ft.

84.
$$6 = -16t^2 + 12t + 60$$

$$16t^2 - 12t - 54 = 0$$

$$2(8t^2 - 6t - 27) = 0$$

$$2(4t - 9)(2t + 3) = 0$$

$$4t - 9 = 0 \quad \text{or} \quad 2t + 3 = 0$$

$$t = \frac{9}{4} \qquad\qquad t = -\frac{3}{2}$$

$$t = 2\frac{1}{4} \text{ sec.} \qquad t \text{ cannot be negative}$$

Chapter 6 Practice Test

1. $14m^7 n^2 = 2 \cdot 7 \cdot m^7 \cdot n^2$

    $21m^3 n^9 = 3 \cdot 7 \cdot m^3 \cdot n^9$

    $\text{GCF} = 7 \cdot m^3 \cdot n^2 = 7m^3 n^2$

2. $3m + 6m^3 - 9m^6 = 3m\left(\dfrac{3m + 6m^3 - 9m^6}{3m}\right)$

    $\qquad = 3m\left(\dfrac{3m}{3m} + \dfrac{6m^3}{3m} - \dfrac{9m^6}{3m}\right)$

    $\qquad = 3m\left(1 + 2m^2 - 3m^5\right)$

3. $3m + 3n - m - n = 2m + 2n$

    $\qquad = 2(m + n)$

4. $9n^2 - 16 = (3n)^2 - (4)^2$

    $\qquad = (3n + 4)(3n - 4)$

5. $8x^3 - 27 = (2x)^3 - (3)^3$

    $\qquad = (2x - 3)\left[(2x)^2 + (2x)(3) + (3)^2\right]$

    $\qquad = (2x - 3)(4x^2 + 6x + 9)$

6. $y^2 + 14y + 49 = (y + 7)^2$

7. $q^2 - 2q - 48 = (q - 8)(q + 6)$

8. $3ab^2 - 30ab + 24a = 3a(b^2 - 10b + 8)$

9. $5 - 125t^2 = 5(1 - 25t^2)$

    $\qquad = 5(1 + 5t)(1 - 5t)$

10. $6d^2 + d - 2 = (3d + 2)(2d - 1)$

11. $8c^3 + 8d^3 = 8(c^3 + d^3)$

    $\qquad = 8(c + d)(c^2 - cd + d^2)$

12. $w^3 + 2w^2 + 3w + 6 = (w^3 + 2w^2) + (3w + 6)$

    $\qquad = w^2(w + 2) + 3(w + 2)$

    $\qquad = (w + 2)(w^2 + 3)$

13. $s^4 - 81 = (s^2 + 9)(s^2 - 9)$

    $\qquad = (s^2 + 9)(s + 3)(s - 3)$

14. $5p^2 - 7pq - 12q^2 = (5p - 12q)(p + q)$

15. $8x^2 - 14x + 5 = 0$

    $(2x - 1)(4x - 5) = 0$

    $2x - 1 = 0 \quad \text{or} \quad 4x - 5 = 0$

    $2x = 1 \qquad\qquad 4x = 5$

    $x = \dfrac{1}{2} \qquad\qquad x = \dfrac{5}{4}$

16. $x(x + 3) - 6 = 12$

    $x^2 + 3x - 18 = 0$

    $(x + 6)(x - 3) = 0$

    $x + 6 = 0 \quad \text{or} \quad x - 3 = 0$

    $x = -6 \qquad\qquad x = 3$

17. $2x^3 + x^2 = 15x$

$2x^3 + x^2 - 15x = 0$

$x(2x^2 + x - 15) = 0$

$x(2x-5)(x+3) = 0$

$x = 0$ or $2x - 5 = 0$ or $x + 3 = 0$

$\qquad\qquad 2x = 5 \qquad\qquad x = -3$

$\qquad\qquad x = \dfrac{5}{2}$

18. $x^2 + (x+7)^2 = (2x+1)^2$

$x^2 + x^2 + 14x + 49 = 4x^2 + 4x + 1$

$2x^2 + 14x + 49 = 4x^2 + 4x + 1$

$-2x^2 + 10x + 48 = 0$

$-2(x^2 - 5x - 24) = 0$

$-2(x-8)(x+3) = 0$

$x - 8 = 0$ or $x + 3 = 0$

$x = 8 \qquad\qquad x = -3$

Because the length of the side cannot be negative, choose $x = 8$ as the length of the shortest side. The other sides are $x + 7 = 8 + 7 = 15$ and $2x + 1 = 2(8) + 1 = 17$.

19. Let $v_0 = 4$ and $h_0 = 56$. Set the formula equal to 0 and solve for t to find the time it takes for the rock to hit the ground.

$0 = -16t^2 + 4t + 56$

$0 = -4(4t^2 - t - 14)$

$0 = -4(4t+7)(t-2)$

$4t + 7 = 0$ or $t - 2 = 0$

$4t = -7 \qquad\qquad t = 2$

$t = -\dfrac{7}{4}$

Disregard the negative solution because time cannot be negative in this example. It takes the rock 2 seconds to hit the ground.

20. To find the x-intercepts, set the function equal to 0 and solve for x.

$x^2 - 9 = 0$

$(x+3)(x-3) = 0$

$x + 3 = 0$ or $x - 3 = 0$

$x = -3 \qquad\qquad x = 3$

The x-intercepts are $(-3, 0)$ and $(3, 0)$.

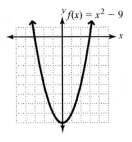

Chapters 1–6 Cumulative Review

1.   True

2.   False; this is an example of a quadratic function.

3.   True

4.   False; $a^2 + b^2$ is prime and cannot be factored further.

5.   the same

6.   coefficients, add

7. $(a+b)(a^2 - ab + b^2)$

8. $\dfrac{4(3^2 - 10) - 4}{3 - \sqrt{25 - 16}} = \dfrac{4(9 - 10) - 4}{3 - \sqrt{9}}$

$= \dfrac{4(-1) - 4}{3 - 3}$

$= \dfrac{-4 - 4}{0}$ is undefined

9. $(2x^3 + 5x^2 - 19x + 14) - (17x^2 + 2x - 6)$

$= (2x^3 + 5x^2 - 19x + 14) + (-17x^2 - 2x + 6)$

$= 2x^3 - 12x^2 - 21x + 20$

10. $(2x+1)(4x^2 - 3x + 2)$

$= 2x(4x^2) + 2x(-3x) + 2x(2) + 1(4x^2)$

$\qquad + 1(-3x) + 1(2)$

$= 8x^3 - 6x^2 + 4x + 4x^2 - 3x + 2$

$= 8x^3 - 2x^2 + x + 2$

11. $\left(\dfrac{2}{x}\right)^{-3} = \left(\dfrac{x}{2}\right)^3 = \dfrac{x^3}{2^3} = \dfrac{x^3}{8}$

12. $4m - 16m^2 = 4m(1 - 4m)$

13. $8x^2 + 29x - 12 = (8x - 3)(x + 4)$

14. $49a^2 + 84a + 36 = (7a + 6)^2$

15. $64p^3 + 8 = 8(8p^3 + 1) = 8(2p + 1)(4p^2 - 2p + 1)$

16. $3x - 6y + ax - 2ay = (3x - 6y) + (ax - 2ay)$
$$= 3(x - 2y) + a(x - 2y)$$
$$= (x - 2y)(3 + a)$$

17. $9x - \left[14 + 2(x - 3)\right] = 8x - 2(3x - 1)$
$$9x - \left[14 + 2x - 6\right] = 8x - 6x + 2$$
$$9x - \left[8 + 2x\right] = 2x + 2$$
$$9x - 8 - 2x = 2x + 2$$
$$5x = 10$$
$$x = 2$$

18. $|2x - 3| - 9 = 16$
$$|2x - 3| = 25$$
$$2x - 3 = -25 \quad \text{or} \quad 2x - 3 = 25$$
$$2x = -22 \qquad\qquad 2x = 28$$
$$x = -11 \qquad\qquad x = 14$$

19. $5x^2 + 26x = 24$
$$5x^2 + 26x - 24 = 0$$
$$(5x - 4)(x + 6) = 0$$
$$5x - 4 = 0 \quad \text{or} \quad x + 6 = 0$$
$$x = \frac{4}{5} \qquad\qquad x = -6$$

20. $\dfrac{3}{4}x + \dfrac{2}{3} \le \dfrac{5}{6}x + \dfrac{1}{2}$
$$12\left(\frac{3}{4}x + \frac{2}{3}\right) \le 12\left(\frac{5}{6}x + \frac{1}{2}\right)$$
$$9x + 8 \le 10x + 6$$
$$-x \le -2$$
$$x \ge 2$$
a) $\{x \mid x \ge 2\}$ b) $[2, \infty)$

c)

21. $2|x - 3| - 2 < 10$
$$2|x - 3| < 12$$
$$|x - 3| < 6$$
$$-6 < x - 3 < 6$$
$$-3 < x < 9$$
a) $\{x \mid -3 < x < 9\}$ b) $(-3, 9)$

c)

22. $\begin{cases} 5x - 2y = -1 \\ 3x + 2y = -7 \end{cases}$

$\begin{aligned} 5x - 2y &= -1 \\ \underline{3x + 2y} &= \underline{-7} \\ 8x &= -8 \\ x &= -1 \end{aligned}$ $\begin{aligned} 5x - 2y &= -1 \\ 5(-1) - 2y &= -1 \\ -5 - 2y &= -1 \\ -2y &= 4 \\ y &= -2 \end{aligned}$

Solution: $(-1, -2)$

23. $\begin{cases} x - 2y + z = 8 \\ x + y + z = 5 \\ 2x + y - 2z = -5 \end{cases}$

Add equation 1 and equation 2 multiplied by -1.
$$\begin{aligned} x - 2y + z &= 8 \\ \underline{-x - y - z} &= \underline{-5} \\ -3y &= 3 \\ y &= -1 \end{aligned}$$

Substitute $y = -1$ into equations 1 and 3.
$$\begin{aligned} x - 2(-1) + z &= 8 \\ x + 2 + z &= 8 \\ x + z &= 6 \end{aligned} \qquad \begin{aligned} 2x - 1 - 2z &= -5 \\ 2x - 2z &= -4 \end{aligned}$$

Solve the new system.
$$\begin{cases} x + z = 6 \\ 2x - 2z = -4 \end{cases}$$
$$\begin{aligned} -2x - 2z &= -12 \\ \underline{2x - 2z} &= \underline{-4} \\ -4z &= -16 \\ z &= 4 \end{aligned} \qquad \begin{aligned} x + z &= 6 \\ x + 4 &= 6 \\ x &= 2 \end{aligned}$$

Solution: $(2, -1, 4)$

24. $\begin{vmatrix} 3 & -2 \\ x & 4 \end{vmatrix} = 16$
$$3(4) - x(-2) = 16$$
$$12 + 2x = 16$$
$$2x = 4$$
$$x = 2$$

25. $\begin{cases} x + y < 6 \\ y \geq 2x \end{cases}$

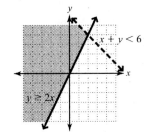

26. $f(x) = x^2 + x - 6$

$x^2 + x - 6 = 0$

$(x+3)(x-2) = 0$

$x + 3 = 0$　or　$x - 2 = 0$

　　$x = -3$　　　$x = 2$

The x-intercepts are $(-3, 0)$ and $(2, 0)$.

$f(0) = (0)^2 + (0) - 6 = -6$

The y-intercept is $(0, -6)$.

x	$f(x)$
-2	-4
-1	-6
1	-4

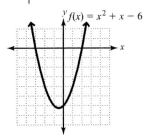

27. a)　$50 \leq 10n < 80$　　　　b)　$n \geq 8$

　　　　$5 \leq n < 8$

28. $\begin{cases} a = 2 + 2t \\ a + t = 77 \end{cases}$　　　　$a + t = 77$

　　　　　　　　　　$(2 + 2t) + t = 77$

　　　　　　　　　　　$2 + 3t = 77$

　　　　　　　　　　　　$3t = 75$

　　　　　　　　　　　　　$t = 25$

Teens drive on average 25 minutes per day. Adults drive on average $2(25) + 2 = 52$ minutes per day.

29. Since the angles are complementary, we know that their sum is 90 degrees.

$\begin{cases} x + y = 90 \\ x = 2y - 6 \end{cases}$

$(2y - 6) + y = 90$

　　　$3y = 96$

　　　　$y = 32$

The angles are $32°$ and $2(32) - 6 = 58°$

30.　$60 = \dfrac{1}{2}(a + 3)(a + a + 5)$

$120 = (a + 3)(2a + 5)$

$120 = 2a^2 + 11a + 15$

$0 = 2a^2 + 11a - 105$

$0 = (2a + 21)(a - 5)$

$2a + 21 = 0$　　　　or　$a - 5 = 0$

　$a = -\dfrac{21}{2}$　　　　　　$a = 5$

a cannot be negative

$a = 5$ in., $a + 3 = 8$ in., $a + 5 = 10$ in.

Chapter 7
Rational Expressions and Equations

Exercise Set 7.1

1. $\dfrac{-28m^3n^5}{16m^7n^6}$

$= \dfrac{-1\cdot 2 \cdot 2 \cdot 7\cdot m\cdot m\cdot m\cdot n\cdot n\cdot n\cdot n\cdot n}{2\cdot 2\cdot 2\cdot 2\cdot m\cdot m\cdot m\cdot m\cdot m\cdot m\cdot m\cdot m\cdot n}$

$= -\dfrac{7}{4m^4n}$

3. $\dfrac{-32x^4y^3z^2}{24x^2y^2z^2}$

$= \dfrac{-2\cdot 2\cdot 2\cdot 2\cdot 2\cdot k\cdot k\cdot x\cdot x\cdot y\cdot y\cdot y\cdot z\cdot z}{2\cdot 2\cdot 2\cdot 3\cdot k\cdot k\cdot y\cdot y\cdot z\cdot z}$

$= -\dfrac{4x^2y}{3}$

5. $\dfrac{2x+12}{3x+18} = \dfrac{2(x+6)}{3(x+6)} = \dfrac{2}{3}$

7. $\dfrac{8x^2+12x}{20x^2+30x} = \dfrac{4x(2x+3)}{10x(2x+3)} = \dfrac{4}{10} = \dfrac{2}{5}$

9. $\dfrac{-3x-15}{4x+20} = \dfrac{-3(x+5)}{4(x+5)} = -\dfrac{3}{4}$

11. $\dfrac{a^2+3a-10}{a^2+2a-15} = \dfrac{(a+5)(a-2)}{(a+5)(a-3)} = \dfrac{a-2}{a-3}$

13. $\dfrac{x^2-6x}{x^2-8x+12} = \dfrac{x(x-6)}{(x-6)(x-2)} = \dfrac{x}{x-2}$

15. $\dfrac{4x^2-9y^2}{6x^2-xy-15y^2} = \dfrac{(2x+3y)(2x-3y)}{(3x-5y)(2x+3y)}$

$= \dfrac{2x-3y}{3x-5y}$

17. $\dfrac{4a^2-4ab-3b^2}{6a^2-ab-2b^2} = \dfrac{(2a+b)(2a-3b)}{(2a+b)(3a-2b)}$

$= \dfrac{2a-3b}{3a-2b}$

19. $\dfrac{x^3-8}{3x^2-2x-8} = \dfrac{(x-2)(x^2+2x+4)}{(3x+4)(x-2)}$

$= \dfrac{x^2+2x+4}{3x+4}$

21. $\dfrac{x^3+27}{3x^2-27} = \dfrac{(x+3)(x^2-3x+9)}{3(x+3)(x-3)}$

$= \dfrac{x^2-3x+9}{3(x-3)}$

23. $\dfrac{xy-4x+3y-12}{xy-6x+3y-18} = \dfrac{x(y-4)+3(y-4)}{x(y-6)+3(y-6)}$

$= \dfrac{(y-4)(x+3)}{(y-6)(x+3)}$

$= \dfrac{y-4}{y-6}$

25. $\dfrac{6a^3+4a^2+9a+6}{6a^2+13a+6} = \dfrac{2a^2(3a+2)+3(3a+2)}{(2a+3)(3a+2)}$

$= \dfrac{(3a+2)(2a^2+3)}{(2a+3)(3a+2)}$

$= \dfrac{2a^2+3}{2a+3}$

27. $\dfrac{2b-a}{a-2b} = \dfrac{-1(a-2b)}{(a-2b)} = -1$

29. $\dfrac{x^2+4x-21}{6-2x} = \dfrac{(x+7)(x-3)}{-2(x-3)} = -\dfrac{x+7}{2}$

31. $\dfrac{8x^2y^4}{12x^3y^2} \cdot \dfrac{4x^3y^3}{6x^4y^3}$

$= \dfrac{\cancel{2} \cdot \cancel{2} \cdot 2 \cdot \cancel{x} \cdot \cancel{x} \cdot \cancel{y} \cdot \cancel{y} \cdot y \cdot y}{\cancel{2} \cdot \cancel{2} \cdot 3 \cdot \cancel{x} \cdot \cancel{x} \cdot x \cdot \cancel{y} \cdot \cancel{y}}$

$\quad \cdot \dfrac{\cancel{2} \cdot 2 \cdot \cancel{x} \cdot \cancel{x} \cdot \cancel{x} \cdot \cancel{y} \cdot \cancel{y} \cdot \cancel{y}}{\cancel{2} \cdot 3 \cdot \cancel{x} \cdot \cancel{x} \cdot \cancel{x} \cdot x \cdot \cancel{y} \cdot \cancel{y} \cdot \cancel{y}}$

$= \dfrac{4y^2}{9x^2}$

33. $-\dfrac{24a^2bc^3}{16ab^3c^2} \cdot \dfrac{32a^3b^2c^3}{18a^4b^3c^3}$

$= -\dfrac{\cancel{2} \cdot \cancel{2} \cdot \cancel{2} \cdot \cancel{3} \cdot \cancel{a} \cdot \cancel{a} \cdot \cancel{b} \cdot \cancel{c} \cdot \cancel{c} \cdot c}{\cancel{2} \cdot \cancel{2} \cdot \cancel{2} \cdot \cancel{2} \cdot \cancel{a} \cdot \cancel{b} \cdot b \cdot b \cdot \cancel{c} \cdot \cancel{c}}$

$\quad \cdot \dfrac{\cancel{2} \cdot \cancel{2} \cdot 2 \cdot 2 \cdot 2 \cdot \cancel{a} \cdot \cancel{a} \cdot \cancel{a} \cdot \cancel{b} \cdot \cancel{b} \cdot \cancel{c} \cdot \cancel{c} \cdot \cancel{c}}{\cancel{2} \cdot \cancel{3} \cdot 3 \cdot \cancel{a} \cdot \cancel{a} \cdot \cancel{a} \cdot \cancel{a} \cdot \cancel{b} \cdot \cancel{b} \cdot b \cdot \cancel{c} \cdot \cancel{c} \cdot \cancel{c}}$

$= -\dfrac{8c}{3b^3}$

35. $\dfrac{3x^2y^4}{15x+10y} \cdot \dfrac{24x+16y}{12xy^2}$

$= \dfrac{\cancel{3} \cdot \cancel{x} \cdot x \cdot \cancel{y} \cdot \cancel{y} \cdot y \cdot y}{5(3x+2y)} \cdot \dfrac{\cancel{2} \cdot \cancel{2} \cdot 2 \cdot \cancel{(3x+2y)}}{\cancel{2} \cdot \cancel{2} \cdot \cancel{3} \cdot \cancel{x} \cdot \cancel{y} \cdot \cancel{y}}$

$= \dfrac{2xy^2}{5}$

37. $\dfrac{8x+12}{18x-24} \cdot \dfrac{12-9x}{10x+15} = \dfrac{4(2x+3)}{6(3x\cancel{-4})} \cdot \dfrac{-3(3x\cancel{-4})}{5(2x\cancel{+3})}$

$= \dfrac{-12}{30} = -\dfrac{2}{5}$

39. $\dfrac{4a-16}{3a-3} \cdot \dfrac{a^2-1}{a^2-16} = \dfrac{4(a\cancel{-4})}{3(a\cancel{-1})} \cdot \dfrac{(a+1)(a\cancel{-1})}{(a+4)(a\cancel{-4})}$

$= \dfrac{4(a+1)}{3(a+4)}$

41. $\dfrac{4x^2y^3}{2x-6} \cdot \dfrac{12-4x}{6x^3y}$

$= \dfrac{\cancel{2} \cdot \cancel{2} \cdot \cancel{x} \cdot \cancel{x} \cdot \cancel{y} \cdot y \cdot y}{\cancel{2}(x\cancel{-3})} \cdot \dfrac{-2 \cdot 2 \cdot \cancel{(x-3)}}{\cancel{2} \cdot 3 \cdot \cancel{x} \cdot \cancel{x} \cdot x \cdot \cancel{y}}$

$= -\dfrac{4y^2}{3x}$

43. $\dfrac{x^2-4}{x^2-3x-10} \cdot \dfrac{x^2-8x+15}{x^2-9}$

$= \dfrac{(x+2)(x-2)}{(x\cancel{-5})(x\cancel{+2})} \cdot \dfrac{(x\cancel{-5})(x\cancel{-3})}{(x+3)(x\cancel{-3})}$

$= \dfrac{x-2}{x+3}$

45. $\dfrac{6x^2+x-12}{2x^2-5x-12} \cdot \dfrac{3x^2-14x+8}{9x^2-18x+8}$

$= \dfrac{(2x\cancel{+3})(3x\cancel{-4})}{(2x\cancel{+3})(x\cancel{-4})} \cdot \dfrac{(3x\cancel{-2})(x\cancel{-4})}{(3x\cancel{-2})(3x\cancel{-4})}$

$= 1$

47. $\dfrac{2x^2-7x+6}{8x^2-6x-9} \cdot \dfrac{3x^2-13x-10}{x^2-7x+10}$

$= \dfrac{(2x\cancel{-3})(x\cancel{-2})}{(4x+3)(2x\cancel{-3})} \cdot \dfrac{(3x+2)(x\cancel{-5})}{(x\cancel{-5})(x\cancel{-2})}$

$= \dfrac{3x+2}{4x+3}$

49. $\dfrac{2x^2-11x+12}{2x^2+11x-21} \cdot \dfrac{3x^2+20x-7}{4-x}$

$= \dfrac{(2x\cancel{-3})(x\cancel{-4})}{(2x\cancel{-3})(x\cancel{+7})} \cdot \dfrac{(3x-1)(x\cancel{+7})}{-1(x\cancel{-4})}$

$= -(3x-1)$

51. $\dfrac{ac+3a+2c+6}{ad+a+2d+2} \cdot \dfrac{ad-5a+2d-10}{bc+3b-4c-12}$

$= \dfrac{a(c+3)+2(c+3)}{a(d+1)+2(d+1)} \cdot \dfrac{a(d-5)+2(d-5)}{b(c+3)-4(c+3)}$

$= \dfrac{(c\cancel{+3})(a\cancel{+2})}{(d+1)(a\cancel{+2})} \cdot \dfrac{(d-5)(a+2)}{(c\cancel{+3})(b-4)}$

$= \dfrac{(d-5)(a+2)}{(d+1)(b-4)}$

53. $\dfrac{4x^3 - 3x^2 + 4x - 3}{4x^3 - 3x^2 - 8x + 6} \cdot \dfrac{2x^3 + x^2 - 4x - 2}{3x^3 + 2x^2 + 3x + 2}$

$= \dfrac{x^2(4x-3) + 1(4x-3)}{x^2(4x-3) - 2(4x-3)} \cdot \dfrac{x^2(2x+1) - 2(2x+1)}{x^2(3x+2) + 1(3x+2)}$

$= \dfrac{(4x-3)(x^2+1)}{(4x-3)(x^2-2)} \cdot \dfrac{(2x+1)(x^2-2)}{(3x+2)(x^2+1)}$

$= \dfrac{2x+1}{3x+2}$

55. $\dfrac{a^3 - b^3}{a^2 + ab + b^2} \cdot \dfrac{2a^2 + ab - b^2}{a^2 - b^2}$

$= \dfrac{(a-b)(a^2+ab+b^2)}{(a^2+ab+b^2)} \cdot \dfrac{(2a-b)(a+b)}{(a+b)(a-b)}$

$= 2a - b$

57. $\dfrac{8x^7 g^3}{15x^2 g^4} \div \dfrac{4xg^2}{3x^2 g^5}$

$= \dfrac{8x^7 g^3}{15x^2 g^4} \cdot \dfrac{3x^2 g^5}{4xg^2}$

$= \dfrac{2 \cdot 2 \cdot 2 \cdot x \cdot x \cdot x \cdot x \cdot x \cdot x \cdot x \cdot g \cdot g \cdot g}{3 \cdot 5 \cdot x \cdot x \cdot g \cdot g \cdot g \cdot g}$

$\cdot \dfrac{3 \cdot x \cdot x \cdot g \cdot g \cdot g \cdot g \cdot g \cdot g}{2 \cdot 2 \cdot x \cdot g \cdot g} = \dfrac{2x^6 g^2}{5}$

59. $\dfrac{4a^3 - 8a^2}{15a} \div \dfrac{3a^2 - 6a}{5a^4}$

$= \dfrac{4a^3 - 8a^2}{15a} \cdot \dfrac{5a^4}{3a^2 - 6a}$

$= \dfrac{2 \cdot 2 \cdot a \cdot a(a-2)}{3 \cdot 5 \cdot a} \cdot \dfrac{5 \cdot a \cdot a \cdot a \cdot a}{3a(a-2)}$

$= \dfrac{4a^4}{9}$

61. $\dfrac{10x^3 + 15x^2}{18x - 6} \div \dfrac{20x + 30}{9x^2 - 3x}$

$= \dfrac{10x^3 + 15x^2}{18x - 6} \cdot \dfrac{9x^2 - 3x}{20x + 30}$

$= \dfrac{5 \cdot x \cdot x(2x+3)}{2 \cdot 3 \cdot (3x-1)} \cdot \dfrac{3x(3x-1)}{2 \cdot 5 \cdot (2x+3)}$

$= \dfrac{x^3}{4}$

63. $\dfrac{3a - 4b}{6a - 9b} \div \dfrac{12b - 9a}{4a - 6b}$

$= \dfrac{3a - 4b}{6a - 9b} \cdot \dfrac{4a - 6b}{12b - 9a}$

$= \dfrac{(3a-4b)}{3(2a-3b)} \cdot \dfrac{2(2a-3b)}{-3(3a-4b)}$

$= -\dfrac{2}{9}$

65. $\dfrac{d^2 - d - 12}{2d^2} \div \dfrac{3d^2 + 13d + 12}{d}$

$= \dfrac{d^2 - d - 12}{2d^2} \cdot \dfrac{d}{3d^2 + 13d + 12}$

$= \dfrac{(d-4)(d+3)}{2 \cdot d \cdot d} \cdot \dfrac{d}{(3d+4)(d+3)}$

$= \dfrac{d-4}{2d(3d+4)}$

67. $\dfrac{m^2 - 25n^2}{2m - 12n} \div \dfrac{3m + 15n}{m^2 - 36n^2}$

$= \dfrac{m^2 - 25n^2}{2m - 12n} \cdot \dfrac{m^2 - 36n^2}{3m + 15n}$

$= \dfrac{(m+5n)(m-5n)}{2(m-6n)} \cdot \dfrac{(m+6n)(m-6n)}{3(m+5n)}$

$= \dfrac{(m-5n)(m+6n)}{6}$

69. $\dfrac{3x + 24}{4x - 32} \div \dfrac{x^2 + 16x + 64}{x^2 - 16x + 64}$

$= \dfrac{3x + 24}{4x - 32} \cdot \dfrac{x^2 - 16x + 64}{x^2 + 16x + 64}$

$= \dfrac{3(x+8)}{4(x-8)} \cdot \dfrac{(x-8)(x-8)}{(x+8)(x+8)}$

$= \dfrac{3(x-8)}{4(x+8)}$

71. $\dfrac{2x^2-7xy+6y^2}{2x^2+7xy+6y^2} \div \dfrac{4x^2-9y^2}{4x^2+12xy+9y^2}$

$= \dfrac{2x^2-7xy+6y^2}{2x^2+7xy+6y^2} \cdot \dfrac{4x^2+12xy+9y^2}{4x^2-9y^2}$

$= \dfrac{(x-2y)\cancel{(2x-3y)}}{(x+2y)\cancel{(2x+3y)}} \cdot \dfrac{\cancel{(2x+3y)}\cancel{(2x+3y)}}{\cancel{(2x+3y)}\cancel{(2x-3y)}}$

$= \dfrac{x-2y}{x+2y}$

73. $\dfrac{2x^2-7x-4}{3x^2-14x+8} \div \dfrac{2x^2+7x+3}{3x^2-8x+4}$

$= \dfrac{2x^2-7x-4}{3x^2-14x+8} \cdot \dfrac{3x^2-8x+4}{2x^2+7x+3}$

$= \dfrac{\cancel{(2x+1)}\cancel{(x-4)}}{\cancel{(3x-2)}\cancel{(x-4)}} \cdot \dfrac{\cancel{(3x-2)}(x-2)}{\cancel{(2x+1)}(x+3)}$

$= \dfrac{x-2}{x+3}$

75. $\dfrac{27a^3-8b^3}{9a^2-4b^2} \div \dfrac{2a^2-5ab-12b^2}{6a^2+13ab+6b^2}$

$= \dfrac{27a^3-8b^3}{9a^2-4b^2} \cdot \dfrac{6a^2+13ab+6b^2}{2a^2-5ab-12b^2}$

$= \dfrac{\cancel{(3a-2b)}(9a^2+6ab+4b^2)}{(3a+2b)\cancel{(3a-2b)}}$

$\cdot \dfrac{(2a+3b)\cancel{(3a+2b)}}{(2a+3b)(a-4b)}$

$= \dfrac{9a^2+6ab+4b^2}{a-4b}$

77. $\dfrac{ab+3a+2b+6}{bc+4b+3c+12} \div \dfrac{ac-3a+2c-6}{bc+4b-4c-16}$

$= \dfrac{ab+3a+2b+6}{bc+4b+3c+12} \cdot \dfrac{bc+4b-4c-16}{ac-3a+2c-6}$

$= \dfrac{a(b+3)+2(b+3)}{b(c+4)+3(c+4)} \cdot \dfrac{b(c+4)-4(c+4)}{a(c-3)+2(c-3)}$

$= \dfrac{\cancel{(b+3)}\cancel{(a+2)}}{\cancel{(c+4)}\cancel{(b+3)}} \cdot \dfrac{\cancel{(c+4)}(b-4)}{(c-3)\cancel{(a+2)}}$

$= \dfrac{b-4}{c-3}$

79. $\dfrac{x^2-3x-18}{2x^2+13x+20} \cdot \dfrac{3x^2+10x-8}{4x^2-24x} \div \dfrac{3x^2+7x-6}{6x^2+11x-10}$

$= \dfrac{\cancel{(x-6)}(x+3)}{\cancel{(2x+5)}\cancel{(x+4)}} \cdot \dfrac{(3x-2)\cancel{(x+4)}}{4x\cancel{(x-6)}}$

$\cdot \dfrac{\cancel{(3x-2)}\cancel{(2x+5)}}{\cancel{(3x-2)}(x+3)} = \dfrac{3x-2}{4x}$

81. $\dfrac{5x^2-17x-12}{12x^2-29x+15}$

$\cdot \left(\dfrac{6x^2+x-12}{3x^2+14x-24} \div \dfrac{2x^2-5x-12}{3x^2+13x-30} \right)$

$= \dfrac{(5x+3)\cancel{(x-4)}}{(4x-3)\cancel{(3x-5)}} \cdot \dfrac{\cancel{(2x+3)}\cancel{(3x-4)}}{\cancel{(3x-4)}\cancel{(x+6)}}$

$\cdot \dfrac{\cancel{(3x-5)}\cancel{(x+6)}}{\cancel{(2x+3)}\cancel{(x-4)}}$

$= \dfrac{5x+3}{4x-3}$

83. a^2+b^2 does not factor.

Correct: $\dfrac{a^2+b^2}{(a+b)\cancel{(a-b)}} \cdot \dfrac{\cancel{(a-b)}\cancel{(a+3b)}}{\cancel{(a+3b)}(a-2b)}$

$= \dfrac{a^2+b^2}{(a+b)(a-2b)}$

85. $f(x) = \dfrac{x+3}{x-4}$

a) $f(3) = \dfrac{3+3}{3-4} = \dfrac{6}{-1} = -6$

b) $f(-3) = \dfrac{-3+3}{-3-4} = \dfrac{0}{-7} = 0$

c) $f(4) = \dfrac{4+3}{4-4} = \dfrac{7}{0}$ is undefined.

87. $f(x) = \dfrac{2x-3}{3x+2}$

   a) $f\left(\dfrac{3}{2}\right) = \dfrac{2\left(\dfrac{3}{2}\right)-3}{3\left(\dfrac{3}{2}\right)+2} = \dfrac{0}{\frac{13}{2}} = 0$

   b) $f(-1) = \dfrac{2(-1)-3}{3(-1)+2} = \dfrac{-5}{-1} = 5$

   c) $f(2) = \dfrac{2(2)-3}{3(2)+2} = \dfrac{1}{8}$

89. $f(x) = \dfrac{3x-4}{2x^2-x-6}$

   a) $f(3) = \dfrac{3(3)-4}{2(3)^2-(3)-6} = \dfrac{5}{9}$

   b) $f\left(\dfrac{4}{3}\right) = \dfrac{3\left(\dfrac{4}{3}\right)-4}{2\left(\dfrac{4}{3}\right)^2-\left(\dfrac{4}{3}\right)-6} = \dfrac{0}{\left(-\dfrac{34}{9}\right)} = 0$

   c) $f(2) = \dfrac{3(2)-4}{2(2)^2-(2)-6} = \dfrac{2}{0}$ is undefined.

91. $f(x) = \dfrac{x^2+x-6}{x^2+x-12}$

   a) $f(-3) = \dfrac{(-3)^2+(-3)-6}{(-3)^2+(-3)-12} = \dfrac{0}{-6} = 0$

   b) $f(5) = \dfrac{(5)^2+(5)-6}{(5)^2+(5)-12} = \dfrac{24}{18} = \dfrac{4}{3}$

   c) $f(3) = \dfrac{(3)^2+(3)-6}{(3)^2+(3)-12} = \dfrac{6}{0}$ is undefined.

93. $f(x) = \dfrac{x+6}{3x+12}$

   $3x+12 = 0$

   $3x = -12$

   $x = -4$

   The domain is $\{x \mid x \neq -4\}$.

95. $f(x) = \dfrac{3x-6}{x^2-25}$

   $x^2-25 = 0$

   $x^2 = 25$

   $x = \pm 5$

   The domain is $\{x \mid x \neq -5, 5\}$.

97. $g(c) = \dfrac{3c-9}{4c^2-81}$

   $4c^2-81 = 0$

   $4c^2 = 81$

   $c^2 = \dfrac{81}{4}$

   $c = \pm\dfrac{9}{2}$

   The domain is $\left\{c \mid c \neq -\dfrac{9}{2}, \dfrac{9}{2}\right\}$.

99. $h(t) = \dfrac{2t^2+4t-8}{6t^2+11t-10}$

   $6t^2+11t-10 = 0$

   $(3t-2)(2t+5) = 0$

   $3t-2 = 0$ or $2t+5 = 0$

   $t = \dfrac{2}{3}$ $t = -\dfrac{5}{2}$

   The domain is $\left\{t \mid t \neq -\dfrac{5}{2}, \dfrac{2}{3}\right\}$.

101. $g(x) = \dfrac{4x-8}{x^3+4x^2-21x}$

   $x^3+4x^2-21x = 0$

   $x(x+7)(x-3) = 0$

   $x = 0$ or $x+7 = 0$ or $x-3 = 0$

                   $x = -7$ $x = 3$

   The domain is $\{x \mid x \neq 0, -7, 3\}$.

103. $f(x) = \dfrac{x-7}{x^3-8}$

$$x^3 - 8 = 0$$

$$(x-2)(x^2+2x+4) = 0$$

$$x - 2 = 0 \quad \text{or} \quad x^2 + 2x + 4 = 0$$

$$x = 2$$

When solved for x, $x^2 + 2x + 4 = 0$ yields complex solutions.

The domain is $\{x \mid x \neq 2\}$.

105. c

107. d

109. $f(x) = \dfrac{1}{x-2}$

Note that $x \neq 2$, so the graph has a vertical asymptote at $x = 2$.

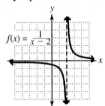

111. $f(x) = \dfrac{-2}{x+1}$

Note that $x \neq -1$, so the graph has a vertical asymptote at $x = -1$.

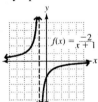

113. Substitute 2 for d and solve for R.

$$R = \frac{5}{d^2} = \frac{5}{2^2} = \frac{5}{4} \text{ ohms}$$

115. Substitute 100 for x and solve for C.

$$C = \frac{15x + 1800}{x}$$

$$= \frac{15(200) + 1800}{200}$$

$$= \frac{4800}{200} = \$24$$

117. Substitute 50 for x and solve for C.

$$C = 200\left(\frac{300}{x^2} + \frac{x}{x+50}\right)$$

$$= 200\left(\frac{300}{50^2} + \frac{50}{50+50}\right)$$

$$= 200\left(\frac{300}{2500} + \frac{50}{100}\right)$$

$$= 200(0.62)$$

$$= 124$$

The cost of the order is \$124,000.

119. Substitute 70 for v and solve for F.

$$F = \frac{130,000}{330 - v} = \frac{130,000}{330 - 70} = 500$$

The frequency is 500 cycles/sec.

Review Exercises

1. $\dfrac{3}{11} + \dfrac{4}{11} = \dfrac{7}{11}$

2. $\dfrac{3}{5} - \dfrac{5}{6} = \dfrac{18}{30} - \dfrac{25}{30} = -\dfrac{7}{30}$

3. $\dfrac{5}{12}y + \dfrac{7}{30}y = \dfrac{25}{60}y + \dfrac{14}{60}y = \dfrac{39}{60}y = \dfrac{13}{20}y$

4. $\dfrac{11}{15}x + 4 - \dfrac{2}{15}x - 3 = \dfrac{9}{15}x + 1$

$$= \dfrac{3}{5}x + 1$$

5. $6x^2 - 15x = 3x(2x - 5)$

6. $x^2 + x - 20 = (x+5)(x-4)$

Exercise Set 7.2

1. $\dfrac{3c}{2d} + \dfrac{5c}{2d} = \dfrac{3c+5c}{2d} = \dfrac{8c}{2d} = \dfrac{4c}{d}$

3. $\dfrac{4a+3b}{2a-5b} + \dfrac{3a-4b}{2a-5b} = \dfrac{4a+3b+3a-4b}{2a-5b} = \dfrac{7a-b}{2a-5b}$

5. $\dfrac{2x-3y}{x+2y} - \dfrac{4x+2y}{x+2y} = \dfrac{2x-3y-(4x+2y)}{x+2y}$

$$= \dfrac{2x-3y-4x-2y}{x+2y}$$

$$= \dfrac{-2x-5y}{x+2y}$$

7. $\dfrac{c^2}{c^2-5c+4} - \dfrac{-2c+24}{c^2-5c+4} = \dfrac{c^2-(-2c+24)}{c^2-5c+4}$

$= \dfrac{c^2+2c-24}{c^2-5c+4}$

$= \dfrac{(c+6)\cancel{(c-4)}}{\cancel{(c-4)}(c-1)}$

$= \dfrac{c+6}{c-1}$

9. $\dfrac{y^2-2y}{y^2-12y+36} + \dfrac{-5y+6}{y^2-12y+36}$

$= \dfrac{y^2-2y-5y+6}{y^2-12y+36}$

$= \dfrac{y^2-7y+6}{y^2-12y+36}$

$= \dfrac{\cancel{(y-6)}(y-1)}{\cancel{(y-6)}(y-6)}$

$= \dfrac{y-1}{y-6}$

11. $\dfrac{h^2-4h}{h^2+10h+21} - \dfrac{-6h+3}{h^2+10h+21}$

$= \dfrac{h^2-4h-(-6h+3)}{h^2+10h+21}$

$= \dfrac{h^2-4h+6h-3}{h^2+10h+21}$

$= \dfrac{h^2+2h-3}{h^2+10h+21}$

$= \dfrac{\cancel{(h+3)}(h-1)}{(h+7)\cancel{(h+3)}}$

$= \dfrac{h-1}{h+7}$

13. $\dfrac{2x^2+x-3}{x^2+6x+5} + \dfrac{x^2-2x+4}{x^2+6x+5} - \dfrac{2x^2+x+4}{x^2+6x+5}$

$= \dfrac{2x^2+x-3+x^2-2x+4-(2x^2+x+4)}{x^2+6x+5}$

$= \dfrac{2x^2+x-3+x^2-2x+4-2x^2-x-4}{x^2+6x+5}$

$= \dfrac{x^2-2x-3}{x^2+6x+5}$

$= \dfrac{(x-3)\cancel{(x+1)}}{(x+5)\cancel{(x+1)}}$

$= \dfrac{x-3}{x+5}$

15. $\dfrac{x^3}{x^2-xy+y^2} + \dfrac{y^3}{x^2-xy+y^2}$

$= \dfrac{x^3+y^3}{x^2-xy+y^2}$

$= \dfrac{(x+y)\cancel{(x^2-xy+y^2)}}{\cancel{(x^2-xy+y^2)}}$

$= x+y$

17. $15a^6b^8 = 3\cdot5\cdot a^6\cdot b^8$

$20a^4b^5 = 2^2\cdot5\cdot a^4\cdot b^5$

$\text{LCD} = 2^2\cdot3\cdot5\cdot a^6\cdot b^8 = 60a^6b^8$

$\dfrac{4}{15a^6b^8} = \dfrac{4\cdot4}{15a^6b^8\cdot4} = \dfrac{16}{60a^6b^8}$

$\dfrac{9}{20a^4b^5} = \dfrac{9\cdot3a^2b^3}{20a^4b^5\cdot3a^2b^3} = \dfrac{27a^2b^3}{60a^6b^8}$

19. $\text{LCD} = (x+7)(x-5)$

$\dfrac{x}{x+7} = \dfrac{x\cdot(x-5)}{(x+7)\cdot(x-5)} = \dfrac{x^2-5x}{(x+7)(x-5)}$

$\dfrac{x}{x-5} = \dfrac{x\cdot(x+7)}{(x-5)\cdot(x+7)} = \dfrac{x^2+7x}{(x+7)(x-5)}$

21. $9r+18 = 3^2(r+2)$

$15r+30 = 3\cdot5(r+2)$

$\text{LCD} = 3^2\cdot5(r+2) = 45(r+2)$

$\dfrac{7r}{9r+18} = \dfrac{7r\cdot5}{9(r+2)\cdot5} = \dfrac{35r}{45(r+2)}$

$\dfrac{8r}{15r+30} = \dfrac{8r\cdot3}{15(r+2)\cdot3} = \dfrac{24r}{45(r+2)}$

23. $b^2 - 1 = (b+1)(b-1)$

$b^2 + 3b - 4 = (b+4)(b-1)$

LCD $= (b+1)(b-1)(b+4)$

$\dfrac{3b}{b^2-1} = \dfrac{3b \cdot (b+4)}{(b+1)(b-1) \cdot (b+4)}$

$\qquad = \dfrac{3b^2 + 12b}{(b+1)(b-1)(b+4)}$

$\dfrac{b}{b^2+3b-4} = \dfrac{b \cdot (b+1)}{(b+4)(b-1) \cdot (b+1)}$

$\qquad = \dfrac{b^2 + b}{(b+1)(b-1)(b+4)}$

25. $c^2 - 2c - 3 = (c-3)(c+1)$

$c^2 - 5c + 6 = (c-2)(c-3)$

LCD $= (c-3)(c+1)(c-2)$

$\dfrac{c-5}{c^2-2c-3} = \dfrac{(c-5) \cdot (c-2)}{(c-3)(c+1) \cdot (c-2)}$

$\qquad = \dfrac{c^2 - 7c + 10}{(c-3)(c+1)(c-2)}$

$\dfrac{3+c}{c^2-5c+6} = \dfrac{(c+3) \cdot (c+1)}{(c-3)(c-2) \cdot (c+1)}$

$\qquad = \dfrac{c^2 + 4c + 3}{(c-3)(c+1)(c-2)}$

27. $n^2 + 8n + 16 = (n+4)^2$

$n^2 + 5n + 4 = (n+4)(n+1)$

LCD $= (n+4)^2 (n+1)$

$\dfrac{n+1}{n^2+8n+16} = \dfrac{(n+1) \cdot (n+1)}{(n+4)^2 \cdot (n+1)}$

$\qquad = \dfrac{n^2 + 2n + 1}{(n+4)^2 (n+1)}$

$\dfrac{n-4}{n^2+5n+4} = \dfrac{(n-4) \cdot (n+4)}{(n+4)(n-1) \cdot (n+4)}$

$\qquad = \dfrac{n^2 - 16}{(n+4)^2 (n-1)}$

29. $x^3 + x^2 - 6x = x(x^2 + x - 6)$

$\qquad = x(x+3)(x-2)$

$x^4 + 7x^3 + 12x^2 = x^2(x^2 + 7x + 12)$

$\qquad = x^2(x+3)(x+4)$

LCD $= x^2(x+3)(x-2)(x+4)$

$\dfrac{x+5}{x^3+x^2-6x} = \dfrac{(x+5) \cdot x(x+4)}{x(x+3)(x-2) \cdot x(x+4)}$

$\qquad = \dfrac{x^3 + 9x^2 + 20x}{x^2(x+3)(x-2)(x+4)}$

$\dfrac{x-3}{x^4+7x^3+12x^2} = \dfrac{(x-3) \cdot (x-2)}{x^2(x+3)(x+4) \cdot (x-2)}$

$\qquad = \dfrac{x^2 - 5x + 6}{x^2(x+3)(x-2)(x+4)}$

31. $x = x$

$x^2 + 4x = x(x+4)$

$x + 4 = (x+4)$

LCD $= x(x+4)$

$\dfrac{3}{x} = \dfrac{3 \cdot (x+4)}{x \cdot (x+4)} = \dfrac{3x+12}{x(x+4)}$

$\dfrac{4}{x^2+4x} = \dfrac{4}{x(x+4)}$

$\dfrac{6}{x+4} = \dfrac{6 \cdot x}{(x+4) \cdot x} = \dfrac{6x}{x(x+4)}$

33. LCD $= 24u$

$\dfrac{5}{8u} - \dfrac{7}{12u} = \dfrac{5(3)}{8u(3)} - \dfrac{7(2)}{12u(2)} = \dfrac{15}{24u} - \dfrac{14}{24u} = \dfrac{1}{24u}$

35. LCD $= 20x$

$\dfrac{3z}{10x} + \dfrac{5z}{4x} = \dfrac{3z(2)}{10x(2)} + \dfrac{5z(5)}{4x(5)} = \dfrac{6z}{20x} + \dfrac{25z}{20x} = \dfrac{31z}{20x}$

37. LCD $= 12y$

$\dfrac{y+6}{4y} + \dfrac{2y-3}{3y} = \dfrac{(y+6)(3)}{4y(3)} + \dfrac{(2y-3)(4)}{3y(4)}$

$\qquad = \dfrac{3y+18}{12y} + \dfrac{8y-12}{12y}$

$\qquad = \dfrac{3y+18+8y-12}{12y}$

$\qquad = \dfrac{11y+6}{12y}$

39. LCD $= 6m$

$$\frac{m+8}{2m} - \frac{m-7}{3m} = \frac{(m+8)(3)}{2m(3)} - \frac{(m-7)(2)}{3m(2)}$$

$$= \frac{3m+24}{6m} - \frac{2m-14}{6m}$$

$$= \frac{3m+24-2m+14}{6m}$$

$$= \frac{m+38}{6m}$$

41. LCD $= 30p^3q^2$

$$\frac{3p+1}{10p^3q^2} + \frac{9p-2}{6p^2q}$$

$$= \frac{(3p+1)(3)}{10p^3q^2(3)} + \frac{(9p-2)(5pq)}{6p^2q(5pq)}$$

$$= \frac{9p+3}{30p^3q^2} + \frac{45p^2q-10pq}{30p^3q^2}$$

$$= \frac{9p+3+45p^2q-10pq}{30p^3q^2}$$

43. LCD $= 24a^3b^4$

$$\frac{2a+b}{8a^2b^4} - \frac{a-3b}{12a^3b^3}$$

$$= \frac{(2a+b)(3a)}{8a^2b^4(3a)} - \frac{(a-3b)(2b)}{12a^3b^3(2b)}$$

$$= \frac{6a^2+3ab}{24a^3b^4} - \frac{2ab-6b^2}{24a^3b^4}$$

$$= \frac{6a^2+3ab-2ab+6b^2}{24a^3b^4}$$

$$= \frac{6a^2+ab+6b^2}{24a^3b^4}$$

45. LCD $= k(k+2)$

$$\frac{4}{k} - \frac{6}{k+2} = \frac{4(k+2)}{k(k+2)} - \frac{6(k)}{(k+2)(k)}$$

$$= \frac{4k+8}{k(k+2)} - \frac{6k}{k(k+2)}$$

$$= \frac{4k+8-6k}{k(k+2)}$$

$$= \frac{8-2k}{k(k+2)}$$

47. LCD $= (w-3)(w+7)$

$$\frac{4}{w-3} + \frac{5}{w+7}$$

$$= \frac{4(w+7)}{(w-3)(w+7)} + \frac{5(w-3)}{(w+7)(w-3)}$$

$$= \frac{4w+28}{(w-3)(w+7)} + \frac{5w-15}{(w-3)(w+7)}$$

$$= \frac{9w+13}{(w-3)(w+7)}$$

49. LCD $= 6(x+3)$

$$\frac{x-4}{3x+9} - \frac{x+5}{6x+18}$$

$$= \frac{(x-4)(2)}{3(x+3)(2)} - \frac{x+5}{6(x+3)}$$

$$= \frac{2x-8}{6(x+3)} - \frac{x+5}{6(x+3)}$$

$$= \frac{2x-8-x-5}{6(x+3)}$$

$$= \frac{x-13}{6(x+3)}$$

51. LCD $= (t-4)^2$

$$\frac{2t}{t^2-8t+16} + \frac{6}{t-4}$$

$$= \frac{2t}{(t-4)^2} + \frac{6(t-4)}{(t-4)(t-4)}$$

$$= \frac{2t}{(t-4)^2} + \frac{6t-24}{(t-4)^2}$$

$$= \frac{8t-24}{(t-4)^2}$$

53. $\text{LCD} = (u-3)^2$

$$\frac{u^2-2u}{u^2-6u+9}+\frac{4}{4u-12}$$

$$=\frac{u^2-2u}{(u-3)^2}+\frac{\cancel{4}}{\cancel{4}(u-3)}$$

$$=\frac{u^2-2u}{(u-3)^2}+\frac{1(u-3)}{(u-3)(u-3)}$$

$$=\frac{u^2-2u}{(u-3)^2}+\frac{u-3}{(u-3)^2}$$

$$=\frac{u^2-2u+u-3}{(u-3)^2}$$

$$=\frac{u^2-u-3}{(u-3)^2}$$

55. $\text{LCD} = (x-6)(x+2)$

$$\frac{x}{x+2}-\frac{16}{x^2-4x-12}$$

$$=\frac{x(x-6)}{(x+2)(x-6)}-\frac{16}{(x+2)(x-6)}$$

$$=\frac{x^2-6x}{(x+2)(x-6)}-\frac{16}{(x+2)(x-6)}$$

$$=\frac{x^2-6x-16}{(x+2)(x-6)}$$

$$=\frac{\cancel{(x+2)}(x-8)}{\cancel{(x+2)}(x-6)}$$

$$=\frac{x-8}{x-6}$$

57. $\text{LCD} = (x-3)(x+2)(x+4)$

$$\frac{x+1}{x^2-x-6}+\frac{2x+8}{x^2+6x+8}$$

$$=\frac{(x+1)(x+4)}{(x-3)(x+2)(x+4)}+\frac{(2x+8)(x-3)}{(x+2)(x+4)(x-3)}$$

$$=\frac{x^2+5x+4}{(x-3)(x+2)(x+4)}+\frac{2x^2+2x-24}{(x-3)(x+2)(x+4)}$$

$$=\frac{x^2+5x+4+2x^2+2x-24}{(x-3)(x+2)(x+4)}$$

$$=\frac{3x^2+7x-20}{(x-3)(x+2)(x+4)}$$

$$=\frac{(3x-5)\cancel{(x+4)}}{(x-3)(x+2)\cancel{(x+4)}}$$

$$=\frac{3x-5}{(x-3)(x+2)}$$

59. $\text{LCD} = (v+4)(v-4)(v+3)$

$$\frac{2v+5}{v^2-16}-\frac{2v+6}{v^2-v-12}$$

$$=\frac{(2v+5)(v+3)}{(v+4)(v-4)(v+3)}-\frac{(2v+6)(v+4)}{(v-4)(v+3)(v+4)}$$

$$=\frac{2v^2+11v+15}{(v+4)(v-4)(v+3)}-\frac{2v^2+14v+24}{(v+4)(v-4)(v+3)}$$

$$=\frac{2v^2+11v+15-2v^2-14v-24}{(v+4)(v-4)(v+3)}$$

$$=\frac{-3v-9}{(v+4)(v-4)(v+3)}$$

$$=\frac{-3\cancel{(v+3)}}{(v+4)(v-4)\cancel{(v+3)}}$$

$$=\frac{-3}{(v+4)(v-4)}$$

61. LCD $= (z+3)^2(z-2)$

$$\frac{z-3}{z^2+6z+9}+\frac{z+1}{z^2+z-6}$$

$$=\frac{(z-3)(z-2)}{(z+3)^2(z-2)}+\frac{(z+1)(z+3)}{(z+3)(z-2)(z+3)}$$

$$=\frac{z^2-5z+6}{(z+3)^2(z-2)}+\frac{z^2+4z+3}{(z+3)^2(z-2)}$$

$$=\frac{z^2-5z+6+z^2+4z+3}{(z+3)^2(z-2)}$$

$$=\frac{2z^2-z+9}{(z+3)^2(z-2)}$$

63. LCD $= x(x+6)(x-6)(x-3)$

$$\frac{x+4}{x^3-36x}-\frac{1}{x^2+3x-18}$$

$$=\frac{(x+4)(x-3)}{x(x+6)(x-6)(x-3)}-\frac{1x(x-6)}{(x+6)(x-3)x(x-6)}$$

$$=\frac{x^2+x-12}{x(x+6)(x-6)(x-3)}-\frac{x^2-6x}{x(x+6)(x-6)(x-3)}$$

$$=\frac{x^2+x-12-x^2+6x}{x(x+6)(x-6)(x-3)}$$

$$=\frac{7x-12}{x(x+6)(x-6)(x-3)}$$

65. LCD $= (x+2)(x-2)$

$$\frac{3x}{x-2}-\frac{4}{x+2}-\frac{3}{x^2-4}$$

$$=\frac{3x(x+2)}{(x-2)(x+2)}-\frac{4(x-2)}{(x+2)(x-2)}-\frac{3}{(x+2)(x-2)}$$

$$=\frac{3x^2+6x}{(x-2)(x+2)}-\frac{4x-8}{(x+2)(x-2)}-\frac{3}{(x+2)(x-2)}$$

$$=\frac{3x^2+6x-4x+8-3}{(x-2)(x+2)}$$

$$=\frac{3x^2+2x+5}{x^2-4}$$

67. LCD $= a(a-4)$

$$\frac{5}{a^2-4a}+\frac{6}{a}+\frac{4}{a-4}$$

$$=\frac{5}{a(a-4)}+\frac{6(a-4)}{a(a-4)}+\frac{4a}{(a-4)a}$$

$$=\frac{5}{a(a-4)}+\frac{6a-24}{a(a-4)}+\frac{4a}{a(a-4)}$$

$$=\frac{5+6a-24+4a}{a(a-4)}$$

$$=\frac{10a-19}{a(a-4)}$$

69. LCD $= x(x-4)$

$$\frac{3x+1}{x-4}-\frac{7}{x}+\frac{2}{x^2-4x}$$

$$=\frac{(3x+1)x}{(x-4)x}-\frac{7(x-4)}{x(x-4)}+\frac{2}{x(x-4)}$$

$$=\frac{3x^2+x}{x(x-4)}-\frac{7x-28}{x(x-4)}+\frac{2}{x(x-4)}$$

$$=\frac{3x^2+x-7x+28+2}{x(x-4)}$$

$$=\frac{3x^2-6x+30}{x(x-4)}$$

71. LCD $= (r+3)(r-3)$

$$\frac{3r}{r+3}-\frac{5r}{r-3}+\frac{2}{r^2-9}$$

$$=\frac{3r(r-3)}{(r+3)(r-3)}-\frac{5r(r+3)}{(r-3)(r+3)}+\frac{2}{(r+3)(r-3)}$$

$$=\frac{3r^2-9r}{(r+3)(r-3)}-\frac{5r^2+15r}{(r+3)(r-3)}+\frac{2}{(r+3)(r-3)}$$

$$=\frac{3r^2-9r-5r^2-15r+2}{(r+3)(r-3)}$$

$$=\frac{-2r^2-24r+2}{(r+3)(r-3)}$$

$$=\frac{-2r^2-24r+2}{r^2-9}$$

73. LCD = $u - v$

$$\frac{v}{u-v} - \frac{3v}{v-u} = \frac{v}{u-v} - \frac{3v}{-1(u-v)}$$

$$= \frac{v}{u-v} + \frac{3v}{(u-v)}$$

$$= \frac{4v}{u-v}$$

75. LCD = $m - n$

$$\frac{2m-n}{m-n} - \frac{m-3n}{n-m} = \frac{2m-n}{m-n} - \frac{m-3n}{-1(m-n)}$$

$$= \frac{2m-n}{m-n} + \frac{m-3n}{(m-n)}$$

$$= \frac{3m-4n}{m-n}$$

77. LCD = $3a - b$

$$\frac{2a-3b}{3a-b} + \frac{3a+2b}{b-3a} = \frac{2a-3b}{3a-b} + \frac{3a+2b}{-1(3a-b)}$$

$$= \frac{2a-3b}{3a-b} - \frac{3a+2b}{(3a-b)}$$

$$= \frac{2a-3b-3a-2b}{3a-b}$$

$$= \frac{-a-5b}{3a-b}$$

79. Mistake: Added denominators instead of finding the LCD.

Correct: $\dfrac{4(5)}{a(5)} + \dfrac{a(a)}{5(a)} = \dfrac{20}{5a} + \dfrac{a^2}{5a} = \dfrac{a^2+20}{5a}$

81. Mistake: Added denominators.

Correct: $\dfrac{9v}{2x}$

83. Mistake: Did not distribute the subtraction sign to both terms in $2c + 3$.

Correct: $\dfrac{5c+2-2c-3}{3c-5} = \dfrac{3c-1}{3c-5}$

85. $f(x) = \dfrac{9x+7}{6x}$ and $g(x) = \dfrac{x-3}{6x}$

a) $f(x) + g(x) = \dfrac{9x+7}{6x} + \dfrac{x-3}{6x}$

$$= \frac{10x+4}{6x}$$

$$= \frac{\cancel{2}(5x+2)}{\cancel{2} \cdot 3x}$$

$$= \frac{5x+2}{3x}$$

b) $f(x) - g(x) = \dfrac{9x+7}{6x} - \dfrac{x-3}{6x}$

$$= \frac{9x+7-x+3}{6x}$$

$$= \frac{8x+10}{6x}$$

$$= \frac{\cancel{2}(4x+5)}{\cancel{2} \cdot 3x}$$

$$= \frac{4x+5}{3x}$$

c) Using part a.

$$(f+g)(4) = \frac{5(4)+2}{3(4)} = \frac{22}{12} = \frac{11}{6}$$

d) Using part b.

$$(f-g)(-2) = \frac{4(-2)+5}{3(-2)} = \frac{-3}{-6} = \frac{1}{2}$$

87. $f(x) = \dfrac{x-14}{x^2-4}$ and $g(x) = \dfrac{x+1}{x-2}$

$$= \frac{x-14}{(x+2)(x-2)}$$

a) $f(x) + g(x) = \dfrac{x-14}{(x+2)(x-2)} + \dfrac{(x+1)(x+2)}{(x-2)(x+2)}$

$$= \frac{x-14}{(x+2)(x-2)} + \frac{x^2+3x+2}{(x-2)(x+2)}$$

$$= \frac{x^2+4x-12}{(x+2)(x-2)}$$

$$= \frac{(x+6)(x-2)}{(x+2)(x-2)}$$

$$= \frac{x+6}{x+2}$$

b) $f(x) - g(x) = \dfrac{x-14}{(x+2)(x-2)} - \dfrac{(x+1)(x+2)}{(x-2)(x+2)}$

$= \dfrac{x-14}{(x+2)(x-2)} - \dfrac{x^2+3x+2}{(x-2)(x+2)}$

$= \dfrac{x-14-x^2-3x-2}{(x+2)(x-2)}$

$= \dfrac{-x^2-2x-16}{(x+2)(x-2)}$

$= -\dfrac{x^2+2x+16}{x^2-4}$ or $\dfrac{x^2+2x+16}{4-x^2}$

c) Using part a.

$(f+g)(1) = \dfrac{1+6}{1+2} = \dfrac{7}{3}$

d) Using part b.

$(f-g)(-1) = -\dfrac{(-1)^2+2(-1)+16}{(-1)^2-4}$

$= -\dfrac{1-2+16}{1-4}$

$= -\dfrac{15}{-3}$

$= 5$

89. $\dfrac{t}{5} + \dfrac{t}{3} = \dfrac{t(3)}{5(3)} + \dfrac{t(5)}{3(5)} = \dfrac{3t}{15} + \dfrac{5t}{15} = \dfrac{8t}{15}$

91. perimeter = 2(length) + 2(width)

$2\left(\dfrac{3}{x+4}\right) + 2\left(\dfrac{1}{x-2}\right)$

$= \dfrac{6}{x+4} + \dfrac{2}{x-2}$

$= \dfrac{6(x-2)}{(x+4)(x-2)} + \dfrac{2(x+4)}{(x-2)(x+4)}$

$= \dfrac{6x-12}{(x+4)(x-2)} + \dfrac{2x+8}{(x-2)(x+4)}$

$= \dfrac{8x-4}{(x+4)(x-2)}$

93. Let x represent the number, then $\left(\dfrac{1}{x}\right)$ represents

the reciprocal.

$3x + 2\left(\dfrac{1}{x}\right) = 3x + \dfrac{2}{x} = \dfrac{3x(x)}{1(x)} + \dfrac{2}{x} = \dfrac{3x^2+2}{x}$

95. $\left(\dfrac{x}{5} + \dfrac{x}{3}\right) \div 2 = \left(\dfrac{x(3)}{5(3)} + \dfrac{x(5)}{3(5)}\right) \div 2$

$= \left(\dfrac{3x}{15} + \dfrac{5x}{15}\right) \cdot \dfrac{1}{2}$

$= \dfrac{8x}{15} \cdot \dfrac{1}{2}$

$= \dfrac{8x}{30}$

$= \dfrac{4x}{15}$

Review Exercises

1. $\dfrac{1}{3} + \dfrac{3}{4} = \dfrac{1(4)}{3(4)} + \dfrac{3(3)}{4(3)} = \dfrac{4}{12} + \dfrac{9}{12} = \dfrac{13}{12}$

2. $\dfrac{5}{12} \div \dfrac{3}{8} = \dfrac{5}{12} \cdot \dfrac{8}{3} = \dfrac{40}{36} = \dfrac{10}{9}$

3. $\left(\dfrac{2}{3} + \dfrac{1}{2}\right) \div \left(\dfrac{3}{4} - \dfrac{1}{3}\right)$

$= \left(\dfrac{2(2)}{3(2)} + \dfrac{1(3)}{2(3)}\right) \div \left(\dfrac{3(3)}{4(3)} - \dfrac{1(4)}{3(4)}\right)$

$= \left(\dfrac{4}{6} + \dfrac{3}{6}\right) \div \left(\dfrac{9}{12} - \dfrac{4}{12}\right)$

$= \dfrac{7}{6} \div \dfrac{5}{12}$

$= \dfrac{7}{\cancel{6}} \cdot \dfrac{\cancel{12}^{2}}{5}$

$= \dfrac{14}{5}$

4. $12\left(\dfrac{2}{3} + \dfrac{5}{6}\right) = \dfrac{12 \cdot 2}{3} + \dfrac{12 \cdot 5}{6}$

$= \dfrac{24}{3} + \dfrac{60}{6}$

$= 8 + 10$

$= 18$

5. $\dfrac{x^2-9x+8}{x^2-6x-16} = \dfrac{\cancel{(x-8)}(x-1)}{\cancel{(x-8)}(x+2)} = \dfrac{x-1}{x+2}$

6. $\dfrac{3y+2}{27y^3+8} = \dfrac{3y+2}{(3y+2)(9y^2-6y+4)}$

$= \dfrac{1}{9y^2-6y+4}$

Exercise Set 7.3

1. $\dfrac{\frac{5}{21}}{\frac{3}{14}} = \dfrac{5}{{}_{3}\cancel{21}} \cdot \dfrac{\cancel{14}^{2}}{3} = \dfrac{10}{9}$

3. $\dfrac{\frac{u}{v^3}}{\frac{w}{v^4}} = \dfrac{u}{\cancel{v^3}_{1}} \cdot \dfrac{\cancel{v^4}^{v}}{w} = \dfrac{uv}{w}$

5. $\dfrac{\frac{u^7 v^2}{w^5}}{\frac{u^3 v^4}{w}} = \dfrac{\cancel{u^7}^{u^4}\ \cancel{v^2}}{w^4\ \cancel{w^5}} \cdot \dfrac{\cancel{w}^{1}}{\cancel{u^3}\ \cancel{v^4}_{v^2}} = \dfrac{u^4}{w^4 v^2}$

7. $\dfrac{\frac{15}{8}}{20} = \dfrac{{}^{3}\cancel{15}}{8} \cdot \dfrac{1}{\cancel{20}_{4}} = \dfrac{3}{32}$

9. $\dfrac{\frac{a}{b}}{\frac{c}{}} = \dfrac{a}{b} \cdot \dfrac{c}{b} = \dfrac{ac}{b}$

11. $\dfrac{\frac{16}{2x-2}}{\frac{8}{x-3}} = \dfrac{\cancel{16}}{2(x-1)} \cdot \dfrac{x-3}{\cancel{8}} = \dfrac{x-3}{x-1}$

13. $\dfrac{\frac{x+4}{6}}{\frac{2x+8}{18}} = \dfrac{(x+4)}{\cancel{6}} \cdot \dfrac{\cancel{18}^{3}}{2(x+4)} = \dfrac{3}{2}$

15. $\dfrac{\frac{3x^2}{x+2}}{\frac{2x}{x+2}} = \dfrac{3x^2}{(x+2)} \cdot \dfrac{(x+2)}{2x} = \dfrac{3x^2}{2x} = \dfrac{3x}{2}$

17. $\dfrac{\frac{4}{9} - \frac{1}{3}}{\frac{7}{12} - \frac{5}{18}} = \dfrac{\frac{4}{9} - \frac{3}{9}}{\frac{21}{36} - \frac{10}{36}} = \dfrac{\frac{1}{9}}{\frac{11}{36}} = \dfrac{1}{\cancel{9}_{1}} \cdot \dfrac{\cancel{36}^{4}}{11} = \dfrac{4}{11}$

19. $\dfrac{1 + \frac{x}{2}}{1 - \frac{x}{2}} = \dfrac{\frac{2}{2} + \frac{x}{2}}{\frac{2}{2} - \frac{x}{2}} = \dfrac{\frac{2+x}{2}}{\frac{2-x}{2}} = \dfrac{2+x}{\cancel{2}} \cdot \dfrac{\cancel{2}}{2-x} = \dfrac{2+x}{2-x}$

21. $\dfrac{\left(\frac{1}{t} - 1\right) \cdot t^3}{\left(\frac{1}{t^3} - 1\right) \cdot t^3} = \dfrac{t^2 - t^3}{1 - t^3}$

$= \dfrac{t^2(1 - t)}{(1 - t)(1 + t + t^2)}$

$= \dfrac{t^2}{1 + t + t^2}$

23. $\dfrac{\left(1 - \frac{2}{v} - \frac{3}{v^2}\right) \cdot v^2}{\left(1 - \frac{4}{v} + \frac{3}{v^2}\right) \cdot v^2} = \dfrac{v^2 - 2v - 3}{v^2 - 4v + 3}$

$= \dfrac{(v+1)(v-3)}{(v-1)(v-3)}$

$= \dfrac{v+1}{v-1}$

25. $\dfrac{\left(\frac{2}{x} - \frac{7}{x^2} - \frac{30}{x^3}\right) \cdot x^3}{\left(2 + \frac{11}{x} + \frac{15}{x^2}\right) \cdot x^3} = \dfrac{2x^2 - 7x - 30}{2x^3 + 11x^2 + 15x}$

$= \dfrac{(2x+5)(x-6)}{x(2x^2 + 11x + 15)}$

$= \dfrac{(2x+5)(x-6)}{x(2x+5)(x+3)}$

$= \dfrac{x-6}{x(x+3)}$

27. $\dfrac{\left(5 + \frac{4}{r+8}\right) \cdot (r+8)}{(r+8) \cdot (r+8)} = \dfrac{5(r+8) + 4}{(r+8)^2}$

$= \dfrac{5r + 40 + 4}{(r+8)^2}$

$= \dfrac{5r + 44}{(r+8)^2}$

29. $\dfrac{\left(t+\dfrac{9}{t-7}\right)\cdot(t-7)}{\left(11-\dfrac{3}{t-7}\right)\cdot(t-7)} = \dfrac{t(t-7)+9}{11(t-7)-3}$

$= \dfrac{t^2-7t+9}{11t-77-3}$

$= \dfrac{t^2-7t+9}{11t-80}$

31. $\dfrac{\left(\dfrac{x+4}{(x+3)(x-3)}\right)\cdot(x+3)(x-3)}{\left(2+\dfrac{1}{x+3}\right)\cdot(x+3)(x-3)}$

$= \dfrac{x+4}{2(x+3)(x-3)+1(x-3)}$

$= \dfrac{x+4}{2x^2-18+x-3}$

$= \dfrac{x+4}{2x^2+x-21}$

33. $\dfrac{\left(\dfrac{5}{a+3}-\dfrac{4}{a-3}\right)\cdot(a+3)(a-3)}{\left(\dfrac{3}{a+3}-\dfrac{2}{a-3}\right)\cdot(a+3)(a-3)}$

$= \dfrac{5(a-3)-4(a+3)}{3(a-3)-2(a+3)}$

$= \dfrac{5a-15-4a-12}{3a-9-2a-6}$

$= \dfrac{a-27}{a-15}$

35. $\dfrac{\left(\dfrac{r+3}{r-3}-\dfrac{r-3}{r+3}\right)\cdot(r-3)(r+3)}{\left(\dfrac{r+3}{r-3}+\dfrac{r-3}{r+3}\right)\cdot(r-3)(r+3)}$

$= \dfrac{(r+3)(r+3)-(r-3)(r-3)}{(r+3)(r+3)+(r-3)(r-3)}$

$= \dfrac{\left(r^2+6r+9\right)-\left(r^2-6r+9\right)}{\left(r^2+6r+9\right)+\left(r^2-6r+9\right)}$

$= \dfrac{r^2+6r+9-r^2+6r-9}{2r^2+18}$

$= \dfrac{12r}{2r^2+18}$

$= \dfrac{\cancel{2}(6r)}{\cancel{2}\left(r^2+9\right)}$

$= \dfrac{6r}{r^2+9}$

37. $\dfrac{\left(\dfrac{3x-9}{x-3}-\dfrac{x-3}{x-5}\right)\cdot(x-3)(x-5)}{\left(\dfrac{x+5}{x-5}+\dfrac{2x-6}{x-3}\right)\cdot(x-3)(x-5)}$

$= \dfrac{(3x-9)(x-5)-(x-3)(x-3)}{(x+5)(x-3)+(2x-6)(x-5)}$

$= \dfrac{\left(3x^2-24x+45\right)-\left(x^2-6x+9\right)}{\left(x^2+2x-15\right)+\left(2x^2-16x+30\right)}$

$= \dfrac{3x^2-24x+45-x^2+6x-9}{3x^2-14x+15}$

$= \dfrac{2x^2-18x+36}{3x^2-14x+15}$

$= \dfrac{2\left(x^2-9x+18\right)}{(3x-5)(x-3)}$

$= \dfrac{2(x-6)\cancel{(x-3)}}{(3x-5)\cancel{(x-3)}}$

$= \dfrac{2(x-6)}{3x-5}$

39. $\dfrac{\left(\dfrac{2}{y^2}-\dfrac{5}{xy}-\dfrac{3}{x^2}\right)\cdot\left(x^2y^2\right)}{\left(\dfrac{2}{y^2}+\dfrac{5}{xy}+\dfrac{2}{x^2}\right)\cdot\left(x^2y^2\right)}=\dfrac{2x^2-5xy-3y^2}{2x^2+5xy+2y^2}$

$\qquad\qquad = \dfrac{\cancel{(2x+y)}\,(x-3y)}{\cancel{(2x+y)}\,(x+2y)}$

$\qquad\qquad = \dfrac{x-3y}{x+2y}$

41. $\dfrac{\left(\dfrac{2a}{a+6}+\dfrac{1}{a}\right)\cdot a(a+6)}{\left(\dfrac{9a}{1}-\dfrac{3}{a+6}\right)\cdot a(a+6)}=\dfrac{2a\cdot a+1(a+6)}{9a\cdot a(a+6)-3a}$

$\qquad\qquad\qquad = \dfrac{2a^2+a+6}{9a^3+54a^2-3a}$

43. $\dfrac{3x^{-1}}{3x^{-1}+1}=\dfrac{\dfrac{3}{x}}{\dfrac{3}{x}+1}=\dfrac{\left(\dfrac{3}{x}\right)\cdot x}{\left(\dfrac{3}{x}+1\right)\cdot x}=\dfrac{3}{3+x}$

45. $\dfrac{1-9x^{-2}}{1+3x^{-1}}=\dfrac{1-\dfrac{9}{x^2}}{1+\dfrac{3}{x}}$

$\qquad = \dfrac{\left(1-\dfrac{9}{x^2}\right)\cdot x^2}{\left(1+\dfrac{3}{x}\right)\cdot x^2}$

$\qquad = \dfrac{x^2-9}{x^2+3x}$

$\qquad = \dfrac{\cancel{(x+3)}\,(x-3)}{x\,\cancel{(x+3)}}$

$\qquad = \dfrac{x-3}{x}$

47. $\dfrac{3x^{-2}+5y^{-1}}{x^{-1}+y^{-1}}=\dfrac{\dfrac{3}{x^2}+\dfrac{5}{y}}{\dfrac{1}{x}+\dfrac{1}{y}}$

$\qquad = \dfrac{\left(\dfrac{3}{x^2}+\dfrac{5}{y}\right)\cdot x^2y}{\left(\dfrac{1}{x}+\dfrac{1}{y}\right)\cdot x^2y}$

$\qquad = \dfrac{3y+5x^2}{xy+x^2}$

49. $\dfrac{x^{-2}y^{-2}}{4x^{-1}+3y^{-1}}=\dfrac{\dfrac{1}{x^2y^2}}{\dfrac{4}{x}+\dfrac{3}{y}}$

$\qquad = \dfrac{\left(\dfrac{1}{x^2y^2}\right)\cdot x^2y^2}{\left(\dfrac{4}{x}+\dfrac{3}{y}\right)\cdot x^2y^2}$

$\qquad = \dfrac{1}{4xy^2+3x^2y}$

51. $\dfrac{36a^{-2}-25b^{-2}}{6a^{-1}+5b^{-1}}=\dfrac{\dfrac{36}{a^2}-\dfrac{25}{b^2}}{\dfrac{6}{a}+\dfrac{5}{b}}$

$\qquad = \dfrac{\left(\dfrac{36}{a^2}-\dfrac{25}{b^2}\right)\cdot a^2b^2}{\left(\dfrac{6}{a}+\dfrac{5}{b}\right)\cdot a^2b^2}$

$\qquad = \dfrac{36b^2-25a^2}{6ab^2+5a^2b}$

$\qquad = \dfrac{\cancel{(6b+5a)}\,(6b-5a)}{ab\,\cancel{(6b+5a)}}$

$\qquad = \dfrac{6b-5a}{ab}$

53. $\dfrac{(4a)^{-1} + 2b^{-2}}{2a^{-1} + b^{-2}} = \dfrac{\dfrac{1}{4a} + \dfrac{2}{b^2}}{\dfrac{2}{a} + \dfrac{1}{b^2}}$

$= \dfrac{\left(\dfrac{1}{4a} + \dfrac{2}{b^2}\right) \cdot 4ab^2}{\left(\dfrac{2}{a} + \dfrac{1}{b^2}\right) \cdot 4ab^2}$

$= \dfrac{b^2 + 8a}{8b^2 + 4a}$

55. Mistake: The +1 was omitted in the
    multiplication of $\dfrac{1}{3} \cdot 3$.

    Correct: $\dfrac{a + \dfrac{1}{3}}{b + \dfrac{1}{3}} = \dfrac{\left(a + \dfrac{1}{3}\right) \cdot 3}{\left(b + \dfrac{1}{3}\right) \cdot 3} = \dfrac{3a + 1}{3b + 1}$

57. Mistake: Did not multiply n by n.

    Correct: $\dfrac{n + \dfrac{3}{n}}{\dfrac{3}{n}} = \dfrac{\left(n + \dfrac{3}{n}\right) \cdot n}{\left(\dfrac{3}{n}\right) \cdot n} = \dfrac{n^2 + 3}{3}$

59. Average rate $= \dfrac{20 + 20}{\dfrac{1}{3} + \dfrac{1}{4}}$

$= \dfrac{(40) \cdot 12}{\left(\dfrac{1}{3} + \dfrac{1}{4}\right) \cdot 12}$

$= \dfrac{480}{4 + 3}$

$= \dfrac{480}{7}$

≈ 68.6 mph

61. $w = \dfrac{A}{l} = \dfrac{\dfrac{3x^2 - 11x - 4}{18}}{\dfrac{x - 4}{2}}$

$= \dfrac{\left(\dfrac{3x^2 - 11x - 4}{18}\right) \cdot 18}{\left(\dfrac{x - 4}{2}\right) \cdot 18}$

$= \dfrac{3x^2 - 11x - 4}{9x - 36}$

$= \dfrac{(3x + 1)\,(\cancel{x - 4})}{9\,(\cancel{x - 4})}$

$= \dfrac{3x + 1}{9}$ in.

63. a) $\dfrac{1}{\dfrac{1}{R_1} + \dfrac{1}{R_2}} = \dfrac{1 \cdot R_1 R_2}{\left(\dfrac{1}{R_1} + \dfrac{1}{R_2}\right) \cdot R_1 R_2} = \dfrac{R_1 R_2}{R_2 + R_1}$

  b) $\dfrac{R_1 R_2}{R_2 + R_1} = \dfrac{60 \cdot 40}{60 + 40} = \dfrac{2400}{100} = 24$ ohms

  c) $\dfrac{R_1 R_2}{R_2 + R_1} = \dfrac{40 \cdot 40}{40 + 40} = \dfrac{1600}{80} = 20$ ohms

Review Exercises

1. $3x + 5 = 2$
     $3x = -3$
     $x = -1$

2. $4x - 6 = 2x - 14$
     $2x - 6 = -14$
     $2x = -8$
     $x = -4$

3. $3x^2 + 7x = 0$
     $x(3x + 7) = 0$
     $x = 0$ or $3x + 7 = 0$
     $x = 0$ $x = -\dfrac{7}{3}$

4. $2x^2 - 7x - 15 = 0$
     $(2x + 3)(x - 5) = 0$
     $2x + 3 = 0$ or $x - 5 = 0$
     $x = -\dfrac{3}{2}$ $x = 5$

5. $d = rt$

$$2.6 = r \cdot \frac{3}{4}$$

$$\frac{4}{3} \cdot \frac{13}{5} = r \cdot \frac{3}{4} \cdot \frac{4}{3}$$

$$\frac{52}{15} = r$$

$$3.4\overline{6} = r$$

6.

	Rate in mph	Time	Distance
Northbound	40	t	$40t$
Southbound	60	t	$60t$

$$40t + 60t = 20$$
$$100t = 20$$
$$t = 0.2 \text{ hr. or } 12 \text{ min.}$$

Exercise Set 7.4

1. The value of x that causes expressions in the equation to be undefined is 4.

3. The values of x that cause expressions in the equation to be undefined are 2 and –6.

5. $x^2 - 4 = (x+2)(x-2)$

$x^2 + 5x + 6 = (x+3)(x+2)$

$x^2 + x - 6 = (x+3)(x-2)$

The values of x that cause expressions in the equation to be undefined are 2, –2, and –3.

7. $$\frac{4}{3u} - \frac{1}{2u} = \frac{5}{6}$$

$$6u \cdot \left(\frac{4}{3u} - \frac{1}{2u} \right) = 6u \cdot \frac{5}{6}$$

$$8 - 3 = 5u$$
$$5 = 5u$$
$$1 = u$$

Check: $\frac{4}{3} - \frac{1}{2} \overset{?}{=} \frac{5}{6}$

$\frac{8}{6} - \frac{3}{6} \overset{?}{=} \frac{5}{6}$

$\frac{5}{6} = \frac{5}{6}$

9. $$\frac{5}{4w} - \frac{3}{5w} = -\frac{13}{40}$$

$$40w \cdot \left(\frac{5}{4w} - \frac{3}{5w} \right) = 40w \cdot \left(-\frac{13}{40} \right)$$

$$50 - 24 = -13w$$
$$26 = -13w$$
$$-2 = w$$

Check: $\frac{5}{-8} - \frac{3}{-10} \overset{?}{=} -\frac{13}{40}$

$-\frac{25}{40} + \frac{12}{40} \overset{?}{=} -\frac{13}{40}$

$-\frac{13}{40} = -\frac{13}{40}$

11. $$\frac{7}{x-3} = 8 - \frac{1}{x-3}$$

$$(x-3)\left(\frac{7}{x-3} \right) = (x-3)\left(8 - \frac{1}{x-3} \right)$$

$$7 = 8(x-3) - 1$$
$$7 = 8x - 24 - 1$$
$$7 = 8x - 25$$
$$32 = 8x$$
$$4 = x$$

Check: $\frac{7}{1} \overset{?}{=} 8 - \frac{1}{1}$

$7 \overset{?}{=} 8 - 1$

$7 = 7$

13. $$\frac{t}{t+4} = 3 - \frac{12}{t+4}$$

$$(t+4)\left(\frac{t}{t+4} \right) = (t+4)\left(3 - \frac{12}{t+4} \right)$$

$$t = 3(t+4) - 12$$
$$t = 3t + 12 - 12$$
$$t - 3t = 0$$
$$-2t = 0$$
$$t = 0$$

Check: $\frac{0}{4} \overset{?}{=} 3 - \frac{12}{4}$

$0 \overset{?}{=} 3 - 3$

$0 = 0$

15.

$$\frac{6a}{a+5} = \frac{3}{a+5} + 3$$

$$(a+5)\left(\frac{6a}{a+5}\right) = (a+5)\left(\frac{3}{a+5} + 3\right)$$

$$6a = 3 + 3(a+5)$$

$$6a = 3 + 3a + 15$$

$$6a = 3a + 18$$

$$3a = 18$$

$$a = 6$$

Check: $\dfrac{36}{11} \overset{?}{=} \dfrac{3}{11} + 3$

$$\frac{36}{11} \overset{?}{=} \frac{3}{11} + \frac{33}{11}$$

$$\frac{36}{11} = \frac{36}{11}$$

17.

$$\frac{4x}{2x-3} = 3 + \frac{6}{2x-3}$$

$$(2x-3)\left(\frac{4x}{2x-3}\right) = (2x-3)\left(3 + \frac{6}{2x-3}\right)$$

$$4x = 3(2x-3) + 6$$

$$4x = 6x - 9 + 6$$

$$4x = 6x - 3$$

$$-2x = -3$$

$$x = \frac{3}{2}$$

Notice that if $x = \dfrac{3}{2}$, then $\dfrac{4x}{2x-3}$ and $\dfrac{6}{2x-3}$ are undefined. Therefore, $x = \dfrac{3}{2}$ is extraneous and the equation has no solution.

19.

$$\frac{3x}{x+3} = \frac{6}{x+7}$$

$$3x(x+7) = 6(x+3)$$

$$3x^2 + 21x = 6x + 18$$

$$3x^2 + 15x - 18 = 0$$

$$3(x^2 + 5x - 6) = 0$$

$$3(x-1)(x+6) = 0$$

$$x - 1 = 0 \quad \text{or} \quad x + 6 = 0$$

$$x = 1 \qquad\qquad x = -6$$

Check $x = 1$:

$$\frac{3(1)}{1+3} \overset{?}{=} \frac{6}{1+7}$$

$$\frac{3}{4} \overset{?}{=} \frac{6}{8}$$

$$\frac{3}{4} = \frac{3}{4}$$

Check $x = -6$:

$$\frac{3(-6)}{-6+3} \overset{?}{=} \frac{6}{-6+7}$$

$$\frac{-18}{-3} \overset{?}{=} \frac{6}{1}$$

$$6 = 6$$

21.

$$\frac{m}{m-1} = \frac{3m}{4m-3}$$

$$(4m-3) \cdot m = (m-1) \cdot 3m$$

$$4m^2 - 3m = 3m^2 - 3m$$

$$m^2 = 0$$

$$m = 0$$

Check: $\dfrac{0}{-1} \overset{?}{=} \dfrac{0}{-3}$

$$0 = 0$$

23.

$$\frac{2y}{3y-6} + \frac{3}{4y-8} = \frac{1}{4}$$

$$12(y-2)\left(\frac{2y}{3(y-2)} + \frac{3}{4(y-2)}\right) = 12(y-2)\left(\frac{1}{4}\right)$$

$$8y + 9 = 3y - 6$$

$$5y + 9 = -6$$

$$5y = -15$$

$$y = -3$$

Check: $\dfrac{-6}{-15} + \dfrac{3}{-20} \overset{?}{=} \dfrac{1}{4}$

$$\frac{24}{60} - \frac{9}{60} \overset{?}{=} \frac{1}{4}$$

$$\frac{15}{60} \overset{?}{=} \frac{1}{4}$$

$$\frac{1}{4} = \frac{1}{4}$$

25.
$$\frac{a^2}{a+2} - 3 = \frac{a+6}{a+2} - 4$$

$$(a+2)\left(\frac{a^2}{a+2} - 3\right) = (a+2)\left(\frac{a+6}{a+2} - 4\right)$$

$$a^2 - 3(a+2) = (a+6) - 4(a+2)$$

$$a^2 - 3a - 6 = a + 6 - 4a - 8$$

$$a^2 - 3a - 6 = -3a - 2$$

$$a^2 - 4 = 0$$

$$(a+2)(a-2) = 0$$

$$a+2 = 0 \quad \text{or} \quad a-2 = 0$$

$$a = -2 \qquad\qquad a = 2$$

Notice that if $a = -2$, then $\dfrac{a^2}{a+2}$ and $\dfrac{a+6}{a+2}$ are undefined. Therefore, $a = -2$ is extraneous, and 2 is the only solution.

Check: $\dfrac{4}{4} - 3 \overset{?}{=} \dfrac{8}{4} - 4$

$$1 - 3 \overset{?}{=} 2 - 4$$

$$-2 = -2$$

27.
$$2x - \frac{12}{x} = 5$$

$$x\left(2x - \frac{12}{x}\right) = x \cdot 5$$

$$2x^2 - 12 = 5x$$

$$2x^2 - 5x - 12 = 0$$

$$(2x+3)(x-4) = 0$$

$$2x+3 = 0 \quad \text{or} \quad x-4 = 0$$

$$2x = -3 \qquad\qquad x = 4$$

$$x = -\frac{3}{2}$$

Check $x = -\dfrac{3}{2}$:

$$2\left(-\frac{3}{2}\right) - \frac{12}{\left(-\dfrac{3}{2}\right)} \overset{?}{=} 5$$

$$-3 - (-8) \overset{?}{=} 5$$

$$5 = 5$$

Check $x = 4$:

$$2(4) - \frac{12}{(4)} \overset{?}{=} 5$$

$$8 - 3 \overset{?}{=} 5$$

$$5 = 5$$

29.
$$\frac{x+1}{x+2} + \frac{5}{2x} = \frac{3x+2}{x+2}$$

$$2x(x+2)\left(\frac{x+1}{x+2} + \frac{5}{2x}\right) = 2x(x+2)\left(\frac{3x+2}{x+2}\right)$$

$$2x(x+1) + 5(x+2) = 2x(3x+2)$$

$$2x^2 + 2x + 5x + 10 = 6x^2 + 4x$$

$$-4x^2 + 3x + 10 = 0$$

$$4x^2 - 3x - 10 = 0$$

$$(4x+5)(x-2) = 0$$

$$4x+5 = 0 \quad \text{or} \quad x-2 = 0$$

$$4x = -5 \qquad\qquad x = 2$$

$$x = -\frac{5}{4}$$

Check $x = -\dfrac{5}{4}$:

$$\frac{-\dfrac{5}{4}+1}{-\dfrac{5}{4}+2} + \frac{5}{2\left(-\dfrac{5}{4}\right)} \overset{?}{=} \frac{-\dfrac{15}{4}+2}{-\dfrac{5}{4}+2}$$

$$\frac{-\dfrac{1}{4}}{\dfrac{3}{4}} + \frac{5}{-\dfrac{5}{2}} \overset{?}{=} \frac{-\dfrac{7}{4}}{\dfrac{3}{4}}$$

$$-\frac{1}{3} - 2 \overset{?}{=} -\frac{7}{3}$$

$$-\frac{1}{3} - \frac{6}{3} \overset{?}{=} -\frac{7}{3}$$

$$-\frac{7}{3} = -\frac{7}{3}$$

Check $x = 2$:

$$\frac{3}{4} + \frac{5}{4} \overset{?}{=} \frac{8}{4}$$

$$2 = 2$$

31. $\dfrac{2x+1}{x+2} - \dfrac{x+1}{3x-2} = \dfrac{x}{x+2}$

$(x+2)(3x-2)\left(\dfrac{2x+1}{x+2} - \dfrac{x+1}{3x-2}\right)$

$= (x+2)(3x-2)\left(\dfrac{x}{x+2}\right)$

$(3x-2)(2x+1) - (x+2)(x+1) = (3x-2)(x)$

$6x^2 - x - 2 - x^2 - 3x - 2 = 3x^2 - 2x$

$2x^2 - 2x - 4 = 0$

$2(x^2 - x - 2) = 0$

$2(x-2)(x+1) = 0$

$x - 2 = 0 \quad \text{or} \quad x + 1 = 0$

$\qquad x = 2 \qquad\qquad x = -1$

Check $x = 2$: $\dfrac{5}{4} - \dfrac{3}{4} \overset{?}{=} \dfrac{2}{4}$

$\dfrac{1}{2} = \dfrac{1}{2}$

Check $x = -1$: $\dfrac{-1}{1} - \dfrac{0}{-5} \overset{?}{=} \dfrac{-1}{1}$

$-1 - 0 \overset{?}{=} -1$

$-1 = -1$

33. $\dfrac{4}{x+5} - \dfrac{2}{x-5} = \dfrac{4x-20}{x^2-25}$

$(x+5)(x-5)\left(\dfrac{4}{x+5} - \dfrac{2}{x-5}\right)$

$= (x+5)(x-5)\left(\dfrac{4x-20}{(x+5)(x-5)}\right)$

$4(x-5) - 2(x+5) = 4x - 20$

$4x - 20 - 2x - 10 = 4x - 20$

$2x - 30 = 4x - 20$

$-2x = 10$

$x = -5$

Notice that if $x = -5$, then $\dfrac{4}{x+5}$ and $\dfrac{4x-20}{x^2-25}$ are undefined. Therefore, $x = -5$ is extraneous and the equation has no solution.

35. $\dfrac{x}{x-2} - \dfrac{4}{x-1} = \dfrac{2}{x^2-3x+2}$

$(x-2)(x-1)\left(\dfrac{x}{x-2} - \dfrac{4}{x-1}\right)$

$= (x-2)(x-1)\left(\dfrac{2}{(x-2)(x-1)}\right)$

$(x-1)(x) - (x-2)(4) = 2$

$x^2 - x - 4x + 8 = 2$

$x^2 - 5x + 6 = 0$

$(x-2)(x-3) = 0$

$x - 2 = 0 \quad \text{or} \quad x - 3 = 0$

$\qquad x = 2 \qquad\qquad x = 3$

Notice that if $x = 2$, then $\dfrac{x}{x-2}$ and $\dfrac{2}{x^2-3x+2}$ are undefined. Therefore, $x = 2$ is extraneous, and the only solution is $x = 3$.

Check: $\dfrac{3}{1} - \dfrac{4}{2} \overset{?}{=} \dfrac{2}{2}$

$3 - 2 \overset{?}{=} 1$

$1 = 1$

37. $\dfrac{6}{t^2+t-12} = \dfrac{4}{t^2-t-6} + \dfrac{1}{t^2+6t+8}$

$(t+4)(t-3)(t+2)\left(\dfrac{6}{(t+4)(t-3)}\right)$

$= (t+4)(t-3)(t+2)$

$\cdot \left(\dfrac{4}{(t-3)(t+2)} + \dfrac{1}{(t+4)(t+2)}\right)$

$6(t+2) = 4(t+4) + 1(t-3)$

$6t + 12 = 4t + 16 + t - 3$

$6t + 12 = 5t + 13$

$t = 1$

Check: $\dfrac{6}{-10} \overset{?}{=} \dfrac{4}{-6} + \dfrac{1}{15}$

$-\dfrac{3}{5} \overset{?}{=} -\dfrac{10}{15} + \dfrac{1}{15}$

$-\dfrac{3}{5} \overset{?}{=} -\dfrac{9}{15}$

$-\dfrac{3}{5} = -\dfrac{3}{5}$

39. $\dfrac{5}{x^2 - x - 6} - \dfrac{8}{x^2 + 2x - 15} = \dfrac{3}{x^2 + 7x + 10}$

$(x-3)(x+2)(x+5)$

$\cdot \left(\dfrac{5}{(x-3)(x+2)} - \dfrac{8}{(x+5)(x-3)} \right)$

$= (x-3)(x+2)(x+5) \left(\dfrac{3}{(x+5)(x+2)} \right)$

$5(x+5) - 8(x+2) = 3(x-3)$

$5x + 25 - 8x - 16 = 3x - 9$

$-3x + 9 = 3x - 9$

$-6x = -18$

$x = 3$

Notice that if $x = 3$, then $\dfrac{5}{x^2 - x - 6}$ and

$\dfrac{8}{x^2 + 2x - 15}$ are undefined. Therefore, $x = 3$ is

extraneous and the equation has no solution.

41. $\dfrac{p+1}{p^2 + 8p + 15} - \dfrac{2}{p^2 + 7p + 10} = \dfrac{1}{p^2 + 5p + 6}$

$(p+3)(p+5)(p+2) \left(\dfrac{\dfrac{p+1}{(p+3)(p+5)}}{-\dfrac{2}{(p+5)(p+2)}} \right)$

$= (p+3)(p+5)(p+2) \left(\dfrac{1}{(p+3)(p+2)} \right)$

$(p+2)(p+1) - 2(p+3) = (p+5)(1)$

$p^2 + 3p + 2 - 2p - 6 = p + 5$

$p^2 + p - 4 = p + 5$

$p^2 = 9$

$p^2 - 9 = 0$

$(p+3)(p-3) = 0$

$p + 3 = 0 \quad \text{or} \quad p - 3 = 0$

$p = -3 \qquad\qquad p = 3$

Notice that if $p = -3$, then $\dfrac{p+1}{p^2 + 8p + 15}$ and

$\dfrac{1}{p^2 + 5p + 6}$ are undefined. Therefore, $p = -3$

is extraneous, and the only solution is $p = 3$.

Check: $\dfrac{4}{48} - \dfrac{2}{40} \overset{?}{=} \dfrac{1}{30}$

$\dfrac{1}{12} - \dfrac{1}{20} \overset{?}{=} \dfrac{1}{30}$

$\dfrac{5}{60} - \dfrac{3}{60} \overset{?}{=} \dfrac{1}{30}$

$\dfrac{2}{60} \overset{?}{=} \dfrac{1}{30}$

$\dfrac{1}{30} = \dfrac{1}{30}$

43. $\dfrac{a-5}{a+5} + \dfrac{a+15}{a-5} = 2 - \dfrac{10}{a^2 - 25}$

$(a+5)(a-5) \left(\dfrac{a-5}{a+5} + \dfrac{a+15}{a-5} \right)$

$= (a+5)(a-5) \left(2 - \dfrac{10}{(a+5)(a-5)} \right)$

$(a-5)^2 + (a+5)(a+15) = 2(a+5)(a-5) - 10$

$a^2 - 10a + 25 + a^2 + 20a + 75 = 2a^2 - 50 - 10$

$2a^2 + 10a + 100 = 2a^2 - 60$

$10a = -160$

$a = -16$

Check: $\dfrac{-21}{-11} + \dfrac{-1}{-21} \overset{?}{=} 2 - \dfrac{10}{231}$

$\dfrac{441}{231} + \dfrac{11}{231} \overset{?}{=} \dfrac{462}{231} - \dfrac{10}{231}$

$\dfrac{452}{231} = \dfrac{452}{231}$

45. $\dfrac{3n}{n^2 - 2n - 15} = \dfrac{2n}{n-5} + \dfrac{n}{n+3}$

$(n-5)(n+3) \left(\dfrac{3n}{(n-5)(n+3)} \right)$

$= (n-5)(n+3) \left(\dfrac{2n}{n-5} + \dfrac{n}{n+3} \right)$

$3n = 2n(n+3) + n(n-5)$

$3n = 2n^2 + 6n + n^2 - 5n$

$0 = 3n^2 - 2n$

$0 = n(3n - 2)$

$n = 0 \quad \text{or} \quad 3n - 2 = 0$

$3n = 2$

$n = \dfrac{2}{3}$

Check $n = 0$: $\dfrac{0}{-15} \stackrel{?}{=} \dfrac{0}{-5} + \dfrac{0}{3}$

$$0 = 0$$

Check $n = \dfrac{2}{3}$:

$$-\dfrac{2}{\dfrac{143}{9}} \stackrel{?}{=} \dfrac{\dfrac{4}{3}}{-\dfrac{13}{3}} + \dfrac{\dfrac{2}{3}}{\dfrac{11}{3}}$$

$$-\dfrac{18}{143} \stackrel{?}{=} -\dfrac{4}{13} + \dfrac{2}{11}$$

$$-\dfrac{18}{143} \stackrel{?}{=} -\dfrac{44}{143} + \dfrac{26}{143}$$

$$-\dfrac{18}{143} = -\dfrac{18}{143}$$

47. $\dfrac{x+5}{x-5} - \dfrac{x-5}{x+5} = \dfrac{40}{x^2 - 25}$

$$(x+5)(x-5)\left(\dfrac{x+5}{x-5} - \dfrac{x-5}{x+5}\right)$$

$$= (x+5)(x-5)\left(\dfrac{40}{(x+5)(x-5)}\right)$$

$$(x+5)^2 - (x-5)^2 = 40$$

$$x^2 + 10x + 25 - x^2 + 10x - 25 = 40$$

$$20x = 40$$

$$x = 2$$

Check: $\dfrac{2+5}{2-5} - \dfrac{2-5}{2+5} \stackrel{?}{=} \dfrac{40}{4-25}$

$$\dfrac{7}{-3} - \dfrac{-3}{7} \stackrel{?}{=} \dfrac{40}{-21}$$

$$-\dfrac{49}{21} + \dfrac{9}{21} \stackrel{?}{=} -\dfrac{40}{21}$$

$$-\dfrac{40}{21} = -\dfrac{40}{21}$$

49. $\dfrac{2}{2x^2 - 4x - 10} = \dfrac{3}{3x^2 + 6x - 3}$

$$\dfrac{2}{2(x^2 - 2x - 5)} = \dfrac{3}{3(x^2 + 2x - 1)}$$

$$\dfrac{1}{x^2 - 2x - 5} = \dfrac{1}{x^2 + 2x - 1}$$

$$x^2 + 2x - 1 = x^2 - 2x - 5$$

$$4x - 1 = -5$$

$$4x = -4$$

$$x = -1$$

Check:

$$\dfrac{2}{2(-1)^2 - 4(-1) - 10} \stackrel{?}{=} \dfrac{3}{3(-1)^2 + 6(-1) - 3}$$

$$\dfrac{2}{2 + 4 - 10} \stackrel{?}{=} \dfrac{3}{3 - 6 - 3}$$

$$\dfrac{2}{-4} \stackrel{?}{=} \dfrac{3}{-6}$$

$$-\dfrac{1}{2} = -\dfrac{1}{2}$$

51. $1 = \dfrac{3}{a-2} - \dfrac{12}{a^2 - 4}$

$$(a+2)(a-2)(1)$$

$$= (a+2)(a-2)\left(\dfrac{3}{a-2} - \dfrac{12}{(a+2)(a-2)}\right)$$

$$a^2 - 4 = 3(a+2) - 12$$

$$a^2 - 4 = 3a + 6 - 12$$

$$a^2 - 3a + 2 = 0$$

$$(a-2)(a-1) = 0$$

$$a - 2 = 0 \quad \text{or} \quad a - 1 = 0$$

$$a = 2 \qquad\qquad a = 1$$

Notice that if $a = 2$, then $\dfrac{3}{a-2}$ and $\dfrac{12}{a^2 - 4}$ are undefined. Therefore, $a = 2$ is extraneous, and the only solution is $a = 1$.

Check: $1 \stackrel{?}{=} \dfrac{3}{-1} - \dfrac{12}{-3}$

$$1 \stackrel{?}{=} -3 - (-4)$$

$$1 = 1$$

53. $\dfrac{w+2}{w^2-5w+6}+\dfrac{2w}{w^2-w-2}=-\dfrac{2}{w^2-2w-3}$

$(w-3)(w-2)(w+1)$

$\quad \cdot\left(\dfrac{w+2}{(w-3)(w-2)}+\dfrac{2w}{(w-2)(w+1)}\right)$

$\quad=(w-3)(w-2)(w+1)\left(-\dfrac{2}{(w-3)(w+1)}\right)$

$(w+2)(w+1)+(2w)(w-3)=-2(w-2)$

$\quad w^2+3w+2+2w^2-6w=-2w+4$

$\quad\quad\quad 3w^2-3w+2=-2w+4$

$\quad\quad\quad\quad 3w^2-w-2=0$

$\quad\quad\quad (3w+2)(w-1)=0$

$3w+2=0 \quad\quad \text{or} \quad w-1=0$

$\quad 3w=-2 \quad\quad\quad\quad\quad w=1$

$\quad\quad w=-\dfrac{2}{3}$

Check $w=-\dfrac{2}{3}$:

$\dfrac{-\dfrac{2}{3}+2}{\dfrac{4}{9}-5\left(-\dfrac{2}{3}\right)+6}+\dfrac{2\left(-\dfrac{2}{3}\right)}{\dfrac{4}{9}+\dfrac{2}{3}-2}\overset{?}{=}-\dfrac{2}{\dfrac{4}{9}-2\left(-\dfrac{2}{3}\right)-3}$

$\dfrac{\dfrac{4}{3}}{\dfrac{88}{9}}+\dfrac{-\dfrac{4}{3}}{-\dfrac{8}{9}}\overset{?}{=}-\dfrac{2}{\dfrac{-11}{9}}$

$\dfrac{3}{22}+\dfrac{3}{2}\overset{?}{=}\dfrac{18}{11}$

$\dfrac{18}{11}=\dfrac{18}{11}$

Check $w=1$:

$\dfrac{3}{2}+\dfrac{2}{-2}\overset{?}{=}-\dfrac{2}{-4}$

$\dfrac{1}{2}=\dfrac{1}{2}$

55. $\dfrac{2p}{2p-3}=\dfrac{15-32p^2}{4p^2-9}+\dfrac{3p}{2p+3}$

$(2p+3)(2p-3)\left(\dfrac{2p}{2p-3}\right)$

$\quad=(2p+3)(2p-3)\cdot\left(\dfrac{15-32p^2}{(2p+3)(2p-3)}+\dfrac{3p}{2p+3}\right)$

$2p(2p+3)=15-32p^2+3p(2p-3)$

$\quad 4p^2+6p=15-32p^2+6p^2-9p$

$30p^2+15p-15=0$

$15\left(2p^2+p-1\right)=0$

$15(2p-1)(p+1)=0$

$2p-1=0 \quad\quad \text{or} \quad p+1=0$

$\quad 2p=1 \quad\quad\quad\quad\quad p=-1$

$\quad p=\dfrac{1}{2}$

Check $p=\dfrac{1}{2}$: $\dfrac{1}{1-3}\overset{?}{=}\dfrac{15-32\left(\dfrac{1}{4}\right)}{4\left(\dfrac{1}{4}\right)-9}+\dfrac{\dfrac{3}{2}}{1+3}$

$\quad\quad -\dfrac{1}{2}\overset{?}{=}\dfrac{7}{-8}+\dfrac{3}{8}$

$\quad\quad -\dfrac{1}{2}\overset{?}{=}-\dfrac{4}{8}$

$\quad\quad -\dfrac{1}{2}=-\dfrac{1}{2}$

Check $p=-1$: $\dfrac{-2}{-2-3}\overset{?}{=}\dfrac{15-32}{4-9}+\dfrac{-3}{-2+3}$

$\quad\quad \dfrac{2}{5}\overset{?}{=}\dfrac{-17}{-5}-3$

$\quad\quad \dfrac{2}{5}\overset{?}{=}\dfrac{17}{5}-\dfrac{15}{5}$

$\quad\quad \dfrac{2}{5}=\dfrac{2}{5}$

57. $\dfrac{4x}{x^2+4x}-\dfrac{2x}{x^2+2x}=\dfrac{3x-2}{x^2+6x+8}$

$x(x+2)(x+4)\left(\dfrac{4x}{x(x+4)}-\dfrac{2x}{x(x+2)}\right)$

$\quad=x(x+2)(x+4)\left(\dfrac{3x-2}{(x+4)(x+2)}\right)$

$4x(x+2)-2x(x+4)=x(3x-2)$

$\quad 4x^2+8x-2x^2-8x=3x^2-2x$

$\quad\quad\quad\quad -x^2+2x=0$

$\quad\quad\quad\quad -x(x-2)=0$

$\quad -x=0 \quad\quad \text{or} \quad x-2=0$

$\quad\quad x=0 \quad\quad\quad\quad\quad x=2$

Notice that if $x=0$, then $\dfrac{4x}{x^2+4x}$ and $\dfrac{2x}{x^2+2x}$ are undefined. Therefore, $x=0$ is extraneous, and the only solution is $x=2$.

Check $x = 2$: $\dfrac{8}{12} - \dfrac{4}{8} \overset{?}{=} \dfrac{4}{24}$

$$\dfrac{16}{24} - \dfrac{12}{24} \overset{?}{=} \dfrac{4}{24}$$

$$\dfrac{4}{24} = \dfrac{4}{24}$$

59. $\dfrac{6}{x^3 + 2x^2 - x - 2} = \dfrac{6}{x+2} - \dfrac{3}{x^2 - 1}$

$$(x+1)(x-1)(x+2)\left(\dfrac{6}{(x+1)(x-1)(x+2)}\right)$$

$$= (x+1)(x-1)(x+2)\left(\dfrac{6}{x+2} - \dfrac{3}{(x+1)(x-1)}\right)$$

$$6 = 6(x+1)(x-1) - 3(x+2)$$

$$6 = 6x^2 - 6 - 3x - 6$$

$$0 = 6x^2 - 3x - 18$$

$$0 = 3(2x^2 - x - 6)$$

$$0 = 3(2x+3)(x-2)$$

$$2x + 3 = 0 \quad \text{or} \quad x - 2 = 0$$

$$2x = -3 \qquad\qquad x = 2$$

$$x = -\dfrac{3}{2}$$

Check $x = -\dfrac{3}{2}$:

$$\dfrac{6}{\left(-\dfrac{3}{2}\right)^3 + 2\left(-\dfrac{3}{2}\right)^2 - \left(-\dfrac{3}{2}\right) - 2}$$

$$\overset{?}{=} \dfrac{6}{\left(-\dfrac{3}{2}\right) + 2} - \dfrac{3}{\left(-\dfrac{3}{2}\right)^2 - 1}$$

$$\dfrac{48}{5} \overset{?}{=} 12 - \dfrac{12}{5}$$

$$\dfrac{48}{5} \overset{?}{=} \dfrac{60}{5} - \dfrac{12}{5}$$

$$\dfrac{48}{5} = \dfrac{48}{5}$$

Check $x = 2$:

$$\dfrac{6}{(2)^3 + 2(2)^2 - (2) - 2} \overset{?}{=} \dfrac{6}{(2)+2} - \dfrac{3}{(2)^2 - 1}$$

$$\dfrac{1}{2} \overset{?}{=} \dfrac{3}{2} - 1$$

$$\dfrac{1}{2} = \dfrac{1}{2}$$

61. $C = \dfrac{90{,}000\,p}{100 - p}$

$$(100 - p)C = (100 - p)\left(\dfrac{90{,}000\,p}{100 - p}\right)$$

$$100C - pC = 90{,}000\,p$$

$$100C = 90{,}000\,p + pC$$

$$100C = p(90{,}000 + C)$$

$$\dfrac{100C}{90{,}000 + C} = p$$

63. $I = \dfrac{2E}{R + 2r}$

$$(R + 2r)I = (R + 2r)\left(\dfrac{2E}{R + 2r}\right)$$

$$RI + 2rI = 2E$$

$$2rI = 2E - RI$$

$$r = \dfrac{2E - RI}{2I}$$

65. $\dfrac{1}{s} + \dfrac{1}{S} = \dfrac{1}{f}$

$$sSf\left(\dfrac{1}{s} + \dfrac{1}{S}\right) = sSf\left(\dfrac{1}{f}\right)$$

$$Sf + sf = sS$$

$$f(S + s) = sS$$

$$f = \dfrac{sS}{S + s}$$

67. $R = \dfrac{1}{\dfrac{1}{R_1} + \dfrac{1}{R_2}}$

$$R = \dfrac{1(R_1 R_2)}{\left(\dfrac{1}{R_1} + \dfrac{1}{R_2}\right)(R_1 R_2)}$$

$$R = \dfrac{R_1 R_2}{R_2 + R_1}$$

$$R(R_2 + R_1) = (R_2 + R_1)\dfrac{R_1 R_2}{R_2 + R_1}$$

$$RR_2 + RR_1 = R_1 R_2$$

$$RR_2 = R_1 R_2 - RR_1$$

$$RR_2 = R_1(R_2 - R)$$

$$\dfrac{RR_2}{R_2 - R} = R_1$$

69. Four is extraneous. There is no solution.

71. Use the formula found in exercise #61.

$$p = \frac{100C}{90,000 + C}$$

$$= \frac{100(22,500)}{90,000 + 22,500}$$

$$= \frac{2,250,000}{112,500}$$

$$= 20\%$$

73.

$$F = \frac{132,000}{330 - s}$$

$$440 = \frac{132,000}{330 - s}$$

$$440(330 - s) = 132,000$$

$$330 - s = 300$$

$$-s = -30$$

$$s = 30$$

The speed is 30 m/sec.

75.

$$T = \frac{80}{x} + \frac{80}{x - 20}$$

$$6 = \frac{80}{x} + \frac{80}{x - 20}$$

$$x(x - 20)(6) = x(x - 20)\left(\frac{80}{x} + \frac{80}{x - 20}\right)$$

$$6x^2 - 120x = 80(x - 20) + 80x$$

$$6x^2 - 120x = 80x - 1600 + 80x$$

$$6x^2 - 280x + 1600 = 0$$

$$2(3x^2 - 140x + 800) = 0$$

$$2(3x - 20)(x - 40) = 0$$

$$3x - 20 = 0 \quad \text{or} \quad x - 40 = 0$$

$$3x = 20 \qquad\qquad x = 40$$

$$x = \frac{20}{3}$$

The solution is 40 mph. Although $\frac{20}{3}$ is also a solution, it is not reasonable because it leads to a time of −6 hours for the return trip.

77.

$$I = \frac{E}{R + r}$$

$$2 = \frac{110}{50 + r}$$

$$2(50 + r) = 110$$

$$100 + 2r = 110$$

$$2r = 10$$

$$r = 5$$

The internal resistance is 5 ohms.

Review Exercises

1. Let n be the number of nickels and d be the number of dimes. Translate to a system of equations.

$$\begin{cases} d + 2 = n \\ 0.05n + 0.10d = 1.30 \end{cases}$$

$$0.05(d + 2) + 0.10d = 1.30$$

$$0.05d + 0.10 + 0.10d = 1.30$$

$$0.15d = 1.20$$

$$d = 8$$

$$8 + 2 = n$$

$$10 = n$$

There are 8 dimes and 10 nickels.

2. Let l be the length of the painting and w be the width of the painting. Translate to a system of equations.

$$\begin{cases} l = 2 + w \\ lw = 80 \end{cases}$$

$$(2 + w)w = 80$$

$$2w + w^2 = 80$$

$$w^2 + 2w - 80 = 0$$

$$(w + 10)(w - 8) = 0$$

$$w = -10 \quad \text{or} \quad w = 8$$

$$l = 2 + 8$$

$$l = 10$$

Since length cannot be negative, −10 is not a reasonable solution. The length is 10 inches and the width is 8 inches.

3. Let one number be x and the other be $(x+3)$.

$$(x+3)^2 - x^2 = 39$$
$$x^2 + 6x + 9 - x^2 = 39$$
$$6x = 30$$
$$x = 5$$

One number is 5 and the other is 8.

4. Let the larger number be x and the smaller be y. Translate to a system of equations.

$$\begin{cases} x + y = 28 \\ x - y = 4 \end{cases}$$

$$\begin{array}{ll} x + y = 28 & 16 + y = 28 \\ \underline{x - y = 4} & y = 12 \\ 2x = 32 & \\ x = 16 & \end{array}$$

The numbers are 16 and 12.

5. Let s be the price of the saw and d be the price of the drill. Translate to a system of equations.

$$\begin{cases} d = s - 54 \\ d + s = 238 \end{cases}$$

$$\begin{array}{ll} (s - 54) + s = 238 & d = 146 - 54 \\ 2s - 54 = 238 & d = 92 \\ 2s = 292 & \\ s = 146 & \end{array}$$

The drill cost \$92 and the saw \$146.

6.

	rate	time	distance
rv	55	x	d
truck	75	$x-4$	d

$$\begin{cases} 55x = d \\ 75(x-4) = d \end{cases}$$

$$55x = 75(x-4)$$
$$55x = 75x - 300$$
$$-20x = -300$$
$$x = 15$$

It will take the truck $15 - 4 = 11$ hours to catch up to the recreational vehicle.

Exercise Set 7.5

1. Complete a table.

Categories	Rate of Work	Time at Work	Amt of Task Completed
Jason	$\dfrac{1}{4}$	t	$\dfrac{t}{4}$
sister	$\dfrac{1}{6}$	t	$\dfrac{t}{6}$

$$\frac{t}{4} + \frac{t}{6} = 1$$
$$24\left(\frac{t}{4} + \frac{t}{6}\right) = 24 \cdot 1$$
$$6t + 4t = 24$$
$$10t = 24$$
$$t = 2\frac{2}{5}$$

Working together, it takes Jason and his sister $2\frac{2}{5}$ hours to wash and wax the car.

3. Complete a table.

Categories	Rate of Work	Time at Work	Amt of Task Completed
Alicia	$\dfrac{1}{25}$	t	$\dfrac{t}{25}$
Geraldine	$\dfrac{1}{35}$	t	$\dfrac{t}{35}$

$$\frac{t}{25} + \frac{t}{35} = 1$$
$$175\left(\frac{t}{25} + \frac{t}{35}\right) = 175 \cdot 1$$
$$7t + 5t = 175$$
$$12t = 175$$
$$t = 14\frac{7}{12}$$

Working together it will take Alicia and Geraldine $14\frac{7}{12}$ days to make the quilt.

5. Complete a table.

Categories	Rate of Work	Time at Work	Amt of Task Completed
1st roofer	$\dfrac{1}{6}$	4	$\dfrac{2}{3}$
2nd roofer	$\dfrac{1}{t}$	4	$\dfrac{4}{t}$

$$\frac{2}{3}+\frac{4}{t}=1$$

$$3t\left(\frac{2}{3}+\frac{4}{t}\right)=3t\cdot 1$$

$$2t+12=3t$$

$$12=t$$

It would take the second roofer 12 hr. to do the job.

7. Complete a table.

Categories	Rate of Work	Time at Work	Amt of Task Completed
cold	$\dfrac{1}{15}$	9	$\dfrac{3}{5}$
hot	$\dfrac{1}{t}$	9	$\dfrac{9}{t}$

$$\frac{3}{5}+\frac{9}{t}=1$$

$$5t\left(\frac{3}{5}+\frac{9}{t}\right)=5t\cdot 1$$

$$3t+45=5t$$

$$45=2t$$

$$22.5=t$$

It would take the hot water faucet 22.5 minutes to fill the tub alone.

9. Complete a table.

Categories	Rate	Time	Distance
northbound bus	63	$\dfrac{x}{63}$	x
southbound car	72	$\dfrac{675-x}{72}$	$675-x$

$$\frac{x}{63}=\frac{675-x}{72} \qquad \text{time}=\frac{x}{63}=\frac{315}{63}=5 \text{ hr.}$$

$$72x=42{,}525-63x$$

$$135x=42{,}525$$

$$x=315$$

They will be 675 miles apart after 5 hours at 3 P.M.

11. Complete a table.

Categories	Rate	Time	Distance
freight train	$\dfrac{3}{5}r$	3	$\dfrac{9}{5}r$
passenger train	r	3	$3r$

$$\frac{9}{5}r+3r=360$$

$$\frac{24}{5}r=360$$

$$r=75$$

The passenger train travels at 75 mph.

13. Complete a table.

Categories	Rate	Time	Distance
Jack	45	$\dfrac{x}{45}$	x
Frances	60	$\dfrac{x}{60}$	x

$$\frac{x}{45}-\frac{x}{60}=2$$

$$180\left(\frac{x}{45}-\frac{x}{60}\right)=180\cdot 2$$

$$4x-3x=360$$

$$x=360$$

Frances' time is $\dfrac{x}{60}=\dfrac{360}{60}=6.$

It takes Frances 6 hours to catch Jack.

15. Complete a table.

Categories	Rate	Time	Distance
against	$r-21$	5.5	$5.5(r-21)$
with	$r+21$	5	$5(r+21)$

$$5.5(r-21)=5(r+21)$$

$$5.5r-115.5=5r+105$$

$$0.5r=220.5$$

$$r=441$$

The speed of the airliner in still air is 441 mph.

17. Translate "a varies directly as b."

$a = k \cdot b$ Now, $a = \dfrac{4}{9}b$

$4 = k \cdot 9$

$\dfrac{4}{9} = k$ $a = \dfrac{4}{9}(27)$

$a = 12$

19. Translate "y varies directly as the square of x."

$y = k \cdot x^2$ Now, $y = 4x^2$

$100 = k \cdot 5^2$ $y = 4(3)^2$

$100 = 25k$ $y = 36$

$4 = k$

21. Translate "m varies directly as n."

$m = k \cdot n$ Now, $m = \dfrac{3}{4}n$

$6 = k \cdot 8$

$\dfrac{3}{4} = k$ $9 = \dfrac{3}{4}n$

$12 = n$

23. Translating "The price increases with the quantity purchased," we write $p = kq$, where p represents purchase price and q represents the quantity.

$p = kq$

$16.25 = k \cdot 2.5$

$6.5 = k$

Now, $p = 6.5q$

$p = 6.5(6)$

$p = 39$

The salmon will cost $39 for 6 pounds.

25. Translating "the volume of a gas varies directly with temperature," we write $v = kt$ where v represents the volume and t represents the temperature.

$v = kt$

$288 = k \cdot 80$

$3.6 = k$

Now, $v = 3.6t$

$v = 3.6(50)$

$v = 180$

The volume is 180 cubic cm.

27. $d = k \cdot t^2$ Now, $d = 16t^2$

$144 = k \cdot 3^2$ $400 = 16t^2$

$\dfrac{144}{9} = k$ $25 = t^2$

$16 = k$ $5 = t$

In 5 seconds it will fall 400 feet.

29. $d = k \cdot t^2$ Now, $d = 1.62t^2$

$6.48 = k \cdot 2^2$ $d = 1.62(5)^2$

$\dfrac{6.48}{4} = k$ $d = 40.5$

$1.62 = k$

In 5 seconds it will fall 40.5 meters.

31. Translating "The circumference varies directly with its diameter," we write $c = kd$, where c represents circumference and d represents the diameter.

$c = k \cdot d$ Now, $c = 3.14d$

$12.56 = k \cdot 4$ $21.98 = 3.14d$

$3.14 = k$ $7 = d$

The diameter will be 7 ft. when the circumference is 21.98 ft.

33. Translate "a varies inversely as b."

$a = \dfrac{k}{b}$ Now, $a = \dfrac{16}{b}$

$3.2 = \dfrac{k}{5}$ $8 = \dfrac{16}{b}$

$16 = k$ $8b = 16$

$b = 2$

35. Translate "m varies inversely as n."

$m = \dfrac{k}{n}$ Now, $m = \dfrac{66}{n}$

$11 = \dfrac{k}{6}$ $8 = \dfrac{66}{n}$

$66 = k$ $8n = 66$

$n = 8.25$

37. Translating "The pressure that a gas exerts against the walls of a container is inversely proportional to the volume of the container," we write $p = \dfrac{k}{v}$, where p represents the pressure the gas exerts and v is the volume of the container.

$$p = \frac{k}{v} \qquad \text{Now,} \quad p = \frac{800}{v}$$

$$40 = \frac{k}{20} \qquad\qquad p = \frac{800}{16}$$

$$800 = k \qquad\qquad p = 50$$

The pressure is 50 psi.

39. Translating "the current I varies inversely as the resistance R," we write $I = \dfrac{k}{R}$, where I represents the current in amperes and R represents the resistance in ohms.

$$I = \frac{k}{R} \qquad \text{Now,} \quad I = \frac{150}{R}$$

$$10 = \frac{k}{15} \qquad\qquad 25 = \frac{150}{R}$$

$$150 = k \qquad\qquad 25R = 150$$

$$\qquad\qquad\qquad\qquad R = 6$$

The resistance is 6 ohms.

41.
$$l = \frac{k}{f} \qquad \text{Now,} \quad l = \frac{360{,}000}{f}$$

$$400 = \frac{k}{900} \qquad\qquad l = \frac{360{,}000}{600}$$

$$360{,}000 = k \qquad\qquad l = 600$$

The wavelength is 600 m.

43.
$$f = \frac{k}{a} \qquad \text{Now,} \quad f = \frac{280}{a}$$

$$5.6 = \frac{k}{50} \qquad\qquad 2.8 = \frac{280}{a}$$

$$280 = k \qquad\qquad 2.8a = 280$$

$$\qquad\qquad\qquad\qquad a = 100$$

The aperture is 100 mm.

45. Translate "a varies jointly with b and c."

$$a = kbc \qquad \text{Now,} \quad a = 4bc$$

$$96 = k \cdot 6 \cdot 4 \qquad\qquad a = 4(2)(8)$$

$$96 = 24k \qquad\qquad a = 64$$

$$4 = k$$

47. Translate "a varies jointly as the square of b and c."

$$a = k \cdot b^2 \cdot c \qquad \text{Now,} \quad a = 4b^2 c$$

$$96 = k \cdot 2^2 \cdot 6 \qquad\qquad a = 4(3)^2(2)$$

$$96 = 24k \qquad\qquad a = 72$$

$$4 = k$$

49. Translating "the volume of a rectangular solid varies jointly with the length and the height," we write $v = klh$, where v represents volume, l represents length, and h represents height.

$$v = k \cdot l \cdot h \qquad \text{Now,} \quad v = 4lh$$

$$192 = k \cdot 8 \cdot 6 \qquad\qquad v = 4(7)(12)$$

$$192 = 48k \qquad\qquad v = 336$$

$$4 = k$$

The volume is 336 cubic inches.

51. Translating "the volume of a right circular cylinder varies jointly as the square of the radius and the height," we write $v = kr^2 h$, where v represents volume, r represents radius, and h represents height.

$$v = k \cdot r^2 \cdot h \qquad \text{Now,} \quad v = 3.14167 \cdot r^2 \cdot h$$

$$301.6 = k \cdot 4^2 \cdot 6 \qquad\qquad v = 3.14167 \cdot 3^2 \cdot 6$$

$$301.6 = 96k \qquad\qquad v \approx 169.65$$

$$3.14167 \approx k$$

The volume is approximately 169.65 cubic cm.

53. Translate "y varies directly with x and inversely as z."

$$y = \frac{kx}{z} \qquad \text{Now,} \quad y = \frac{12x}{z}$$

$$8 = \frac{k \cdot 4}{6} \qquad\qquad y = \frac{12 \cdot 5}{10}$$

$$48 = 4k \qquad\qquad y = 6$$

$$12 = k$$

55. Translate "y varies jointly with x and z and inversely with n."

$$y = \frac{kxz}{n} \qquad \text{Now,} \quad y = \frac{18xz}{n}$$

$$81 = \frac{k(4)(9)}{8} \qquad\qquad 8 = \frac{18(6)(12)}{n}$$

$$648 = 36k \qquad\qquad 8n = 1296$$

$$18 = k \qquad\qquad n = 162$$

57. Translating "the resistance of a wire varies directly with the length and inversely with the square of the diameter," we write $R = \dfrac{kl}{d^2}$, where R represents resistance, l represents length, and d represents diameter.

$$R = \frac{kl}{d^2} \qquad \text{Now,} \qquad R = \frac{0.0005l}{d^2}$$

$$7.5 = \frac{k \cdot 6}{0.02^2} \qquad\qquad R = \frac{0.0005 \cdot 10}{0.04^2}$$

$$0.003 = 6k \qquad\qquad 0.0016R = 0.005$$

$$0.0005 = k \qquad\qquad R = 3.125$$

The resistance is 3.125 ohms.

59. Translating "the force of attraction between two bodies is jointly proportional to their masses and inversely proportional to the square of the distance between them," we write

$f = \dfrac{k \cdot m_1 \cdot m_2}{d^2}$, where f is the force of attraction, m_1 is the mass of one body, m_2 is the mass of the other body, and d is the distance between them.

$$f = \frac{k \cdot m_1 \cdot m_2}{d^2} \qquad \text{Now,} \quad f = \frac{18 m_1 m_2}{d^2}$$

$$48 = \frac{k \cdot 4 \cdot 6}{3^2} \qquad\qquad f = \frac{18(2)(12)}{6^2}$$

$$432k = 24k \qquad\qquad f = 12$$

$$18 = k$$

The force of attraction is 12 dynes.

61. a)

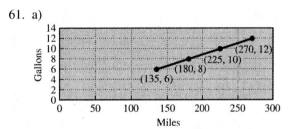

b) They are directly proportional. As the number of miles increases, so does the number of gallons.

c) $d = k \cdot g = 135 = k \cdot 6 = 22.5 = k$, which represents the miles per gallon the car gets.

d) Yes, the data represent a function because for any given number of miles, there will be one quantity of gasoline (assuming that the miles per gallon stays constant).

Review Exercises

1. $2^3 = 8$

2. $-12^2 = -144$

3. $\left(6a^3 b^2\right)\left(-5ab^4\right) = -30 a^4 b^6$

4. $\dfrac{-48x^3 y^6}{-12x^7 y^4} = \dfrac{4y^2}{x^4}$

5. $-3(x+4) - 2(x-5) = -3x - 12 - 2x + 10$
$$= -5x - 2$$

6. $3x^2 - 13x - 10 = 0$
$$(3x+2)(x-5) = 0$$
$$3x + 2 = 0 \qquad \text{or} \qquad x - 5 = 0$$
$$x = -\frac{2}{3} \qquad\qquad\qquad x = 5$$

Chapter 7 Review Exercises

1. $\dfrac{32x^2 y^5}{8xy^3} = 4xy^2$

2. $\dfrac{-42 p^6 q^2}{27 p^3 q^4} = \dfrac{-2 \cdot \cancel{3} \cdot 7 \cdot p^6 q^2}{\cancel{3} \cdot 3 \cdot 3 \cdot p^3 q^4} = -\dfrac{14 p^3}{9q^2}$

3. $-\dfrac{18 m^3 n^2}{54 m^5 n^5} = -\dfrac{1}{3m^2 n^3}$

4. $\dfrac{14x+63}{10x+45} = \dfrac{7\cancel{(2x+9)}}{5\cancel{(2x+9)}} = \dfrac{7}{5}$

5. $\dfrac{4a-20}{7a-35} = \dfrac{4\cancel{(a-5)}}{7\cancel{(a-5)}} = \dfrac{4}{7}$

6. $\dfrac{a^2+4a-21}{a^2+9a+14} = \dfrac{\cancel{(a+7)}(a-3)}{\cancel{(a+7)}(a+2)} = \dfrac{a-3}{a+2}$

7. $\dfrac{2x-3}{6x^2-x-12} = \dfrac{\cancel{(2x-3)}}{\cancel{(2x-3)}(3x+4)} = \dfrac{1}{3x+4}$

8. $\dfrac{25x^2-9}{10x^2+x-3} = \dfrac{\cancel{(5x+3)}(5x-3)}{\cancel{(5x+3)}(2x-1)} = \dfrac{5x-3}{2x-1}$

9. $\dfrac{2c+3d}{8c^2+26cd+21d^2} = \dfrac{\cancel{(2c+3d)}}{\cancel{(2c+3d)}(4c+7d)}$
$$= \dfrac{1}{4c+7d}$$

10. $\dfrac{5-2x}{4x^2-25} = \dfrac{-1\cancel{(2x-5)}}{(2x+5)\cancel{(2x-5)}} = -\dfrac{1}{2x+5}$

11. $\dfrac{8ac-2a+12c-3}{6ac-8a+9c-12} = \dfrac{2a(4c-1)+3(4c-1)}{2a(3c-4)+3(3c-4)}$

$\qquad = \dfrac{(4c-1)\cancel{(2a+3)}}{(3c-4)\cancel{(2a+3)}}$

$\qquad = \dfrac{4c-1}{3c-4}$

12. $\dfrac{8x^3+27}{2x^2-7x-15} = \dfrac{\cancel{(2x+3)}(4x^2-6x+9)}{\cancel{(2x+3)}(x-5)}$

$\qquad = \dfrac{4x^2-6x+9}{x-5}$

13. $\dfrac{\overset{2}{\cancel{28}}m^2n^3}{\underset{3}{\cancel{15}}p^2q^6} \cdot \dfrac{\overset{5}{\cancel{25}}p^4q^3}{\cancel{14}mn} = \dfrac{10m^2n^3p^4q^3}{3mnp^2q^6}$

$\qquad = \dfrac{10mn^2p^2}{3q^3}$

14. $\dfrac{8m-12n}{32m^3n} \cdot \dfrac{36mn^3}{6m-9n} = \dfrac{\cancel{4}\cancel{(2m-3n)}}{_2\cancel{8}\,32m^3n} \cdot \dfrac{\overset{12}{\cancel{36}}\,mn^3}{\cancel{3}\cancel{(2m-3n)}}$

$\qquad = \dfrac{3mn^3}{2m^3n}$

$\qquad = \dfrac{3n^2}{2m^2}$

15. $\dfrac{14x-21}{30x-40} \cdot \dfrac{15x-20}{8x-12} = \dfrac{7\cancel{(2x-3)}}{_2\cancel{10}\cancel{(3x-4)}} \cdot \dfrac{\cancel{5}\cancel{(3x-4)}}{4\cancel{(2x-3)}}$

$\qquad = \dfrac{7}{8}$

16. $\dfrac{25x^2-16}{16x+24} \cdot \dfrac{12x+18}{5x-4}$

$\qquad = \dfrac{(5x+4)\cancel{(5x-4)}}{_4\cancel{8}\cancel{(2x+3)}} \cdot \dfrac{\overset{3}{\cancel{6}}\cancel{(2x+3)}}{\cancel{(5x-4)}}$

$\qquad = \dfrac{3(5x+4)}{4}$

17. $\dfrac{\cancel{(2x-7)}}{_6\cancel{12}} \cdot \dfrac{\overset{5}{\cancel{10}}}{-1\cancel{(2x-7)}} = -\dfrac{5}{6}$

18. $\dfrac{x^2-16}{x^2+6x+8} \cdot \dfrac{x^2-3x-10}{x^2-25}$

$\qquad = \dfrac{\cancel{(x+4)}(x-4)}{\cancel{(x+4)}\cancel{(x+2)}} \cdot \dfrac{\cancel{(x-5)}\cancel{(x+2)}}{(x+5)\cancel{(x-5)}}$

$\qquad = \dfrac{x-4}{x+5}$

19. $\dfrac{y^2-9}{y^2-3y-18} \cdot \dfrac{y^2-4y-12}{y^2-6y+9}$

$\qquad = \dfrac{\cancel{(y+3)}\cancel{(y-3)}}{\cancel{(y-6)}\cancel{(y+3)}} \cdot \dfrac{\cancel{(y-6)}(y+2)}{\cancel{(y-3)}(y-3)}$

$\qquad = \dfrac{y+2}{y-3}$

20. $\dfrac{8x^2+2x-15}{6x^2+x-12} \cdot \dfrac{3x^2-13x+12}{3x^2-7x-6}$

$\qquad = \dfrac{(4x-5)\cancel{(2x+3)}}{\cancel{(3x-4)}\cancel{(2x+3)}} \cdot \dfrac{\cancel{(3x-4)}\cancel{(x-3)}}{(3x+2)\cancel{(x-3)}}$

$\qquad = \dfrac{4x-5}{3x+2}$

21. $\dfrac{ab+2ad-3bc-6cd}{ab-4ad+2bc-8cd} \cdot \dfrac{ab+5ad+2bc+10cd}{ab+5ad-3bc-15cd}$

$\qquad = \dfrac{a(b+2d)-3c(b+2d)}{a(b-4d)+2c(b-4d)} \cdot \dfrac{a(b+5d)+2c(b+5d)}{a(b+5d)-3c(b+5d)}$

$\qquad = \dfrac{(b+2d)\cancel{(a-3c)}}{(b-4d)\cancel{(a+2c)}} \cdot \dfrac{\cancel{(b+5d)}\cancel{(a+2c)}}{\cancel{(b+5d)}\cancel{(a-3c)}}$

$\qquad = \dfrac{b+2d}{b-4d}$

22. $\dfrac{8x^3+y^3}{4x^2-y^2} \cdot \dfrac{6x^2+5xy-4y^2}{4x^2-2xy+y^2}$

$\qquad = \dfrac{\cancel{(2x+y)}\cancel{(4x^2-2xy+y^2)}}{\cancel{(2x+y)}\cancel{(2x-y)}} \cdot \dfrac{(3x+4y)\cancel{(2x-y)}}{\cancel{(4x^2-2xy+y^2)}}$

$\qquad = 3x+4y$

23. $\dfrac{39a^2b^4}{27x^3y^2} \div \dfrac{26a^3b}{28xy^4} = \dfrac{\overset{13}{\cancel{39}}a^2b^4}{_9\cancel{27}x^3y^2} \cdot \dfrac{\overset{14}{\cancel{28}}xy^4}{\cancel{26}^{13}a^3b}$

$\qquad = \dfrac{14a^2b^4xy^4}{9a^3bx^3y^2}$

$\qquad = \dfrac{14b^3y^2}{9ax^2}$

24. $\dfrac{8y^3 - 20y^2}{9y} \div \dfrac{6y^4 - 15y^3}{10y^2}$

$= \dfrac{4y^2\left(2y - 5\right)}{9y} \cdot \dfrac{10y^2}{3y^3\left(2y - 5\right)}$

$= \dfrac{40y^4}{27y^4} = \dfrac{40}{27}$

25. $\dfrac{21a + 7b}{16a - 24b} \div \dfrac{42a + 14b}{24a - 36b}$

$= \dfrac{\overset{1}{\cancel{7}}\left(3a + b\right)}{\underset{2}{\cancel{8}}\left(2a - 3b\right)} \cdot \dfrac{\overset{3}{\cancel{12}}\left(2a - 3b\right)}{\underset{2}{\cancel{14}}\left(3a + b\right)}$

$= \dfrac{3}{4}$

26. $\dfrac{z^2}{z^2 - 2z - 8} \div \dfrac{9z^3 + 3z^4}{z^2 + 5z + 6}$

$= \dfrac{z^2}{\left(z - 4\right)\left(z + 2\right)} \cdot \dfrac{\left(z + 3\right)\left(z + 2\right)}{3z^3\left(z + 3\right)}$

$= \dfrac{z^2}{3z^3\left(z - 4\right)}$

$= \dfrac{1}{3z\left(z - 4\right)}$

27. $\dfrac{4p^2 - 9}{24p + 28} \div \dfrac{10p - 15}{36p^2 - 49}$

$= \dfrac{\left(2p + 3\right)\left(2p - 3\right)}{4\left(6p + 7\right)} \cdot \dfrac{\left(6p + 7\right)\left(6p - 7\right)}{5\left(2p - 3\right)}$

$= \dfrac{\left(2p + 3\right)\left(6p - 7\right)}{20}$

28. $\dfrac{16p^2 - 8pq - 3q^2}{8p^2 + 22pq + 5q^2} \div \dfrac{8p^2 - 10pq + 3q^2}{10p^2 + pq - 3q^2}$

$= \dfrac{\left(4p + q\right)\left(4p - 3q\right)}{\left(4p + q\right)\left(2p + 5q\right)} \cdot \dfrac{\left(5p + 3q\right)\left(2p - q\right)}{\left(4p - 3q\right)\left(2p - q\right)}$

$= \dfrac{5p + 3q}{2p + 5q}$

29. $\dfrac{x^3 - 8y^3}{x^2 - 36y^2} \div \dfrac{x^2 + 2xy - 8y^2}{x^2 - 2xy - 24y^2}$

$= \dfrac{\left(x - 2y\right)\left(x^2 + 2xy + 4y^2\right)}{\left(x + 6y\right)\left(x - 6y\right)} \cdot \dfrac{\left(x - 6y\right)\left(x + 4y\right)}{\left(x + 4y\right)\left(x - 2y\right)}$

$= \dfrac{x^2 + 2xy + 4y^2}{x + 6y}$

30. $\dfrac{xz + 2xw - 3yz - 6yw}{4xz - 2xw + 6yz - 3yw} \div \dfrac{xz - 2xw - 3yz + 6yw}{2xz - 4xw + 3yz - 6yw}$

$= \dfrac{x\left(z + 2w\right) - 3y\left(z + 2w\right)}{2x\left(2z - w\right) + 3y\left(2z - w\right)} \cdot \dfrac{2x\left(z - 2w\right) + 3y\left(z - 2w\right)}{x\left(z - 2w\right) - 3y\left(z - 2w\right)}$

$= \dfrac{\left(z + 2w\right)\left(x - 3y\right)}{\left(2z - w\right)\left(2x + 3y\right)} \cdot \dfrac{\left(z - 2w\right)\left(2x + 3y\right)}{\left(z - 2w\right)\left(x - 3y\right)}$

$= \dfrac{z + 2w}{2z - w}$

31. $f\left(x\right) = \dfrac{2x}{5 - x}$

   a) $f\left(3\right) = \dfrac{2\left(3\right)}{5 - 3} = \dfrac{6}{2} = 3$

   b) $f\left(0\right) = \dfrac{2\left(0\right)}{5 - 0} = \dfrac{0}{5} = 0$

   c) $f\left(-1\right) = \dfrac{2\left(-1\right)}{5 - \left(-1\right)} = \dfrac{-2}{6} = -\dfrac{1}{3}$

32. $g\left(x\right) = \dfrac{x + 2}{x^2 - 8x + 15}$

   a) $g\left(4\right) = \dfrac{4 + 2}{4^2 - 8\left(4\right) + 15} = \dfrac{6}{-1} = -6$

   b) $g\left(-2\right) = \dfrac{-2 + 2}{\left(-2\right)^2 - 8\left(-2\right) + 15} = \dfrac{0}{35} = 0$

   c) $g\left(5\right) = \dfrac{5 + 2}{5^2 - 8\left(5\right) + 15} = \dfrac{7}{0}$ is undefined

33. $f\left(x\right) = \dfrac{2x + 4}{3x - 5}$

$3x - 5 = 0$

$3x = 5$

$x = \dfrac{5}{3}$

The domain is $\left\{ x \,\middle|\, x \neq \dfrac{5}{3} \right\}$.

34. $f(x) = \dfrac{3x-4}{x^2+4x-12}$

$x^2 + 4x - 12 = 0$

$(x+6)(x-2) = 0$

$x + 6 = 0$ or $\begin{aligned}x - 2 &= 0 \\ x &= 2\end{aligned}$

$x = -6$

The domain is $\{x \mid x \neq -6, 2\}$.

35.

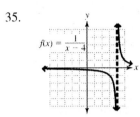
$f(x) = \dfrac{1}{x-4}$

36.

$f(x) = \dfrac{2}{x+2}$

37. $\dfrac{5r}{14x} + \dfrac{3r}{14x} = \dfrac{8r}{14x} = \dfrac{4r}{7x}$

38. $\dfrac{3a+4b}{2a-3b} - \dfrac{a-2b}{2a-3b} = \dfrac{3a+4b-(a-2b)}{2a-3b}$

$\qquad = \dfrac{3a+4b-a+2b}{2a-3b}$

$\qquad = \dfrac{2a+6b}{2a-3b}$

39. $\dfrac{2p^2-p+2}{p^2-16} + \dfrac{p^2+4p-8}{p^2-16} = \dfrac{3p^2+3p-6}{p^2-16}$

40. $\dfrac{x^2+2x-5}{x^2-3x-18} - \dfrac{3x+7}{x^2-3x-18}$

$\qquad = \dfrac{x^2+2x-5-(3x+7)}{x^2-3x-18}$

$\qquad = \dfrac{x^2+2x-5-3x-7}{x^2-3x-18}$

$\qquad = \dfrac{x^2-x-12}{x^2-3x-18}$

$\qquad = \dfrac{(x-4)\cancel{(x+3)}}{(x-6)\cancel{(x+3)}}$

$\qquad = \dfrac{x-4}{x-6}$

41. $9p^3q^4 = 3^2 \cdot p^3 \cdot q^4$

$12p^2q^5 = 2^2 \cdot 3 \cdot p^2 \cdot q^5$

$\text{LCD} = 2^2 \cdot 3^2 \cdot p^3 \cdot q^5$

$\qquad = 36p^3q^5$

$\dfrac{8b}{9p^3q^4} = \dfrac{8b(4q)}{9p^3q^4(4q)} = \dfrac{32bq}{36p^3q^5}$

$\dfrac{11c}{12p^2q^5} = \dfrac{11c(3p)}{12p^2q^5(3p)} = \dfrac{33cp}{36p^3q^5}$

42. $\text{LCD} = (t-4)(t+2)$

$\dfrac{7}{t-4} = \dfrac{7(t+2)}{(t-4)(t+2)} = \dfrac{7t+14}{(t-4)(t+2)}$

$\dfrac{9}{t+2} = \dfrac{9(t-4)}{(t+2)(t-4)} = \dfrac{9t-36}{(t-4)(t+2)}$

43. $8u+12 = 4(2u+3)$

$14u+21 = 7(2u+3)$

$\text{LCD} = 4 \cdot 7 \cdot (2u+3)$

$\qquad = 28(2u+3)$

$\dfrac{5u}{8u+12} = \dfrac{5u(7)}{4(2u+3)(7)} = \dfrac{35u}{28(2u+3)}$

$\dfrac{9u}{14u+21} = \dfrac{9u(4)}{7(2u+3)(4)} = \dfrac{36u}{28(2u+3)}$

44. $w^2-16 = (w+4)(w-4)$

$w^2+5w+4 = (w+4)(w+1)$

$\text{LCD} = (w+4)(w-4)(w+1)$

$\dfrac{2w}{w^2-16} = \dfrac{2w(w+1)}{(w+4)(w-4)(w+1)}$

$\qquad = \dfrac{2w^2+2w}{(w+4)(w-4)(w+1)}$

$\dfrac{6w}{w^2+5w+4} = \dfrac{6w(w-4)}{(w+4)(w+1)(w-4)}$

$\qquad = \dfrac{6w^2-24w}{(w+4)(w-4)(w+1)}$

45. $a^2 + 6a + 9 = (a+3)^2$

$a^2 - 2a - 15 = (a-5)(a+3)$

$\text{LCD} = (a-5)(a+3)^2$

$$\frac{3a}{a^2+6a+9} = \frac{3a(a-5)}{(a+3)^2(a-5)} = \frac{3a^2-15a}{(a+3)^2(a-5)}$$

$$\frac{10a}{a^2-2a-15} = \frac{10a(a+3)}{(a+3)(a-5)(a+3)}$$

$$= \frac{10a^2+30a}{(a+3)^2(a-5)}$$

46. $m^2 - 2m - 8 = (m-4)(m+2)$

$m^2 - 3m - 4 = (m-4)(m+1)$

$\text{LCD} = (m-4)(m+2)(m+1)$

$$\frac{m+1}{m^2-2m-8} = \frac{(m+1)(m+1)}{(m-4)(m+2)(m+1)}$$

$$= \frac{m^2+2m+1}{(m-4)(m+2)(m+1)}$$

$$\frac{m-1}{m^2-3m-4} = \frac{(m-1)(m+2)}{(m-4)(m+1)(m+2)}$$

$$= \frac{m^2+m-2}{(m-4)(m+2)(m+1)}$$

47. $x^3 + 6x^2 + 8x = x(x+4)(x+2)$

$x^5 + 2x^4 - 8x^3 = x^3(x+4)(x-2)$

$\text{LCD} = x^3(x+4)(x-2)(x+2)$

$$\frac{2x-3}{x^3+6x^2+8x} = \frac{(2x-3)x^2(x-2)}{x(x+4)(x+2)x^2(x-2)}$$

$$= \frac{2x^4-7x^3+6x^2}{x^3(x+4)(x+2)(x-2)}$$

$$\frac{6x+3}{x^5+2x^4-8x^3} = \frac{(6x+3)(x+2)}{x^3(x+4)(x-2)(x+2)}$$

$$= \frac{6x^2+15x+6}{x^3(x+4)(x+2)(x-2)}$$

48. $x^3 = x^3$

$x^2 + 10 + 24 = (x+6)(x+4)$

$4x^3 + 16x^2 = 4x^2(x+4)$

$\text{LCD} = 4x^3(x+4)(x+6)$

$$\frac{6}{x^3} = \frac{6(4)(x+6)(x+4)}{x^3(4)(x+6)(x+4)} = \frac{24x^2+240x+576}{4x^3(x+4)(x+6)}$$

$$\frac{4x-3}{x^2+10x+24} = \frac{(4x-3)4x^3}{(x+6)(x+4)4x^3}$$

$$= \frac{16x^4-12x^3}{4x^3(x+4)(x+6)}$$

$$\frac{3x}{4x^3+16x^2} = \frac{3x \cdot x(x+6)}{4x^2(x+4)\cdot x(x+6)}$$

$$= \frac{3x^3+18x^2}{4x^3(x+6)(x+4)}$$

49. $\text{LCD} = 45x$

$$\frac{8a}{15x} - \frac{4a}{9x} = \frac{(8a)\cdot 3}{(15x)\cdot 3} - \frac{(4a)\cdot 5}{(9x)\cdot 5}$$

$$= \frac{24a}{45x} - \frac{20a}{45x}$$

$$= \frac{4a}{45x}$$

50. $\text{LCD} = 20y$

$$\frac{y-3}{4y} - \frac{y+2}{5y} = \frac{(y-3)\cdot 5}{(4y)\cdot 5} - \frac{(y+2)\cdot 4}{(5y)\cdot 4}$$

$$= \frac{5y-15}{20y} - \frac{4y+8}{20y}$$

$$= \frac{5y-15-4y-8}{20y}$$

$$= \frac{y-23}{20y}$$

51. $\text{LCD} = 36t^4u^5$

$$\frac{2t-3}{18t^4u^2} + \frac{5t+1}{12t^3u^5}$$

$$= \frac{(2t-3)\cdot 2u^3}{18t^4u^2 \cdot 2u^3} + \frac{(5t+1)\cdot 3t}{12t^3u^5 \cdot 3t}$$

$$= \frac{4tu^3-6u^3}{36t^4u^5} + \frac{15t^2+3t}{36t^4u^5}$$

$$= \frac{4tu^3-6u^3+15t^2+3t}{36t^4u^5}$$

52. LCD = $w(w-3)$

$$\frac{8}{w-3} - \frac{-3}{w} = \frac{8 \cdot w}{(w-3) \cdot w} - \frac{-3 \cdot (w-3)}{w \cdot (w-3)}$$

$$= \frac{8w}{w(w-3)} - \frac{-3w+9}{w(w-3)}$$

$$= \frac{8w+3w-9}{w(w-3)}$$

$$= \frac{11w-9}{w(w-3)}$$

53. LCD = $4(v+2)(v-3)$

$$\frac{v+4}{4v+8} - \frac{v-2}{2v-6}$$

$$= \frac{(v+4) \cdot (v-3)}{4(v+2) \cdot (v-3)} - \frac{(v-2) \cdot 2(v+2)}{2(v-3) \cdot 2(v+2)}$$

$$= \frac{v^2+v-12}{4(v+2)(v-3)} - \frac{2v^2-8}{4(v+2)(v-3)}$$

$$= \frac{v^2+v-12-2v^2+8}{4(v+2)(v-3)}$$

$$= \frac{-v^2+v-4}{4(v+2)(v-3)}$$

54. LCD = $(t+4)^2$

$$\frac{t^2-5t}{t^2+8t+16} + \frac{6}{t+4} = \frac{t^2-5t}{(t+4)^2} + \frac{6(t+4)}{(t+4) \cdot (t+4)}$$

$$= \frac{t^2-5t}{(t+4)^2} + \frac{6t+24}{(t+4)^2}$$

$$= \frac{t^2+t+24}{(t+4)^2}$$

55. LCD = $(w+5)(w-5)(w+2)$

$$\frac{2w+5}{w^2-25} + \frac{6w}{w^2-3w-10}$$

$$= \frac{(2w+5) \cdot (w+2)}{(w+5)(w-5) \cdot (w+2)} + \frac{6w \cdot (w+5)}{(w-5)(w+2) \cdot (w+5)}$$

$$= \frac{2w^2+9w+10}{(w+5)(w-5)(w+2)} + \frac{6w^2+30w}{(w+5)(w-5)(w+2)}$$

$$= \frac{8w^2+39w+10}{(w+5)(w-5)(w+2)}$$

56. LCD = $(4z+5)(z+7)^2$

$$\frac{z+9}{4z^2+33z+35} - \frac{z-12}{z^2+14z+49}$$

$$= \frac{(z+9) \cdot (z+7)}{(4z+5)(z+7) \cdot (z+7)} - \frac{(z-12) \cdot (4z+5)}{(z+7)^2 \cdot (4z+5)}$$

$$= \frac{z^2+16z+63}{(4z+5)(z+7)^2} - \frac{4z^2-43z-60}{(4z+5)(z+7)^2}$$

$$= \frac{z^2+16z+63-4z^2+43z+60}{(4z+5)(z+7)^2}$$

$$= \frac{-3z^2+59z+123}{(4z+5)(z+7)^2}$$

57. LCD = $a(a+3)$

$$\frac{3}{a} - \frac{3a+5}{a+3} + \frac{5a-3}{a^2+3a}$$

$$= \frac{3 \cdot (a+3)}{a \cdot (a+3)} - \frac{(3a+5) \cdot a}{(a+3) \cdot a} + \frac{5a-3}{a(a+3)}$$

$$= \frac{3a+9}{a(a+3)} - \frac{3a^2+5a}{a(a+3)} + \frac{5a-3}{a(a+3)}$$

$$= \frac{3a+9-3a^2-5a+5a-3}{a(a+3)}$$

$$= \frac{-3a^2+3a+6}{a(a+3)}$$

58. LCD = $3x-y$

$$\frac{4x-3y}{3x-y} - \frac{3x-4y}{-1(3x-y)} = \frac{4x-3y}{3x-y} - \frac{-1(3x-4y)}{3x-y}$$

$$= \frac{4x-3y}{3x-y} - \frac{-3x+4y}{3x-y}$$

$$= \frac{4x-3y+3x-4y}{3x-y}$$

$$= \frac{7x-7y}{3x-y}$$

59. $$\frac{\dfrac{4}{5}}{\dfrac{3}{10}} = \frac{4}{5} \cdot \frac{10}{3} = \frac{40}{15} = \frac{8}{3}$$

60. $$\frac{\dfrac{4}{3} - \dfrac{8}{9}}{\dfrac{5}{6} - \dfrac{4}{12}} = \frac{\left(\dfrac{4}{3} - \dfrac{8}{9}\right) \cdot 36}{\left(\dfrac{5}{6} - \dfrac{4}{12}\right) \cdot 36} = \frac{48-32}{30-12} = \frac{16}{18} = \frac{8}{9}$$

61. $\dfrac{\dfrac{u^4v^2}{w}}{\dfrac{uv^3}{w^2}} = \dfrac{u^4v^2}{w} \cdot \dfrac{w^2}{uv^3} = \dfrac{u^4v^2w^2}{uv^3w} = \dfrac{u^3w}{v}$

62. $\dfrac{\dfrac{x^6y^3}{t^4}}{\dfrac{x^2y^2}{t}} = \dfrac{x^6y^3}{t^4} \cdot \dfrac{t}{x^2y^2} = \dfrac{x^6y^3t}{x^2y^2t^4} = \dfrac{x^4y}{t^3}$

63. $\dfrac{2w - \dfrac{w}{4}}{6 - \dfrac{w}{4}} = \dfrac{\left(2w - \dfrac{w}{4}\right) \cdot 4}{\left(6 - \dfrac{w}{4}\right) \cdot 4} = \dfrac{8w - w}{24 - w} = \dfrac{7w}{24 - w}$

64. $\dfrac{\dfrac{7}{b-15} - 6}{b - 15} = \dfrac{\left(\dfrac{7}{b-15} - 6\right) \cdot (b-15)}{(b-15) \cdot (b-15)}$

$= \dfrac{7 - 6(b-15)}{(b-15)^2}$

$= \dfrac{7 - 6b + 90}{(b-15)^2}$

$= \dfrac{-6b + 97}{(b-15)^2}$

65. $\dfrac{1 - \dfrac{1}{y} - \dfrac{12}{y}}{1 - \dfrac{6}{y} + \dfrac{8}{y}} = \dfrac{\left(1 - \dfrac{1}{y} - \dfrac{12}{y}\right) \cdot y}{\left(1 - \dfrac{6}{y} + \dfrac{8}{y}\right) \cdot y}$

$= \dfrac{y - 1 - 12}{y - 6 + 8}$

$= \dfrac{y - 13}{y + 2}$

66. $\dfrac{x + \dfrac{25}{x+10}}{1 - \dfrac{5}{x+10}} = \dfrac{\left(x + \dfrac{25}{x+10}\right) \cdot (x+10)}{\left(1 - \dfrac{5}{x+10}\right) \cdot (x+10)}$

$= \dfrac{x(x+10) + 25}{x + 10 - 5}$

$= \dfrac{x^2 + 10x + 25}{x + 5}$

$= \dfrac{\cancel{(x+5)}(x+5)}{\cancel{(x+5)}}$

$= x + 5$

67. $\dfrac{\dfrac{6}{y^2} - \dfrac{1}{xy} - \dfrac{12}{x^2}}{\dfrac{4}{y^2} - \dfrac{4}{xy} - \dfrac{3}{x^2}} = \dfrac{\left(\dfrac{6}{y^2} - \dfrac{1}{xy} - \dfrac{12}{x^2}\right) \cdot x^2y^2}{\left(\dfrac{4}{y^2} - \dfrac{4}{xy} - \dfrac{3}{x^2}\right) \cdot x^2y^2}$

$= \dfrac{6x^2 - xy - 12y^2}{4x^2 - 4xy - 3y^2}$

$= \dfrac{(3x + 4y)\cancel{(2x - 3y)}}{\cancel{(2x - 3y)}(2x + y)}$

$= \dfrac{3x + 4y}{2x + y}$

68. $\dfrac{\dfrac{x+4}{x-4} - \dfrac{x-4}{x+4}}{\dfrac{x+4}{x-4} + \dfrac{x-4}{x+4}} = \dfrac{\left(\dfrac{x+4}{x-4} - \dfrac{x-4}{x+4}\right) \cdot (x-4)(x+4)}{\left(\dfrac{x+4}{x-4} + \dfrac{x-4}{x+4}\right) \cdot (x-4)(x+4)}$

$= \dfrac{(x+4)^2 - (x-4)^2}{(x+4)^2 + (x-4)^2}$

$= \dfrac{x^2 + 8x + 16 - x^2 + 8x - 16}{x^2 + 8x + 16 + x^2 - 8x + 16}$

$= \dfrac{16x}{2x^2 + 32}$

$= \dfrac{2 \cdot 8x}{2 \cdot (x^2 + 16)}$

$= \dfrac{8x}{x^2 + 16}$

69. $\dfrac{5}{4t} - \dfrac{3}{8t} = \dfrac{1}{2}$

$8t \cdot \left(\dfrac{5}{4t} - \dfrac{3}{8t}\right) = 8t \cdot \dfrac{1}{2}$

$10 - 3 = 4t$

$7 = 4t$

$\dfrac{7}{4} = t$

Check: $\dfrac{5}{4\left(\dfrac{7}{4}\right)} - \dfrac{3}{8\left(\dfrac{7}{4}\right)} \overset{?}{=} \dfrac{1}{2}$

$\dfrac{5}{7} - \dfrac{3}{14} \overset{?}{=} \dfrac{1}{2}$

$\dfrac{10}{14} - \dfrac{3}{14} \overset{?}{=} \dfrac{1}{2}$

$\dfrac{1}{2} = \dfrac{1}{2}$

70.
$$\frac{12}{m-2} = 9 + \frac{m}{m-2}$$

$$(m-2)\left(\frac{12}{m-2}\right) = (m-2)\left(9 + \frac{m}{m-2}\right)$$

$$12 = 9(m-2) + m$$

$$12 = 9m - 18 + m$$

$$30 = 10m$$

$$3 = m$$

Check: $\dfrac{12}{3-2} \overset{?}{=} 9 + \dfrac{3}{3-2}$

$$12 \overset{?}{=} 9 + 3$$

$$12 = 12$$

71.
$$\frac{x^2}{x-4} = \frac{3x+4}{x-4} - 3$$

$$(x-4)\left(\frac{x^2}{x-4}\right) = (x-4)\left(\frac{3x+4}{x-4} - 3\right)$$

$$x^2 = 3x + 4 - 3(x-4)$$

$$x^2 = 3x + 4 - 3x + 12$$

$$x^2 = 16$$

$$x^2 - 16 = 0$$

$$(x+4)(x-4) = 0$$

$$x + 4 = 0 \quad \text{or} \quad x - 4 = 0$$

$$x = -4 \qquad\qquad x = 4$$

Notice that if $x = 4$, then $\dfrac{x^2}{x-4}$ and $\dfrac{3x+4}{x-4}$ are undefined. Therefore, $x = 4$ is extraneous, and the only solution is $x = -4$.

Check: $\dfrac{(-4)^2}{-4-4} \overset{?}{=} \dfrac{3(-4)+4}{-4-4} - 3$

$$\frac{16}{-8} \overset{?}{=} \frac{-8}{-8} - 3$$

$$-2 \overset{?}{=} 1 - 3$$

$$-2 = -2$$

72. $\dfrac{2a-2}{a+1} - \dfrac{a}{2a-3} = \dfrac{a-2}{2a-3}$

$$(a+1)(2a-3)\left(\frac{2a-2}{a+1} - \frac{a}{2a-3}\right)$$

$$= (a+1)(2a-3)\left(\frac{a-2}{2a-3}\right)$$

$$(2a-2)(2a-3) - a(a+1) = (a-2)(a+1)$$

$$4a^2 - 10a + 6 - a^2 - a = a^2 - a - 2$$

$$3a^2 - 11a + 6 = a^2 - a - 2$$

$$2a^2 - 10a + 8 = 0$$

$$2(a-1)(a-4) = 0$$

$$a = 1 \quad \text{or} \quad a = 4$$

Check: $a = 1$

$$\frac{2(1)-2}{1+1} - \frac{(1)}{2(1)-3} \overset{?}{=} \frac{1-2}{2(1)-3}$$

$$0 - \frac{1}{-1} \overset{?}{=} \frac{-1}{-1}$$

$$0 + 1 \overset{?}{=} 1$$

$$1 = 1$$

$a = 4$

$$\frac{2(4)-2}{4+1} - \frac{(4)}{2(4)-3} \overset{?}{=} \frac{4-2}{2(4)-3}$$

$$\frac{6}{5} - \frac{4}{5} \overset{?}{=} \frac{2}{5}$$

$$\frac{2}{5} = \frac{2}{5}$$

73. $\dfrac{1}{q+4} = \dfrac{q}{3q+2}$

$$3q + 2 = q^2 + 4q$$

$$0 = q^2 + q - 2$$

$$0 = (q+2)(q-1)$$

$$q = -2 \quad \text{or} \quad q = 1$$

Check: $q = -2$

$$\frac{1}{-2+4} \overset{?}{=} \frac{-2}{3(-2)+2}$$

$$\frac{1}{2} \overset{?}{=} \frac{-2}{-4}$$

$$\frac{1}{2} = \frac{1}{2}$$

Check: $q = 1$

$$\frac{1}{1+4} \overset{?}{=} \frac{1}{3(1)+2}$$

$$\frac{1}{5} = \frac{1}{5}$$

74. $\dfrac{5}{v^2+5v+6} - \dfrac{2}{v^2-2v-8} = \dfrac{3}{v^2-16}$

$(v+2)(v+3)(v-4)(v+4)$

$\qquad \cdot \left(\dfrac{5}{(v+2)(v+3)} - \dfrac{2}{(v-4)(v+2)} \right)$

$\qquad = (v+2)(v+3)(v-4)(v+4)\left(\dfrac{3}{(v-4)(v+4)} \right)$

$5(v-4)(v+4) - 2(v+3)(v+4) = 3(v+2)(v+3)$

$5(v^2-16) - 2(v^2+7v+12) = 3(v^2+5v+6)$

$5v^2 - 80 - 2v^2 - 14v - 24 = 3v^2 + 15v + 18$

$\qquad\qquad\qquad -29v = 122$

$\qquad\qquad\qquad\qquad v = -\dfrac{122}{29}$

Check:

$$\dfrac{5}{\left(-\dfrac{122}{29}+2\right)\left(-\dfrac{122}{29}+3\right)} - \dfrac{2}{\left(-\dfrac{122}{29}-4\right)\left(-\dfrac{122}{29}+2\right)}$$

$$\overset{?}{=} \dfrac{3}{\left(-\dfrac{122}{29}-4\right)\left(-\dfrac{122}{29}+4\right)}$$

$$\dfrac{5}{\left(-\dfrac{64}{29}\right)\left(-\dfrac{35}{29}\right)} - \dfrac{2}{\left(\dfrac{-238}{29}\right)\left(\dfrac{-64}{29}\right)} \overset{?}{=} \dfrac{3}{\left(\dfrac{-238}{29}\right)\left(\dfrac{-6}{29}\right)}$$

$$\dfrac{14,297}{7616} - \dfrac{841}{7616} \overset{?}{=} \dfrac{841}{476}$$

$$\dfrac{841}{476} = \dfrac{841}{476}$$

75. $\dfrac{3}{x^2+5x+6} = \dfrac{2}{x+3} + \dfrac{x-3}{x^2+x-2}$

$(x+2)(x+3)(x-1)\cdot\left(\dfrac{3}{(x+2)(x+3)} \right)$

$\qquad = (x+2)(x+3)(x-1)\cdot\left(\dfrac{2}{x+3} + \dfrac{x-3}{(x+2)(x-1)} \right)$

$3(x-1) = 2(x+2)(x-1) + (x-3)(x+3)$

$3x-3 = 2x^2 + 2x - 4 + x^2 - 9$

$\qquad 0 = 3x^2 - x - 10$

$\qquad 0 = (3x+5)(x-2)$

$\qquad x = -\dfrac{5}{3}$ or $x = 2$

Check: $x = 2$

$$\dfrac{3}{(2+2)(2+3)} \overset{?}{=} \dfrac{2}{2+3} + \dfrac{2-3}{(2+2)(2-1)}$$

$$\dfrac{3}{20} \overset{?}{=} \dfrac{2}{5} + \dfrac{-1}{4}$$

$$\dfrac{3}{20} \overset{?}{=} \dfrac{8}{20} - \dfrac{5}{20}$$

$$\dfrac{3}{20} = \dfrac{3}{20}$$

Check: $x = -\dfrac{5}{3}$

$$\dfrac{3}{\left(-\dfrac{5}{3}+2\right)\left(-\dfrac{5}{3}+3\right)} \overset{?}{=} \dfrac{2}{-\dfrac{5}{3}+3} + \dfrac{-\dfrac{5}{3}-3}{\left(-\dfrac{5}{3}+2\right)\left(-\dfrac{5}{3}-1\right)}$$

$$\dfrac{3}{\left(\dfrac{1}{3}\right)\left(\dfrac{4}{3}\right)} \overset{?}{=} \dfrac{2}{\left(\dfrac{4}{3}\right)} + \dfrac{-\dfrac{14}{3}}{\left(\dfrac{1}{3}\right)\left(\dfrac{-8}{3}\right)}$$

$$\dfrac{27}{4} \overset{?}{=} \dfrac{6}{4} + \dfrac{21}{4}$$

$$\dfrac{27}{4} = \dfrac{27}{4}$$

76. $\dfrac{7x}{x^2-2x-8}-\dfrac{4x}{x^2-5x+4}=\dfrac{3}{x-4}$

$(x-4)(x+2)(x-1)$

$\quad\cdot\left(\dfrac{7x}{(x-4)(x+2)}-\dfrac{4x}{(x-4)(x-1)}\right)$

$\quad=(x-4)(x+2)(x-1)\left(\dfrac{3}{x-4}\right)$

$7x(x-1)-4x(x+2)=3(x+2)(x-1)$

$7x^2-7x-4x^2-8x=3x^2+3x-6$

$3x^2-15x=3x^2+3x-6$

$-18x=-6$

$x=\dfrac{1}{3}$

Check:

$\dfrac{7\left(\dfrac{1}{3}\right)}{\left(\dfrac{1}{3}-4\right)\left(\dfrac{1}{3}+2\right)}-\dfrac{4\left(\dfrac{1}{3}\right)}{\left(\dfrac{1}{3}-4\right)\left(\dfrac{1}{3}-1\right)}\overset{?}{=}\dfrac{3}{\dfrac{1}{3}-4}$

$\dfrac{\dfrac{7}{3}}{\left(-\dfrac{11}{3}\right)\left(\dfrac{7}{3}\right)}-\dfrac{\dfrac{4}{3}}{\left(-\dfrac{11}{3}\right)\left(-\dfrac{2}{3}\right)}\overset{?}{=}\dfrac{3}{\dfrac{-11}{3}}$

$-\dfrac{3}{11}-\dfrac{6}{11}\overset{?}{=}-\dfrac{9}{11}$

$-\dfrac{9}{11}=-\dfrac{9}{11}$

77. Complete a table.

Categories	Rate of Work	Time at Work	Amt of Task Completed
George	$\dfrac{1}{30}$	t	$\dfrac{t}{30}$
Lucille	$\dfrac{1}{45}$	t	$\dfrac{t}{45}$

$\dfrac{t}{30}+\dfrac{t}{45}=1$

$90\left(\dfrac{t}{30}+\dfrac{t}{45}\right)=90\cdot1$

$3t+2t=90$

$5t=90$

$t=18$

George and Lucille can write a chapter in 18 days working together.

78. Complete a table.

Categories	Rate	Time	Distance
against current	$r-3$	$\dfrac{200}{r-3}$	200
with current	$r+3$	$\dfrac{260}{r+3}$	260

$\dfrac{200}{r-3}=\dfrac{260}{r+3}$

$260(r-3)=200(r+3)$

$260r-780=200r+600$

$60r=1380$

$r=23$

The cutter can sail in still water at 23 mph.

79. Complete a table.

Categories	Rate	Time	Distance
truck	45	t	$45t$
car	55	t	$55t$

$45t+55t=510$

$100t=510$

$t=5.1$

$0.1(60)=6$ min.

It will be 5 hours and 6 minutes before they meet at 9:06 am.

80. Translate "*p* varies directly as *q*."

$p=kq$ \quad\quad Now, $p=6q$

$24=k\cdot4$ \quad\quad\quad $p=6(7)$

$6=k$ \quad\quad\quad\quad $p=42$

81. Translate "*s* varies directly as the square of *t*."

$s=k\cdot t^2$ \quad Now, $s=8t^2$

$72=k\cdot9$ \quad\quad\quad $s=8(5)^2$

$8=k$ \quad\quad\quad\quad $s=200$

82. Translate "*y* varies inversely as *x*."

$y=\dfrac{k}{x}$ \quad\quad Now, $y=\dfrac{24}{x}$

$8=\dfrac{k}{3}$ \quad\quad\quad $4=\dfrac{24}{x}$

$24=k$ \quad\quad\quad $4x=24$

$\quad\quad\quad\quad\quad x=6$

83. Translate "suppose m varies inversely as the square of n."

$$m = \frac{k}{n^2} \qquad \text{Now, } m = \frac{144}{n^2}$$

$$16 = \frac{k}{3^2} \qquad\qquad m = \frac{144}{4^2}$$

$$144 = k \qquad\qquad m = 9$$

84. Translate "y varies jointly with x and z."

$$y = kxz \qquad \text{Now, } y = 4xz$$

$$40 = k \cdot 2 \cdot 5 \qquad y = 4 \cdot 4 \cdot 2$$

$$4 = k \qquad\qquad y = 32$$

85. Translate "m varies directly with n and inversely with p."

$$m = \frac{kn}{p} \qquad \text{Now, } m = \frac{6n}{p}$$

$$2 = \frac{k \cdot 3}{9} \qquad\qquad m = \frac{6 \cdot 6}{4}$$

$$18 = 3k \qquad\qquad m = 9$$

$$6 = k$$

86. Translating "the distance a car can travel varies directly with the amount of gas it carries," we write $d = kg$ where d is the distance traveled and g is the amount of gas.

$$d = kg \qquad \text{Now,} \qquad d = 26g$$

$$156 = k \cdot 6 \qquad\qquad 234 = 26g$$

$$26 = k \qquad\qquad\quad 9 = g$$

Therefore, 9 gallons are required to travel 234 miles.

87.
$$v = \frac{k}{T} \qquad \text{Now, } v = \frac{28}{T}$$

$$4 = \frac{k}{7} \qquad\qquad v = \frac{28}{12}$$

$$28 = k \qquad\qquad v = \frac{7}{3}$$

The velocity is $\frac{7}{3}$ cm/sec.

88.
$$v = kr^2 h \qquad\qquad \text{Now, } v = 3.14 r^2 h$$

$$62.8 = k \cdot 2^2 \cdot 5 \qquad\qquad v = 3.14(4)^2(2)$$

$$62.8 = 20k \qquad\qquad\qquad v = 100.48$$

$$3.14 = k$$

The volume is 100.48 in.^{3}.

Chapter 7 Practice Test

1.
$$\frac{42a^3 b^4}{16ab^6} = \frac{\cancel{2} \cdot 3 \cdot 7 \cdot \cancel{a} \cdot a \cdot a \cdot \cancel{b} \cdot \cancel{b} \cdot \cancel{b} \cdot \cancel{b}}{\cancel{2} \cdot 2 \cdot 2 \cdot 2 \cdot \cancel{a} \cdot \cancel{b} \cdot \cancel{b} \cdot \cancel{b} \cdot \cancel{b} \cdot b \cdot b}$$

$$= \frac{21a^2}{8b^2}$$

2.
$$\frac{6x^2 + 11x - 10}{4x^2 + 4x - 15} = \frac{(3x-2)\cancel{(2x+5)}}{(2x-3)\cancel{(2x+5)}}$$

$$= \frac{3x-2}{2x-3}$$

3.
$$\frac{m^3 - 64n^3}{3m^2 - 14mn + 8n^2}$$

$$= \frac{\cancel{(m-4n)}\left(m^2 + 4mn + 16n^2\right)}{(3m-2n)\cancel{(m-4n)}}$$

$$= \frac{m^2 + 4mn + 16n^2}{3m - 2n}$$

4.
$$\frac{2bn - 4bm - 3cn + 6cm}{2b^2 + 7bc - 15c^2}$$

$$= \frac{2b(n-2m) - 3c(n-2m)}{(2b-3c)(b+5c)}$$

$$= \frac{\cancel{(2b-3c)}(n-2m)}{\cancel{(2b-3c)}(b+5c)}$$

$$= \frac{n-2m}{b+5c}$$

5.
$$\frac{27a^2 b^4}{14x^4 y} \cdot \frac{35xy^3}{18a^5 b^2}$$

$$= \frac{\cancel{3} \cdot \cancel{3} \cdot 3 \cdot \cancel{a} \cdot \cancel{a} \cdot \cancel{b} \cdot \cancel{b} \cdot b \cdot b}{2 \cdot \cancel{7} \cdot \cancel{x} \cdot x \cdot x \cdot x \cdot \cancel{y}}$$

$$\cdot \frac{5 \cdot \cancel{7} \cdot \cancel{x} \cdot \cancel{y} \cdot y \cdot y}{2 \cdot \cancel{3} \cdot \cancel{3} \cdot \cancel{a} \cdot \cancel{a} \cdot a \cdot a \cdot a \cdot \cancel{b} \cdot \cancel{b}}$$

$$= \frac{15b^2 y^2}{4a^3 x^3}$$

6.
$$\frac{16y^2 - 25}{6y^2 - 17y - 14} \cdot \frac{3y^2 + 2y}{8y^2 - 2y - 15}$$

$$= \frac{\cancel{(4y+5)}(4y-5)}{\cancel{(3y+2)}(2y-7)} \cdot \frac{y\cancel{(3y+2)}}{\cancel{(4y+5)}(2y-3)}$$

$$= \frac{y(4y-5)}{(2y-7)(2y-3)}$$

7. $\dfrac{6q-8p}{8p-2q} \div \dfrac{4p-3q}{12p-3q} = \dfrac{6q-8p}{8p-2q} \cdot \dfrac{12p-3q}{4p-3q}$

$\qquad\qquad = \dfrac{-2\cancel{(4p-3q)}}{2\cancel{(4p-q)}} \cdot \dfrac{3\cancel{(4p-q)}}{\cancel{(4p-3q)}}$

$\qquad\qquad = \dfrac{-6}{2} = -3$

8. $\dfrac{2n^3-5n^2-6n+15}{3n^3+2n^2-9n-6} \div \dfrac{2n^3-5n^2-8n+20}{3n^3+2n^2+12n+8}$

$= \dfrac{2n^3-5n^2-6n+15}{3n^3+2n^2-9n-6} \cdot \dfrac{3n^3+2n^2+12n+8}{2n^3-5n^2-8n+20}$

$= \dfrac{n^2(2n-5)-3(2n-5)}{n^2(3n+2)-3(3n+2)} \cdot \dfrac{n^2(3n+2)+4(3n+2)}{n^2(2n-5)-4(2n-5)}$

$= \dfrac{\cancel{(2n-5)}\cancel{(n^2-3)}}{\cancel{(3n+2)}\cancel{(n^2-3)}} \cdot \dfrac{\cancel{(3n+2)}(n^2+4)}{\cancel{(2n-5)}(n^2-4)}$

$= \dfrac{n^2+4}{n^2-4}$

9. $f(x) = \dfrac{2x-4}{3x^2-7x-6}$

a) $f(-1) = \dfrac{2(-1)-4}{3(-1)^2-7(-1)-6}$

$\qquad = \dfrac{-2-4}{3+7-6}$

$\qquad = \dfrac{-6}{4}$

$\qquad = -\dfrac{3}{2}$

b) $f(2) = \dfrac{2(2)-4}{3(2)^2-7(2)-6}$

$\qquad = \dfrac{4-4}{12-14-6}$

$\qquad = \dfrac{0}{-8}$

$\qquad = 0$

c) $f(3) = \dfrac{2(3)-4}{3(3)^2-7(3)-6}$

$\qquad = \dfrac{6-4}{27-21-6}$

$\qquad = \dfrac{2}{0}$ is undefined

10. To find the domain of the function
$f(x) = \dfrac{3x-4}{2x^2-x-10}$, set the denominator equal
to zero and solve for x to find any values that
should be excluded from the domain.

$\quad 2x^2-x-10 = 0$

$(2x-5)(x+2) = 0$

$2x-5 = 0 \qquad$ or $\qquad x+2 = 0$

$\quad 2x = 5 \qquad\qquad\qquad x = -2$

$\quad x = \dfrac{5}{2}$

The domain of $f(x) = \dfrac{3x-4}{2x^2-x-10}$ is

$\left\{ x \mid x \ne -2, \dfrac{5}{2} \right\}$.

11. $\qquad a^2-9 = (a+3)(a-3)$

$2a^2+13a+21 = (2a+7)(a+3)$

$\qquad \text{LCD} = (a+3)(a-3)(2a+7)$

$\dfrac{3a}{a^2-9} = \dfrac{3a}{(a+3)(a-3)}$

$\qquad = \dfrac{3a(2a+7)}{(a+3)(a-3)(2a+7)}$

$\qquad = \dfrac{6a^2+21a}{(a+3)(a-3)(2a+7)}$

$\dfrac{6a}{2a^2+13a+21} = \dfrac{6a}{(a+3)(2a+7)}$

$\qquad = \dfrac{6a(a-3)}{(a+3)(a-3)(2a+7)}$

$\qquad = \dfrac{6a^2-18a}{(a+3)(a-3)(2a+7)}$

12. LCD = $36y^2$

$$\frac{2y^2+3}{9y^2} - \frac{y-6}{12y} = \frac{(2y^2+3)(4)}{9y^2(4)} - \frac{(y-6)(3y)}{12y(3y)}$$

$$= \frac{8y^2+12}{36y^2} - \frac{3y^2-18y}{36y^2}$$

$$= \frac{(8y^2+12)-(3y^2-18y)}{36y^2}$$

$$= \frac{(8y^2+12)+(-3y^2+18y)}{36y^2}$$

$$= \frac{5y^2+18y+12}{36y^2}$$

13. $\frac{2r^2-2r-5}{r^2-16} - \frac{-7r+7}{r^2-16}$

$$= \frac{(2r^2-2r-5)-(-7r+7)}{r^2-16}$$

$$= \frac{(2r^2-2r-5)+(7r-7)}{r^2-16}$$

$$= \frac{2r^2+5r-12}{r^2-16}$$

$$= \frac{(2r-3)\cancel{(r+4)}}{(r-4)\cancel{(r+4)}}$$

$$= \frac{2r-3}{r-4}$$

14. LCD = $2(t+5)(t-5)^2$

$$\frac{t+3}{t^2-10t+25} - \frac{t-4}{2t^2-50}$$

$$= \frac{t+3}{(t-5)^2} - \frac{t-4}{2(t+5)(t-5)}$$

$$= \frac{(t+3)\cdot 2(t+5)}{(t-5)^2\cdot 2(t+5)} - \frac{(t-4)\cdot(t-5)}{2(t+5)(t-5)\cdot(t-5)}$$

$$= \frac{2t^2+16t+30}{2(t+5)(t-5)^2} - \frac{t^2-9t+20}{2(t+5)(t-5)^2}$$

$$= \frac{(2t^2+16t+30)-(t^2-9t+20)}{2(t+5)(t-5)^2}$$

$$= \frac{(2t^2+16t+30)+(-t^2+9t-20)}{2(t+5)(t-5)^2}$$

$$= \frac{t^2+25t+10}{2(t+5)(t-5)^2}$$

15. LCD = $3a-5b$

$$\frac{3a+4b}{3a-5b} + \frac{2a-b}{5b-3a} = \frac{3a+4b}{3a-5b} + \frac{2a-b}{-1(3a-5b)}$$

$$= \frac{3a+4b}{3a-5b} + \frac{-2a+b}{3a-5b}$$

$$= \frac{3a+4b-2a+b}{3a-5b}$$

$$= \frac{a+5b}{3a-5b}$$

16. LCD = $x(x+5)$

$$\frac{5}{x} - \frac{3x-1}{x^2+5x} - \frac{2x-5}{x+5}$$

$$= \frac{5(x+5)}{x(x+5)} - \frac{3x-1}{x(x+5)} - \frac{(2x-5)x}{(x+5)x}$$

$$= \frac{5x+25}{x(x+5)} - \frac{3x-1}{x(x+5)} - \frac{2x^2-5x}{x(x+5)}$$

$$= \frac{(5x+25)-(3x-1)-(2x^2-5x)}{x(x+5)}$$

$$= \frac{(5x+25)+(-3x+1)+(-2x^2+5x)}{x(x+5)}$$

$$= \frac{5x+25-3x+1-2x^2+5x}{x(x+5)}$$

$$= \frac{-2x^2+7x+26}{x(x+5)}$$

17. $\dfrac{\dfrac{6a^3b^2}{c^3}}{\dfrac{9a^2b^4}{c^6}} = \dfrac{6a^3b^2}{c^3} \div \dfrac{9a^2b^4}{c^6}$

$= \dfrac{6a^3b^2}{c^3} \cdot \dfrac{c^6}{9a^2b^4}$

$= \dfrac{6a^3b^2c^6}{9a^2b^4c^3}$

$= \dfrac{2ac^3}{3b^2}$

18. $\dfrac{6 - \dfrac{1}{x} - \dfrac{15}{x^2}}{4 + \dfrac{4}{x} - \dfrac{3}{x^2}} = \dfrac{\left(6 - \dfrac{1}{x} - \dfrac{15}{x^2}\right) \cdot x^2}{\left(4 + \dfrac{4}{x} - \dfrac{3}{x^2}\right) \cdot x^2}$

$= \dfrac{6 \cdot x^2 - \dfrac{1}{x} \cdot x^2 - \dfrac{15}{x^2} \cdot x^2}{4 \cdot x^2 + \dfrac{4}{x} \cdot x^2 - \dfrac{3}{x^2} \cdot x^2}$

$= \dfrac{6x^2 - x - 15}{4x^2 + 4x - 3}$

$= \dfrac{(3x - 5)(2x + 3)}{(2x - 1)(2x + 3)}$

$= \dfrac{3x - 5}{2x - 1}$

19. $\dfrac{t - \dfrac{14}{t - 5}}{2 - \dfrac{4}{t - 5}} = \dfrac{\left(t - \dfrac{14}{t - 5}\right) \cdot (t - 5)}{\left(2 - \dfrac{4}{t - 5}\right) \cdot (t - 5)}$

$= \dfrac{t \cdot (t - 5) - \dfrac{14}{t - 5} \cdot (t - 5)}{2 \cdot (t - 5) - \dfrac{4}{t - 5} \cdot (t - 5)}$

$= \dfrac{t^2 - 5t - 14}{2t - 10 - 4}$

$= \dfrac{t^2 - 5t - 14}{2t - 14}$

$= \dfrac{(t - 7)(t + 2)}{2(t - 7)}$

$= \dfrac{t + 2}{2}$

20. $\dfrac{\dfrac{9}{a^2} - \dfrac{4}{b^2}}{\dfrac{3}{a} + \dfrac{2}{b}} = \dfrac{\left(\dfrac{9}{a^2} - \dfrac{4}{b^2}\right) \cdot a^2b^2}{\left(\dfrac{3}{a} + \dfrac{2}{b}\right) \cdot a^2b^2}$

$= \dfrac{\dfrac{9}{a^2} \cdot a^2b^2 - \dfrac{4}{b^2} \cdot a^2b^2}{\dfrac{3}{a} \cdot a^2b^2 + \dfrac{2}{b} \cdot a^2b^2}$

$= \dfrac{9b^2 - 4a^2}{3ab^2 + 2a^2b}$

$= \dfrac{(3b + 2a)(3b - 2a)}{ab(3b + 2a)}$

$= \dfrac{3b - 2a}{ab}$

21. Multiply both sides by the LCD, $18x$.

$18x\left(\dfrac{8}{6x} + \dfrac{5}{2x} = \dfrac{13}{9}\right)$

$24 + 45 = 26x$

$69 = 26x$

$\dfrac{69}{26} = x$

22. Multiply both sides by the LCD, $(x - 4)$. Notice that if $x = 4$, then parts of the equation are undefined, so 4 cannot be a solution.

$\dfrac{3x}{x - 4} = 6 + \dfrac{12}{x - 4}$

$(x - 4)\left(\dfrac{3x}{x - 4}\right) = (x - 4)\left(6 + \dfrac{12}{x - 4}\right)$

$3x = (x - 4)(6) + (x - 4)\dfrac{12}{x - 4}$

$3x = 6x - 24 + 12$

$-3x = -12$

$x = 4$

We already noted that 4 causes expressions in the equation to be undefined, so 4 is extraneous. Since 4 was the only possible solution, this equation has no solution.

23. Multiply both sides by the LCD, $(x+1)(x+2)$.

$$\frac{x}{x+1}+\frac{9x}{x^2+3x+2}=\frac{12}{x+2}$$

$$\frac{x}{x+1}+\frac{9x}{(x+2)(x+1)}=\frac{12}{x+2}$$

$$(x+1)(x+2)\left(\frac{x}{x+1}+\frac{9x}{(x+2)(x+1)}\right)$$

$$=(x+1)(x+2)\left(\frac{12}{x+2}\right)$$

$$x(x+2)+9x=12(x+1)$$

$$x^2+2x+9x=12x+12$$

$$x^2+11x=12x+12$$

$$x^2-x-12=0$$

$$(x-4)(x+3)=0$$

$$x-4=0 \quad\text{or}\quad x+3=0$$

$$x=4 \qquad\qquad x=-3$$

24. Multiply both sides by the LCD, $(n+1)(n+2)$.

Notice that if $n=-1$ or if $n=-2$, then parts of the equation are undefined, so neither of these values can be a solution.

$$\frac{n}{n+1}+\frac{n+1}{n+2}=\frac{6n+5}{n^2+3n+2}$$

$$\frac{n}{n+1}+\frac{n+1}{n+2}=\frac{6n+5}{(n+1)(n+2)}$$

$$(n+1)(n+2)\left(\frac{n}{n+1}+\frac{n+1}{n+2}\right)$$

$$=(n+1)(n+2)\left(\frac{6n+5}{(n+1)(n+2)}\right)$$

$$n(n+2)+(n+1)^2=6n+5$$

$$n^2+2n+n^2+2n+1=6n+5$$

$$2n^2+4n+1=6n+5$$

$$2n^2-2n-4=0$$

$$2(n^2-n-2)=0$$

$$2(n-2)(n+1)=0$$

$$n-2=0 \quad\text{or}\quad n+1=0$$

$$n=2 \qquad\qquad n=-1$$

We already noted that -1 causes expressions in the equation to be undefined, so -1 is extraneous. The only solution is 2.

25. Let t represent the number of days it would take Eduardo to do the project if he worked by himself. Complete a table.

Categories	Rate of Work	Time at Work	Part of Task Completed
Elena	$\dfrac{1}{15}$	10	$\dfrac{10}{15}$
Eduardo	$\dfrac{1}{t}$	10	$\dfrac{10}{t}$

The total job in this case is 1 project completed, so we can write an equation that combines Elena's part of the task and Eduardo's part of the task and set this sum equal to 1.

$$\frac{10}{15}+\frac{10}{t}=1$$

$$15t\left(\frac{10}{15}+\frac{10}{t}\right)=15t\cdot 1$$

$$10t+150=15t$$

$$150=5t$$

$$30=t$$

It would take Eduardo 30 days to complete the project by himself.

26. Let d represent the distance flown by the cargo plane. Because the sum of the distances flown by each plane is 1900, the distance flown by the airliner is $1900-d$. The times that the planes are flying are equal. Using the relationship $rt=d$, complete a table.

Categories	Rate	Time	Distance
cargo plane	420	$\dfrac{d}{420}$	d
airliner	530	$\dfrac{1900-d}{530}$	$1900-d$

$$\frac{d}{420}=\frac{1900-d}{530}$$

$$530d=420(1900-d)$$

$$530d=798{,}000-420d$$

$$950d=798{,}000$$

$$d=840$$

Time is $\dfrac{d}{420}=\dfrac{840}{420}=2$ hr.

It takes 2 hours for the airplanes to be 1900 miles apart.

27. $r=ks$ \qquad so, $42=7s$

$14=k\cdot 2$ \qquad\qquad $6=s$

$7=k$

28. $p=kqr$ \qquad so, $p=2\cdot 3\cdot 5$

$48=k\cdot 4\cdot 6$ \qquad $p=30$

$2=k$

29. Translate "intensity of light is inversely proportional to the square of the distance from the source" as $I = \dfrac{k}{d^2}$, where I is the intensity of light and d is the distance from the source.

$$I = \frac{k}{d^2} \quad \text{so,} \quad I = \frac{576}{20^2}$$

$$16 = \frac{k}{6^2} \qquad I = 1.44 \text{ foot-candles}$$

$$576 = k$$

The intensity 20 feet from the bulb is 1.44 foot-candles.

30. $d = kg \quad$ so, $\quad d = kg$

$255 = k \cdot 5 \qquad d = 51 \cdot 8$

$51 = k \qquad\quad d = 408$ mi.

Chapters 1–7 Cumulative Review

1. False; the given statement is an example of the commutative property.

2. False; the statement is false for $x = -2$.

3. True

4. True

5. $\dfrac{1}{12}$

6. no

7. -2

8. $\dfrac{2(-4)^2 - 7^2}{5^2 - 3(-3)^2} = \dfrac{2(16) - 49}{25 - 3(9)} = \dfrac{32 - 49}{25 - 27} = \dfrac{-17}{-2} = \dfrac{17}{2}$

9. $-5 - 2 \cdot 6^2 \div (-36)(-4) + 8$

$$= -5 - 2 \cdot 36 \div (-36)(-4) + 8$$
$$= -5 - 72 \div (-36)(-4) + 8$$
$$= -5 - (-2)(-4) + 8$$
$$= -5 - 8 + 8$$
$$= -13 + 8$$
$$= -5$$

10. $x - y\left(x^4 - 2z\right) = (-2) - (-3)\left((-2)^4 - 2(6)\right)$

$$= (-2) - (-3)(16 - 2(6))$$
$$= (-2) - (-3)(16 - 12)$$
$$= (-2) - (-3)(4)$$
$$= (-2) - (-12)$$
$$= -2 + 12$$
$$= 10$$

11. $\left(3x^3 - 2x^2 + 4\right) - \left(-4x^3 + 3x^2 - 5x + 6\right)$

$$= \left(3x^3 - 2x^2 + 4\right) + \left(4x^3 - 3x^2 + 5x - 6\right)$$
$$= 7x^3 - 5x^2 + 5x - 2$$

12. $\left(2a - 3b\right)\left(4a^2 + 6ab + 9b^2\right)$

$$= 2a \cdot 4a^2 + 2a \cdot 6ab + 2a \cdot 9b^2 - 3b \cdot 4a^2$$
$$\quad - 3b \cdot 6ab - 3b \cdot 9b^2$$
$$= 8a^3 + 12a^2b + 18ab^2 - 12a^2b - 18ab^2 - 27b^3$$
$$= 8a^3 - 27b^3$$

13. $\dfrac{12x^3y^4 - 16x^4y^3 + 3x^2y^5}{4x^3y^4}$

$$= \frac{12x^3y^4}{4x^3y^4} - \frac{16x^4y^3}{4x^3y^4} + \frac{3x^2y^5}{4x^3y^4}$$
$$= 3 - \frac{4x}{y} + \frac{3y}{4x}$$

14. $\dfrac{6x^2 + x - 12}{2x^2 - 5x - 12} \cdot \dfrac{3x^2 - 14x + 8}{9x^2 - 18x + 8}$

$$= \frac{\cancel{(3x-4)}\,\cancel{(2x+3)}}{\cancel{(x-4)}\,\cancel{(2x+3)}} \cdot \frac{\cancel{(3x-2)}\,\cancel{(x-4)}}{\cancel{(3x-4)}\,\cancel{(3x-2)}}$$
$$= 1$$

15. $\dfrac{2x+5}{x^2 - 16} - \dfrac{x-9}{x^2 - x - 12}$

$$= \frac{(2x+5)(x+3)}{(x+4)(x-4)(x+3)} - \frac{(x-9)(x+4)}{(x+3)(x-4)(x+4)}$$
$$= \frac{2x^2 + 11x + 15}{(x+4)(x-4)(x+3)} - \frac{x^2 - 5x - 36}{(x+4)(x-4)(x+3)}$$
$$= \frac{2x^2 + 11x + 15 - x^2 + 5x + 36}{(x+4)(x-4)(x+3)}$$
$$= \frac{x^2 + 16x + 51}{(x+4)(x-4)(x+3)}$$

16. $\dfrac{1-\dfrac{3}{x}-\dfrac{4}{x^2}}{\dfrac{3}{x}-\dfrac{13}{x^2}+\dfrac{4}{x^3}} = \dfrac{\left(1-\dfrac{3}{x}-\dfrac{4}{x^2}\right)\cdot x^3}{\left(\dfrac{3}{x}-\dfrac{13}{x^2}+\dfrac{4}{x^3}\right)\cdot x^3}$

$$= \frac{x^3-3x^2-4x}{3x^2-13x+4}$$

$$= \frac{x(x+1)\,(x-4)}{(3x-1)\,(x-4)}$$

$$= \frac{x(x+1)}{3x-1}$$

17. $36a^4b^3 - 39a^3b^4 - 12a^2b^5$

$$= 3a^2b^3\left(12a^2 - 13ab - 4b^2\right)$$

$$= 3a^2b^3\left(4a+b\right)\left(3a-4b\right)$$

18. $108c^3 - 32d^3 = 4\left(27c^3 - 8d^3\right)$

$$= 4\left(3c - 2d\right)\left(9c^2 + 6cd + 4d^2\right)$$

19. $\dfrac{3}{4}x + 3 = \dfrac{2}{3}(x - 6)$

$$12\cdot\left(\dfrac{3}{4}x + 3\right) = 12\cdot\left(\dfrac{2}{3}(x-6)\right)$$

$$9x + 36 = 8(x - 6)$$

$$9x + 36 = 8x - 48$$

$$x + 36 = -48$$

$$x = -84$$

20. $8 + |3 + 2x| \ge 4$

$$|3 + 2x| \ge -4$$

This inequality indicates that the absolute value is greater than a negative number. Because the absolute value of every real number is either positive or 0, the solution set is $\mathbb{R}$.

$\{x \mid x \text{ is a real number}\}$ or $(-\infty, \infty)$

21. $\begin{cases} 2x + y - z = 2 & \text{Eqtn. 1} \\ x + 3y + 2z = 1 & \text{Eqtn. 2} \\ x + y + z = 2 & \text{Eqtn. 3} \end{cases}$

Multiply equation 1 by 2 and add to equation 2. This makes equation 4.

$$4x + 2y - 2z = 4$$
$$\underline{x + 3y + 2z = 1}$$
$$5x + 5y = 5 \qquad \rightarrow \quad x + y = 1 \quad \text{Eqtn. 4}$$

Add equation 1 to equation 3. This makes equation 5.

$$2x + y - z = 2$$
$$\underline{x + y + z = 2}$$
$$3x + 2y = 4 \qquad \text{Eqtn. 5}$$

Use equations 4 and 5 to make a system of equations in two variables. Solve for y.

$$x + y = 1 \qquad \text{Multiply by } -3$$
$$\underline{3x + 2y = 4}$$

$$-3x - 3y = -3$$
$$\underline{3x + 2y = 4}$$
$$-y = 1$$
$$y = -1$$

Substitute the value for y into equation 4 and solve for x.

$$x + y = 1$$
$$x + (-1) = 1$$
$$x = 2$$

Substitute the values for x and y into equation 3 to solve for z.

$$x + y + z = 2$$
$$2 + (-1) + z = 2$$
$$1 + z = 2$$
$$z = 1$$

Solution: $(2, -1, 1)$

22. $6x^2 - 10 = 11x$

$$6x^2 - 11x - 10 = 0$$

$$(3x + 2)(2x - 5) = 0$$

$$3x + 2 = 0 \qquad \text{or} \qquad 2x - 5 = 0$$

$$3x = -2 \qquad\qquad\qquad 2x = 5$$

$$x = -\frac{2}{3} \qquad\qquad\qquad x = \frac{5}{2}$$

23. $\dfrac{3}{x^2+2x-24}+\dfrac{x-5}{x^2-16}=\dfrac{x}{x^2+10x+24}$

$(x+4)(x-4)(x+6)$

$\qquad \cdot\left(\dfrac{3}{(x+6)(x-4)}+\dfrac{x-5}{(x+4)(x-4)}\right)$

$=(x+4)(x-4)(x+6)\cdot\left(\dfrac{x}{(x+6)(x+4)}\right)$

$3(x+4)+(x-5)(x+6)=x(x-4)$

$3x+12+x^2+x-30=x^2-4x$

$x^2+4x-18=x^2-4x$

$4x-18=-4x$

$8x=18$

$x=\dfrac{18}{8}=\dfrac{9}{4}$

24. $2x-5y=-10$

x	y	Ordered Pair
0	2	$(0,2)$
-5	0	$(-5,0)$
5	4	$(5,4)$

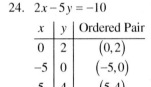

25. $3x-y\le-6$

$\qquad -y\le-6-3x$

$\qquad y\ge 3x+6$

Begin by graphing the related equation
$y=3x+6$ with a solid line. Now choose $(0,0)$
as a test point.

$3x-y \quad\le\quad -6$

$3(0)-0 \quad\overset{?}{\le}\quad -6$

$0-0 \quad\overset{?}{\le}\quad -6$

$0 \quad\le\quad -6$

Because $(0,0)$ does not satisfy the inequality,
shade the side of the line on the opposite side
from $(0,0)$.

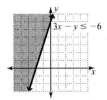

26. Find the slope: $m=\dfrac{1-(-3)}{-1-(-3)}=\dfrac{1+3}{-1+3}=\dfrac{4}{2}=2$

$y-y_1=m(x-x_1)$

$y-1=2(x-(-1))$

$y-1=2(x+1)$

$y-1=2x+2$

$y=2x+3$

27. If two angles are supplementary their sum is 180
degrees. Let one of the angles be x and the other
angle be $3x+20$.

$x+3x+20=180$

$4x+20=180$

$4x=160$

$x=40$

The angles are 40° and $3(40)+20=140°$.

28. Complete a table.

Categories	Selling Price	Number of pounds	Revenue
peppermint	1.80	x	$1.80x$
butterscotch	2.30	15	$2.30(15)$
mixture	2.10	$x+15$	$2.10(x+15)$

$1.80x+2.30(15)=2.10(x+15)$

$1.8x+34.5=2.1x+31.5$

$34.5=0.3x+31.5$

$3=0.3x$

$10=x$

A total of 10 pounds of butterscotch candy will
be needed for the mixture.

29. Let x be the width and $2x-3$ be the length.

$x(2x-3)=54$

$2x^2-3x=54$

$2x^2-3x-54=0$

$(2x+9)(x-6)=0$

$$2x + 9 = 0 \qquad \text{or} \qquad x - 6 = 0$$
$$2x = -9 \qquad\qquad\qquad x = 6$$
$$x = -\frac{9}{2}$$

x cannot be negative

The width is 6 feet and the length is $2(6) - 3 = 9$
feet.

30. Complete a table.

Categories	Rate of Work	Time at Work	Amt of Task Completed
Hal	$\dfrac{1}{7}$	t	$\dfrac{t}{7}$
Frank	$\dfrac{1}{10}$	t	$\dfrac{t}{10}$

$$\frac{t}{7} + \frac{t}{10} = 1$$
$$70\left(\frac{t}{7} + \frac{t}{10}\right) = 70 \cdot 1$$
$$10t + 7t = 70$$
$$17t = 70$$
$$t = \frac{70}{17}$$

Working together it will take Hal and Frank
$\dfrac{70}{17}$ hours to paint the room.

Chapter 8
Rational Exponents, Radicals, and Complex Numbers

Exercise Set 8.1

1. The square roots of 36 are ± 6.

3. The square roots of 121 are ± 11.

5. The square roots of 196 are ± 14.

7. The square roots of 225 are ± 15.

9. $\sqrt{25} = 5$

11. $\sqrt{-64}$ is not a real number.

13. $-\sqrt{25} = -5$

15. $\pm\sqrt{25} = \pm 5$

17. $\sqrt{1.44} = 1.2$

19. $\sqrt{-10.64}$ is not a real number.

21. $-\sqrt{0.0121} = -0.11$

23. $\sqrt{\dfrac{49}{81}} = \dfrac{7}{9}$

25. $-\sqrt{\dfrac{144}{169}} = -\dfrac{12}{13}$

27. $\sqrt[3]{27} = 3$

29. $\sqrt[3]{-64} = -4$

31. $-\sqrt[3]{-216} = -(-6) = 6$

33. $\sqrt[4]{256} = 4$

35. $\sqrt[4]{-625}$ is not a real number.

37. $-\sqrt[4]{16} = -2$

39. $\sqrt[5]{32} = 2$

41. $\sqrt[5]{-243} = -3$

43. $-\sqrt[5]{-32} = -(-2) = 2$

45. $\sqrt[6]{64} = 2$

47. $\sqrt[3]{-\dfrac{8}{27}} = -\dfrac{2}{3}$

49. $\sqrt[4]{\dfrac{16}{81}} = \dfrac{2}{3}$

51. $\sqrt{7} \approx 2.646$

53. $-\sqrt{11} \approx -3.317$

55. $\sqrt[3]{50} \approx 3.684$

57. $\sqrt[3]{-53} \approx -3.756$

59. $\sqrt[4]{189} \approx 3.708$

61. $-\sqrt[4]{85} \approx -3.036$

63. $\sqrt[5]{89} \approx 2.454$

65. $\sqrt[6]{146} \approx 2.295$

67. $\sqrt{b^4} = b^2$

69. $\sqrt{16x^2} = 4x$

71. $\sqrt{100r^8 s^6} = 10r^4 s^3$

73. $\sqrt{0.25a^6 b^{12}} = 0.5a^3 b^6$

75. $\sqrt[3]{m^3} = m$

77. $\sqrt[3]{27a^9 b^6} = 3a^3 b^2$

79. $\sqrt[3]{-64a^3 b^{12}} = -4ab^4$

81. $\sqrt[3]{0.008x^{18}} = 0.2x^6$

83. $\sqrt[4]{a^4} = a$

85. $\sqrt[4]{16x^{16}} = 2x^4$

87. $\sqrt[5]{32x^{10}} = 2x^2$

89. $\sqrt[6]{x^{12} y^6} = x^2 y$

91. $\sqrt{36m^2} = 6|m|$

93. $\sqrt{(r-1)^2} = |r-1|$

95. $\sqrt[4]{256y^{12}} = 4|y^3|$

97. $\sqrt[3]{27y^3} = 3y$

99. $\sqrt{(y-3)^4} = (y-3)^2$

101. $\sqrt[3]{(y-4)^6} = (y-4)^2$

103. $f(x) = \sqrt{2x+4}$

    $f(0) = \sqrt{2 \cdot 0 + 4}$

         $= \sqrt{0+4}$

         $= \sqrt{4}$

         $= 2$

105. $f(x) = \sqrt{4x+3}$

    $f(3) = \sqrt{4 \cdot 3 + 3}$

         $= \sqrt{12+3}$

         $= \sqrt{15}$

107. Because the index is even, the radicand must be nonnegative.

    $2x - 8 \geq 0$

      $2x \geq 8$

       $x \geq 4$

    Domain: $\{x | x \geq 4\}$, or $[4, \infty)$

109. Because the index is even, the radicand must be nonnegative.

    $-4x + 16 \geq 0$

      $-4x \geq -16$

       $x \leq 4$

    Domain: $\{x | x \leq 4\}$, or $(-\infty, 4]$

111. a)

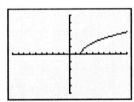

    b) $\{x | x \geq 2\}$, or $[2, \infty)$

113. a)

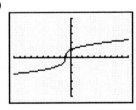

    b) $\mathbb{R}$, or $(-\infty, \infty)$

115. $v = -\sqrt{19.6h}$

    $v = -\sqrt{19.6(16)}$

      $= -\sqrt{313.6}$

      ≈ -17.709 m/sec.

117. $T = 2\pi \sqrt{\dfrac{L}{9.8}}$

    $T = 2\pi \sqrt{\dfrac{3}{9.8}}$

      $\approx 2\pi \sqrt{0.306}$

      ≈ 3.476 sec.

119. $S = \dfrac{7}{2}\sqrt{2D}$

    $S = \dfrac{7}{2}\sqrt{2 \cdot 15}$

      $= \dfrac{7}{2}\sqrt{30}$

      $\approx \dfrac{7}{2} \cdot 5.477$

      ≈ 19.170 mph

121. $c = \sqrt{a^2 + b^2}$

    $c = \sqrt{5^2 + 12^2}$

      $= \sqrt{25 + 144}$

      $= \sqrt{169}$

      $= 13$ ft.

123. $R = \sqrt{F_1^2 + F_2^2}$

    $R = \sqrt{9^2 + 12^2}$

      $= \sqrt{81 + 144}$

      $= \sqrt{225}$

      $= 15$ N

125. a) $f(x) = 10.5\sqrt{x} + 21$

        $f(2) = 10.5\sqrt{2} + 21$

           ≈ 35.849

           ≈ 36

    Approximately 36 earthquakes with that magnitude occurred in 2010.

    b) $f(x) = 10.5\sqrt{x} + 21$

        $f(3) = 10.5\sqrt{3} + 21$

           ≈ 39.187

           ≈ 39

    Approximately 39 earthquakes with that magnitude occurred in 2011.

Review Exercises

1. a) $\sqrt{16 \cdot 9} = \sqrt{144}$
    $$= 12$$
    b) $\sqrt{16} \cdot \sqrt{9} = 4 \cdot 3$
    $$= 12$$

2. $x^5 \cdot x^3 = x^{5+3} = x^8$

3. $\left(-9m^3 n\right)\left(5mn^2\right) = -9 \cdot 5m^{3+1}n^{1+2} = -45m^4 n^3$

4. $4x^2 \left(3x^2 - 5x + 1\right)$
    $$= 4x^2 \cdot 3x^2 + 4x^2 \cdot \left(-5x\right) + 4x^2 \cdot 1$$
    $$= 12x^4 - 20x^3 + 4x^2$$

5. $\left(7y - 4\right)\left(3y + 5\right)$
    $$= 7y \cdot 3y + 7y \cdot 5 - 4 \cdot 3y - 4 \cdot 5$$
    $$= 21y^2 + 35y - 12y - 20$$
    $$= 21y^2 + 23y - 20$$

6. $A = \left(x - 9\right)\left(2x + 1\right)$
    $$= x \cdot 2x + x \cdot 1 - 9 \cdot 2x - 9 \cdot 1$$
    $$= 2x^2 + x - 18x - 9$$
    $$= 2x^2 - 17x - 9$$

Exercise Set 8.2

1. $25^{1/2} = \sqrt{25} = 5$

3. $-100^{1/2} = -\sqrt{100} = -10$

5. $27^{1/3} = \sqrt[3]{27} = 3$

7. $\left(-64\right)^{1/3} = \sqrt[3]{-64} = -4$

9. $y^{1/4} = \sqrt[4]{y}$

11. $\left(144x^8\right)^{1/2} = \sqrt{144x^8} = 12x^4$

13. $18r^{1/2} = 18\sqrt{r}$

15. $\left(\dfrac{x^4}{81}\right)^{1/2} = \sqrt{\dfrac{x^4}{81}} = \dfrac{x^2}{9}$

17. $64^{2/3} = \left(\sqrt[3]{64}\right)^2 = \left(4\right)^2 = 16$

19. $-81^{3/4} = -\left(\sqrt[4]{81}\right)^3 = -\left(3\right)^3 = -27$

21. $\left(-8\right)^{4/3} = \left(\sqrt[3]{-8}\right)^4 = \left(-2\right)^4 = 16$

23. $16^{-3/4} = \dfrac{1}{16^{3/4}} = \dfrac{1}{\left(\sqrt[4]{16}\right)^3} = \dfrac{1}{2^3} = \dfrac{1}{8}$

25. $x^{4/5} = \sqrt[5]{x^4}$

27. $8n^{2/3} = 8\sqrt[3]{n^2}$

29. $\left(-32\right)^{-2/5} = \dfrac{1}{\left(-32\right)^{2/5}}$
    $$= \dfrac{1}{\left(\sqrt[5]{-32}\right)^2}$$
    $$= \dfrac{1}{\left(-2\right)^2}$$
    $$= \dfrac{1}{4}$$

31. $\left(\dfrac{1}{25}\right)^{3/2} = \left(\sqrt{\dfrac{1}{25}}\right)^3 = \left(\dfrac{1}{5}\right)^3 = \dfrac{1}{125}$

33. $\left(2a + 4\right)^{5/6} = \sqrt[6]{\left(2a + 4\right)^5}$

35. $\sqrt[4]{25} = 25^{1/4}$

37. $\sqrt[6]{z^5} = z^{5/6}$

39. $\dfrac{1}{\sqrt[6]{5^5}} = \dfrac{1}{5^{5/6}} = 5^{-5/6}$

41. $\dfrac{5}{\sqrt[5]{x^4}} = \dfrac{5}{x^{4/5}} = 5x^{-4/5}$

43. $\left(\sqrt[3]{5}\right)^7 = 5^{7/3}$

45. $\left(\sqrt[7]{x}\right)^2 = x^{2/7}$

47. $\sqrt[4]{\left(4a - 5\right)^7} = \left(4a - 7\right)^{7/4}$

49. $\left(\sqrt[5]{2r - 5}\right)^8 = \left(2r - 5\right)^{8/5}$

51. $x^{1/5} \cdot x^{3/5} = x^{1/5 + 3/5} = x^{4/5}$

53. $x^{3/2} \cdot x^{-1/3} = x^{3/2 + (-1/3)} = x^{9/6 - 2/6} = x^{7/6}$

55. $a^{2/3} \cdot a^{3/4} = a^{2/3 + 3/4} = a^{8/12 + 9/12} = a^{17/12}$

57. $\left(3w^{1/7}\right)\left(7w^{3/7}\right) = 21w^{1/7 + 3/7} = 21w^{4/7}$

59. $\left(-3a^{2/3}\right)\left(4a^{3/4}\right) = -12a^{2/3+3/4}$
$$= -12a^{8/12+9/12}$$
$$= -12a^{17/12}$$

61. $\dfrac{7^{7/3}}{7^{2/3}} = 7^{7/3-2/3} = 7^{5/3}$

63. $\dfrac{x^{1/6}}{x^{5/6}} = x^{1/6-5/6} = x^{-4/6} = x^{-2/3} = \dfrac{1}{x^{2/3}}$

65. $\dfrac{x^{3/4}}{x^{1/2}} = x^{3/4-1/2} = x^{3/4-2/4} = x^{1/4}$

67. $\dfrac{r^{3/4}}{r^{2/3}} = r^{3/4-2/3} = r^{9/12-8/12} = r^{1/12}$

69. $\dfrac{x^{-3/7}}{x^{2/7}} = x^{-3/7-2/7} = x^{-5/7} = \dfrac{1}{x^{5/7}}$

71. $\dfrac{a^{3/4}}{a^{-3/2}} = a^{3/4-(-3/2)} = a^{3/4+6/4} = a^{9/4}$

73. $\left(5s^{-2/7}\right)\left(4s^{5/7}\right) = 20s^{-2/7+5/7} = 20s^{3/7}$

75. $\left(-6b^{-5/4}\right)\left(4b^{3/2}\right) = -24b^{-5/4+3/2}$
$$= -24b^{-5/4+6/4}$$
$$= -24b^{1/4}$$

77. $\left(x^{2/3}\right)^3 = x^{(2/3)\cdot 3} = x^2$

79. $\left(a^{5/6}\right)^2 = a^{(5/6)\cdot 2} = a^{10/6} = a^{5/3}$

81. $\left(b^{2/3}\right)^{3/5} = b^{(2/3)\cdot(3/5)} = b^{6/15} = b^{2/5}$

83. $\left(2x^{2/3}y^{1/2}\right)^6 = 2^6\, x^{(2/3)\cdot 6}\, y^{(1/2)\cdot 6}$
$$= 64x^{12/3}y^{6/2}$$
$$= 64x^4 y^3$$

85. $\left(8q^{3/2}t^{3/4}\right)^{1/3} = 8^{1/3}q^{(3/2)\cdot(1/3)}t^{(3/4)\cdot(1/3)}$
$$= 2q^{3/6}t^{3/12}$$
$$= 2q^{1/2}t^{1/4}$$

87. $\dfrac{\left(3a^{5/4}\right)^4}{a^2} = \dfrac{3^4\, a^{(5/4)\cdot 4}}{a^2}$
$$= \dfrac{81a^{20/4}}{a^2}$$
$$= \dfrac{81a^5}{a^2}$$
$$= 81a^{5-2}$$
$$= 81a^3$$

89. $\dfrac{\left(9z^{7/3}\right)^{1/2}}{z^{5/6}} = \dfrac{9^{1/2}\, z^{(7/3)\cdot(1/2)}}{z^{5/6}}$
$$= \dfrac{3z^{7/6}}{z^{5/6}}$$
$$= 3z^{7/6-5/6}$$
$$= 3z^{2/6}$$
$$= 3z^{1/3}$$

91. $\sqrt[4]{4} = 4^{1/4} = \left(2^2\right)^{1/4} = 2^{2\,(1/4)} = 2^{1/2} = \sqrt{2}$

93. $\sqrt[6]{49} = \left(7^2\right)^{1/6} = 7^{2\cdot(1/6)} = 7^{1/3} = \sqrt[3]{7}$

95. $\sqrt[4]{x^2} = \left(x^2\right)^{1/4} = x^{2\cdot(1/4)} = x^{1/2} = \sqrt{x}$

97. $\sqrt[8]{r^6} = \left(r^6\right)^{1/8} = r^{3/4} = \sqrt[4]{r^3}$

99. $\sqrt[8]{x^6 y^2} = \left(x^6 y^2\right)^{1/8}$
$$= x^{6\cdot(1/8)}y^{2\cdot(1/8)}$$
$$= x^{3/4}y^{1/4}$$
$$= \left(x^3 y\right)^{1/4}$$
$$= \sqrt[4]{x^3 y}$$

101. $\sqrt[10]{m^4 n^6} = \left(m^4 n^6\right)^{1/10}$
$$= m^{4\cdot(1/10)}n^{6\cdot(1/10)}$$
$$= m^{2/5}n^{3/5}$$
$$= \left(m^2 m^3\right)^{1/5}$$
$$= \sqrt[5]{m^2 n^3}$$

103. $\sqrt[3]{x}\cdot\sqrt{x} = x^{1/3}\cdot x^{1/2}$
$$= x^{1/3+1/2}$$
$$= x^{2/6+3/6}$$
$$= x^{5/6}$$
$$= \sqrt[6]{x^5}$$

105. $\sqrt[4]{y^2} \cdot \sqrt[3]{y^2} = y^{2/4} \cdot y^{2/3}$

$\qquad = y^{1/2 + 2/3}$

$\qquad = y^{3/6 + 4/6}$

$\qquad = y^{7/6}$

$\qquad = \sqrt[6]{y^7}$

107. $\dfrac{\sqrt[3]{x^4}}{\sqrt[4]{x^2}} = \dfrac{x^{4/3}}{x^{1/2}} = x^{4/3 - 1/2} = x^{8/6 - 3/6} = x^{5/6} = \sqrt[6]{x^5}$

109. $\dfrac{\sqrt[5]{n^4}}{\sqrt[3]{n^2}} = \dfrac{n^{4/5}}{n^{2/3}} = n^{4/5 - 2/3} = n^{12/15 - 10/15} = n^{2/15} = \sqrt[15]{n^2}$

111. $\sqrt{5} \cdot \sqrt[3]{3} = 5^{1/2} \cdot 3^{1/3}$

$\qquad = 5^{3/6} \cdot 3^{2/6}$

$\qquad = \left(5^3 \cdot 3^2\right)^{1/6}$

$\qquad = \left(125 \cdot 9\right)^{1/6}$

$\qquad = \left(1125\right)^{1/6}$

$\qquad = \sqrt[6]{1125}$

113. $\sqrt[4]{6} \cdot \sqrt[3]{2} = 6^{1/4} \cdot 2^{1/3}$

$\qquad = 6^{3/12} \cdot 2^{4/12}$

$\qquad = \left(6^3 \cdot 2^4\right)^{1/12}$

$\qquad = \left(216 \cdot 16\right)^{1/12}$

$\qquad = \left(3456\right)^{1/12}$

$\qquad = \sqrt[12]{3456}$

115. $\sqrt[3]{\sqrt[3]{x}} = \left(x^{1/3}\right)^{1/3} = x^{(1/3)\cdot(1/3)} = x^{1/9} = \sqrt[9]{x}$

117. $\sqrt{\sqrt[3]{n}} = \left(n^{1/3}\right)^{1/2} = n^{(1/3)\cdot(1/2)} = n^{1/6} = \sqrt[6]{n}$

Review Exercises

1. $2 \cdot 2 \cdot 2 \cdot 2 \cdot x \cdot x \cdot x \cdot y \cdot y = 2^4 x^3 y^2$

2. $\sqrt{16} \cdot \sqrt{9} = 4 \cdot 3 = 12$

3. $\sqrt[3]{27} \cdot \sqrt[3]{125} = 3 \cdot 5 = 15$

4. $\left(2.5 \times 10^6\right)\left(3.2 \times 10^5\right) = 8 \times 10^{6+5} = 8 \times 10^{11}$

5. $\left(\dfrac{3}{4} x^3 y\right)\left(-\dfrac{5}{6} xyz^2\right) = -\dfrac{5}{8} x^{3+1} y^{1+1} z^2$

$\qquad = -\dfrac{5}{8} x^4 y^2 z^2$

6.
$$\begin{array}{r} 2x^2 - 8x + 5 \\ x+3{\overline{\smash{\big)}\,2x^3 - 2x^2 - 19x + 18}} \\ \underline{2x^3 + 6x^2} \\ -8x^2 - 19x \\ \underline{-8x^2 - 24x} \\ 5x + 18 \\ \underline{5x + 15} \\ 3 \end{array}$$

Answer: $2x^2 - 8x + 5 + \dfrac{3}{x+3}$

Exercise Set 8.3

1. $\sqrt{2} \cdot \sqrt{32} = \sqrt{64} = 8$

3. $\sqrt{3x} \cdot \sqrt{27x^5} = \sqrt{3x \cdot 27x^5} = \sqrt{81x^6} = 9x^3$

5. $\sqrt{6xy^3} \cdot \sqrt{24xy} = \sqrt{6xy^3 \cdot 24xy}$

$\qquad = \sqrt{144x^2 y^4}$

$\qquad = 12xy^2$

7. $\sqrt{5} \cdot \sqrt{13} = \sqrt{5 \cdot 13} = \sqrt{65}$

9. $\sqrt{15} \cdot \sqrt{x} = \sqrt{15 \cdot x} = \sqrt{15x}$

11. $\sqrt[3]{3} \cdot \sqrt[3]{9} = \sqrt[3]{3 \cdot 9} = \sqrt[3]{27} = 3$

13. $\sqrt[3]{5y} \cdot \sqrt[3]{2y} = \sqrt[3]{5y \cdot 2y} = \sqrt[3]{10y^2}$

15. $\sqrt[4]{3} \cdot \sqrt[4]{7} = \sqrt[4]{3 \cdot 7} = \sqrt[4]{21}$

17. $\sqrt[4]{12w^3} \cdot \sqrt[4]{6w} = \sqrt[4]{12w^3 \cdot 6w}$

$\qquad = \sqrt[4]{72w^4}$

$\qquad = \sqrt[4]{w^4 \cdot 72}$

$\qquad = \sqrt[4]{w^4} \cdot \sqrt[4]{72}$

$\qquad = w\sqrt[4]{72}$

19. $\sqrt[4]{3x^2 y} \cdot \sqrt[4]{5xy^2} = \sqrt[4]{3x^2 y \cdot 5xy^2} = \sqrt[4]{15x^3 y^3}$

21. $\sqrt[5]{6x^3} \cdot \sqrt[5]{5x^4} = \sqrt[5]{6x^3 \cdot 5x^4}$

$\qquad = \sqrt[5]{30x^7}$

$\qquad = \sqrt[5]{x^5 \cdot 30x^2}$

$\qquad = \sqrt[5]{x^5} \cdot \sqrt[5]{30x^2}$

$\qquad = x\sqrt[5]{30x^2}$

23. $\sqrt[6]{4x^2 y^3} \cdot \sqrt[6]{2x^3 y} = \sqrt[6]{4x^2 y^3 \cdot 2x^3 y} = \sqrt[6]{8x^5 y^4}$

25. $\sqrt{\dfrac{7}{2}} \cdot \sqrt{\dfrac{3}{5}} = \sqrt{\dfrac{7}{2} \cdot \dfrac{3}{5}} = \sqrt{\dfrac{21}{10}}$

27. $\sqrt{\dfrac{6}{x}} \cdot \sqrt{\dfrac{y}{5}} = \sqrt{\dfrac{6}{x} \cdot \dfrac{y}{5}} = \sqrt{\dfrac{6y}{5x}}$

29. $\sqrt{\dfrac{25}{36}} = \dfrac{\sqrt{25}}{\sqrt{36}} = \dfrac{5}{6}$

31. $\sqrt{\dfrac{10}{9}} = \dfrac{\sqrt{10}}{\sqrt{9}} = \dfrac{\sqrt{10}}{3}$

33. $\dfrac{\sqrt{180}}{\sqrt{5}} = \sqrt{\dfrac{180}{5}} = \sqrt{36} = 6$

35. $\dfrac{\sqrt{15}}{\sqrt{5}} = \sqrt{\dfrac{15}{5}} = \sqrt{3}$

37. $\sqrt[3]{\dfrac{4}{w^6}} = \dfrac{\sqrt[3]{4}}{\sqrt[3]{w^6}} = \dfrac{\sqrt[3]{4}}{w^2}$

39. $\sqrt[3]{\dfrac{5y^2}{27x^9}} = \dfrac{\sqrt[3]{5y^2}}{\sqrt[3]{27x^9}} = \dfrac{\sqrt[3]{5y^2}}{3x^3}$

41. $\dfrac{\sqrt[3]{320}}{\sqrt[3]{5}} = \sqrt[3]{\dfrac{320}{5}} = \sqrt[3]{64} = 4$

43. $\sqrt[4]{\dfrac{3u^3}{16x^8}} = \dfrac{\sqrt[4]{3u^3}}{\sqrt[4]{16x^8}} = \dfrac{\sqrt[4]{3u^3}}{2x^2}$

45. $\sqrt{98} = \sqrt{49 \cdot 2} = \sqrt{49} \cdot \sqrt{2} = 7\sqrt{2}$

47. $\sqrt{128} = \sqrt{64 \cdot 2} = \sqrt{64} \cdot \sqrt{2} = 8\sqrt{2}$

49. $6\sqrt{80} = 6\sqrt{16 \cdot 5} = 6\sqrt{16} \cdot \sqrt{5} = 6 \cdot 4\sqrt{5} = 24\sqrt{5}$

51. $5\sqrt{112} = 5\sqrt{16 \cdot 7} = 5\sqrt{16} \cdot \sqrt{7} = 5 \cdot 4\sqrt{7} = 20\sqrt{7}$

53. $\sqrt{a^7} = \sqrt{a^6 \cdot a} = \sqrt{a^6} \cdot \sqrt{a} = a^3\sqrt{a}$

55. $\sqrt{x^2 y^4} = xy^2$

57. $\sqrt{x^6 y^8 z^{10}} = x^3 y^4 z^5$

59. $rs^2 \sqrt{r^9 s^5} = rs^2 \sqrt{r^8 s^4 \cdot rs}$
$= rs^2 \sqrt{r^8 s^4} \cdot \sqrt{rs}$
$= rs^2 \cdot r^4 s^2 \cdot \sqrt{rs}$
$= r^5 s^4 \sqrt{rs}$

61. $3\sqrt{72x^5} = 3\sqrt{36x^4 \cdot 2x}$
$= 3\sqrt{36x^4} \cdot \sqrt{2x}$
$= 3 \cdot 6x^2 \cdot \sqrt{2x}$
$= 18x^2 \sqrt{2x}$

63. $\sqrt[3]{32} = \sqrt[3]{8 \cdot 4} = \sqrt[3]{8} \cdot \sqrt[3]{4} = 2\sqrt[3]{4}$

65. $\sqrt[3]{x^7} = \sqrt[3]{x^6 \cdot x} = \sqrt[3]{x^6} \cdot \sqrt[3]{x} = x^2 \sqrt[3]{x}$

67. $\sqrt[3]{x^6 y^5} = \sqrt[3]{x^6 y^3 \cdot y^2} = \sqrt[3]{x^6 y^3} \cdot \sqrt[3]{y^2} = x^2 y\sqrt[3]{y^2}$

69. $\sqrt[3]{128z^8} = \sqrt[3]{64z^6 \cdot 2z^2}$
$= \sqrt[3]{64z^6} \cdot \sqrt[3]{2z^2}$
$= 4z^2 \sqrt[3]{2z^2}$

71. $2\sqrt[3]{40} = 2\sqrt[3]{8 \cdot 5} = 2\sqrt[3]{8} \cdot \sqrt[3]{5} = 2 \cdot 2 \cdot \sqrt[3]{5} = 4\sqrt[3]{5}$

73. $\sqrt[4]{80} = \sqrt[4]{16 \cdot 5} = \sqrt[4]{16} \cdot \sqrt[4]{5} = 2\sqrt[4]{5}$

75. $3x^2 \sqrt[4]{243x^9} = 3x^2 \sqrt[4]{81x^8 \cdot 3x}$
$= 3x^2 \sqrt[4]{81x^8} \cdot \sqrt[4]{3x}$
$= 3x^2 \cdot 3x^2 \sqrt[4]{3x}$
$= 9x^4 \sqrt[4]{3x}$

77. $\sqrt[5]{486x^{16}} = \sqrt[5]{243x^{15} \cdot 2x}$
$= \sqrt[5]{243x^{15}} \cdot \sqrt[5]{2x}$
$= 3x^3 \sqrt[5]{2x}$

79. $\sqrt[6]{x^8 y^{14} z^{11}} = \sqrt[6]{x^6 y^{12} z^6 \cdot x^2 y^2 z^5}$
$= \sqrt[6]{x^6 y^{12} z^6} \cdot \sqrt[6]{x^2 y^2 z^5}$
$= xy^2 z\sqrt[6]{x^2 y^2 z^5}$

81. $\sqrt{3} \cdot \sqrt{21} = \sqrt{63} = \sqrt{9 \cdot 7} = 3\sqrt{7}$

83. $5\sqrt{10} \cdot 3\sqrt{6} = 15\sqrt{60}$
$= 15\sqrt{4 \cdot 15}$
$= 15 \cdot 2\sqrt{15}$
$= 30\sqrt{15}$

85. $\sqrt{y^3} \cdot \sqrt{y^2} = \sqrt{y^5} = \sqrt{y^4 \cdot y} = y^2 \sqrt{y}$

87. $x\sqrt{x^2 y^3} \cdot y^2 \sqrt{x^4 y^4} = xy^2 \sqrt{x^6 y^7}$
$= xy^2 \sqrt{x^6 y^6 \cdot y}$
$= xy^2 \cdot x^3 y^3 \sqrt{y}$
$= x^4 y^5 \sqrt{y}$

89. $4\sqrt{6c^3} \cdot 3\sqrt{10c^5} = 12\sqrt{60c^8}$
$$= 12\sqrt{4c^8 \cdot 15}$$
$$= 12 \cdot 2c^4\sqrt{15}$$
$$= 24c^4\sqrt{15}$$

91. $4\sqrt{3} \cdot 5\sqrt{6} = 20\sqrt{18}$
$$= 20\sqrt{9 \cdot 2}$$
$$= 20 \cdot 3\sqrt{2}$$
$$= 60\sqrt{2}$$

93. $\dfrac{\sqrt{48}}{\sqrt{6}} = \sqrt{\dfrac{48}{6}} = \sqrt{8} = \sqrt{4 \cdot 2} = 2\sqrt{2}$

95. $\dfrac{9\sqrt{160}}{3\sqrt{8}} = 3\sqrt{\dfrac{160}{8}}$
$$= 3\sqrt{20}$$
$$= 3\sqrt{4 \cdot 5}$$
$$= 3 \cdot 2\sqrt{5}$$
$$= 6\sqrt{5}$$

97. $\dfrac{\sqrt{c^5 d^6}}{\sqrt{cd^3}} = \sqrt{\dfrac{c^5 d^6}{cd^3}} = \sqrt{c^4 d^3} = \sqrt{c^4 d^2 \cdot d} = c^2 d\sqrt{d}$

99. $\dfrac{8\sqrt{45a^5}}{2\sqrt{5a}} = 4\sqrt{\dfrac{45a^5}{5a}}$
$$= 4\sqrt{9a^4}$$
$$= 4 \cdot 3a^2$$
$$= 12a^2$$

101. $\dfrac{12\sqrt{72c^5}}{4\sqrt{6c^2}} = 3\sqrt{\dfrac{72c^5}{6c^2}}$
$$= 3\sqrt{12c^3}$$
$$= 3\sqrt{4c^2 \cdot 3c}$$
$$= 3 \cdot 2c\sqrt{3c}$$
$$= 6c\sqrt{3c}$$

103. $\dfrac{36\sqrt{96x^6 y^{11}}}{4\sqrt{3x^2 y^4}} = 9\sqrt{\dfrac{96x^6 y^{11}}{3x^2 y^4}}$
$$= 9\sqrt{32x^4 y^7}$$
$$= 9\sqrt{16x^4 y^6 \cdot 2y}$$
$$= 9 \cdot 4x^2 y^3 \sqrt{2y}$$
$$= 36x^2 y^3 \sqrt{2y}$$

105. $\sqrt{\dfrac{3}{7}} \cdot \sqrt{\dfrac{8}{7}} = \sqrt{\dfrac{24}{49}} = \dfrac{\sqrt{24}}{7} = \dfrac{\sqrt{4 \cdot 6}}{7} = \dfrac{2\sqrt{6}}{7}$

107. $\sqrt{\dfrac{a^3}{2}} \cdot \sqrt{\dfrac{a^5}{2}} = \sqrt{\dfrac{a^8}{4}} = \dfrac{a^4}{2}$

109. $\sqrt{\dfrac{3x^5}{2}} \cdot \sqrt{\dfrac{15x^5}{8}} = \sqrt{\dfrac{45x^{10}}{16}}$
$$= \dfrac{\sqrt{45x^{10}}}{\sqrt{16}}$$
$$= \dfrac{\sqrt{9x^{10} \cdot 5}}{4}$$
$$= \dfrac{3x^5 \sqrt{5}}{4}$$

111. $\dfrac{1}{2} \cdot \sqrt{\dfrac{3}{8}} \cdot \sqrt{\dfrac{15}{2}} = \dfrac{1}{2}\sqrt{\dfrac{45}{16}}$
$$= \dfrac{1}{2} \cdot \dfrac{\sqrt{45}}{\sqrt{16}}$$
$$= \dfrac{1}{2} \cdot \dfrac{\sqrt{9 \cdot 5}}{4}$$
$$= \dfrac{1}{2} \cdot \dfrac{3\sqrt{5}}{4}$$
$$= \dfrac{3}{8}\sqrt{5}$$

Review Exercises

1. No   2. Yes

3. $6x^2 - 4x - 3x + 2x^2 = 8x^2 - 7x$

4. $(2a - 3b)(4a + 3b)$
$$= 2a \cdot 4a + 2a \cdot 3b - 3b \cdot 4a - 3b \cdot 3b$$
$$= 8a^2 + 6ab - 12ab - 9b^2$$
$$= 8a^2 - 6ab - 9b^2$$

5. $(3m + 5n)(3m - 5n) = (3m)^2 - (5n)^2$
$$= 9m^2 - 25n^2$$

6. $(2x - 3y)^2 = (2x)^2 - 2(2x)(3y) + (3y)^2$
$$= 4x^2 - 12xy + 9y^2$$

Exercise Set 8.4

1. $9\sqrt{6} - 15\sqrt{6} = (9 - 15)\sqrt{6} = -6\sqrt{6}$

3. $7\sqrt{a} + 2\sqrt{a} = (7 + 2)\sqrt{a} = 9\sqrt{a}$

5. $4\sqrt{5} - 2\sqrt{6} + 8\sqrt{5} - 6\sqrt{6}$
    $= (4+8)\sqrt{5} + (-2-6)\sqrt{6}$
    $= 12\sqrt{5} - 8\sqrt{6}$

7. $3a\sqrt{5a} - 4b\sqrt{7b} + 8a\sqrt{5a} + 2b\sqrt{7b}$
    $= (3a+8a)\sqrt{5a} + (-4b+2b)\sqrt{7b}$
    $= 11a\sqrt{5a} - 2b\sqrt{7b}$

9. $6x\sqrt[3]{9} - 3x\sqrt[3]{9} = (6x-3x)\sqrt[3]{9} = 3x\sqrt[3]{9}$

11. $6x^2\sqrt[4]{5x} - 12x^2\sqrt[4]{5x} = (6x^2-12x^2)\sqrt[4]{5x}$
    $= -6x^2\sqrt[4]{5x}$

13. $3x\sqrt{5x} + 4x\sqrt[3]{5x}$
    Cannot combine because the radicals are not like.

15. $\sqrt{48} - \sqrt{75} = \sqrt{16\cdot3} - \sqrt{25\cdot3}$
    $= 4\sqrt{3} - 5\sqrt{3}$
    $= -\sqrt{3}$

17. $\sqrt{80y} - \sqrt{125y} = \sqrt{16\cdot5y} - \sqrt{25\cdot5y}$
    $= 4\sqrt{5y} - 5\sqrt{5y}$
    $= -\sqrt{5y}$

19. $\sqrt{80} - 4\sqrt{45} = \sqrt{16\cdot5} - 4\sqrt{9\cdot5}$
    $= 4\sqrt{5} - 4\cdot3\sqrt{5}$
    $= 4\sqrt{5} - 12\sqrt{5}$
    $= -8\sqrt{5}$

21. $3\sqrt{96} - 2\sqrt{54} = 3\sqrt{16\cdot6} - 2\sqrt{9\cdot6}$
    $= 3\cdot4\sqrt{6} - 2\cdot3\sqrt{6}$
    $= 12\sqrt{6} - 6\sqrt{6}$
    $= 6\sqrt{6}$

23. $6\sqrt{48a^3} - 2\sqrt{75a^3}$
    $= 6\sqrt{16a^2\cdot3a} - 2\sqrt{25a^2\cdot3a}$
    $= 6\cdot4a\sqrt{3a} - 2\cdot5a\sqrt{3a}$
    $= 24a\sqrt{3a} - 10a\sqrt{3a}$
    $= 14a\sqrt{3a}$

25. $\sqrt{150} - \sqrt{54} + \sqrt{24} = \sqrt{25\cdot6} - \sqrt{9\cdot6} + \sqrt{4\cdot6}$
    $= 5\sqrt{6} - 3\sqrt{6} + 2\sqrt{6}$
    $= 4\sqrt{6}$

27. $2\sqrt{8} - 3\sqrt{48} + 2\sqrt{98} - \sqrt{75}$
    $= 2\sqrt{4\cdot2} - 3\sqrt{16\cdot3} + 2\sqrt{49\cdot2} - \sqrt{25\cdot3}$
    $= 2\cdot2\sqrt{2} - 3\cdot4\sqrt{3} + 2\cdot7\sqrt{2} - 5\sqrt{3}$
    $= 4\sqrt{2} - 12\sqrt{3} + 14\sqrt{2} - 5\sqrt{3}$
    $= 18\sqrt{2} - 17\sqrt{3}$

29. $\sqrt[3]{128} + \sqrt[3]{54} = \sqrt[3]{64\cdot2} + \sqrt[3]{27\cdot2}$
    $= 4\sqrt[3]{2} + 3\sqrt[3]{2}$
    $= 7\sqrt[3]{2}$

31. $4\sqrt[3]{135x^5} - 6x\sqrt[3]{320x^2}$
    $= 4\sqrt[3]{27x^3\cdot5x^2} - 6x\sqrt[3]{64\cdot5x^2}$
    $= 4\cdot3x\sqrt[3]{5x^2} - 6x\cdot4\sqrt[3]{5x^2}$
    $= 12x\sqrt[3]{5x^2} - 24x\sqrt[3]{5x^2}$
    $= -12x\sqrt[3]{5x^2}$

33. $-4\sqrt[4]{32x^9} + 2x\sqrt[4]{162x^5}$
    $= -4\sqrt[4]{16x^8\cdot2x} + 2x\sqrt[4]{81x^4\cdot2x}$
    $= -4\cdot2x^2\sqrt[4]{2x} + 2x\cdot3x\sqrt[4]{2x}$
    $= -8x^2\sqrt[4]{2x} + 6x^2\sqrt[4]{2x}$
    $= -2x^2\sqrt[4]{2x}$

35. $\sqrt{2}(3+\sqrt{2}) = \sqrt{2}\cdot3 + \sqrt{2}\cdot\sqrt{2}$
    $= 3\sqrt{2} + \sqrt{4}$
    $= 3\sqrt{2} + 2$

37. $\sqrt{3}(\sqrt{3}-\sqrt{15}) = \sqrt{3}\cdot\sqrt{3} - \sqrt{3}\cdot\sqrt{15}$
    $= \sqrt{9} - \sqrt{45}$
    $= 3 - \sqrt{9\cdot5}$
    $= 3 - 3\sqrt{5}$

39. $\sqrt{5}(\sqrt{6}+2\sqrt{10}) = \sqrt{5}\cdot\sqrt{6} + \sqrt{5}\cdot2\sqrt{10}$
    $= \sqrt{30} + 2\sqrt{50}$
    $= \sqrt{30} + 2\sqrt{25\cdot2}$
    $= \sqrt{30} + 2\cdot5\sqrt{2}$
    $= \sqrt{30} + 10\sqrt{2}$

41. $4\sqrt{3x}\left(2\sqrt{3x}-4\sqrt{6x}\right)$

$= 4\sqrt{3x}\cdot 2\sqrt{3x}-4\sqrt{3x}\cdot 4\sqrt{6x}$

$= 8\sqrt{9x^2}-16\sqrt{18x^2}$

$= 8\cdot 3x-16\sqrt{9x^2\cdot 2}$

$= 24x-16\cdot 3x\sqrt{2}$

$= 24x-48x\sqrt{2}$

43. $\left(3+\sqrt{5}\right)\left(4-\sqrt{2}\right)$

$= 3\cdot 4-3\cdot\sqrt{2}+\sqrt{5}\cdot 4-\sqrt{5}\cdot\sqrt{2}$

$= 12-3\sqrt{2}+4\sqrt{5}-\sqrt{10}$

45. $\left(3+\sqrt{x}\right)\left(2+\sqrt{x}\right)$

$= 3\cdot 2+3\cdot\sqrt{x}+\sqrt{x}\cdot 2+\sqrt{x}\cdot\sqrt{x}$

$= 6+3\sqrt{x}+2\sqrt{x}+\sqrt{x^2}$

$= 6+5\sqrt{x}+x$

47. $\left(2+3\sqrt{3}\right)\left(3+5\sqrt{2}\right)$

$= 2\cdot 3+2\cdot 5\sqrt{2}+3\sqrt{3}\cdot 3+3\sqrt{3}\cdot 5\sqrt{2}$

$= 6+10\sqrt{2}+9\sqrt{3}+15\sqrt{6}$

49. $\left(\sqrt{3}+\sqrt{5}\right)\left(\sqrt{5}+\sqrt{7}\right)$

$= \sqrt{3}\cdot\sqrt{5}+\sqrt{3}\cdot\sqrt{7}+\sqrt{5}\cdot\sqrt{5}+\sqrt{5}\cdot\sqrt{7}$

$= \sqrt{15}+\sqrt{21}+\sqrt{25}+\sqrt{35}$

$= \sqrt{15}+\sqrt{21}+5+\sqrt{35}$

51. $\left(\sqrt{x}+3\sqrt{y}\right)\left(\sqrt{x}-2\sqrt{y}\right)$

$= \sqrt{x}\cdot\sqrt{x}-\sqrt{x}\cdot 2\sqrt{y}+3\sqrt{y}\cdot\sqrt{x}-3\sqrt{y}\cdot 2\sqrt{y}$

$= \sqrt{x^2}-2\sqrt{xy}+3\sqrt{xy}-6\sqrt{y^2}$

$= x+\sqrt{xy}-6y$

53. $\left(4\sqrt{2}+2\sqrt{5}\right)\left(3\sqrt{7}-3\sqrt{3}\right)$

$= 4\sqrt{2}\cdot 3\sqrt{7}-4\sqrt{2}\cdot 3\sqrt{3}+2\sqrt{5}\cdot 3\sqrt{7}-2\sqrt{5}\cdot 3\sqrt{3}$

$= 12\sqrt{14}-12\sqrt{6}+6\sqrt{35}-6\sqrt{15}$

55. $\left(2\sqrt{a}+3\sqrt{b}\right)\left(4\sqrt{a}-\sqrt{b}\right)$

$= 2\sqrt{a}\cdot 4\sqrt{a}-2\sqrt{a}\cdot\sqrt{b}+3\sqrt{b}\cdot 4\sqrt{a}-3\sqrt{b}\cdot\sqrt{b}$

$= 8\sqrt{a^2}-2\sqrt{ab}+12\sqrt{ab}-3\sqrt{b^2}$

$= 8a+10\sqrt{ab}-3b$

57. $\left(\sqrt[3]{4}+5\right)\left(\sqrt[3]{4}-8\right)$

$= \sqrt[3]{4}\cdot\sqrt[3]{4}-\sqrt[3]{4}\cdot 8+5\sqrt[3]{4}-5\cdot 8$

$= \sqrt[3]{16}-8\sqrt[3]{4}+5\sqrt[3]{4}-40$

$= \sqrt[3]{8\cdot 2}-3\sqrt[3]{4}-40$

$= 2\sqrt[3]{2}-3\sqrt[3]{4}-40$

59. $\left(\sqrt[3]{9}+\sqrt[3]{4}\right)\left(\sqrt[3]{3}-\sqrt[3]{2}\right)$

$= \sqrt[3]{9}\cdot\sqrt[3]{3}-\sqrt[3]{9}\cdot\sqrt[3]{2}+\sqrt[3]{4}\cdot\sqrt[3]{3}-\sqrt[3]{4}\cdot\sqrt[3]{2}$

$= \sqrt[3]{27}-\sqrt[3]{18}+\sqrt[3]{12}-\sqrt[3]{8}$

$= 3-\sqrt[3]{18}+\sqrt[3]{12}-2$

$= 1-\sqrt[3]{18}+\sqrt[3]{12}$

61. $\left(\sqrt[3]{x}+2\right)\left(\sqrt[3]{x^2}-2\sqrt[3]{x}+4\right)$

$= \sqrt[3]{x}\cdot\sqrt[3]{x^2}-\sqrt[3]{x}\cdot 2\sqrt[3]{x}+\sqrt[3]{x}\cdot 4+2\sqrt[3]{x^2}$

$\quad -2\cdot 2\sqrt[3]{x}+2\cdot 4$

$= \sqrt[3]{x^3}-2\sqrt[3]{x^2}+4\sqrt[3]{x}+2\sqrt[3]{x^2}$

$\quad -4\sqrt[3]{x}+8$

$= x-2\sqrt[3]{x^2}+4\sqrt[3]{x}+2\sqrt[3]{x^2}-4\sqrt[3]{x}+8$

$= x+8$

63. $\left(4+\sqrt{6}\right)^2 = 4^2+2\cdot 4\sqrt{6}+\left(\sqrt{6}\right)^2$

$= 16+8\sqrt{6}+6$

$= 22+8\sqrt{6}$

65. $\left(4-\sqrt{2}\right)^2 = 4^2-2\cdot 4\sqrt{2}+\left(\sqrt{2}\right)^2$

$= 16-8\sqrt{2}+2$

$= 18-8\sqrt{2}$

67. $\left(2+2\sqrt{3}\right)^2 = 2^2+2\cdot 4\sqrt{3}+\left(2\sqrt{3}\right)^2$

$= 4+8\sqrt{3}+4\cdot 3$

$= 4+8\sqrt{3}+12$

$= 16+8\sqrt{3}$

69. $\left(2\sqrt{3}+3\sqrt{2}\right)^2 = \left(2\sqrt{3}\right)^2+2\cdot 6\sqrt{6}+\left(3\sqrt{2}\right)^2$

$= 4\cdot 3+12\sqrt{6}+9\cdot 2$

$= 12+12\sqrt{6}+18$

$= 30+12\sqrt{6}$

71. $\left(4+\sqrt{3}\right)\left(4-\sqrt{3}\right) = 4^2-\left(\sqrt{3}\right)^2 = 16-3 = 13$

73. $\left(\sqrt{2}+4\right)\left(\sqrt{2}-4\right)=\left(\sqrt{2}\right)^2-4^2=2-16=-14$

75. $\left(6+\sqrt{x}\right)\left(6-\sqrt{x}\right)=6^2-\left(\sqrt{x}\right)^2=36-x$

77. $\left(\sqrt{3}+\sqrt{2}\right)\left(\sqrt{3}-\sqrt{2}\right)=\left(\sqrt{3}\right)^2-\left(\sqrt{2}\right)^2=3-2=1$

79. $\left(\sqrt{x}+\sqrt{y}\right)\left(\sqrt{x}-\sqrt{y}\right)=\left(\sqrt{x}\right)^2-\left(\sqrt{y}\right)^2=x-y$

81. $\left(4+2\sqrt{3}\right)\left(4-2\sqrt{3}\right)=4^2-\left(2\sqrt{3}\right)^2$
$$=16-4\cdot3$$
$$=16-12$$
$$=4$$

83. $\left(3\sqrt{7}+\sqrt{13}\right)\left(3\sqrt{7}-\sqrt{13}\right)=\left(3\sqrt{7}\right)^2-\left(\sqrt{13}\right)^2$
$$=9\cdot7-13$$
$$=63-13$$
$$=50$$

85. $\sqrt{3}\cdot\sqrt{15}+\sqrt{8}\cdot\sqrt{10}=\sqrt{3\cdot15}+\sqrt{8\cdot10}$
$$=\sqrt{45}+\sqrt{80}$$
$$=\sqrt{9\cdot5}+\sqrt{16\cdot5}$$
$$=3\sqrt{5}+4\sqrt{5}$$
$$=7\sqrt{5}$$

87. $3\sqrt{3}\cdot\sqrt{18}-4\sqrt{18}\cdot\sqrt{12}=3\sqrt{3\cdot18}-4\sqrt{18\cdot12}$
$$=3\sqrt{54}-4\sqrt{216}$$
$$=3\sqrt{9\cdot6}-4\sqrt{36\cdot6}$$
$$=3\cdot3\sqrt{6}-4\cdot6\sqrt{6}$$
$$=9\sqrt{6}-24\sqrt{6}$$
$$=-15\sqrt{6}$$

89. $\dfrac{\sqrt{54}}{\sqrt{3}}+\sqrt{72}=\sqrt{\dfrac{54}{3}}+\sqrt{72}$
$$=\sqrt{18}+\sqrt{36\cdot2}$$
$$=\sqrt{9\cdot2}+6\sqrt{2}$$
$$=3\sqrt{2}+6\sqrt{2}$$
$$=9\sqrt{2}$$

91. $\dfrac{\sqrt{540}}{\sqrt{3}}-4\sqrt{125}=\sqrt{\dfrac{540}{3}}-4\sqrt{125}$
$$=\sqrt{180}-4\sqrt{125}$$
$$=\sqrt{36\cdot5}-4\sqrt{25\cdot5}$$
$$=6\sqrt{5}-4\cdot5\sqrt{5}$$
$$=6\sqrt{5}-20\sqrt{5}$$
$$=-14\sqrt{5}$$

93. $5\sqrt{3}+4\sqrt{12}+4\sqrt{27}$
$$=5\sqrt{3}+4\sqrt{4\cdot3}+4\sqrt{9\cdot3}$$
$$=5\sqrt{3}+4\cdot2\sqrt{3}+4\cdot3\sqrt{3}$$
$$=5\sqrt{3}+8\sqrt{3}+12\sqrt{3}$$
$$=25\sqrt{3}$$

95. a) $13+10+\sqrt{5}+9+\sqrt{5}+10=\left(42+2\sqrt{5}\right)$ ft.

b) 46.5 ft.   c) $46.5\left(\$1.89\right)=\$87.89$

Review Exercises

1. $2x+5$

2. $\left(4x+3\right)\left(4x-3\right)=\left(4x\right)^2-3^2=16x^2-9$

3. $\sqrt{2}$ because $\sqrt{8}\cdot\sqrt{2}=\sqrt{8\cdot2}=\sqrt{16}$

4. $\sqrt[3]{4}$ because $\sqrt[3]{2}\cdot\sqrt[3]{4}=\sqrt[3]{2\cdot4}=\sqrt[3]{8}$

5. $5y-2x=10$
$$5y=2x+10$$
$$\dfrac{5y}{5}=\dfrac{2x+10}{5}$$
$$y=\dfrac{2}{5}x+2$$

slope is $\dfrac{2}{5}$; y-intercept is $(0,2)$

6. $y=-\dfrac{1}{3}x+4$

Exercise Set 8.5

1. $\dfrac{1}{\sqrt{3}} = \dfrac{1}{\sqrt{3}} \cdot \dfrac{\sqrt{3}}{\sqrt{3}} = \dfrac{\sqrt{3}}{\sqrt{9}} = \dfrac{\sqrt{3}}{3}$

3. $\dfrac{3}{\sqrt{8}} = \dfrac{3}{\sqrt{8}} \cdot \dfrac{\sqrt{2}}{\sqrt{2}} = \dfrac{3\sqrt{2}}{\sqrt{16}} = \dfrac{3\sqrt{2}}{4}$

5. $\sqrt{\dfrac{36}{7}} = \dfrac{\sqrt{36}}{\sqrt{7}} = \dfrac{6}{\sqrt{7}} \cdot \dfrac{\sqrt{7}}{\sqrt{7}} = \dfrac{6\sqrt{7}}{\sqrt{49}} = \dfrac{6\sqrt{7}}{7}$

7. $\sqrt{\dfrac{5}{12}} = \dfrac{\sqrt{5}}{\sqrt{12}} = \dfrac{\sqrt{5}}{\sqrt{12}} \cdot \dfrac{\sqrt{3}}{\sqrt{3}} = \dfrac{\sqrt{15}}{\sqrt{36}} = \dfrac{\sqrt{15}}{6}$

9. $\dfrac{\sqrt{7x^2}}{\sqrt{50}} = \dfrac{\sqrt{x^2 \cdot 7}}{\sqrt{25 \cdot 2}} = \dfrac{x\sqrt{7}}{5\sqrt{2}} \cdot \dfrac{\sqrt{2}}{\sqrt{2}}$

$= \dfrac{x\sqrt{14}}{5\sqrt{4}} = \dfrac{x\sqrt{14}}{5 \cdot 2} = \dfrac{x\sqrt{14}}{10}$

11. $\dfrac{\sqrt{8}}{\sqrt{56}} = \sqrt{\dfrac{8}{56}} = \sqrt{\dfrac{1}{7}} = \dfrac{\sqrt{1}}{\sqrt{7}} = \dfrac{1}{\sqrt{7}} \cdot \dfrac{\sqrt{7}}{\sqrt{7}}$

$= \dfrac{\sqrt{7}}{\sqrt{49}} = \dfrac{\sqrt{7}}{7}$

13. $\dfrac{5}{\sqrt{3a}} = \dfrac{5}{\sqrt{3a}} \cdot \dfrac{\sqrt{3a}}{\sqrt{3a}} = \dfrac{5\sqrt{3a}}{\sqrt{9a^2}} = \dfrac{5\sqrt{3a}}{3a}$

15. $\sqrt{\dfrac{3m}{11n}} = \dfrac{\sqrt{3m}}{\sqrt{11n}} = \dfrac{\sqrt{3m}}{\sqrt{11n}} \cdot \dfrac{\sqrt{11n}}{\sqrt{11n}} = \dfrac{\sqrt{33mn}}{\sqrt{121n^2}} = \dfrac{\sqrt{33mn}}{11n}$

17. $\dfrac{10}{\sqrt{5x}} = \dfrac{10}{\sqrt{5x}} \cdot \dfrac{\sqrt{5x}}{\sqrt{5x}} = \dfrac{10\sqrt{5x}}{\sqrt{25x^2}} = \dfrac{10\sqrt{5x}}{5x} = \dfrac{2\sqrt{5x}}{x}$

19. $\dfrac{\sqrt{6x}}{\sqrt{32x}} = \sqrt{\dfrac{6x}{32x}} = \sqrt{\dfrac{3}{16}} = \dfrac{\sqrt{3}}{\sqrt{16}} = \dfrac{\sqrt{3}}{4}$

21. $\dfrac{3}{\sqrt{x^3}} = \dfrac{3}{\sqrt{x^2 \cdot x}}$

$= \dfrac{3}{x\sqrt{x}} \cdot \dfrac{\sqrt{x}}{\sqrt{x}}$

$= \dfrac{3\sqrt{x}}{x\sqrt{x^2}}$

$= \dfrac{3\sqrt{x}}{x \cdot x}$

$= \dfrac{3\sqrt{x}}{x^2}$

23. $\dfrac{8x^2}{\sqrt{2x}} = \dfrac{8x^2}{\sqrt{2x}} \cdot \dfrac{\sqrt{2x}}{\sqrt{2x}}$

$= \dfrac{8x^2\sqrt{2x}}{\sqrt{4x^2}}$

$= \dfrac{8x^2\sqrt{2x}}{2x}$

$= 4x\sqrt{2x}$

25. Mistake: The product of $\sqrt{2}$ and 2 is not 2.

Correct: $\dfrac{\sqrt{3}}{\sqrt{2}} \cdot \dfrac{\sqrt{2}}{\sqrt{2}} = \dfrac{\sqrt{6}}{\sqrt{4}} = \dfrac{\sqrt{6}}{2}$

27. $\dfrac{5}{\sqrt[3]{3}} = \dfrac{5}{\sqrt[3]{3}} \cdot \dfrac{\sqrt[3]{9}}{\sqrt[3]{9}} = \dfrac{5\sqrt[3]{9}}{\sqrt[3]{27}} = \dfrac{5\sqrt[3]{9}}{3}$

29. $\sqrt[3]{\dfrac{5}{2}} = \dfrac{\sqrt[3]{5}}{\sqrt[3]{2}} \cdot \dfrac{\sqrt[3]{4}}{\sqrt[3]{4}} = \dfrac{\sqrt[3]{20}}{\sqrt[3]{8}} = \dfrac{\sqrt[3]{20}}{2}$

31. $\dfrac{6}{\sqrt[3]{4}} = \dfrac{6}{\sqrt[3]{4}} \cdot \dfrac{\sqrt[3]{2}}{\sqrt[3]{2}} = \dfrac{6\sqrt[3]{2}}{\sqrt[3]{8}} = \dfrac{6\sqrt[3]{2}}{2} = 3\sqrt[3]{2}$

33. $\dfrac{m}{\sqrt[3]{n}} = \dfrac{m}{\sqrt[3]{n}} \cdot \dfrac{\sqrt[3]{n^2}}{\sqrt[3]{n^2}} = \dfrac{m\sqrt[3]{n^2}}{\sqrt[3]{n^3}} = \dfrac{m\sqrt[3]{n^2}}{n}$

35. $\sqrt[3]{\dfrac{a}{b^2}} = \dfrac{\sqrt[3]{a}}{\sqrt[3]{b^2}} \cdot \dfrac{\sqrt[3]{b}}{\sqrt[3]{b}} = \dfrac{\sqrt[3]{ab}}{\sqrt[3]{b^3}} = \dfrac{\sqrt[3]{ab}}{b}$

37. $\dfrac{4}{\sqrt[3]{2x}} = \dfrac{4}{\sqrt[3]{2x}} \cdot \dfrac{\sqrt[3]{4x^2}}{\sqrt[3]{4x^2}}$

$= \dfrac{4\sqrt[3]{4x^2}}{\sqrt[3]{8x^3}}$

$= \dfrac{4\sqrt[3]{4x^2}}{2x}$

$= \dfrac{2\sqrt[3]{4x^2}}{x}$

39. $\sqrt[3]{\dfrac{6}{25a^2}} = \dfrac{\sqrt[3]{6}}{\sqrt[3]{25a^2}} \cdot \dfrac{\sqrt[3]{5a}}{\sqrt[3]{5a}} = \dfrac{\sqrt[3]{30a}}{\sqrt[3]{125a^3}} = \dfrac{\sqrt[3]{30a}}{5a}$

41. $\dfrac{5}{\sqrt[4]{4}} = \dfrac{5}{\sqrt[4]{4}} \cdot \dfrac{\sqrt[4]{4}}{\sqrt[4]{4}} = \dfrac{5\sqrt[4]{4}}{\sqrt[4]{16}} = \dfrac{5\sqrt[4]{4}}{2} = \dfrac{5\sqrt[4]{2^2}}{2} = \dfrac{5\sqrt{2}}{2}$

43. $\sqrt[4]{\dfrac{3}{x^2}} = \dfrac{\sqrt[4]{3}}{\sqrt[4]{x^2}} \cdot \dfrac{\sqrt[4]{x^2}}{\sqrt[4]{x^2}} = \dfrac{\sqrt[4]{3x^2}}{\sqrt[4]{x^4}} = \dfrac{\sqrt[4]{3x^2}}{x}$

45. $\dfrac{9}{\sqrt[4]{3x^3}} = \dfrac{9}{\sqrt[4]{3x^3}} \cdot \dfrac{\sqrt[4]{27x}}{\sqrt[4]{27x}}$

$= \dfrac{9\sqrt[4]{27x}}{\sqrt[4]{81x^4}}$

$= \dfrac{9\sqrt[4]{27x}}{3x}$

$= \dfrac{3\sqrt[4]{27x}}{x}$

47. $\dfrac{3}{\sqrt{2}+1} \cdot \dfrac{\sqrt{2}-1}{\sqrt{2}-1} = \dfrac{3\left(\sqrt{2}-1\right)}{\left(\sqrt{2}\right)^2 - 1^2}$

$= \dfrac{3\sqrt{2}-3}{2-1}$

$= \dfrac{3\sqrt{2}-3}{1}$

$= 3\sqrt{2}-3$

49. $\dfrac{4}{2-\sqrt{3}} \cdot \dfrac{2+\sqrt{3}}{2+\sqrt{3}} = \dfrac{4\left(2+\sqrt{3}\right)}{2^2 - \left(\sqrt{3}\right)^2}$

$= \dfrac{8+4\sqrt{3}}{4-3}$

$= \dfrac{8+4\sqrt{3}}{1}$

$= 8+4\sqrt{3}$

51. $\dfrac{5}{\sqrt{2}+\sqrt{3}} \cdot \dfrac{\sqrt{2}-\sqrt{3}}{\sqrt{2}-\sqrt{3}} = \dfrac{5\left(\sqrt{2}-\sqrt{3}\right)}{\left(\sqrt{2}\right)^2 - \left(\sqrt{3}\right)^2}$

$= \dfrac{5\sqrt{2}-5\sqrt{3}}{2-3}$

$= \dfrac{5\sqrt{2}-5\sqrt{3}}{-1}$

$= 5\sqrt{3}-5\sqrt{2}$

53. $\dfrac{4}{1-\sqrt{5}} \cdot \dfrac{1+\sqrt{5}}{1+\sqrt{5}} = \dfrac{4\left(1+\sqrt{5}\right)}{1^2 - \left(\sqrt{5}\right)^2}$

$= \dfrac{4\left(1+\sqrt{5}\right)}{1-5}$

$= \dfrac{4\left(1+\sqrt{5}\right)}{-4}$

$= -1\left(1+\sqrt{5}\right)$

$= -1-\sqrt{5}$

55. $\dfrac{\sqrt{3}}{\sqrt{3}-1} \cdot \dfrac{\sqrt{3}+1}{\sqrt{3}+1} = \dfrac{\sqrt{3}\left(\sqrt{3}+1\right)}{\left(\sqrt{3}\right)^2 - 1^2}$

$= \dfrac{3+\sqrt{3}}{3-1}$

$= \dfrac{3+\sqrt{3}}{2}$

57. $\dfrac{2\sqrt{3}}{\sqrt{3}-4} \cdot \dfrac{\sqrt{3}+4}{\sqrt{3}+4} = \dfrac{2\sqrt{3}\left(\sqrt{3}+4\right)}{\left(\sqrt{3}\right)^2 - 4^2}$

$= \dfrac{2\cdot 3 + 8\sqrt{3}}{3-16}$

$= \dfrac{6+8\sqrt{3}}{-13}$

$= \dfrac{-6-8\sqrt{3}}{13}$

59. $\dfrac{4\sqrt{3}}{\sqrt{7}+\sqrt{2}} \cdot \dfrac{\sqrt{7}-\sqrt{2}}{\sqrt{7}-\sqrt{2}} = \dfrac{4\sqrt{3}\left(\sqrt{7}-\sqrt{2}\right)}{\left(\sqrt{7}\right)^2 - \left(\sqrt{2}\right)^2}$

$= \dfrac{4\sqrt{21}-4\sqrt{6}}{7-2}$

$= \dfrac{4\sqrt{21}-4\sqrt{6}}{5}$

61. $\dfrac{8\sqrt{2}}{4\sqrt{2}-\sqrt{6}} \cdot \dfrac{4\sqrt{2}+\sqrt{6}}{4\sqrt{2}+\sqrt{6}} = \dfrac{32\sqrt{4}+8\sqrt{12}}{16\cdot 2 - 6}$

$= \dfrac{32\cdot 2 + 8\sqrt{4\cdot 3}}{32-6}$

$= \dfrac{64+8\cdot 2\sqrt{3}}{26}$

$= \dfrac{64+16\sqrt{3}}{26}$

$= \dfrac{\cancel{2}\left(32+8\sqrt{3}\right)}{\cancel{2}\cdot 13}$

$= \dfrac{32+8\sqrt{3}}{13}$

63. $\dfrac{6\sqrt{y}}{\sqrt{y}+1} \cdot \dfrac{\sqrt{y}-1}{\sqrt{y}-1} = \dfrac{6\sqrt{y}\left(\sqrt{y}-1\right)}{\left(\sqrt{y}\right)^2 - 1^2} = \dfrac{6y-6\sqrt{y}}{y-1}$

65. $\dfrac{3\sqrt{t}}{\sqrt{t}+2\sqrt{u}}\cdot\dfrac{\sqrt{t}-2\sqrt{u}}{\sqrt{t}-2\sqrt{u}}=\dfrac{3\sqrt{t}\left(\sqrt{t}-2\sqrt{u}\right)}{\left(\sqrt{t}\right)^2-\left(2\sqrt{u}\right)^2}$

$\qquad\qquad\qquad\qquad=\dfrac{3t-6\sqrt{tu}}{t-4u}$

67. $\dfrac{\sqrt{2y}}{\sqrt{x}-\sqrt{6y}}\cdot\dfrac{\sqrt{x}+\sqrt{6y}}{\sqrt{x}+\sqrt{6y}}=\dfrac{\sqrt{2y}\left(\sqrt{x}+\sqrt{6y}\right)}{\left(\sqrt{x}\right)^2-\left(\sqrt{6y}\right)^2}$

$\qquad\qquad\qquad\qquad=\dfrac{\sqrt{2xy}+\sqrt{12y^2}}{x-6y}$

$\qquad\qquad\qquad\qquad=\dfrac{\sqrt{2xy}+\sqrt{4y^2\cdot 3}}{x-6y}$

$\qquad\qquad\qquad\qquad=\dfrac{\sqrt{2xy}+2y\sqrt{3}}{x-6y}$

69. $\dfrac{\sqrt{3}}{2}\cdot\dfrac{\sqrt{3}}{\sqrt{3}}=\dfrac{\sqrt{9}}{2\sqrt{3}}=\dfrac{3}{2\sqrt{3}}$

71. $\dfrac{\sqrt{2x}}{5}\cdot\dfrac{\sqrt{2x}}{\sqrt{2x}}=\dfrac{\sqrt{4x^2}}{5\sqrt{2x}}=\dfrac{2x}{5\sqrt{2x}}$

73. $\dfrac{\sqrt{8n}}{6}=\dfrac{\sqrt{4\cdot 2n}}{6}=\dfrac{2\sqrt{2n}}{6}$

$\qquad\quad=\dfrac{\sqrt{2n}}{3}\cdot\dfrac{\sqrt{2n}}{\sqrt{2n}}$

$\qquad\quad=\dfrac{\sqrt{4n^2}}{3\sqrt{2n}}$

$\qquad\quad=\dfrac{2n}{3\sqrt{2n}}$

75. $\dfrac{2+\sqrt{3}}{5}\cdot\dfrac{2-\sqrt{3}}{2-\sqrt{3}}=\dfrac{2^2-\left(\sqrt{3}\right)^2}{5\left(2-\sqrt{3}\right)}$

$\qquad\qquad\qquad\qquad=\dfrac{4-3}{10-5\sqrt{3}}$

$\qquad\qquad\qquad\qquad=\dfrac{1}{10-5\sqrt{3}}$

77. $\dfrac{\sqrt{5x}-6}{9}\cdot\dfrac{\sqrt{5x}+6}{\sqrt{5x}+6}=\dfrac{\left(\sqrt{5x}\right)^2-6^2}{9\left(\sqrt{5x}+6\right)}$

$\qquad\qquad\qquad\qquad=\dfrac{5x-36}{9\sqrt{5x}+54}$

79. $\dfrac{5\sqrt{n}+\sqrt{6n}}{2n}\cdot\dfrac{5\sqrt{n}-\sqrt{6n}}{5\sqrt{n}-\sqrt{6n}}=\dfrac{\left(5\sqrt{n}\right)^2-\left(\sqrt{6n}\right)^2}{2n\left(5\sqrt{n}-\sqrt{6n}\right)}$

$\qquad\qquad\qquad\qquad=\dfrac{25n-6n}{10n\sqrt{n}-2n\sqrt{6n}}$

$\qquad\qquad\qquad\qquad=\dfrac{19n}{10n\sqrt{n}-2n\sqrt{6n}}$

$\qquad\qquad\qquad\qquad=\dfrac{19\not{n}}{\not{n}\left(10\sqrt{n}-2\sqrt{6n}\right)}$

$\qquad\qquad\qquad\qquad=\dfrac{19}{10\sqrt{n}-2\sqrt{6n}}$

81. $f(x)=\dfrac{5\sqrt{2}}{x}$

a) $f\left(\sqrt{6}\right)=\dfrac{5\sqrt{2}}{\sqrt{6}}=5\sqrt{\dfrac{2}{6}}=5\sqrt{\dfrac{1}{3}}$

$\qquad\qquad=\dfrac{5}{\sqrt{3}}\cdot\dfrac{\sqrt{3}}{\sqrt{3}}=\dfrac{5\sqrt{3}}{3}$

b) $f\left(\sqrt{10}\right)=\dfrac{5\sqrt{2}}{\sqrt{10}}=5\sqrt{\dfrac{2}{10}}=5\sqrt{\dfrac{1}{5}}$

$\qquad\qquad=\dfrac{5}{\sqrt{5}}\cdot\dfrac{\sqrt{5}}{\sqrt{5}}=\dfrac{5\sqrt{5}}{\sqrt{25}}=\dfrac{5\sqrt{5}}{5}$

$\qquad\qquad=\sqrt{5}$

c) $f\left(\sqrt{22}\right)=\dfrac{5\sqrt{2}}{\sqrt{22}}=5\sqrt{\dfrac{2}{22}}=5\sqrt{\dfrac{1}{11}}$

$\qquad\qquad=\dfrac{5}{\sqrt{11}}\cdot\dfrac{\sqrt{11}}{\sqrt{11}}=\dfrac{5\sqrt{11}}{\sqrt{121}}$

$\qquad\qquad=\dfrac{5\sqrt{11}}{11}$

83. a) The graphs are identical. The functions are identical.

b) $f(x)=g(x)$

85. a) $T=\dfrac{2\pi\sqrt{L}}{\sqrt{9.8}}\cdot\dfrac{\sqrt{9.8}}{\sqrt{9.8}}$

$\qquad\quad=\dfrac{2\pi\sqrt{9.8L}}{9.8}$

$\qquad\quad=\dfrac{\pi\sqrt{9.8L}}{4.9}$

b) $T=\dfrac{2\pi\sqrt{L}}{\sqrt{9.8}}\cdot\dfrac{\sqrt{L}}{\sqrt{L}}$

$\qquad\quad=\dfrac{2\pi L}{\sqrt{9.8L}}$

87. a) $s = \dfrac{\sqrt{3V}}{\sqrt{h}} \cdot \dfrac{\sqrt{h}}{\sqrt{h}} = \dfrac{\sqrt{3Vh}}{\sqrt{h^2}} = \dfrac{\sqrt{3Vh}}{h}$

    b) $s = \dfrac{\sqrt{3(83,068,742)449}}{449} \approx 745$ ft.

89. a) $V_{rms} = \dfrac{V_m}{\sqrt{2}} \cdot \dfrac{\sqrt{2}}{\sqrt{2}} = \dfrac{\sqrt{2}\,V_m}{\sqrt{4}} = \dfrac{\sqrt{2}\,V_m}{2}$

    b) $V_{rms} = \dfrac{163\sqrt{2}}{2}$ c) $V_{rms} \approx 115.3$

91. $\dfrac{5\sqrt{2}}{3+\sqrt{6}} \cdot \dfrac{3-\sqrt{6}}{3-\sqrt{6}} = \dfrac{5\sqrt{2}\left(3-\sqrt{6}\right)}{3^2 - \left(\sqrt{6}\right)^2}$

$$= \dfrac{15\sqrt{2} - 5\sqrt{12}}{9-6}$$

$$= \dfrac{15\sqrt{2} - 5 \cdot 2\sqrt{3}}{3}$$

$$= \dfrac{15\sqrt{2} - 10\sqrt{3}}{3}\,\Omega$$

Review Exercises

1. $\pm\sqrt{28} = \pm\sqrt{4 \cdot 7} = \pm 2\sqrt{7}$

2. $x^2 - 6x + 9 = (x-3)(x-3) = (x-3)^2$

3. $2x - 3 = 5$ 4. $2x - 3 = -5$
      $2x = 8$ $2x = -2$
        $x = 4$ $x = -1$

5. $x^2 - 36 = 0$
     $(x+6)(x-6) = 0$
    $x + 6 = 0$ or $x - 6 = 0$
        $x = -6$ $x = 6$

6. $x^2 - 5x + 6 = 0$
    $(x-2)(x-3) = 0$
       $x - 2 = 0$ or $x - 3 = 0$
           $x = 2$ $x = 3$

Exercise Set 8.6

1. $\sqrt{x} = 2$
    $\left(\sqrt{x}\right)^2 = 2^2$
        $x = 4$

3. $\sqrt{k} = -4$ has no real-number solution

5. $\sqrt[3]{y} = 3$ 7. $\sqrt[3]{z} = -2$
    $\left(\sqrt[3]{y}\right)^3 = 3^3$ $\left(\sqrt[3]{z}\right)^3 = (-2)^3$
         $y = 27$ $z = -8$

9. $\sqrt{n-1} = 4$ 11. $\sqrt{t+5} = 4$
   $\left(\sqrt{n-1}\right)^2 = 4^2$ $\left(\sqrt{t+5}\right)^2 = 4^2$
      $n - 1 = 16$ $t + 5 = 16$
         $n = 17$ $t = 11$

13. $\sqrt{3x-2} = 4$ 15. $\sqrt{2x+24} = 4$
   $\left(\sqrt{3x-2}\right)^2 = 4^2$ $\left(\sqrt{2x+24}\right)^2 = 4^2$
      $3x - 2 = 16$ $2x + 24 = 16$
         $3x = 18$ $2x = -8$
          $x = 6$ $x = -4$

17. $\sqrt{2n-8} = -3$ has no real-number solution.

19. $\sqrt[3]{x-3} = 2$ 21. $\sqrt[3]{3y-2} = -2$
   $\left(\sqrt[3]{x-3}\right)^3 = 2^3$ $\left(\sqrt[3]{3y-2}\right)^3 = (-2)^3$
      $x - 3 = 8$ $3y - 2 = -8$
         $x = 11$ $3y = -6$
                           $y = -2$

23. $\sqrt{u-3} - 10 = 1$ 25. $\sqrt{y-6} + 2 = 9$
      $\sqrt{u-3} = 11$ $\sqrt{y-6} = 7$
   $\left(\sqrt{u-3}\right)^2 = 11^2$ $\left(\sqrt{y-6}\right)^2 = 7^2$
        $u - 3 = 121$ $y - 6 = 49$
          $u = 124$ $y = 55$

27. $\sqrt{6x-5} - 2 = 3$ 29. $\sqrt[3]{n+3} - 2 = -4$
      $\sqrt{6x-5} = 5$ $\sqrt[3]{n+3} = -2$
   $\left(\sqrt{6x-5}\right)^2 = 5^2$ $\left(\sqrt[3]{n+3}\right)^3 = (-2)^3$
        $6x - 5 = 25$ $n + 3 = -8$
          $6x = 30$ $n = -11$
           $x = 5$

31. $\sqrt[4]{x-2} - 2 = -4$
       $\sqrt[4]{x-2} = -2$

This equation has no real-number solution.

33. $\sqrt{3x-2} = \sqrt{8-2x}$

$\left(\sqrt{3x-2}\right)^2 = \left(\sqrt{8-2x}\right)^2$

$3x-2 = 8-2x$

$5x = 10$

$x = 2$

Check:

$\sqrt{3(2)-2} \overset{?}{=} \sqrt{8-2(2)}$

$\sqrt{6-2} \overset{?}{=} \sqrt{8-4}$

$\sqrt{4} \overset{?}{=} \sqrt{4}$

$2 = 2$

35. $\sqrt{4x-5} = \sqrt{6x+5}$

$\left(\sqrt{4x-5}\right)^2 = \left(\sqrt{6x+5}\right)^2$

$4x-5 = 6x+5$

$-10 = 2x$

$-5 = x$

Check:

$\sqrt{4(-5)-5} \overset{?}{=} \sqrt{6(-5)+5}$

$\sqrt{-20-5} \overset{?}{=} \sqrt{-30+5}$

$\sqrt{-25} \overset{?}{=} \sqrt{-25}$

$\sqrt{-25}$ is not a real number.

No real-number solution. (-5 is an extraneous solution.)

37. $\sqrt[3]{2r+2} = \sqrt[3]{3r-1}$

$\left(\sqrt[3]{2r+2}\right)^3 = \left(\sqrt[3]{3r-1}\right)^3$

$2r+2 = 3r-1$

$3 = r$

Check:

$\sqrt[3]{2(3)+2} \overset{?}{=} \sqrt[3]{3(3)-1}$

$\sqrt[3]{6+2} \overset{?}{=} \sqrt[3]{9-1}$

$\sqrt[3]{8} \overset{?}{=} \sqrt[3]{8}$

$2 = 2$

39. $\sqrt[4]{4x+4} = \sqrt[4]{5x+1}$

$\left(\sqrt[4]{4x+4}\right)^4 = \left(\sqrt[4]{5x+1}\right)^4$

$4x+4 = 5x+1$

$3 = x$

Check:

$\sqrt[4]{4(3)+4} \overset{?}{=} \sqrt[4]{5(3)+1}$

$\sqrt[4]{12+4} \overset{?}{=} \sqrt[4]{15+1}$

$\sqrt[4]{16} \overset{?}{=} \sqrt[4]{16}$

$2 = 2$

41. $\sqrt{2x+24} = x+8$

$\left(\sqrt{2x+24}\right)^2 = \left(x+8\right)^2$

$2x+24 = x^2+16x+64$

$0 = x^2+14x+40$

$0 = (x+4)(x+10)$

$x+4 = 0$ or $x+10 = 0$

$x = -4$ $x = -10$

Check $x = -4$:

$\sqrt{2(-4)+24} \overset{?}{=} (-4)+8$

$\sqrt{-8+24} \overset{?}{=} 4$

$\sqrt{16} \overset{?}{=} 4$

$4 = 4$

Check $x = -10$:

$\sqrt{2(-10)+24} \overset{?}{=} (-10)+8$

$\sqrt{-20+24} \overset{?}{=} -2$

$\sqrt{4} \overset{?}{=} -2$

$2 \neq -2$

-4 is the only solution. (-10 is an extraneous solution.)

43.
$$y - 1 = \sqrt{2y - 2}$$
$$(y - 1)^2 = \left(\sqrt{2y - 2}\right)^2$$
$$y^2 - 2y + 1 = 2y - 2$$
$$y^2 - 4y + 3 = 0$$
$$(y - 3)(y - 1) = 0$$

$$y - 3 = 0 \quad \text{or} \quad y - 1 = 0$$
$$y = 3 \qquad\qquad y = 1$$

Check $y = 3$:
$$3 - 1 \stackrel{?}{=} \sqrt{2(3) - 2}$$
$$2 \stackrel{?}{=} \sqrt{6 - 2}$$
$$2 \stackrel{?}{=} \sqrt{4}$$
$$2 = 2$$

Check $y = 1$:
$$1 - 1 \stackrel{?}{=} \sqrt{2(1) - 2}$$
$$0 \stackrel{?}{=} \sqrt{2 - 2}$$
$$0 \stackrel{?}{=} \sqrt{0}$$
$$0 = 0$$

45. $\sqrt{3x + 10} - 4 = x$
$$\sqrt{3x + 10} = x + 4$$
$$\left(\sqrt{3x + 10}\right)^2 = (x + 4)^2$$
$$3x + 10 = x^2 + 8x + 16$$
$$0 = x^2 + 5x + 6$$
$$0 = (x + 2)(x + 3)$$
$$x + 2 = 0 \quad \text{or} \quad x + 3 = 0$$
$$x = -2 \qquad\qquad x = -3$$

Check $x = -2$:
$$\sqrt{3(-2) + 10} - 4 \stackrel{?}{=} -2$$
$$\sqrt{-6 + 10} - 4 \stackrel{?}{=} -2$$
$$\sqrt{4} - 4 \stackrel{?}{=} -2$$
$$2 - 4 \stackrel{?}{=} -2$$
$$-2 = -2$$

Check $x = -3$:
$$\sqrt{3(-3) + 10} - 4 \stackrel{?}{=} -3$$
$$\sqrt{-9 + 10} - 4 \stackrel{?}{=} -3$$
$$\sqrt{1} - 4 \stackrel{?}{=} -3$$
$$1 - 4 \stackrel{?}{=} -3$$
$$-3 = -3$$

47. $\sqrt{10n+4} - 3n = n+1$

$\sqrt{10n+4} = 4n+1$

$\left(\sqrt{10n+4}\right)^2 = \left(4n+1\right)^2$

$10n+4 = 16n^2 + 8n + 1$

$0 = 16n^2 - 2n - 3$

$0 = \left(2n-1\right)\left(8n+3\right)$

$2n-1=0 \quad \text{or} \quad 8n+3=0$

$2n=1 \qquad\qquad 8n=-3$

$n=\dfrac{1}{2} \qquad\qquad n=-\dfrac{3}{8}$

Check $n=\dfrac{1}{2}$:

$\sqrt{10\left(\dfrac{1}{2}\right)+4} - 3\left(\dfrac{1}{2}\right) \overset{?}{=} \left(\dfrac{1}{2}\right)+1$

$\sqrt{5+4} - \dfrac{3}{2} \overset{?}{=} \left(\dfrac{1}{2}\right)+\dfrac{2}{2}$

$\sqrt{9} - \dfrac{3}{2} \overset{?}{=} \dfrac{3}{2}$

$3 - \dfrac{3}{2} \overset{?}{=} \dfrac{3}{2}$

$\dfrac{3}{2} = \dfrac{3}{2}$

Check $n=-\dfrac{3}{8}$:

$\sqrt{10\left(-\dfrac{3}{8}\right)+4} - 3\left(-\dfrac{3}{8}\right) \overset{?}{=} \left(-\dfrac{3}{8}\right)+1$

$\sqrt{-\dfrac{30}{8}+\dfrac{32}{8}} + \dfrac{9}{8} \overset{?}{=} \left(-\dfrac{3}{8}\right)+\dfrac{8}{8}$

$\sqrt{\dfrac{2}{8}} + \dfrac{9}{8} \overset{?}{=} \dfrac{5}{8}$

$\sqrt{\dfrac{1}{4}} + \dfrac{9}{8} \overset{?}{=} \dfrac{5}{8}$

$\dfrac{1}{2} + \dfrac{9}{8} \overset{?}{=} \dfrac{5}{8}$

$\dfrac{13}{8} \neq \dfrac{5}{8}$

$\dfrac{1}{2}$ is the only solution.

$\left(-\dfrac{3}{8} \text{ is an extraneous solution.}\right)$

49. $\sqrt[3]{5x+2} + 2 = 5$

$\sqrt[3]{5x+2} = 3$

$\left(\sqrt[3]{5x+2}\right)^3 = 3^3$

$5x+2 = 27$

$5x = 25$

$x = 5$

Check:

$\sqrt[3]{5(5)+2} + 2 \overset{?}{=} 5$

$\sqrt[3]{25+2} + 2 \overset{?}{=} 5$

$\sqrt[3]{27} + 2 \overset{?}{=} 5$

$3 + 2 \overset{?}{=} 5$

$5 = 5$

51. $\sqrt[3]{n^2 - 2n + 5} = 2$

$\left(\sqrt[3]{n^2 - 2n + 5}\right)^3 = \left(2\right)^3$

$n^2 - 2n + 5 = 8$

$n^2 - 2n - 3 = 0$

$\left(n-3\right)\left(n+1\right) = 0$

$n-3=0 \quad \text{or} \quad n+1=0$

$n=3 \qquad\qquad n=-1$

Check $n=3$:

$\sqrt[3]{3^2 - 2(3) + 5} \overset{?}{=} 2$

$\sqrt[3]{9 - 6 + 5} \overset{?}{=} 2$

$\sqrt[3]{8} \overset{?}{=} 2$

$2 = 2$

Check $n=-1$:

$\sqrt[3]{(-1)^2 - 2(-1) + 5} \overset{?}{=} 2$

$\sqrt[3]{1 + 2 + 5} \overset{?}{=} 2$

$\sqrt[3]{8} \overset{?}{=} 2$

$2 = 2$

53. $1 + \sqrt{x} = \sqrt{2x+1}$

$$\left(1 + \sqrt{x}\right)^2 = \left(\sqrt{2x+1}\right)^2$$

$$1 + 2\sqrt{x} + x = 2x + 1$$

$$2\sqrt{x} = x$$

$$\left(2\sqrt{x}\right)^2 = x^2$$

$$4x = x^2$$

$$0 = x^2 - 4x$$

$$0 = x(x - 4)$$

$$x = 0 \quad \text{or} \quad x - 4 = 0$$

$$x = 4$$

Check $x = 0$:

$$1 + \sqrt{0} \overset{?}{=} \sqrt{2(0)+1}$$

$$1 + 0 \overset{?}{=} \sqrt{0+1}$$

$$1 \overset{?}{=} \sqrt{1}$$

$$1 = 1$$

Check $x = 4$:

$$1 + \sqrt{4} \overset{?}{=} \sqrt{2(4)+1}$$

$$1 + 2 \overset{?}{=} \sqrt{8+1}$$

$$3 \overset{?}{=} \sqrt{9}$$

$$3 = 3$$

55. $\sqrt{3x+1} + \sqrt{3x} = 2$

$$\left(\sqrt{3x+1}\right)^2 = \left(2 - \sqrt{3x}\right)^2$$

$$3x + 1 = 4 - 4\sqrt{3x} + 3x$$

$$\left(-3\right)^2 = \left(-4\sqrt{3x}\right)^2$$

$$9 = 16 \cdot 3x$$

$$9 = 48x$$

$$\frac{3}{16} = x$$

Check:

$$\sqrt{3\left(\frac{3}{16}\right)+1} + \sqrt{3\left(\frac{3}{16}\right)} \overset{?}{=} 2$$

$$\sqrt{\frac{9}{16}+1} + \sqrt{\frac{9}{16}} \overset{?}{=} 2$$

$$\sqrt{\frac{9}{16}+\frac{16}{16}} + \sqrt{\frac{9}{16}} \overset{?}{=} 2$$

$$\sqrt{\frac{25}{16}} + \sqrt{\frac{9}{16}} \overset{?}{=} 2$$

$$\frac{5}{4} + \frac{3}{4} \overset{?}{=} 2$$

$$\frac{8}{4} \overset{?}{=} 2$$

$$2 = 2$$

57. $\sqrt{6x+7} - 2 = \sqrt{2x+3}$

$$\left(\sqrt{6x+7} - 2\right)^2 = \left(\sqrt{2x+3}\right)^2$$

$$6x + 7 - 4\sqrt{6x+7} + 4 = 2x + 3$$

$$6x + 11 - 4\sqrt{6x+7} = 2x + 3$$

$$4x + 8 = 4\sqrt{6x+7}$$

$$x + 2 = \sqrt{6x+7}$$

$$\left(x+2\right)^2 = \left(\sqrt{6x+7}\right)^2$$

$$x^2 + 4x + 4 = 6x + 7$$

$$x^2 - 2x - 3 = 0$$

$$\left(x+1\right)\left(x-3\right) = 0$$

$$x + 1 = 0 \quad \text{or} \quad x - 3 = 0$$

$$x = -1 \qquad\qquad x = 3$$

Check $x = -1$:

$$\sqrt{6(-1)+7} - 2 \overset{?}{=} \sqrt{2(-1)+3}$$

$$\sqrt{-6+7} - 2 \overset{?}{=} \sqrt{-2+3}$$

$$\sqrt{1} - 2 \overset{?}{=} \sqrt{1}$$

$$1 - 2 \overset{?}{=} 1$$

$$-1 \ne 1$$

Check $x = 3$:

$$\sqrt{6(3)+7} - 2 \overset{?}{=} \sqrt{2(3)+3}$$

$$\sqrt{18+7} - 2 \overset{?}{=} \sqrt{6+3}$$

$$\sqrt{25} - 2 \overset{?}{=} \sqrt{9}$$

$$5 - 2 \overset{?}{=} 3$$

$$3 = 3$$

3 is the only solution (-1 is an extraneous solution).

59. Mistake: You cannot take the principal square root of a number and get a negative.
Correct: No real-number solution.

61. Mistake: The binomial $x - 3$ was not squared correctly.
Correct:

$$\sqrt{x+3} = x - 3$$

$$\left(\sqrt{x+3}\right)^2 = (x-3)^2$$

$$x + 3 = x^2 - 6x + 9$$

$$0 = x^2 - 7x + 6$$

$$0 = (x-6)(x-1)$$

$$x - 6 = 0 \quad \text{or} \quad x - 1 = 0$$

$$x = 6 \qquad x = 1 \text{ is extraneous}$$

63. Substitute and solve.

$$T = 2\pi\sqrt{\frac{L}{9.8}}$$

$$2\pi = 2\pi\sqrt{\frac{L}{9.8}}$$

$$1 = \sqrt{\frac{L}{9.8}}$$

$$1^2 = \left(\sqrt{\frac{L}{9.8}}\right)^2$$

$$1 = \frac{L}{9.8}$$

$$9.8 = L$$

The length of the pendulum is 9.8 m.

65. Substitute and solve.

$$T = 2\pi\sqrt{\frac{L}{9.8}}$$

$$\frac{\pi}{2} = 2\pi\sqrt{\frac{L}{9.8}}$$

$$\frac{1}{4} = \sqrt{\frac{L}{9.8}}$$

$$\left(\frac{1}{4}\right)^2 = \left(\sqrt{\frac{L}{9.8}}\right)^2$$

$$\frac{1}{16} = \frac{L}{9.8}$$

$$\frac{9.8}{16} = L$$

$$0.6125 = L$$

The length of the pendulum is 0.6125 m.

67.

$$0.3 = \sqrt{\frac{h}{16}}$$

$$(0.3)^2 = \left(\sqrt{\frac{h}{16}}\right)^2$$

$$0.09 = \frac{h}{16}$$

$$1.44 = h$$

The distance is 1.44 ft.

69.

$$\frac{1}{4} = \sqrt{\frac{h}{16}}$$

$$\left(\frac{1}{4}\right)^2 = \left(\sqrt{\frac{h}{16}}\right)^2$$

$$\frac{1}{16} = \frac{h}{16}$$

$$1 = h$$

The distance is 1 ft.

71.

$$30 = \frac{7}{2}\sqrt{2D}$$

$$60 = 7\sqrt{2D}$$

$$\frac{60}{7} = \sqrt{2D}$$

$$\left(\frac{60}{7}\right)^2 = \left(\sqrt{2D}\right)^2$$

$$73.469 \approx 2D$$

$$36.73 \approx D$$

The skid distance is approximately 36.73 ft.

73. $45 = \dfrac{7}{2}\sqrt{2D}$

$90 = 7\sqrt{2D}$

$\dfrac{90}{7} = \sqrt{2D}$

$\left(\dfrac{90}{7}\right)^2 = \left(\sqrt{2D}\right)^2$

$165.306 \approx 2D$

$82.65 \text{ ft.} \approx D$

The skid distance is approximately 82.65 ft.

75. $5 = \sqrt{F_1^2 + 3^2}$

$5^2 = \left(\sqrt{F_1^2 + 9}\right)^2$

$25 = F_1^2 + 9$

$16 = F_1^2$

$\sqrt{16} = F_1$

$4 = F_1$

Force 1 has a value of 4 N.

77. $3\sqrt{5} = \sqrt{3^2 + F_2^2}$

$\left(3\sqrt{5}\right)^2 = \left(\sqrt{9 + F_2^2}\right)^2$

$9 \cdot 5 = 9 + F_2^2$

$45 - 9 = F_2^2$

$36 = F_2^2$

$\sqrt{36} = F_2$

$6 = F_2$

Force 2 has a value of 6 N.

79. a) $x = 9$ $x = 16$ $y = 5$

$y = \sqrt{9}$ $y = \sqrt{16}$ $5 = \sqrt{x}$

$y = 3$ $y = 4$ $5^2 = \left(\sqrt{x}\right)^2$

$25 = x$

b)

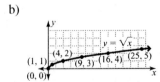

c) No. The x-values must be 0 or positive because real square roots exist only when $x \geq 0$. The y-values must be 0 or positive because, by definition, the principal square root is 0 or positive.

d) Yes, because it passes the vertical line test.

81. The graph becomes steeper from left to right.

83. The graph rises or lowers according to the value of the constant.

Review Exercises

1. $3^4 = 81$

2. $(-0.2)^3 = -0.008$

3. $\left(\dfrac{2}{5}\right)^{-4} = \left(\dfrac{5}{2}\right)^4 = \dfrac{625}{16}$ or 39.0625

4. $x^3 \cdot x^5 = x^{3+5} = x^8$

5. $\left(n^4\right)^6 = n^{4 \cdot 6} = n^{24}$

6. $\dfrac{y^7}{y^3} = y^{7-3} = y^4$

Exercise Set 8.7

1. $\sqrt{-36} = \sqrt{-1 \cdot 36} = \sqrt{-1} \cdot \sqrt{36} = i\sqrt{36} = 6i$

3. $\sqrt{-5} = \sqrt{-1 \cdot 5} = \sqrt{-1} \cdot \sqrt{5} = i\sqrt{5}$

5. $\sqrt{-8} = \sqrt{-1 \cdot 8} = \sqrt{-1} \cdot \sqrt{8} = i\sqrt{4 \cdot 2} = 2i\sqrt{2}$

7. $\sqrt{-18} = \sqrt{-1 \cdot 18} = \sqrt{-1} \cdot \sqrt{18} = i\sqrt{9 \cdot 2} = 3i\sqrt{2}$

9. $\sqrt{-27} = \sqrt{-1 \cdot 27} = \sqrt{-1} \cdot \sqrt{27} = i\sqrt{9 \cdot 3} = 3i\sqrt{3}$

11. $\sqrt{-125} = \sqrt{-1 \cdot 125}$

$= \sqrt{-1} \cdot \sqrt{125}$

$= i\sqrt{25 \cdot 5}$

$= 5i\sqrt{5}$

13. $\sqrt{-63} = \sqrt{-1 \cdot 63} = \sqrt{-1} \cdot \sqrt{63} = i\sqrt{9 \cdot 7} = 3i\sqrt{7}$

15. $\sqrt{-245} = \sqrt{-1 \cdot 245}$

$= \sqrt{-1} \cdot \sqrt{245}$

$= i\sqrt{49 \cdot 5}$

$= 7i\sqrt{5}$

17. $(9 + 3i) + (-3 + 4i) = 6 + 7i$

19. $(6 + 2i) + (5 - 8i) = 11 - 6i$

21. $(-4 + 6i) - (3 + 5i) = (-4 + 6i) + (-3 - 5i)$

$= -7 + i$

23. $(8 - 5i) - (-3i) = (8 - 5i) + (3i)$

$= 8 - 2i$

25. $(12+3i)+(-15-13i)=-3-10i$

27. $(-5-9i)-(-5-9i)=(-5-9i)+(5+9i)=0$

29. $(10+i)-(2-13i)+(6-5i)$
$=(10+i)+(-2+13i)+(6-5i)$
$=14+9i$

31. $(5-2i)-(9-14i)+(16i)$
$=(5-2i)+(-9+14i)+(16i)$
$=-4+28i$

33. $(8i)(3i)=24i^2=24(-1)=-24$

35. $(-8i)(5i)=-40i^2=-40(-1)=40$

37. $2i(6-7i)=12i-14i^2=12i-14(-1)=14+12i$

39. $-8i(4-9i)=-32i+72i^2$
$=-32i+72(-1)$
$=-72-32i$

41. $(6+i)(3-i)=18-6i+3i-i^2$
$=18-3i-(-1)$
$=18-3i+1$
$=19-3i$

43. $(8+5i)(5-2i)=40-16i+25i-10i^2$
$=40+9i-10(-1)$
$=40+9i+10$
$=50+9i$

45. $(8+i)(8-i)=64-i^2$
$=64-(-1)$
$=64+1$
$=65$

47. $(3-4i)^2=3^2-2\cdot3\cdot4i+(4i)^2$
$=9-24i+16i^2$
$=9-24i+16(-1)$
$=9-24i-16$
$=-7-24i$

49. $\dfrac{2}{i}=\dfrac{2}{i}\cdot\dfrac{i}{i}=\dfrac{2i}{i^2}=\dfrac{2i}{-1}=-2i$

51. $\dfrac{4}{5i}=\dfrac{4}{5i}\cdot\dfrac{i}{i}=\dfrac{4i}{5i^2}=\dfrac{4i}{5(-1)}=-\dfrac{4i}{5}$

53. $\dfrac{6}{2i}=\dfrac{3}{i}\cdot\dfrac{i}{i}=\dfrac{3i}{i^2}=\dfrac{3i}{-1}=-3i$

55. $\dfrac{2+i}{2i}=\dfrac{2+i}{2i}\cdot\dfrac{i}{i}$
$=\dfrac{2i+i^2}{2i^2}$
$=\dfrac{2i+(-1)}{2(-1)}$
$=\dfrac{-1+2i}{-2}$
$=\dfrac{1}{2}-i$

57. $\dfrac{4+2i}{4i}=\dfrac{2(2+i)}{4i}$
$=\dfrac{2+i}{2i}\cdot\dfrac{i}{i}$
$=\dfrac{2i+i^2}{2i^2}$
$=\dfrac{-1+2i}{-2}$
$=\dfrac{1}{2}-i$

59. $\dfrac{7}{2+i}=\dfrac{7}{2+i}\cdot\dfrac{2-i}{2-i}$
$=\dfrac{14-7i}{4-i^2}$
$=\dfrac{14-7i}{4-(-1)}$
$=\dfrac{14-7i}{4+1}$
$=\dfrac{14-7i}{5}$
$=\dfrac{14}{5}-\dfrac{7}{5}i$

61. $\dfrac{2i}{3-7i} = \dfrac{2i}{3-7i} \cdot \dfrac{3+7i}{3+7i}$

$ = \dfrac{6i + 14i^2}{9 - 49i^2}$

$ = \dfrac{6i + 14(-1)}{9 - 49(-1)}$

$ = \dfrac{6i - 14}{9 + 49}$

$ = \dfrac{-14 + 6i}{58}$

$ = -\dfrac{14}{58} + \dfrac{6}{58}i$

$ = -\dfrac{7}{29} + \dfrac{3}{29}i$

63. $\dfrac{5-9i}{1-i} = \dfrac{5-9i}{1-i} \cdot \dfrac{1+i}{1+i}$

$ = \dfrac{5 + 5i - 9i - 9i^2}{1 - i^2}$

$ = \dfrac{5 - 4i - 9(-1)}{1 - (-1)}$

$ = \dfrac{5 - 4i + 9}{1 + 1}$

$ = \dfrac{14 - 4i}{2}$

$ = 7 - 2i$

65. $\dfrac{3+i}{2+3i} = \dfrac{3+i}{2+3i} \cdot \dfrac{2-3i}{2-3i}$

$ = \dfrac{6 - 9i + 2i - 3i^2}{4 - 9i^2}$

$ = \dfrac{6 - 7i - 3(-1)}{4 - 9(-1)}$

$ = \dfrac{6 - 7i + 3}{4 + 9}$

$ = \dfrac{9 - 7i}{13}$

$ = \dfrac{9}{13} - \dfrac{7}{13}i$

67. $\dfrac{1+6i}{4+5i} = \dfrac{1+6i}{4+5i} \cdot \dfrac{4-5i}{4-5i}$

$ = \dfrac{4 - 5i + 24i - 30i^2}{16 - 25i^2}$

$ = \dfrac{4 + 19i - 30(-1)}{16 - 25(-1)}$

$ = \dfrac{4 + 19i + 30}{16 + 25}$

$ = \dfrac{34 + 19i}{41}$

$ = \dfrac{34}{41} + \dfrac{19}{41}i$

69. $i^{19} = i^{16} \cdot i^3 = \left(i^4\right)^4 \cdot i^3 = (1)^4 \cdot (-i) = 1 \cdot (-i) = -i$

71. $i^{42} = i^{40} \cdot i^2 = \left(i^4\right)^{10} \cdot i^2 = 1^{10} \cdot (-1) = 1(-1) = -1$

73. $i^{38} = i^{36} \cdot i^2 = \left(i^4\right)^9 \cdot i^2 = 1^9 \cdot (-1) = 1(-1) = -1$

75. $i^{60} = \left(i^4\right)^{15} = 1^{15} = 1$

77. $i^{-20} = \dfrac{1}{i^{20}} = \dfrac{1}{\left(i^4\right)^5} = \dfrac{1}{1^5} = \dfrac{1}{1} = 1$

79. $i^{-30} = \dfrac{1}{i^{30}} = \dfrac{1}{i^{28} \cdot i^2} = \dfrac{1}{\left(i^4\right)^7 \cdot i^2} = \dfrac{1}{1^7 \cdot (-1)}$

$ = \dfrac{1}{1(-1)} = \dfrac{1}{-1} = -1$

81. $i^{-21} = \dfrac{1}{i^{21}} = \dfrac{1}{i^{20} \cdot i} = \dfrac{1}{\left(i^4\right)^5 \cdot i} = \dfrac{1}{1^5 \cdot i} = \dfrac{1}{1 \cdot i}$

$ = \dfrac{1}{i} \cdot \dfrac{i}{i} = \dfrac{i}{i^2} = \dfrac{i}{-1} = -i$

83. $i^{-35} = \dfrac{1}{i^{35}} = \dfrac{1}{i^{32} \cdot i^3} = \dfrac{1}{\left(i^4\right)^8 \cdot i^3} = \dfrac{1}{1^8 \cdot (-i)} = \dfrac{1}{1(-i)}$

$ = \dfrac{1}{-i} \cdot \dfrac{i}{i} = \dfrac{i}{-i^2} = \dfrac{i}{-(-1)} = \dfrac{i}{1} = i$

Review Exercises

1. $4x^2 - 12x + 9 = (2x - 3)(2x - 3) = (2x - 3)^2$

2. $\pm\sqrt{48} = \pm\sqrt{16 \cdot 3} = \pm 4\sqrt{3}$

3. $3x - 2 = 4$

$ \quad 3x = 6$

$ \quad\; x = 2$

4. $2x^2 - x - 6 = 0$

       $(2x + 3)(x - 2) = 0$

           $2x + 3 = 0$ or $x - 2 = 0$

             $2x = -3$ $x = 2$

               $x = -\dfrac{3}{2}$

5. $2x + 3y = 6$

      $y = 0 : 2x + 3 \cdot 0 = 6$

            $2x = 6$

            $x = 3$

             $(3, 0)$

      $x = 0 : 2 \cdot 0 + 3y = 6$

            $3y = 6$

            $y = 2$

             $(0, 2)$

6. $y = x^2 - 3$

x	y	(x, y)
-2	1	$(-2, 1)$
-1	-2	$(-1, -2)$
0	-3	$(0, -3)$
1	-2	$(1, -2)$
2	1	$(2, 1)$

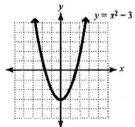

Chapter 8 Review Exercises

1.   The square roots of 121 are ± 11.

2.   The square roots of 49 are ± 7.

3. $\sqrt{169} = 13$

4. $-\sqrt{49} = -7$

5. $\sqrt{-36}$ is not a real number.

6. $\sqrt{\dfrac{1}{25}} = \dfrac{1}{5}$

7. $c = \sqrt{4^2 + 3^2}$

     $c = \sqrt{16 + 9}$

     $c = \sqrt{25}$

     $c = 5$

   The connecting piece should be 5 ft.

8. $T = 2\pi \sqrt{\dfrac{2.45}{9.8}}$

       $= 2\pi \sqrt{0.25}$

       $= 2\pi \cdot 0.5$

       $= \pi$

   The period is π sec.

9. $\sqrt{7} \approx 2.646$

10. $\sqrt{90} \approx 9.487$

11.   a) $S = 2\sqrt{2(40)} = 2\sqrt{80} = 2 \cdot 4\sqrt{5} = 8\sqrt{5}$

        The exact speed is $8\sqrt{5}$ mph.

     b) $8\sqrt{5} \approx 17.9$

        The speed is approximately 17.9 mph.

12.   a) $t = \sqrt{\dfrac{40}{16}} = \dfrac{\sqrt{40}}{\sqrt{16}} = \dfrac{2\sqrt{10}}{4} = \dfrac{\sqrt{10}}{2}$

        The exact time is $\dfrac{\sqrt{10}}{2}$ sec.

     b) $\dfrac{\sqrt{10}}{2} \approx 1.58$

        The time is approximately 1.58 sec.

13. $\sqrt{49x^8} = 7x^4$

14. $\sqrt{144a^6 b^{12}} = 12a^3 b^6$

15. $\sqrt{0.16m^2 n^{10}} = 0.4mn^5$

16. $\sqrt[3]{x^{15}} = x^5$

17. $\sqrt[3]{-64r^9 s^3} = -4r^3 s$

18. $\sqrt[4]{81x^{12}} = 3x^3$

19. $\sqrt[5]{32x^{15} y^{20}} = 2x^3 y^4$

20. $\sqrt[7]{x^{14} y^7} = x^2 y$

21. $\sqrt{81x^2} = 9|x|$

22. $\sqrt[4]{(x - 1)^8} = (x - 1)^2$

23. $f(x) = \sqrt{4x-4}$
$$f(5) = \sqrt{4(5)-4}$$
$$= \sqrt{20-4}$$
$$= \sqrt{16}$$
$$= 4$$

24. $f(x) = \sqrt[3]{5x+2}$
$$f(-2) = \sqrt[3]{5(-2)+2}$$
$$= \sqrt[3]{-10+2}$$
$$= \sqrt[3]{-8}$$
$$= -2$$

25. The index is even, so the radicand must be nonnegative.
$$2x+10 \geq 0$$
$$2x \geq -10$$
$$x \geq -5$$
Domain: $\{x \mid x \geq -5\}$, or $[-5, \infty)$.

26. The index is even, so the radicand must be nonnegative.
$$9-3x \geq 0$$
$$-3x \geq -9$$
$$x \leq 3$$
Domain: $\{x \mid x \leq 3\}$, or $(-\infty, 3]$.

27. $(-64)^{1/3} = \sqrt[3]{-64} = -4$

28. $\left(24a^4\right)^{1/2} = \sqrt{24a^4} = \sqrt{4a^4 \cdot 6} = 2a^2\sqrt{6}$

29. $\left(\dfrac{1}{32}\right)^{3/5} = \left(\sqrt[5]{\dfrac{1}{32}}\right)^3 = \left(\dfrac{1}{2}\right)^3 = \dfrac{1}{8}$

30. $(5r-2)^{5/7} = \sqrt[7]{(5r-2)^5}$

31. $121^{-1/2} = \dfrac{1}{121^{1/2}} = \dfrac{1}{\sqrt{121}} = \dfrac{1}{11}$

32. $81^{-3/4} = \dfrac{1}{81^{3/4}} = \dfrac{1}{\left(\sqrt[4]{81}\right)^3} = \dfrac{1}{3^3} = \dfrac{1}{27}$

33. $\left(\dfrac{16}{49}\right)^{3/2} = \left(\dfrac{16^{1/2}}{49^{1/2}}\right)^3 = \left(\dfrac{\sqrt{16}}{\sqrt{49}}\right)^3 = \left(\dfrac{4}{7}\right)^3 = \dfrac{64}{343}$

34. $81x^{3/4} = 81\sqrt[4]{x^3}$

35. $\sqrt[8]{33} = 33^{1/8}$

36. $\dfrac{8}{\sqrt[7]{n^3}} = \dfrac{8}{n^{3/7}} = 8n^{-3/7}$

37. $\left(\sqrt[5]{8}\right)^3 = 8^{3/5}$

38. $\left(\sqrt[3]{m}\right)^8 = m^{8/3}$

39. $\left(\sqrt[4]{3xw}\right)^3 = \left(3xw\right)^{3/4}$

40. $\sqrt[3]{(a+b)^4} = (a+b)^{4/3}$

41. $x^{2/3} \cdot x^{4/3} = x^{2/3+4/3} = x^{6/3} = x^2$

42. $\left(4m^{1/4}\right)\left(8m^{5/4}\right) = 32m^{1/4+5/4} = 32m^{6/4} = 32m^{3/2}$

43. $\dfrac{y^{3/5}}{y^{4/5}} = y^{3/5-4/5} = y^{-1/5} = \dfrac{1}{y^{1/5}}$

44. $\dfrac{b^{2/5}}{b^{-3/5}} = b^{2/5-(-3/5)} = b^{2/5+3/5} = b^{5/5} = b$

45. $\left(k^{2/3}\right)^{3/4} = k^{(2/3)\cdot(3/4)} = k^{6/12} = k^{1/2}$

46. $\left(2xy^{1/5}\right)^{3/4} = 2^{3/4}x^{3/4}y^{(1/5)\cdot(3/4)} = 2^{3/4}x^{3/4}y^{3/20}$

47. $\sqrt[4]{\sqrt[3]{x^2}} = \sqrt[4]{x^{2/3}} = \left(x^{2/3}\right)^{1/4} = x^{(2/3)\cdot(1/4)}$
$$= x^{2/12} = x^{1/6} = \sqrt[6]{x}$$

48. $\sqrt{3} \cdot \sqrt{27} = \sqrt{81} = 9$

49. $\sqrt{5x^5} \cdot \sqrt{20x^3} = \sqrt{100x^8} = 10x^4$

50. $\sqrt[3]{2} \cdot \sqrt[3]{4} = \sqrt[3]{8} = 2$

51. $\sqrt[4]{7} \cdot \sqrt[4]{6} = \sqrt[4]{42}$

52. $\sqrt[5]{3x^2y^3} \cdot \sqrt[5]{5x^2y} = \sqrt[5]{15x^4y^4}$

53. $4\sqrt{6} \cdot 7\sqrt{15} = 28\sqrt{90}$
$$= 28\sqrt{9 \cdot 10}$$
$$= 28 \cdot 3\sqrt{10}$$
$$= 84\sqrt{10}$$

54. $\sqrt{x^9} \cdot \sqrt{x^6} = \sqrt{x^{15}} = \sqrt{x^{14} \cdot x} = x^7\sqrt{x}$

55. $4\sqrt{10c} \cdot 2\sqrt{6c^4} = 8\sqrt{60c^5}$
$$= 8\sqrt{4c^4 \cdot 15c}$$
$$= 8 \cdot 2c^2\sqrt{15c}$$
$$= 16c^2\sqrt{15c}$$

56. $a\sqrt{a^3b^2} \cdot b^2\sqrt{a^5b^3} = ab^2\sqrt{a^8b^5}$
$$= ab^2\sqrt{a^8b^4 \cdot b}$$
$$= ab^2 \cdot a^4b^2\sqrt{b}$$
$$= a^5b^4\sqrt{b}$$

57. $\left(\sqrt[4]{6ab^2}\right)^4 = \left(\left(6ab^2\right)^{1/4}\right)^4 = 6ab^2$

58. $\sqrt{\dfrac{49}{121}} = \dfrac{7}{11}$

59. $\sqrt[3]{-\dfrac{27}{8}} = -\dfrac{3}{2}$

60. $\dfrac{9\sqrt{160}}{3\sqrt{8}} = 3\sqrt{\dfrac{160}{8}}$
$$= 3\sqrt{20}$$
$$= 3\sqrt{4 \cdot 5}$$
$$= 3 \cdot 2\sqrt{5}$$
$$= 6\sqrt{5}$$

61. $\dfrac{36\sqrt{96x^6y^{11}}}{4\sqrt{3x^2y^4}} = 9\sqrt{\dfrac{96x^6y^{11}}{3x^2y^4}}$
$$= 9\sqrt{32x^4y^7}$$
$$= 9\sqrt{16x^4y^6 \cdot 2y}$$
$$= 9 \cdot 4x^2y^3\sqrt{2y}$$
$$= 36x^2y^3\sqrt{2y}$$

62. $4b\sqrt{27b^7} = 4b\sqrt{9b^6 \cdot 3b}$
$$= 4b \cdot 3b^3\sqrt{3b}$$
$$= 12b^4\sqrt{3b}$$

63. $5\sqrt[3]{108} = 5\sqrt[3]{27 \cdot 4} = 5 \cdot 3\sqrt[3]{4} = 15\sqrt[3]{4}$

64. $2\sqrt[3]{40x^{10}} = 2\sqrt[3]{8x^9 \cdot 5x} = 2 \cdot 2x^3\sqrt[3]{5x} = 4x^3\sqrt[3]{5x}$

65. $2x^4\sqrt[4]{162x^7} = 2x^4\sqrt[4]{81x^4 \cdot 2x^3}$
$$= 2x^4 \cdot 3x\sqrt[4]{2x^3}$$
$$= 6x^5\sqrt[4]{2x^3}$$

66. $-5\sqrt{n} + 2\sqrt{n} = -3\sqrt{n}$

67. $3y^3\sqrt[4]{8y} - 9y^3\sqrt[4]{8y} = -6y^3\sqrt[4]{8y}$

68. $\sqrt{45} + \sqrt{20} = \sqrt{9 \cdot 5} + \sqrt{4 \cdot 5} = 3\sqrt{5} + 2\sqrt{5} = 5\sqrt{5}$

69. $4\sqrt{24} - 6\sqrt{54} = 4\sqrt{4 \cdot 6} - 6\sqrt{9 \cdot 6}$
$$= 4 \cdot 2\sqrt{6} - 6 \cdot 3\sqrt{6}$$
$$= 8\sqrt{6} - 18\sqrt{6}$$
$$= -10\sqrt{6}$$

70. $\sqrt{150} - \sqrt{54} + \sqrt{24} = \sqrt{25 \cdot 6} - \sqrt{9 \cdot 6} + \sqrt{4 \cdot 6}$
$$= 5\sqrt{6} - 3\sqrt{6} + 2\sqrt{6}$$
$$= 4\sqrt{6}$$

71. $4\sqrt{72x^2y} - 2x\sqrt{128y} + 5\sqrt{32x^2y}$
$$= 4\sqrt{36x^2 \cdot 2y} - 2x\sqrt{64 \cdot 2y} + 5\sqrt{16x^2 \cdot 2y}$$
$$= 24x\sqrt{2y} - 16x\sqrt{2y} + 20x\sqrt{2y}$$
$$= 28x\sqrt{2y}$$

72. $\sqrt[3]{250x^4y^5} + \sqrt[3]{128x^4y^5}$
$$= \sqrt[3]{125x^3y^3 \cdot 2xy^2} + \sqrt[3]{64x^3y^3 \cdot 2xy^2}$$
$$= 5xy\sqrt[3]{2xy^2} + 4xy\sqrt[3]{2xy^2}$$
$$= 9xy\sqrt[3]{2xy^2}$$

73. $\sqrt[4]{48} + \sqrt[4]{243} = \sqrt[4]{16 \cdot 3} + \sqrt[4]{81 \cdot 3}$
$$= 2\sqrt[4]{3} + 3\sqrt[4]{3}$$
$$= 5\sqrt[4]{3}$$

74. $\sqrt{5}\left(\sqrt{3} + \sqrt{2}\right) = \sqrt{5} \cdot \sqrt{3} + \sqrt{5} \cdot \sqrt{2} = \sqrt{15} + \sqrt{10}$

75. $\sqrt[3]{7}\left(\sqrt[3]{3} + 2\sqrt[3]{7}\right) = \sqrt[3]{7} \cdot \sqrt[3]{3} + \sqrt[3]{7} \cdot 2\sqrt[3]{7}$
$$= \sqrt[3]{21} + 2\sqrt[3]{49}$$

76. $3\sqrt{6}\left(2 - 3\sqrt{6}\right) = 3\sqrt{6} \cdot 2 + 3\sqrt{6} \cdot \left(-3\sqrt{6}\right)$
$$= 6\sqrt{6} - 9\sqrt{36}$$
$$= 6\sqrt{6} - 9 \cdot 6$$
$$= 6\sqrt{6} - 54$$

77. $\left(\sqrt{2} - \sqrt{3}\right)\left(\sqrt{5} + \sqrt{7}\right)$
$$= \sqrt{2} \cdot \sqrt{5} + \sqrt{2} \cdot \sqrt{7} - \sqrt{3} \cdot \sqrt{5} - \sqrt{3} \cdot \sqrt{7}$$
$$= \sqrt{10} + \sqrt{14} - \sqrt{15} - \sqrt{21}$$

78. $\left(\sqrt[3]{2} - 4\right)\left(\sqrt[3]{4} + 2\right)$
$$= \sqrt[3]{2} \cdot \sqrt[3]{4} + \sqrt[3]{2} \cdot 2 - 4 \cdot \sqrt[3]{4} - 4 \cdot 2$$
$$= \sqrt[3]{8} + 2\sqrt[3]{2} - 4\sqrt[3]{4} - 8$$
$$= 2 + 2\sqrt[3]{2} - 4\sqrt[3]{4} - 8$$
$$= -6 + 2\sqrt[3]{2} - 4\sqrt[3]{4}$$

79. $\left(\sqrt[4]{6x}+2\right)\left(\sqrt[4]{2x}-1\right)$

$= \sqrt[4]{6x}\cdot\sqrt[4]{2x}-\sqrt[4]{6x}\cdot 1+2\cdot\sqrt[4]{2x}-2\cdot 1$

$= \sqrt[4]{12x^2}-\sqrt[4]{6x}+2\sqrt[4]{2x}-2$

80. $\left(\sqrt{5a}+\sqrt{3b}\right)\left(\sqrt{5a}-\sqrt{3b}\right)=\left(\sqrt{5a}\right)^2-\left(\sqrt{3b}\right)^2$

$= 5a-3b$

81. $\left(2\sqrt{3}-\sqrt{5}\right)\left(2\sqrt{3}+\sqrt{5}\right)=\left(2\sqrt{3}\right)^2-\left(\sqrt{5}\right)^2$

$= 4\cdot 3-5$

$= 12-5$

$= 7$

82. $\left(\sqrt{2}-\sqrt{5}\right)^2=\left(\sqrt{2}\right)^2-2\cdot\sqrt{2}\cdot\sqrt{5}+\left(\sqrt{5}\right)^2$

$= 2-2\sqrt{10}+5$

$= 7-2\sqrt{10}$

83. $\dfrac{1}{\sqrt{2}}=\dfrac{1}{\sqrt{2}}\cdot\dfrac{\sqrt{2}}{\sqrt{2}}=\dfrac{\sqrt{2}}{\sqrt{4}}=\dfrac{\sqrt{2}}{2}$

84. $\dfrac{3}{\sqrt[3]{3}}=\dfrac{3}{\sqrt[3]{3}}\cdot\dfrac{\sqrt[3]{9}}{\sqrt[3]{9}}=\dfrac{3\sqrt[3]{9}}{\sqrt[3]{27}}=\dfrac{3\sqrt[3]{9}}{3}=\sqrt[3]{9}$

85. $\sqrt{\dfrac{4}{7}}=\dfrac{\sqrt{4}}{\sqrt{7}}=\dfrac{2}{\sqrt{7}}\cdot\dfrac{\sqrt{7}}{\sqrt{7}}=\dfrac{2\sqrt{7}}{\sqrt{49}}=\dfrac{2\sqrt{7}}{7}$

86. $\dfrac{\sqrt[4]{5x^2}}{\sqrt[4]{2}}=\dfrac{\sqrt[4]{5x^2}}{\sqrt[4]{2}}\cdot\dfrac{\sqrt[4]{8}}{\sqrt[4]{8}}=\dfrac{\sqrt[4]{40x^2}}{\sqrt[4]{16}}=\dfrac{\sqrt[4]{40x^2}}{2}$

87. $\sqrt[3]{\dfrac{17}{3y^2}}=\dfrac{\sqrt[3]{17}}{\sqrt[3]{3y^2}}\cdot\dfrac{\sqrt[3]{9y}}{\sqrt[3]{9y}}=\dfrac{\sqrt[3]{153y}}{\sqrt[3]{27y^3}}=\dfrac{\sqrt[3]{153y}}{3y}$

88. $\dfrac{\sqrt[4]{9}}{\sqrt[4]{27}}=\dfrac{\sqrt[4]{9}}{\sqrt[4]{27}}\cdot\dfrac{\sqrt[4]{3}}{\sqrt[4]{3}}=\dfrac{\sqrt[4]{27}}{\sqrt[4]{81}}=\dfrac{\sqrt[4]{27}}{3}$

89. $\dfrac{4}{\sqrt{2}-\sqrt{3}}=\dfrac{4}{\sqrt{2}-\sqrt{3}}\cdot\dfrac{\sqrt{2}+\sqrt{3}}{\sqrt{2}+\sqrt{3}}$

$= \dfrac{4\sqrt{2}+4\sqrt{3}}{\left(\sqrt{2}\right)^2-\left(\sqrt{3}\right)^2}$

$= \dfrac{4\sqrt{2}+4\sqrt{3}}{2-3}$

$= \dfrac{4\sqrt{2}+4\sqrt{3}}{-1}$

$= -4\sqrt{2}-4\sqrt{3}$

90. $\dfrac{1}{4+\sqrt{3}}=\dfrac{1}{4+\sqrt{3}}\cdot\dfrac{4-\sqrt{3}}{4-\sqrt{3}}$

$= \dfrac{4-\sqrt{3}}{\left(4\right)^2-\left(\sqrt{3}\right)^2}$

$= \dfrac{4-\sqrt{3}}{16-3}$

$= \dfrac{4-\sqrt{3}}{13}$

91. $\dfrac{1}{2-\sqrt{n}}=\dfrac{1}{2-\sqrt{n}}\cdot\dfrac{2+\sqrt{n}}{2+\sqrt{n}}$

$= \dfrac{2+\sqrt{n}}{2^2-\left(\sqrt{n}\right)^2}$

$= \dfrac{2+\sqrt{n}}{4-n}$

92. $\dfrac{2\sqrt{3}}{3\sqrt{2}-2\sqrt{3}}=\dfrac{2\sqrt{3}}{3\sqrt{2}-2\sqrt{3}}\cdot\dfrac{3\sqrt{2}+2\sqrt{3}}{3\sqrt{2}+2\sqrt{3}}$

$= \dfrac{6\sqrt{6}+4\sqrt{9}}{\left(3\sqrt{2}\right)^2-\left(2\sqrt{3}\right)^2}$

$= \dfrac{6\sqrt{6}+4\cdot 3}{9\cdot 2-4\cdot 3}$

$= \dfrac{6\sqrt{6}+12}{18-12}$

$= \dfrac{6\sqrt{6}+12}{6}$

$= \sqrt{6}+2$

93. $\dfrac{\sqrt{10}}{6}=\dfrac{\sqrt{10}}{6}\cdot\dfrac{\sqrt{10}}{\sqrt{10}}=\dfrac{\sqrt{100}}{6\sqrt{10}}=\dfrac{10}{6\sqrt{10}}=\dfrac{5}{3\sqrt{10}}$

94. $\dfrac{\sqrt{3x}}{5}=\dfrac{\sqrt{3x}}{5}\cdot\dfrac{\sqrt{3x}}{\sqrt{3x}}=\dfrac{\sqrt{9x^2}}{5\sqrt{3x}}=\dfrac{3x}{5\sqrt{3x}}$

95. $\dfrac{2-\sqrt{3}}{8}=\dfrac{2-\sqrt{3}}{8}\cdot\dfrac{2+\sqrt{3}}{2+\sqrt{3}}$

$= \dfrac{2^2-\left(\sqrt{3}\right)^2}{8\left(2+\sqrt{3}\right)}$

$= \dfrac{4-3}{16+8\sqrt{3}}$

$= \dfrac{1}{16+8\sqrt{3}}$

96. $\dfrac{2\sqrt{t}+\sqrt{3t}}{5t}=\dfrac{2\sqrt{t}+\sqrt{3t}}{5t}\cdot\dfrac{2\sqrt{t}-\sqrt{3t}}{2\sqrt{t}-\sqrt{3t}}$

$\qquad = \dfrac{\left(2\sqrt{t}\right)^2-\left(\sqrt{3t}\right)^2}{5t\left(2\sqrt{t}-\sqrt{3t}\right)}$

$\qquad = \dfrac{4t-3t}{10t\sqrt{t}-5t\sqrt{3t}}$

$\qquad = \dfrac{t}{10t\sqrt{t}-5t\sqrt{3t}}$

$\qquad = \dfrac{\cancel{t}}{\cancel{t}\left(10\sqrt{t}-5\sqrt{3t}\right)}$

$\qquad = \dfrac{1}{10\sqrt{t}-5\sqrt{3t}}$

97. $\sqrt{x}=9$

$\qquad \left(\sqrt{x}\right)^2=9^2$

$\qquad x=81$

Check:

$\qquad \sqrt{81}\overset{?}{=}9$

$\qquad 9=9$

98. $\sqrt{y}=-3$ has no real-number solution.

99. $\sqrt{w-1}=3$

$\qquad \left(\sqrt{w-1}\right)^2=3^2$

$\qquad w-1=9$

$\qquad w=10$

Check:

$\qquad \sqrt{10-1}\overset{?}{=}3$

$\qquad \sqrt{9}\overset{?}{=}3$

$\qquad 3=3$

100. $\sqrt[3]{3x-2}=-2$

$\qquad \left(\sqrt[3]{3x-2}\right)^3=(-2)^3$

$\qquad 3x-2=-8$

$\qquad 3x=-6$

$\qquad x=-2$

Check:

$\qquad \sqrt[3]{3(-2)-2}\overset{?}{=}-2$

$\qquad \sqrt[3]{-6-2}\overset{?}{=}-2$

$\qquad \sqrt[3]{-8}\overset{?}{=}-2$

$\qquad -2=-2$

101. $\sqrt[4]{x-2}-3=-1$

$\qquad \sqrt[4]{x-2}=2$

$\qquad \left(\sqrt[4]{x-2}\right)^4=2^4$

$\qquad x-2=16$

$\qquad x=18$

Check:

$\qquad \sqrt[4]{18-2}-3\overset{?}{=}-1$

$\qquad \sqrt[4]{16}-3\overset{?}{=}-1$

$\qquad 2-3\overset{?}{=}-1$

$\qquad -1=-1$

102. $\sqrt{y+1}=\sqrt{2y-4}$

$\qquad \left(\sqrt{y+1}\right)^2=\left(\sqrt{2y-4}\right)^2$

$\qquad y+1=2y-4$

$\qquad 5=y$

Check:

$\qquad \sqrt{5+1}\overset{?}{=}\sqrt{2(5)-4}$

$\qquad \sqrt{6}\overset{?}{=}\sqrt{10-4}$

$\qquad \sqrt{6}=\sqrt{6}$

103. $\sqrt{x-6}=x+2$

$\qquad \left(\sqrt{x-6}\right)^2=(x+2)^2$

$\qquad x-6=x^2+4x+4$

$\qquad 0=x^2+3x+10$

This polynomial cannot be factored. The equation has no real-number solution.

104. $\sqrt[3]{3x+10} - 4 = 5$

$\sqrt[3]{3x+10} = 9$

$\left(\sqrt[3]{3x+10}\right)^3 = 9^3$

$3x + 10 = 729$

$3x = 719$

$x = \dfrac{719}{3}$

Check:

$\sqrt[3]{3\left(\dfrac{719}{3}\right)+10} - 4 \overset{?}{=} 5$

$\sqrt[3]{719+10} - 4 \overset{?}{=} 5$

$\sqrt[3]{729} - 4 \overset{?}{=} 5$

$9 - 4 \overset{?}{=} 5$

$5 = 5$

105. $\sqrt[4]{x+8} = \sqrt[4]{2x+1}$

$\left(\sqrt[4]{x+8}\right)^4 = \left(\sqrt[4]{2x+1}\right)^4$

$x + 8 = 2x + 1$

$7 = x$

Check:

$\sqrt[4]{7+8} \overset{?}{=} \sqrt[4]{2(7)+1}$

$\sqrt[4]{15} \overset{?}{=} \sqrt[4]{14+1}$

$\sqrt[4]{15} = \sqrt[4]{15}$

106. $\sqrt{5n-1} = 4 - 2n$

$\left(\sqrt{5n-1}\right)^2 = (4-2n)^2$

$5n - 1 = 16 - 16n + 4n^2$

$0 = 4n^2 - 21n + 17$

$0 = (4n-17)(n-1)$

$4n - 17 = 0 \quad$ or $\quad n - 1 = 0$

$4n = 17 \qquad\qquad n = 1$

$n = \dfrac{17}{4}$

Check $n = \dfrac{17}{4}$:

$\sqrt{5\left(\dfrac{17}{4}\right)-1} \overset{?}{=} 4 - 2\left(\dfrac{17}{4}\right)$

$\sqrt{\dfrac{85}{4} - \dfrac{4}{4}} \overset{?}{=} 4 - \dfrac{17}{2}$

$\sqrt{\dfrac{81}{4}} \overset{?}{=} \dfrac{8}{2} - \dfrac{17}{2}$

$\dfrac{9}{2} \neq -\dfrac{9}{2}$

Check $n = 1$:

$\sqrt{5(1)-1} \overset{?}{=} 4 - 2(1)$

$\sqrt{5-1} \overset{?}{=} 4 - 2$

$\sqrt{4} \overset{?}{=} 2$

$2 = 2$

The only solution is $n = 1$ ($n = \dfrac{17}{4}$ is an

extraneous solution).

107. $1 + \sqrt{x} = \sqrt{2x+1}$

$\left(1+\sqrt{x}\right)^2 = \left(\sqrt{2x+1}\right)^2$

$1 + 2\sqrt{x} + x = 2x + 1$

$2\sqrt{x} = x$

$\left(2\sqrt{x}\right)^2 = x^2$

$4x = x^2$

$0 = x^2 - 4x$

$0 = x(x-4)$

$x = 0 \quad$ or $\quad x - 4 = 0$

$x = 4$

Check $x = 0$:

$1 + \sqrt{0} \overset{?}{=} \sqrt{2(0)+1}$

$1 + 0 \overset{?}{=} \sqrt{0+1}$

$1 \overset{?}{=} \sqrt{1}$

$1 = 1$

Check $x = 4$:

$$1 + \sqrt{4} \stackrel{?}{=} \sqrt{2(4) + 1}$$

$$1 + 2 \stackrel{?}{=} \sqrt{8 + 1}$$

$$3 \stackrel{?}{=} \sqrt{9}$$

$$3 = 3$$

108. $\sqrt{3x + 1} = 2 - \sqrt{3x}$

$$\left(\sqrt{3x + 1}\right)^2 = \left(2 - \sqrt{3x}\right)^2$$

$$3x + 1 = 4 - 4\sqrt{3x} + 3x$$

$$4\sqrt{3x} = 3$$

$$\left(4\sqrt{3x}\right)^2 = 3^2$$

$$16 \cdot 3x = 9$$

$$48x = 9$$

$$x = \frac{9}{48}$$

$$x = \frac{3}{16}$$

Check:

$$\sqrt{3\left(\frac{3}{16}\right) + 1} \stackrel{?}{=} 2 - \sqrt{3\left(\frac{3}{16}\right)}$$

$$\sqrt{\frac{9}{16} + \frac{16}{16}} \stackrel{?}{=} 2 - \sqrt{\frac{9}{16}}$$

$$\sqrt{\frac{25}{16}} \stackrel{?}{=} 2 - \frac{3}{4}$$

$$\frac{5}{4} \stackrel{?}{=} 2 - \frac{3}{4}$$

$$\frac{5}{4} \stackrel{?}{=} \frac{8}{4} - \frac{3}{4}$$

$$\frac{5}{4} = \frac{5}{4}$$

109. $S = 2\sqrt{2L}$

$$50 = 2\sqrt{2L}$$

$$25 = \sqrt{2L}$$

$$(25)^2 = \left(\sqrt{2L}\right)^2$$

$$625 = 2L$$

$$312.5 = L$$

The length of the skid marks is 312.5 ft.

110. $T = 2\pi\sqrt{\dfrac{L}{9.8}}$

$$\frac{\pi}{3} = 2\pi\sqrt{\frac{L}{9.8}}$$

$$\frac{1}{6} = \sqrt{\frac{L}{9.8}}$$

$$\left(\frac{1}{6}\right)^2 = \left(\sqrt{\frac{L}{9.8}}\right)^2$$

$$\frac{1}{36} = \frac{L}{9.8}$$

$$0.272 \approx L$$

The length of the pendulum is approximately 0.272 m.

111. $t = \sqrt{\dfrac{h}{16}}$

$$0.3 = \frac{\sqrt{h}}{\sqrt{16}}$$

$$0.3 = \frac{\sqrt{h}}{4}$$

$$1.2 = \sqrt{h}$$

$$(1.2)^2 = \left(\sqrt{h}\right)^2$$

$$1.44 = h$$

The distance the object falls is 1.44 ft.

112. $\sqrt{-9} = \sqrt{-1 \cdot 9} = \sqrt{-1} \cdot \sqrt{9} = 3i$

113. $\sqrt{-20} = \sqrt{-1 \cdot 20}$

$$= \sqrt{-1} \cdot \sqrt{20}$$

$$= i\sqrt{20}$$

$$= i\sqrt{4 \cdot 5}$$

$$= 2i\sqrt{5}$$

114. $(3 + 2i) + (5 - 8i) = 8 - 6i$

115. $(7 - 3i) - (-2 + 4i) = (7 - 3i) + (2 - 4i)$
$$= 9 - 7i$$

116. $(3i)(4i) = 12i^2 = 12(-1) = -12$

117. $2i(4 - i) = 8i - 2i^2 = 8i - 2(-1) = 2 + 8i$

118. $(6+2i)(4-i) = 6\cdot4 - 6\cdot i + 2i\cdot4 - 2i\cdot i$
$$= 24 - 6i + 8i - 2i^2$$
$$= 24 + 2i - 2(-1)$$
$$= 24 + 2i + 2$$
$$= 26 + 2i$$

119. $(5-i)^2 = 5^2 - 2(5)(i) + (-i)^2$
$$= 25 - 10i + i^2$$
$$= 25 - 10i - 1$$
$$= 24 - 10i$$

120. $\dfrac{5}{i} = \dfrac{5}{i}\cdot\dfrac{i}{i} = \dfrac{5i}{i^2} = \dfrac{5i}{-1} = -5i$

121. $\dfrac{3}{-i} = \dfrac{3}{-i}\cdot\dfrac{i}{i} = \dfrac{3i}{-i^2} = \dfrac{3i}{-(-1)} = \dfrac{3i}{1} = 3i$

122. $\dfrac{4}{3i} = \dfrac{4}{3i}\cdot\dfrac{i}{i} = \dfrac{4i}{3i^2} = \dfrac{4i}{3(-1)} = -\dfrac{4i}{3}$

123. $\dfrac{7+i}{5i}\cdot\dfrac{i}{i} = \dfrac{7i+i^2}{5i^2} = \dfrac{7i-1}{5(-1)} = \dfrac{-1+7i}{-5} = \dfrac{1}{5} - \dfrac{7}{5}i$

124. $\dfrac{3}{2+i}\cdot\dfrac{2-i}{2-i} = \dfrac{3(2-i)}{2^2-i^2}$
$$= \dfrac{6-3i}{4-(-1)}$$
$$= \dfrac{6-3i}{5}$$
$$= \dfrac{6}{5} - \dfrac{3}{5}i$$

125. $\dfrac{5+i}{2-3i} = \dfrac{5+i}{2-3i}\cdot\dfrac{2+3i}{2+3i} = \dfrac{(5+i)(2+3i)}{2^2-(3i)^2}$
$$= \dfrac{10+15i+2i+3i^2}{4-9i^2} = \dfrac{10+17i+3(-1)}{4-9(-1)}$$
$$= \dfrac{10+17i-3}{4+9} = \dfrac{7+17i}{13}$$
$$= \dfrac{7}{13} + \dfrac{17}{13}i$$

126. $i^{20} = (i^4)^5 = 1^5 = 1$

127. $i^{15} = i^{12}\cdot i^3 = (i^4)^3\cdot i^3 = 1^3\cdot(-i) = 1(-i) = -i$

Chapter 8 Practice Test

1. $\sqrt{36} = 6$

2. $\sqrt{-49} = \sqrt{-1\cdot49}$
$$= \sqrt{-1}\cdot\sqrt{49}$$
$$= i\cdot7$$
$$= 7i$$

3. $\sqrt{81x^2y^5} = \sqrt{81x^2y^4\cdot y}$
$$= 9xy^2\sqrt{y}$$

4. $\sqrt[3]{54} = \sqrt[3]{27\cdot2} = 3\sqrt[3]{2}$

5. $\sqrt[4]{4x}\cdot\sqrt[4]{4x^5} = \sqrt[4]{16x^6}$
$$= \sqrt[4]{16x^4\cdot x^2}$$
$$= 2x\sqrt[4]{x^2}\text{ or }2x\sqrt{x}$$

6. $-\sqrt[3]{-27r^{15}} = -(-3r^5) = 3r^5$

7. $\dfrac{\sqrt{5}}{\sqrt{45}} = \sqrt{\dfrac{5}{45}} = \sqrt{\dfrac{1}{9}} = \dfrac{\sqrt{1}}{\sqrt{9}} = \dfrac{1}{3}$

8. $\dfrac{\sqrt[4]{1}}{\sqrt[4]{81}} = \dfrac{1}{3}$

9. $6\sqrt{7} - \sqrt{7} = (6-1)\sqrt{7}$
$$= 5\sqrt{7}$$

10. $(\sqrt{3}-1)^2 = (\sqrt{3})^2 - 2(\sqrt{3})(1) + (-1)^2$
$$= 3 - 2\sqrt{3} + 1$$
$$= 4 - 2\sqrt{3}$$

11. $x^{2/3}\cdot x^{-4/3} = x^{2/3+(-4/3)}$
$$= x^{-2/3}$$
$$= \dfrac{1}{x^{2/3}}$$

12. $(\sqrt[3]{2}-4)(\sqrt[3]{4}+2)$
$$= \sqrt[3]{2}\cdot\sqrt[3]{4} + \sqrt[3]{2}\cdot2 - 4\cdot\sqrt[3]{4} - 4\cdot2$$
$$= \sqrt[3]{8} + 2\sqrt[3]{2} - 4\sqrt[3]{4} - 8$$
$$= 2 + 2\sqrt[3]{2} - 4\sqrt[3]{4} - 8$$
$$= -6 + 2\sqrt[3]{2} - 4\sqrt[3]{4}$$

13. $\sqrt[5]{8x^3} = \left(8x^3\right)^{1/5}$

$= \left(8\right)^{1/5}\left(x^3\right)^{1/5}$

$= 8^{1/5} x^{3\cdot1/5}$

$= 8^{1/5} x^{3/5}$

14. $\sqrt[3]{\left(2x+5\right)^2} = \left(2x+5\right)^{2/3}$

15. $\dfrac{1}{\sqrt[3]{4}} = \dfrac{1}{\sqrt[3]{4}}\cdot\dfrac{\sqrt[3]{2}}{\sqrt[3]{2}} = \dfrac{\sqrt[3]{2}}{\sqrt[3]{8}} = \dfrac{\sqrt[3]{2}}{2}$

16. $\dfrac{\sqrt{x}}{\sqrt{x}+\sqrt{y}} = \dfrac{\sqrt{x}}{\sqrt{x}+\sqrt{y}}\cdot\dfrac{\sqrt{x}-\sqrt{y}}{\sqrt{x}-\sqrt{y}}$

$= \dfrac{\sqrt{x}\left(\sqrt{x}-\sqrt{y}\right)}{\left(\sqrt{x}\right)^2-\left(\sqrt{y}\right)^2}$

$= \dfrac{\sqrt{x^2}-\sqrt{xy}}{x-y}$

$= \dfrac{x-\sqrt{xy}}{x-y}$

17. $\sqrt{3x-2} = 8$

$\left(\sqrt{3x-2}\right)^2 = 8^2$

$3x-2 = 64$

$3x = 66$

$x = 22$

18. $\sqrt[4]{x+8} = \sqrt[4]{2x+1}$

$\left(\sqrt[4]{x+8}\right)^4 = \left(\sqrt[4]{2x+1}\right)^4$

$x+8 = 2x+1$

$8 = x+1$

$7 = x$

19. $\sqrt{2x+1}-\sqrt{x+1} = 2$

$\sqrt{2x+1} = 2+\sqrt{x+1}$

$\left(\sqrt{2x+1}\right)^2 = \left(2+\sqrt{x+1}\right)^2$

$2x+1 = 4+4\sqrt{x+1}+x+1$

$2x+1 = 5+x+4\sqrt{x+1}$

$x-4 = 4\sqrt{x+1}$

$\left(x-4\right)^2 = \left(4\sqrt{x+1}\right)^2$

$x^2-8x+16 = 16\left(x+1\right)$

$x^2-8x+16 = 16x+16$

$x^2-24x = 0$

$x\left(x-24\right) = 0$

$x = \cancel{0}$ or $x-24 = 0$

$x = 24$

0 is an extraneous solution.

20. $\left(2-i\right)-\left(4+3i\right) = \left(2-i\right)+\left(-4-3i\right)$

$= -2-4i$

21. $\left(4-i\right)\left(4+i\right) = 4^2-i^2$

$= 16-\left(-1\right)$

$= 16+1$

$= 17$ or $17+0i$

22. $\dfrac{2}{4-3i} = \dfrac{2}{4-3i}\cdot\dfrac{4+3i}{4+3i}$

$= \dfrac{2\left(4+3i\right)}{4^2-\left(3i\right)^2}$

$= \dfrac{8+6i}{16-9i^2}$

$= \dfrac{8+6i}{16-9\left(-1\right)}$

$= \dfrac{8+6i}{16+9}$

$= \dfrac{8+6i}{25}$

$= \dfrac{8}{25}+\dfrac{6}{25}i$

23. Area of a parallelogram is base $\times$ height .

Area $= 5\sqrt{12}\cdot2\sqrt{3} = 10\sqrt{36} = 10\cdot6 = 60$ m^2

24. a) Let $h = 12$.

$$t = \sqrt{\frac{h}{16}} = \sqrt{\frac{12}{16}} = \sqrt{\frac{3}{4}} = \frac{\sqrt{3}}{\sqrt{4}} = \frac{\sqrt{3}}{2} \text{ seconds.}$$

b) Let $t = 2$.

$$2 = \sqrt{\frac{h}{16}}$$

$$2 = \frac{\sqrt{h}}{\sqrt{16}}$$

$$2 = \frac{\sqrt{h}}{4}$$

$$8 = \sqrt{h}$$

$$8^2 = \left(\sqrt{h}\right)^2$$

$$64 \text{ ft.} = h$$

25. a) Let $D = 40$ feet.

$$S = \frac{7}{4}\sqrt{D}$$

$$= \frac{7}{4}\sqrt{40}$$

$$= \frac{7}{4}\sqrt{4 \cdot 10}$$

$$= \frac{7}{4} \cdot 2\sqrt{10}$$

$$= \frac{7}{2}\sqrt{10}$$

$$= 3.5\sqrt{10}$$

The speed is $3.5\sqrt{10}$ mph.

b) Let $S = 30$ mph.

$$S = \frac{7}{4}\sqrt{D}$$

$$30 = \frac{7}{4}\sqrt{D}$$

$$\frac{120}{7} = \sqrt{D}$$

$$\left(\frac{120}{7}\right)^2 = \left(\sqrt{D}\right)^2$$

$$\frac{14,400}{49} = D$$

The length of the skid marks is approximately 293.88 ft.

Chapters 1–8 Cumulative Review

1. False; if $|x| > 3$, then $x > 3$ or $x < -3$.

2. False; $(2, -3)$ is not a solution because it does not satisfy both equations.

Equation 1	Equation 2
$4x + 3y = -1$	$2x - 5y = -11$

$$4(2) + 3(-3) \overset{?}{=} -1 \qquad 2(2) - 5(-3) \overset{?}{=} -11$$

$$8 - 9 \overset{?}{=} -1 \qquad 4 + 15 \overset{?}{=} -11$$

$$-1 = -1 \qquad 19 \neq -11$$

3. True

4. True

5. x-intercept

6. inconsistent

7. $24x^5 y^2$

8. $4 - 6\left(5 - 3^2\right) + 12 \div 2 - 6$

$$= 4 - 6(5 - 9) + 12 \div 2 - 6$$

$$= 4 - 6(-4) + 12 \div 2 - 6$$

$$= 4 + 24 + 12 \div 2 - 6$$

$$= 4 + 24 + 6 - 6$$

$$= 28$$

9. $\dfrac{7.44 \times 10^{-2}}{3.1 \times 10^4} = 2.4 \times 10^{-2-4}$

$$= 2.4 \times 10^{-6}$$

$$= 0.0000024$$

10. $\dfrac{\left(x^{-2}\right)^3 \left(x^3\right)^4}{\left(x^{-3}\right)^3} = \dfrac{x^{-6} x^{12}}{x^{-9}}$

$$= \frac{x^{-6+12}}{x^{-9}}$$

$$= \frac{x^6}{x^{-9}}$$

$$= x^{6-(-9)}$$

$$= x^{6+9}$$

$$= x^{15}$$

11. $\left(3x^2 - 4y\right)^2$

$$= \left(3x^2\right)^2 - 2 \cdot 3x^2 \cdot 4y + \left(4y\right)^2$$

$$= 9x^4 - 24x^2 y + 16y^2$$

12.

$$2x-3 \overline{\smash{\big)}\,8x^3 + 0x^2 - 22x + 8} \quad \underset{4x^2+6x-2}{}$$

$$\underline{8x^3 - 12x^2}$$
$$12x^2 - 22x$$
$$\underline{12x^2 - 18x}$$
$$-4x + 8$$
$$\underline{-4x + 6}$$
$$2$$

Solution: $\dfrac{8x^3 - 22x + 8}{2x-3} = 4x^2 + 6x - 2 + \dfrac{2}{2x-3}$

13. $\dfrac{2x^2 + 7x + 3}{x^2 - 9} \div \dfrac{2x^2 + 11x + 5}{x^2 - 3x}$

$= \dfrac{2x^2 + 7x + 3}{x^2 - 9} \cdot \dfrac{x^2 - 3x}{2x^2 + 11x + 5}$

$= \dfrac{(2x+1)(x+3)}{(x+3)(x-3)} \cdot \dfrac{x(x-3)}{(2x+1)(x+5)}$

$= \dfrac{x}{x+5}$

14. $2\sqrt{5a^2} \cdot 3\sqrt{10a^5} = 6\sqrt{50a^7}$

$= 6\sqrt{25a^6 \cdot 2a}$

$= 6 \cdot 5a^3 \sqrt{2a}$

$= 30a^3 \sqrt{2a}$

15. $3x\sqrt[3]{24x^4} - 4x^2 \sqrt[3]{81x}$

$= 3x\sqrt[3]{8x^3 \cdot 3x} - 4x^2 \sqrt[3]{27 \cdot 3x}$

$= 3x \cdot 2x\sqrt[3]{3x} - 4x^2 \cdot 3\sqrt[3]{3x}$

$= 6x^2 \sqrt[3]{3x} - 12x^2 \sqrt[3]{3x}$

$= -6x^2 \sqrt[3]{3x}$

16. $32^{-3/5} = \dfrac{1}{32^{3/5}} = \dfrac{1}{\left(\sqrt[5]{32}\right)^3} = \dfrac{1}{(2)^3} = \dfrac{1}{8}$

17. $(4+5i)(2-3i) = 8 - 12i + 10i - 15i^2$

$= 8 - 2i - 15(-1)$

$= 8 - 2i + 15$

$= 23 - 2i$

18. $\dfrac{2+\sqrt{2}}{4-\sqrt{2}} = \dfrac{2+\sqrt{2}}{4-\sqrt{2}} \cdot \dfrac{4+\sqrt{2}}{4+\sqrt{2}}$

$= \dfrac{8 + 2\sqrt{2} + 4\sqrt{2} + \left(\sqrt{2}\right)^2}{4^2 - \left(\sqrt{2}\right)^2}$

$= \dfrac{8 + 6\sqrt{2} + 2}{16 - 2}$

$= \dfrac{10 + 6\sqrt{2}}{14}$

$= \dfrac{2\left(5 + 3\sqrt{2}\right)}{2 \cdot 7}$

$= \dfrac{5 + 3\sqrt{2}}{7}$

19. $A = \dfrac{1}{2}h(B+b)$

$2A = h(B+b)$

$\dfrac{2A}{h} = B + b$

$\dfrac{2A}{h} - B = b$

$\dfrac{2A}{h} - \dfrac{Bh}{h} = b$

$\dfrac{2A - Bh}{h} = b$

20. $(x+2)(x+4) = 63$

$x^2 + 6x + 8 = 63$

$x^2 + 6x - 55 = 0$

$(x-5)(x+11) = 0$

$x - 5 = 0 \quad \text{or} \quad x + 11 = 0$

$x = 5 \qquad\qquad x = -11$

21. $\dfrac{3y}{y-4} - \dfrac{12}{y-4} = 6$

$(y-4)\left(\dfrac{3y}{y-4} - \dfrac{12}{y-4}\right) = (y-4)(6)$

$3y - 12 = 6y - 24$

$-12 = 3y - 24$

$12 = 3y$

$4 = y$

Notice that if $y = 4$, then $\dfrac{3y}{y-4}$ and $\dfrac{12}{y-4}$ are undefined. Therefore, $y = 4$ is extraneous and the equation has no solution.

22. $\sqrt{3x+10} = x+2$

$$\left(\sqrt{3x+10}\right)^2 = \left(x+2\right)^2$$

$$3x+10 = x^2+4x+4$$

$$0 = x^2+x-6$$

$$0 = \left(x+3\right)\left(x-2\right)$$

$x+3 = 0$ or $x-2 = 0$

$x = -3$ $x = 2$

Check $x = -3$:

$$\sqrt{3(-3)+10} \overset{?}{=} -3+2$$

$$\sqrt{-9+10} \overset{?}{=} -1$$

$$\sqrt{1} \overset{?}{=} -1$$

$$1 \neq -1$$

Check $x = 2$:

$$\sqrt{3(2)+10} \overset{?}{=} 2+2$$

$$\sqrt{6+10} \overset{?}{=} 4$$

$$\sqrt{16} \overset{?}{=} 4$$

$$4 = 4$$

The only solution is $x = 2$ ($x = -3$ is an extraneous solution).

23. $5x-2y = -15$

$$-2y = -5x-15$$

$$y = \frac{5}{2}x + \frac{15}{2}$$

The slope is $m = \frac{5}{2}$ and the y-intercept is $\frac{15}{2}$.

24.   The slope of the given line is $m = 3$. Use $m = 3$ as the slope.

$$y - y_1 = m\left(x - x_1\right)$$

$$y - 2 = 3\left(x - \left(-4\right)\right)$$

$$y - 2 = 3\left(x + 4\right)$$

$$y - 2 = 3x + 12$$

$$y = 3x + 14$$

25. $y = -\dfrac{5}{3}x + 2$

$m = -\dfrac{5}{3}$; y-intercept: $\left(0,2\right)$

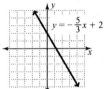

26. $\begin{cases} 2x - y > -4 \\ y > -\dfrac{2}{3}x - 2 \end{cases}$

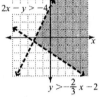

27.   Create a table.

Categories	Hourly wage	Hours worked	Pay
0 – 40 hours worked	w	40	$40w$
hours worked in excess of 40	$1.25w$	6	$6\left(1.25w\right)$

Translate the information in the table to an equation and solve.

$$40w + 6\left(1.25w\right) = 782.80$$

$$40w + 7.5w = 782.80$$

$$47.5w = 782.80$$

$$w = 16.48$$

Hernando's normal hourly wage is \$16.48 per hour.

28. Complete a table.

Categories	Rate	Time	Distance
against current	$r-5$	$\dfrac{20}{r-5}$	20
with current	$r+5$	$\dfrac{30}{r+5}$	30

$$\frac{20}{r-5} = \frac{30}{r+5}$$

$$30(r-5) = 20(r+5)$$

$$30r - 150 = 20r + 100$$

$$10r = 250$$

$$r = 25$$

The speed of the boat in still water is 25 mph.

29. Translate "y varies jointly as x and the square of z."

$y = k \cdot x \cdot z^2$ Now, $y = 2xz^2$

$72 = k \cdot 4 \cdot (3)^2$ $y = 2(3)(4)^2$

$72 = k \cdot 4 \cdot 9$ $y = 2(3)(16)$

$72 = 36k$ $y = 96$

$2 = k$

30. Substitute and solve.

$$T = 2\pi\sqrt{\frac{L}{9.8}}$$

$$4\pi = 2\pi\sqrt{\frac{L}{9.8}}$$

$$2 = \sqrt{\frac{L}{9.8}}$$

$$2^2 = \left(\sqrt{\frac{L}{9.8}}\right)^2$$

$$4 = \frac{L}{9.8}$$

$$39.2 = L$$

The length of the pendulum is 39.2 m.

Chapter 9
Quadratic Equations and Functions

Exercise Set 9.1

1. $x^2 = 49$

   $x = \pm\sqrt{49}$

   $x = \pm 7$

3. $y^2 = \dfrac{4}{25}$

   $y = \pm\sqrt{\dfrac{4}{25}}$

   $y = \pm\dfrac{2}{5}$

5. $n^2 = 1.44$

   $n = \pm\sqrt{1.44}$

   $n = \pm 1.2$

7. $z^2 = 45$

   $z = \pm\sqrt{45}$

   $z = \pm\sqrt{9 \cdot 5}$

   $z = \pm 3\sqrt{5}$

9. $w^2 = -25$

   $w = \pm\sqrt{-25}$

   $w = \pm 5i$

11. $n^2 - 7 = 42$

    $n^2 = 49$

    $n = \pm\sqrt{49}$

    $n = \pm 7$

13. $y^2 - 16 = 65$

    $y^2 = 81$

    $y = \pm\sqrt{81}$

    $y = \pm 9$

15. $4n^2 = 36$

    $n^2 = 9$

    $n = \pm\sqrt{9}$

    $n = \pm 3$

17. $25t^2 = 9$

    $t^2 = \dfrac{9}{25}$

    $t = \pm\sqrt{\dfrac{9}{25}}$

    $t = \pm\dfrac{3}{5}$

19. $4h^2 = -16$

    $h^2 = -4$

    $h = \pm\sqrt{-4}$

    $h = \pm 2i$

21. $\dfrac{5}{6}x^2 = \dfrac{24}{5}$

    $x^2 = \dfrac{6}{5} \cdot \dfrac{24}{5}$

    $x^2 = \dfrac{144}{25}$

    $x = \pm\sqrt{\dfrac{144}{25}}$

    $x = \pm\dfrac{12}{5}$

23. $2x^2 + 5 = 21$

    $2x^2 = 16$

    $x^2 = 8$

    $x = \pm\sqrt{8}$

    $x = \pm\sqrt{4 \cdot 2}$

    $x = \pm 2\sqrt{2}$

25. $5y^2 - 7 = -97$

    $5y^2 = -90$

    $y^2 = -18$

    $y = \pm\sqrt{-18}$

    $y = \pm\sqrt{-9 \cdot 2}$

    $y = \pm 3i\sqrt{2}$

27. $\dfrac{3}{4}y^2 - 5 = 3$

$$\dfrac{3}{4}y^2 = 8$$

$$y^2 = 8 \cdot \dfrac{4}{3}$$

$$y^2 = \dfrac{32}{3}$$

$$y = \pm\sqrt{\dfrac{32}{3}} = \pm\dfrac{\sqrt{32}}{\sqrt{3}} \cdot \dfrac{\sqrt{3}}{\sqrt{3}}$$

$$= \pm\dfrac{\sqrt{96}}{3} = \pm\dfrac{\sqrt{16 \cdot 6}}{3} = \pm\dfrac{4\sqrt{6}}{3}$$

29. $0.2t^2 - 0.5 = 0.012$

$$0.2t^2 = 0.512$$

$$t^2 = 2.56$$

$$t = \pm\sqrt{2.56}$$

$$t = \pm 1.6$$

31. $(x+8)^2 = 49$

$$x + 8 = \pm\sqrt{49}$$

$$x + 8 = \pm 7$$

$$x = -8 \pm 7$$

$$x = -8 + 7 = -1$$

$$x = -8 - 7 = -15$$

33. $(5n-3)^2 = 16$

$$5n - 3 = \pm\sqrt{16}$$

$$5n - 3 = \pm 4$$

$$5n = 3 \pm 4$$

$$n = \dfrac{3 \pm 4}{5}$$

$$n = \dfrac{3+4}{5} = \dfrac{7}{5}$$

$$n = \dfrac{3-4}{5} = -\dfrac{1}{5}$$

35. $(m-8)^2 = -16$

$$m - 8 = \pm\sqrt{-16}$$

$$m - 8 = \pm 4i$$

$$m = 8 \pm 4i$$

37. $(4k-1)^2 = 40$

$$4k - 1 = \pm\sqrt{40}$$

$$4k - 1 = \pm\sqrt{4 \cdot 10}$$

$$4k - 1 = \pm 2\sqrt{10}$$

$$4k = 1 \pm 2\sqrt{10}$$

$$k = \dfrac{1 \pm 2\sqrt{10}}{4}$$

39. $(m-7)^2 = -12$

$$m - 7 = \pm\sqrt{-12}$$

$$m - 7 = \pm\sqrt{-4 \cdot 3}$$

$$m - 7 = \pm 2i\sqrt{3}$$

$$m = 7 \pm 2i\sqrt{3}$$

41. $\left(y - \dfrac{3}{4}\right)^2 = \dfrac{9}{16}$

$$y - \dfrac{3}{4} = \pm\sqrt{\dfrac{9}{16}}$$

$$y - \dfrac{3}{4} = \pm\dfrac{3}{4}$$

$$y = \dfrac{3}{4} \pm \dfrac{3}{4}$$

$$y = \dfrac{3}{4} + \dfrac{3}{4} = \dfrac{3}{2}$$

$$y = \dfrac{3}{4} - \dfrac{3}{4} = 0$$

43. $\left(\dfrac{5}{9}d - \dfrac{1}{2}\right)^2 = \dfrac{1}{36}$

$$\dfrac{5}{9}d - \dfrac{1}{2} = \pm\sqrt{\dfrac{1}{36}}$$

$$\dfrac{5}{9}d - \dfrac{1}{2} = \pm\dfrac{1}{6}$$

$$\dfrac{5}{9}d = \dfrac{1}{2} \pm \dfrac{1}{6}$$

$$d = \dfrac{9}{5} \cdot \left(\dfrac{1}{2} \pm \dfrac{1}{6}\right)$$

$$d = \dfrac{9}{5} \cdot \left(\dfrac{1}{2} + \dfrac{1}{6}\right) = \dfrac{6}{5}$$

$$d = \dfrac{9}{5} \cdot \left(\dfrac{1}{2} - \dfrac{1}{6}\right) = \dfrac{3}{5}$$

45. $(0.4x + 3.8)^2 = 2.56$

$$0.4x + 3.8 = \pm\sqrt{2.56}$$
$$0.4x + 3.8 = \pm 1.6$$
$$0.4x = -3.8 \pm 1.6$$
$$x = \frac{-3.8 \pm 1.6}{0.4}$$
$$x = \frac{-3.8 + 1.6}{0.4} = -5.5$$
$$x = \frac{-3.8 - 1.6}{0.4} = -13.5$$

47. Mistake: Gave only the positive solution.
    Correct: ± 7

49. Mistake: Changed -6 to 6.
    Correct: $5 \pm i\sqrt{6}$

51. $x^2 = 96$

$$x = \pm\sqrt{96}$$
$$x = \pm\sqrt{16 \cdot 6}$$
$$x = \pm 4\sqrt{6} \approx \pm 9.798$$

53. $y^2 - 15 = 5$

$$y^2 = 20$$
$$y = \pm\sqrt{20}$$
$$y = \pm\sqrt{4 \cdot 5}$$
$$y = \pm 2\sqrt{5} \approx \pm 4.472$$

55. $(n-6)^2 = 15$

$$n - 6 = \pm\sqrt{15}$$
$$n = 6 \pm \sqrt{15} \approx 9.873,\ 2.127$$

57. a) $\left(\dfrac{14}{2}\right)^2 = 7^2 = 49$

$$x^2 + 14x + 49$$

b) $(x+7)^2$

59. a) $\left(\dfrac{-10}{2}\right)^2 = (-5)^2 = 25$

$$n^2 - 10n + 25$$

b) $(n-5)^2$

61. a) $\left(\dfrac{-9}{2}\right)^2 = \dfrac{81}{4}$

$$y^2 - 9y + \frac{81}{4}$$

b) $\left(y - \dfrac{9}{2}\right)^2$

63. a) $\left(\dfrac{-\frac{2}{3}}{2}\right)^2 = \left(-\dfrac{1}{3}\right)^2 = \dfrac{1}{9}$

$$s^2 - \frac{2}{3}s + \frac{1}{9}$$

b) $\left(s - \dfrac{1}{3}\right)^2$

65. $\left(\dfrac{2}{2}\right)^2 = 1^2 = 1$

$$w^2 + 2w + 1 = 15 + 1$$
$$(w+1)^2 = 16$$
$$w + 1 = \pm 4$$
$$w = -1 \pm 4$$
$$w = -1 + 4 = 3$$
$$= -1 - 4 = -5$$

67. $\left(\dfrac{-2}{2}\right)^2 = (-1)^2 = 1$

$$r^2 - 2r + 1 = -50 + 1$$
$$(r-1)^2 = -49$$
$$r - 1 = \pm 7i$$
$$r = 1 \pm 7i$$

69.
$$\left(\frac{-9}{2}\right)^2 = \frac{81}{4}$$

$$k^2 - 9k + \frac{81}{4} = -18 + \frac{81}{4}$$

$$\left(k - \frac{9}{2}\right)^2 = \frac{9}{4}$$

$$k - \frac{9}{2} = \pm\frac{3}{2}$$

$$k = \frac{9}{2} \pm \frac{3}{2}$$

$$k = \frac{9}{2} + \frac{3}{2} = \frac{12}{2} = 6$$

$$k = \frac{9}{2} - \frac{3}{2} = \frac{6}{2} = 3$$

71.
$$\left(\frac{-2}{2}\right)^2 = (-1)^2 = 1$$

$$b^2 - 2b + 1 = 16 + 1$$

$$(b-1)^2 = 17$$

$$b - 1 = \pm\sqrt{17}$$

$$b = 1 \pm \sqrt{17}$$

73.
$$h^2 - 6h = -29$$

$$\left(-\frac{6}{2}\right)^2 = (-3)^2 = 9$$

$$h^2 - 6h + 9 = -29 + 9$$

$$(h-3)^2 = -20$$

$$h - 3 = \pm\sqrt{-20}$$

$$h - 3 = \pm 2i\sqrt{5}$$

$$h = 3 \pm 2i\sqrt{5}$$

75.
$$\left(\frac{\frac{1}{2}}{2}\right)^2 = \left(\frac{1}{4}\right)^2 = \frac{1}{16}$$

$$u^2 + \frac{1}{2}u + \frac{1}{16} = \frac{3}{2} + \frac{1}{16}$$

$$\left(u + \frac{1}{4}\right)^2 = \frac{25}{16}$$

$$u + \frac{1}{4} = \pm\frac{5}{4}$$

$$u = -\frac{1}{4} \pm \frac{5}{4}$$

$$u = -\frac{1}{4} + \frac{5}{4} = 1$$

$$u = -\frac{1}{4} - \frac{5}{4} = -\frac{3}{2}$$

77.
$$\frac{4x^2}{4} + \frac{16x}{4} = \frac{9}{4}$$

$$x^2 + 4x = \frac{9}{4}$$

$$\left(\frac{b}{2}\right)^2 = \left(\frac{4}{2}\right)^2 = 2^2 = 4$$

$$x^2 + 4x + 4 = \frac{9}{4} + 4$$

$$(x+2)^2 = \frac{25}{4}$$

$$x + 2 = \pm\sqrt{\frac{25}{4}}$$

$$x + 2 = \pm\frac{5}{2}$$

$$x = -2 + \frac{5}{2} = \frac{1}{2}$$

$$= -2 - \frac{5}{2} = -\frac{9}{2}$$

79.
$$\frac{9x^2}{9} - \frac{18x}{9} = -\frac{5}{9}$$

$$x^2 - 2x = -\frac{5}{9}$$

$$x^2 - 2x + 1 = -\frac{5}{9} + 1$$

$$(x-1)^2 = \frac{4}{9}$$

$$x - 1 = \pm\sqrt{\frac{4}{9}}$$

$$x = 1 \pm \frac{2}{3}$$

$$x = 1 + \frac{2}{3} = \frac{5}{3}$$

$$x = 1 - \frac{2}{3} = \frac{1}{3}$$

81. $n^2 - \dfrac{1}{2}n = \dfrac{3}{2}$

$$\left(\dfrac{\frac{-1}{2}}{2}\right)^2 = \left(-\dfrac{1}{4}\right)^2 = \dfrac{1}{16}$$

$$n^2 - \dfrac{1}{2}n + \dfrac{1}{16} = \dfrac{3}{2} + \dfrac{1}{16}$$

$$\left(n - \dfrac{1}{4}\right)^2 = \dfrac{25}{16}$$

$$n - \dfrac{1}{4} = \pm\dfrac{5}{4}$$

$$n = \dfrac{1}{4} \pm \dfrac{5}{4}$$

$$n = \dfrac{1}{4} + \dfrac{5}{4} = \dfrac{3}{2}$$

$$n = \dfrac{1}{4} - \dfrac{5}{4} = -1$$

83. $\dfrac{4x^2}{4} + \dfrac{16x}{4} = \dfrac{7}{4}$

$$x^2 + 4x = \dfrac{7}{4}$$

$$x^2 + 4x + 4 = \dfrac{7}{4} + 4$$

$$(x + 2)^2 = \dfrac{23}{4}$$

$$x + 2 = \pm\sqrt{\dfrac{23}{4}}$$

$$x + 2 = \pm\dfrac{\sqrt{23}}{2}$$

$$x = -2 \pm \dfrac{\sqrt{23}}{2}$$

85. $k^2 + \dfrac{1}{5}k = \dfrac{2}{5}$

$$\left(\dfrac{\frac{1}{5}}{2}\right)^2 = \left(\dfrac{1}{10}\right)^2 = \dfrac{1}{100}$$

$$k^2 + \dfrac{1}{5}k + \dfrac{1}{100} = \dfrac{2}{5} + \dfrac{1}{100}$$

$$\left(k + \dfrac{1}{10}\right)^2 = \dfrac{41}{100}$$

$$k + \dfrac{1}{10} = \pm\sqrt{\dfrac{41}{100}}$$

$$k + \dfrac{1}{10} = \pm\dfrac{\sqrt{41}}{10}$$

$$k = \dfrac{-1 \pm \sqrt{41}}{10}$$

87. $3a^2 + 4a = 8$

$$a^2 + \dfrac{4}{3}a = \dfrac{8}{3}$$

$$\left(\dfrac{4}{3} \cdot \dfrac{1}{2}\right)^2 = \left(\dfrac{2}{3}\right)^2 = \dfrac{4}{9}$$

$$a^2 + \dfrac{4}{3}a + \dfrac{4}{9} = \dfrac{8}{3} + \dfrac{4}{9}$$

$$\left(a + \dfrac{2}{3}\right)^2 = \dfrac{28}{9}$$

$$a + \dfrac{2}{3} = \pm\sqrt{\dfrac{28}{9}}$$

$$a + \dfrac{2}{3} = \pm\dfrac{2\sqrt{7}}{3}$$

$$a = -\dfrac{2}{3} \pm \dfrac{2\sqrt{7}}{3}$$

$$a = \dfrac{-2 \pm 2\sqrt{7}}{3}$$

89. Mistake: Did not divide by 3 so that x^2 has a coefficient of 1. Then wrote an incorrect factored form.
Correct:

$$3x^2 + 4x = 2$$

$$x^2 + \dfrac{4}{3}x = \dfrac{2}{3}$$

$$x^2 + \dfrac{4}{3}x + \dfrac{4}{9} = \dfrac{2}{3} + \dfrac{4}{9}$$

$$\left(x + \dfrac{2}{3}\right)^2 = \dfrac{10}{9}$$

$$x + \dfrac{2}{3} = \pm\sqrt{\dfrac{10}{9}}$$

$$x = -\dfrac{2}{3} \pm \dfrac{\sqrt{10}}{3}$$

$$x = \dfrac{-2 \pm \sqrt{10}}{3}$$

91.
$$A = s^2$$
$$\sqrt{A} = \sqrt{s^2}$$
$$\sqrt{A} = s$$
$$\sqrt{196} = s$$
$$14 = s$$
The length of each side is 14 inches.

93.
$$V = lwh$$
$$144 = 8 \cdot h \cdot 2h$$
$$144 = 16h^2$$
$$9 = h^2$$
$$\sqrt{9} = \sqrt{h^2}$$
$$3 = h$$
height: 3 ft.

width: $2(3) = 6$ ft.

95.
$$A = \pi r^2$$
$$23,256.25\pi = \pi r^2$$
$$23,256.25 = r^2$$
$$\sqrt{23,256.25} = \sqrt{r^2}$$
$$152.5 = r$$
So, the diameter is $2(152.5) = 305$ m

97. Translate to the equation $(x-2)^2 = 256$ and solve.
$$(x-2)^2 = 256$$
$$\sqrt{(x-2)^2} = \sqrt{256}$$
$$x - 2 = 16$$
$$x = 18$$
The length x is 18 inches.

99. Let the length be x and the width be $x - 4$.
$$x^2 + (x-4)^2 = 20^2$$
$$x^2 + x^2 - 8x + 16 = 400$$
$$2x^2 - 8x = 384$$
$$x^2 - 4x = 192$$
$$x^2 - 4x + 4 = 192 + 4$$
$$(x-2)^2 = 196$$
$$x - 2 = \pm\sqrt{196}$$
$$x - 2 = \pm 14$$
$$x = 2 \pm 14$$
$$x = 2 - 14 = -12$$
$$x = 2 + 14 = 16$$
Length must be positive, so disregard the negative solution. The length is 16 ft. and the width is $16 - 4 = 12$ ft.

101. a) Subtract the area of the interior rectangle from the area of the exterior rectangle.
$$3l \cdot 2l - 8l = 1230$$
$$6l^2 - 8l = 1230$$
$$l^2 - \frac{4}{3}l = 205$$
$$l^2 - \frac{4}{3}l + \frac{4}{9} = 205 + \frac{4}{9}$$
$$\left(l - \frac{2}{3}\right)^2 = \frac{1849}{9}$$
$$l - \frac{2}{3} = \pm\sqrt{\frac{1849}{9}}$$
$$l - \frac{2}{3} = \pm\frac{43}{3}$$
$$l = \frac{2}{3} \pm \frac{43}{3}$$
$$l = \frac{2}{3} \pm \frac{43}{3} = \frac{45}{3} = 15$$
$$l = \frac{2}{3} \pm \frac{43}{3} = -\frac{41}{3}$$
Because lengths cannot be negative, disregard the negative solution. The value of l is 15 cm.

b) length: $3 \cdot 15 = 45$ cm

width: $2 \cdot 15 = 30$ cm

103. $9 = 16t^2$

$$\frac{9}{16} = t^2$$

$$\sqrt{\frac{9}{16}} = t$$

$$\frac{3}{4} = t$$

The cover takes $\dfrac{3}{4}$ sec. to hit the floor.

105. $400 = \dfrac{1}{2} \cdot 50 \cdot v^2$

$$400 = 25v^2$$

$$16 = v^2$$

$$\sqrt{16} = \sqrt{v^2}$$

$$4 = v$$

The velocity is 4 m/sec.

Review Exercises

1. $x^2 - 10x + 25 = (x - 5)^2$

2. $x^2 + 6x + 9 = (x + 3)^2$

3. $\sqrt{36x^2} = 6x$

4. $\sqrt{(2x+3)^2} = 2x + 3$

5. $\dfrac{-4 + \sqrt{8^2 - 4(2)(5)}}{2(2)} = \dfrac{-4 + \sqrt{64 - 40}}{4}$

$$= \frac{-4 + \sqrt{24}}{4}$$

$$= \frac{-4 + \sqrt{4 \cdot 6}}{4}$$

$$= \frac{-4 + 2\sqrt{6}}{4}$$

$$= \frac{-2 + \sqrt{6}}{2} \text{ or } -1 + \frac{\sqrt{6}}{2}$$

6. $\dfrac{-6 - \sqrt{6^2 - 4(5)(2)}}{2(5)} = \dfrac{-6 - \sqrt{36 - 40}}{10}$

$$= \frac{-6 - \sqrt{-4}}{10}$$

$$= \frac{-6 - 2i}{10}$$

$$= \frac{-3 - i}{5} \text{ or } -\frac{3}{5} - \frac{1}{5}i$$

Exercise Set 9.2

1. $x^2 - 3x + 7 = 0$
     $a = 1$
     $b = -3$
     $c = 7$

3. $3x^2 - 9x - 4 = 0$
     $a = 3$
     $b = -9$
     $c = -4$

5. $1.5x^2 - x + 0.2 = 0$
     $a = 1.5$
     $b = -1$
     $c = 0.2$

7. $\dfrac{1}{2}x^2 + \dfrac{3}{4}x - 6 = 0$

     $a = \dfrac{1}{2}$

     $b = \dfrac{3}{4}$

     $c = -6$

9. $x^2 + 9x + 20 = 0$
     $a = 1, b = 9, c = 20$

$$x = \frac{-(9) \pm \sqrt{(9)^2 - 4(1)(20)}}{2(1)}$$

$$= \frac{-9 \pm \sqrt{81 - 80}}{2}$$

$$= \frac{-9 \pm \sqrt{1}}{2}$$

$$= \frac{-9 \pm 1}{2}$$

$$= -4, -5$$

11. $4x^2 + 5x - 6 = 0$
$a = 4, b = 5, c = -6$

$$x = \frac{-(5) \pm \sqrt{(5)^2 - 4(4)(-6)}}{2(4)}$$

$$= \frac{-5 \pm \sqrt{25 + 96}}{8}$$

$$= \frac{-5 \pm \sqrt{121}}{8}$$

$$= \frac{-5 \pm 11}{8}$$

$$= \frac{-5 + 11}{8} = \frac{3}{4}$$

$$= \frac{-5 - 11}{8} = -2$$

13. $x^2 - 9x = 0$
$a = 1, b = -9, c = 0$

$$x = \frac{-(-9) \pm \sqrt{(-9)^2 - 4(1)(0)}}{2(1)}$$

$$= \frac{9 \pm \sqrt{81}}{2}$$

$$= \frac{9 \pm 9}{2}$$

$$= \frac{9 + 9}{2} = 9$$

$$= \frac{9 - 9}{2} = 0$$

15. $x^2 - 8x + 16 = 0$
$a = 1, b = -8, c = 16$

$$x = \frac{-(-8) \pm \sqrt{(-8)^2 - 4(1)(16)}}{2(1)}$$

$$= \frac{8 \pm \sqrt{64 - 64}}{2}$$

$$= \frac{8 \pm \sqrt{0}}{2}$$

$$= \frac{8}{2}$$

$$= 4$$

17. $3x^2 + 4x - 4 = 0$
$a = 3, b = 4, c = -4$

$$x = \frac{-(4) \pm \sqrt{(4)^2 - 4(3)(-4)}}{2(3)}$$

$$= \frac{-4 \pm \sqrt{16 + 48}}{6}$$

$$= \frac{-4 \pm \sqrt{64}}{6}$$

$$= \frac{-4 \pm 8}{6}$$

$$= \frac{-4 + 8}{6} = \frac{2}{3}$$

$$= \frac{-4 - 8}{6} = -2$$

19. $x^2 - 2x + 2 = 0$
$a = 1, b = -2, c = 2$

$$x = \frac{-(-2) \pm \sqrt{(-2)^2 - 4(1)(2)}}{2(1)}$$

$$= \frac{2 \pm \sqrt{4 - 8}}{2}$$

$$= \frac{2 \pm \sqrt{-4}}{2}$$

$$= \frac{2 \pm 2i}{2}$$

$$= 1 \pm i$$

21. $x^2 - x - 1 = 0$
$a = 1, b = -1, c = -1$

$$x = \frac{-(-1) \pm \sqrt{(-1)^2 - 4(1)(-1)}}{2(1)}$$

$$= \frac{1 \pm \sqrt{1 + 4}}{2}$$

$$= \frac{1 \pm \sqrt{5}}{2}$$

23. $3x^2 + 10x + 5 = 0$
    $a = 3, b = 10, c = 5$

$$x = \frac{-(10) \pm \sqrt{(10)^2 - 4(3)(5)}}{2(3)}$$

$$= \frac{-10 \pm \sqrt{100 - 60}}{6}$$

$$= \frac{-10 \pm \sqrt{40}}{6}$$

$$= \frac{-10 \pm 2\sqrt{10}}{6}$$

$$= \frac{-5 \pm \sqrt{10}}{3}$$

25. $-4x^2 + 6x - 5 = 0$
    $a = -4, b = 6, c = -5$

$$x = \frac{-(6) \pm \sqrt{(6)^2 - 4(-4)(-5)}}{2(-4)}$$

$$= \frac{-6 \pm \sqrt{36 - 80}}{-8}$$

$$= \frac{-6 \pm \sqrt{-44}}{-8}$$

$$= \frac{-6 \pm 2i\sqrt{11}}{-8}$$

$$= \frac{3 \pm i\sqrt{11}}{4}$$

27. $18x^2 + 15x + 2 = 0$
    $a = 18, b = 15, c = 2$

$$x = \frac{-(15) \pm \sqrt{(15)^2 - 4(18)(2)}}{2(18)}$$

$$= \frac{-15 \pm \sqrt{225 - 144}}{36}$$

$$= \frac{-15 \pm \sqrt{81}}{36}$$

$$= \frac{-15 \pm 9}{36}$$

$$= \frac{-15 + 9}{36} = -\frac{1}{6}$$

$$= \frac{-15 - 9}{36} = -\frac{2}{3}$$

29. $3x^2 - 4x + 3 = 0$
    $a = 3, b = -4, c = 3$

$$x = \frac{-(-4) \pm \sqrt{(-4)^2 - 4(3)(3)}}{2(3)}$$

$$= \frac{4 \pm \sqrt{16 - 36}}{6}$$

$$= \frac{4 \pm \sqrt{-20}}{6}$$

$$= \frac{4 \pm 2i\sqrt{5}}{6}$$

$$= \frac{2 \pm i\sqrt{5}}{3}$$

31. $6x^2 - 3x - 4 = 0$
    $a = 6, b = -3, c = -4$

$$x = \frac{-(-3) \pm \sqrt{(-3)^2 - 4(6)(-4)}}{2(6)}$$

$$= \frac{3 \pm \sqrt{9 + 96}}{12}$$

$$= \frac{3 \pm \sqrt{105}}{12}$$

33. $2x^2 + 0.1x - 0.03 = 0$
    $a = 2, b = 0.1, c = -0.03$

$$x = \frac{-(0.1) \pm \sqrt{(0.1)^2 - 4(2)(-0.03)}}{2(2)}$$

$$= \frac{-0.1 \pm \sqrt{0.01 + 0.24}}{4}$$

$$= \frac{-0.1 \pm \sqrt{0.25}}{4}$$

$$= \frac{-0.1 \pm 0.5}{4}$$

$$= \frac{-0.1 + 0.5}{4} = 0.1$$

$$= \frac{-0.1 - 0.5}{4} = -0.15$$

35. $x^2 + \dfrac{1}{2}x - 3 = 0$

$a = 1, b = \dfrac{1}{2}, c = -3$

$x = \dfrac{-\left(\dfrac{1}{2}\right) \pm \sqrt{\left(\dfrac{1}{2}\right)^2 - 4(1)(-3)}}{2(1)}$

$= \dfrac{-\dfrac{1}{2} \pm \sqrt{\dfrac{1}{4} + 12}}{2}$

$= \dfrac{-\dfrac{1}{2} \pm \sqrt{\dfrac{49}{4}}}{2}$

$= \dfrac{-\dfrac{1}{2} \pm \dfrac{7}{2}}{2}$

$= \dfrac{-\dfrac{1}{2} + \dfrac{7}{2}}{2} = \dfrac{3}{2}$

$= \dfrac{-\dfrac{1}{2} - \dfrac{7}{2}}{2} = -2$

37. Use: $36x^2 - 49 = 0$
    $a = 36, b = 0, c = -49$

$x = \dfrac{-(0) \pm \sqrt{(0)^2 - 4(36)(-49)}}{2(36)}$

$= \dfrac{0 \pm \sqrt{0 + 7056}}{72}$

$x = \dfrac{0 \pm 84}{72}$

$= \dfrac{\pm 84}{72}$

$= \pm \dfrac{7}{6}$

39. Use: $x^2 - 2x + 3 = 0$
    $a = 1, b = -2, c = 3$

$x = \dfrac{-(-2) \pm \sqrt{(-2)^2 - 4(1)(3)}}{2(1)}$

$= \dfrac{2 \pm \sqrt{4 - 12}}{2}$

$= \dfrac{2 \pm \sqrt{-8}}{2}$

$= \dfrac{2 \pm 2i\sqrt{2}}{2} = 1 \pm i\sqrt{2}$

41. Use: $100x^2 + 6x - 50 = 0$
    $a = 100, b = 6, c = -50$

$x = \dfrac{-(6) \pm \sqrt{(6)^2 - 4(100)(-50)}}{2(100)}$

$= \dfrac{-6 \pm \sqrt{36 + 20,000}}{200}$

$= \dfrac{-6 \pm \sqrt{20,036}}{200}$

$= \dfrac{-6 \pm 2\sqrt{5009}}{200}$

$= \dfrac{-3 \pm \sqrt{5009}}{100}$

$\approx -0.738, 0.678$

43. Use: $12x^2 - 6x + 5 = 0$
    $a = 12, b = -6, c = 5$

$x = \dfrac{-(-6) \pm \sqrt{(-6)^2 - 4(12)(5)}}{2(12)}$

$= \dfrac{6 \pm \sqrt{36 - 240}}{24}$

$= \dfrac{6 \pm \sqrt{-204}}{24}$

$= \dfrac{6 \pm 2i\sqrt{51}}{24}$

$= \dfrac{3 \pm i\sqrt{51}}{12}$

45. Mistake: Did not evaluate $-b$.
    Correct:

$\dfrac{-(-7) \pm \sqrt{(-7)^2 - (4)(3)(1)}}{2(3)} = \dfrac{7 \pm \sqrt{49 - 12}}{6}$

$= \dfrac{7 \pm \sqrt{37}}{6}$

47. Mistake: The result was not completely simplified.

    Correct: $\dfrac{2\pm\sqrt{-8}}{2}=\dfrac{2\pm 2i\sqrt{2}}{2}=1+i\sqrt{2}$

49. $x^2+10x+25=0$

    $b^2-4ac=(10)^2-4(1)(25)=100-100=0$

    1 rational solution

51. $\dfrac{1}{4}x^2-4x+4=0$

    $b^2-4ac=(-4)^2-4\left(\dfrac{1}{4}\right)(4)=16-4=12$

    $12>0$ and 12 is not a perfect square
    2 irrational solutions

53. $3x^2+10x-8=0$

    $b^2-4ac=(10)^2-4(3)(-8)=100+96=196$

    $196>0$ and 196 is a perfect square $\left(196=14^2\right)$

    2 rational solutions

55. $x^2-x+3=0$

    $b^2-4ac=(-1)^2-4(1)(3)=1-12=-11$

    $-11<0$
    2 nonreal complex solutions

57. $x^2-6x+6=0$

    $b^2-4ac=(-6)^2-4(1)(6)=36-24=12$

    $12>0$ and 12 is not a perfect square
    2 irrational solutions

59. Square root principle or factoring

    $x^2-81=0$

    $\quad x^2=81$

    $\quad\; x=\pm\sqrt{81}$

    $\quad\; x=\pm 9$

61. Quadratic formula

    $2x^2-4x-3=0$

    $x=\dfrac{-(-4)\pm\sqrt{(-4)^2-4(2)(-3)}}{2(2)}$

    $=\dfrac{4\pm\sqrt{16+24}}{4}$

    $=\dfrac{4\pm\sqrt{40}}{4}$

    $=\dfrac{4\pm 2\sqrt{10}}{4}$

    $=\dfrac{2\pm\sqrt{10}}{2}$

63. Factoring

    $x^2+6x=0$

    $x(x+6)=0$

    $x=0\;$ or $\;x+6=0$

    $x=-6$

65. Square root principle

    $(x+7)^2=40$

    $\quad x+7=\pm\sqrt{40}$

    $\quad x+7=\pm 2\sqrt{10}$

    $\qquad x=-7\pm 2\sqrt{10}$

67. Quadratic formula

    $x^2-8x+19=0$

    $x=\dfrac{-(-8)\pm\sqrt{(-8)^2-4(1)(19)}}{2(1)}$

    $x=\dfrac{8\pm\sqrt{64-76}}{2}$

    $x=\dfrac{8\pm\sqrt{-12}}{2}$

    $x=\dfrac{8\pm 2i\sqrt{3}}{2}$

    $x=4\pm i\sqrt{3}$

69. x-intercepts: $\quad x^2 - x - 2 = 0$

$$(x-2)(x+1) = 0$$

$$x - 2 = 0 \quad \text{or} \quad x + 1 = 0$$

$$x = 2 \qquad x = -1$$

$$(2,0),(-1,0)$$

y-intercept: $y = 0^2 - 0 - 2$

$$y = -2$$

$$(0,-2)$$

71. x-intercept: $4x^2 - 12x + 9 = 0$

$$(2x-3)^2 = 0$$

$$2x - 3 = \pm\sqrt{0}$$

$$2x - 3 = 0$$

$$2x = 3$$

$$x = \frac{3}{2}$$

$$\left(\frac{3}{2}, 0\right)$$

y-intercept: $y = 4(0)^2 - 12(0) + 9$

$$y = 9$$

$$(0,9)$$

73. x-intercepts: $\quad 2x^2 + 15x - 8 = 0$

$$(2x-1)(x+8) = 0$$

$$2x - 1 = 0 \quad \text{or} \quad x + 8 = 0$$

$$x = \frac{1}{2} \qquad x = -8$$

$$\left(\frac{1}{2}, 0\right),(-8,0)$$

y-intercept: $y = 2 \cdot 0^2 + 15 \cdot 0 - 8$

$$y = -8$$

$$(0,-8)$$

75. x-intercepts: $-2x^2 + 3x - 6 = 0$

$$x = \frac{-3 \pm \sqrt{3^2 - 4(-2)(-6)}}{2(-2)}$$

$$= \frac{-3 \pm \sqrt{9 - 48}}{-4}$$

$$= \frac{-3 \pm \sqrt{-39}}{-4}$$

not real: no x-intercepts

y-intercept: $y = -2 \cdot 0^2 + 3 \cdot 0 - 6$

$$y = -6$$

$$(0,-6)$$

77. Let x represent the first positive integer and $x + 1$ represent the next consecutive integer.

$$x^2 + 5(x+1) = 71$$

$$x^2 + 5x + 5 = 71$$

$$x^2 + 5x - 66 = 0$$

$$(x+11)(x-6) = 0$$

$$x + 11 = 0 \qquad \text{or} \quad x - 6 = 0$$

$$x = \cancel{-11} \qquad\qquad x = 6$$

The integers are 6 and 6 + 1 = 7.

79. Let x be the first integer, $x + 1$ be the next consecutive integer, and $x + 2$ be the third consecutive integer.

$$x^2 + (x+1)^2 = (x+2)^2$$

$$x^2 + x^2 + 2x + 1 = x^2 + 4x + 4$$

$$2x^2 + 2x + 1 = x^2 + 4x + 4$$

$$x^2 - 2x - 3 = 0$$

$$(x-3)(x+1) = 0$$

$$x - 3 = 0 \quad \text{or} \quad x + 1 = 0$$

$$x = 3 \qquad\qquad x = \cancel{-1}$$

The sides are 3, 3 + 1 = 4, and 3 + 2 = 5.

81. If the width is w, then the length is $(3w - 3.5)$.

$$w(3w - 3.5) = 34$$

$$3w^2 - 3.5w - 34 = 0$$

$$w = \frac{-(-3.5) \pm \sqrt{(-3.5)^2 - 4(3)(-34)}}{2(3)}$$

$$w = \frac{3.5 \pm \sqrt{12.25 + 408}}{6}$$

$$w = \frac{3.5 \pm \sqrt{420.25}}{6}$$

$$w = \frac{3.5 \pm 20.5}{6}$$

$$w = \frac{3.5 + 20.5}{6} = 4$$

$$w = \frac{3.5 - 20.5}{6} = \cancel{-2.83}$$

The width is 4 ft. and the length is 3(4) − 3.5 = 8.5 ft.

83. a) $22x = 1.5x \cdot x + 0.5x \cdot 8$

$22x = 1.5x^2 + 4x$

$0 = 1.5x^2 - 18x$

$0 = x(1.5x - 18)$

$0 = 1.5x - 18$ or $\cancel{0} = x$

$18 = 1.5x$

$12 \text{ ft.} = x$

b)

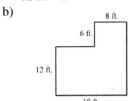

85. $460 = \dfrac{1}{2}(-32.2)t^2 + 48t + 485$

$0 = 16.1t^2 - 48t - 25$

$t = \dfrac{-(-48) \pm \sqrt{(-48)^2 - 4(16.1)(-25)}}{2(16.1)}$

$= \dfrac{48 \pm \sqrt{2304 + 1610}}{32.2}$

$= \dfrac{48 + \sqrt{3914}}{32.2} \approx 3.43$

$= \dfrac{48 - \sqrt{3914}}{32.2} \approx \cancel{-0.452}$

He was in flight for approximately 3.43 seconds.

87. a) $h = \dfrac{1}{2}(-9.8)t^2 + 0t + 10$

$h = -4.9t^2 + 10$

b) $5 = -4.9t^2 + 10$

$4.9t^2 = 5$

$t^2 = \dfrac{5}{4.9}$

$t \approx \pm\sqrt{1.02}$

$t \approx 1.01,\ \cancel{-1.01}$

It takes approximately 1.01 second.

c) $0 = -4.9t^2 + 10$

$4.9t^2 = 10$

$t^2 = \dfrac{10}{4.9}$

$t \approx \pm\sqrt{2.04}$

$t \approx 1.43,\ \cancel{-1.43}$

It takes approximately 1.43 seconds.

89. $0.5n^2 + 2.5n = 4.5n + 16$

$0.5n^2 - 2n - 16 = 0$

$5n^2 - 20n - 160 = 0$

$n^2 - 4n - 32 = 0$

$(n - 8)(n + 4) = 0$

$n = 8,\ \cancel{-4}$

The break-even point is 8000 units.

91. a) $b^2 - 4ac = 0$

$(-5)^2 - 4(2)c = 0$

$25 - 8c = 0$

$-8c = -25$

$c = \dfrac{25}{8}$

b) $b^2 - 4ac > 0$

$(-5)^2 - 4(2)c > 0$

$25 - 8c > 0$

$-8c > -25$

$c < \dfrac{25}{8}$

c) $b^2 - 4ac < 0$

$(-5)^2 - 4(2)c < 0$

$25 - 8c < 0$

$-8c < -25$

$c > \dfrac{25}{8}$

93. a) $b^2 - 4ac = 0$

$12^2 - 4a(8) = 0$

$144 - 32a = 0$

$-32a = -144$

$a = \dfrac{-144}{-32} = \dfrac{9}{2}$

b) $b^2 - 4ac > 0$

$12^2 - 4a(8) > 0$

$144 - 32a > 0$

$-32a > -144$

$a < \dfrac{-144}{-32}$

$a < \dfrac{9}{2}$

c) $b^2 - 4ac < 0$

$12^2 - 4a(8) < 0$

$144 - 32a < 0$

$-32a < -144$

$a > \dfrac{-144}{-32}$

$a > \dfrac{9}{2}$

Review Exercises

1. $u^2 - 9u + 14 = (u - 7)(u - 2)$

2. $3u^2 - 2u - 16 = (3u - 8)(u + 2)$

3. $\left(x^2\right)^2 = x^{2 \cdot 2} = x^4$

4. $\left(x^{1/3}\right)^2 = x^{(1/3) \cdot 2} = x^{2/3}$

5. $5u^2 + 13u - 6 = 0$

$(5u - 2)(u + 3) = 0$

$5u - 2 = 0 \quad \text{or} \quad u + 3 = 0$

$\quad 5u = 2 \qquad\qquad u = -3$

$\quad u = \dfrac{2}{5}$

6. $\dfrac{7}{3u} = \dfrac{5}{u} - \dfrac{1}{u-5}$

$3u(u-5) \cdot \dfrac{7}{3u} = 3u(u-5) \cdot \left[\dfrac{5}{u} - \dfrac{1}{u-5}\right]$

$(u-5)7 = 3(u-5) \cdot 5 - 3u \cdot 1$

$7u - 35 = 15u - 75 - 3u$

$7u - 35 = 12u - 75$

$-5u = -40$

$u = 8$

Exercise Set 9.3

1. $\dfrac{60}{x+2} = \dfrac{60}{x} - 5$

$x(x+2)\left(\dfrac{60}{x+2}\right) = x(x+2)\left(\dfrac{60}{x} - 5\right)$

$60x = 60(x+2) - 5x(x+2)$

$60x = 60x + 120 - 5x^2 - 10x$

$5x^2 + 10x - 120 = 0$

$x^2 + 2x - 24 = 0$

$(x+6)(x-4) = 0$

$x + 6 = 0 \quad \text{or} \quad x - 4 = 0$

$\quad x = -6 \qquad\qquad x = 4$

3. $\dfrac{1}{x} + \dfrac{1}{x+2} = \dfrac{3}{4}$

$4x(x+2)\left(\dfrac{1}{x} + \dfrac{1}{x+2}\right) = 4x(x+2)\left(\dfrac{3}{4}\right)$

$4(x+2) + 4x = x(x+2) \cdot 3$

$4x + 8 + 4x = 3x^2 + 6x$

$-3x^2 + 2x + 8 = 0$

$3x^2 - 2x - 8 = 0$

$(3x+4)(x-2) = 0$

$3x + 4 = 0 \quad \text{or} \quad x - 2 = 0$

$\quad x = -\dfrac{4}{3} \qquad\qquad x = 2$

5. $\dfrac{1}{p-4}+\dfrac{1}{4}=\dfrac{8}{p^2-16}$

$4(p-4)(p+4)\left(\dfrac{1}{p-4}+\dfrac{1}{4}\right)$

$\qquad =4(p-4)(p+4)\left(\dfrac{8}{(p+4)(p-4)}\right)$

$4(p+4)+(p-4)(p+4)=4\cdot 8$

$4p+16+p^2-16=32$

$p^2+4p-32=0$

$(p+8)(p-4)=0$

$p+8=0\quad$ or $\quad p-4=0$

$p=-8\qquad\qquad p=4$

Note that $p=4$ makes expressions in the original equation undefined. Therefore, $p=4$ is an extraneous solution.

7. $\dfrac{6}{2y+5}=\dfrac{2}{y+5}+\dfrac{1}{5}$

$5(2y+5)(y+5)\left(\dfrac{6}{2y+5}\right)$

$\qquad =5(2y+5)(y+5)\left(\dfrac{2}{y+5}+\dfrac{1}{5}\right)$

$5(y+5)\cdot 6=5(2y+5)\cdot 2+(2y+5)(y+5)$

$30y+150=20y+50+2y^2+15y+25$

$0=2y^2+5y-75$

$0=(y-5)(2y+15)$

$y-5=0\quad$ or $\quad 2y+15=0$

$y=5\qquad\qquad y=-7.5$

9. $1+2x^{-1}-8x^{-2}=0$

$1+\dfrac{2}{x}-\dfrac{8}{x^2}=0$

$x^2\left(1+\dfrac{2}{x}-\dfrac{8}{x^2}\right)=x^2\cdot 0$

$x^2+2x-8=0$

$(x+4)(x-2)=0$

$x+4=0\quad$ or $\quad x-2=0$

$x=-4\qquad\qquad x=2$

11. $3+13x^{-1}-10x^{-2}=0$

$3+\dfrac{13}{x}-\dfrac{10}{x^2}=0$

$x^2\left(3+\dfrac{13}{x}-\dfrac{10}{x^2}\right)=x^2\cdot 0$

$3x^2+13x-10=0$

$(3x-2)(x+5)=0$

$3x-2=0\quad$ or $\quad x+5=0$

$x=\dfrac{2}{3}\qquad\qquad x=-5$

13. $x-8\sqrt{x}+15=0$

$-8\sqrt{x}=-x-15$

$8\sqrt{x}=x+15$

$\left(8\sqrt{x}\right)^2=(x+15)^2$

$64x=x^2+30x+225$

$0=x^2-34x+225$

$0=(x-9)(x-25)$

$x-9=0\quad$ or $\quad x-25=0$

$x=9\qquad\qquad x=25$

15. $2x-5\sqrt{x}-7=0$

$2x-7=5\sqrt{x}$

$(2x-7)^2=\left(5\sqrt{x}\right)^2$

$4x^2-28x+49=25x$

$4x^2-53x+49=0$

$x=\dfrac{53\pm\sqrt{53^2-4(4)(49)}}{2(4)}$

$x=\dfrac{53\pm\sqrt{2025}}{8}=\dfrac{53\pm 45}{8}=\dfrac{49}{4},\cancel{1}$

17. $\sqrt{2a+5}=3a-3$

$\left(\sqrt{2a+5}\right)^2=(3a-3)^2$

$2a+5=9a^2-18a+9$

$0=9a^2-20a+4$

$0=(a-2)(9a-2)$

$a-2=0\quad$ or $\quad 9a-2=0$

$a=2\qquad\qquad a=\cancel{\dfrac{2}{9}}$

19. $\sqrt{2m-8} - m - 1 = 0$

$\sqrt{2m-8} = m+1$

$\left(\sqrt{2m-8}\right)^2 = \left(m+1\right)^2$

$2m-8 = m^2 + 2m + 1$

$0 = m^2 + 9$

$-9 = m^2$

$\pm 3i = m$

21. $\sqrt{4x+1} = \sqrt{x+2} + 1$

$\left(\sqrt{4x+1}\right)^2 = \left(\sqrt{x+2} + 1\right)^2$

$4x+1 = x+2 + 2\sqrt{x+2} + 1$

$3x-2 = 2\sqrt{x+2}$

$\left(3x-2\right)^2 = \left(2\sqrt{x+2}\right)^2$

$9x^2 - 12x + 4 = 4\left(x+2\right)$

$9x^2 - 12x + 4 = 4x + 8$

$9x^2 - 16x - 4 = 0$

$\left(x-2\right)\left(9x+2\right) = 0$

$x-2 = 0$ or $9x+2 = 0$

$x = 2$ $\qquad x = \cancel{-\dfrac{2}{9}}$

23. $\sqrt{2x+1} - \sqrt{3x+4} = -1$

$\left(\sqrt{2x+1}\right)^2 = \left(\sqrt{3x+4} - 1\right)^2$

$2x+1 = 3x+4 - 2\sqrt{3x+4} + 1$

$2\sqrt{3x+4} = x+4$

$\left(2\sqrt{3x+4}\right)^2 = \left(x+4\right)^2$

$4\left(3x+4\right) = x^2 + 8x + 16$

$12x+16 = x^2 + 8x + 16$

$0 = x^2 - 4x$

$0 = x\left(x-4\right)$

$x-4 = 0$ or $x = 0$

$x = 4$

25. Let $u = x^2$.

$u^2 - 10u + 9 = 0$

$\left(u-9\right)\left(u-1\right) = 0$

$u-9 = 0$ or $u-1 = 0$

$u = 9$ $\qquad u = 1$

$x^2 = 9$ $\qquad x^2 = 1$

$x = \pm 3$ $\qquad x = \pm 1$

27. Let $u = x^2$.

$4u^2 - 13u + 9 = 0$

$\left(4u-9\right)\left(u-1\right) = 0$

$4u-9 = 0$ or $u-1 = 0$

$u = \dfrac{9}{4}$ $\qquad u = 1$

$x^2 = \dfrac{9}{4}$ $\qquad x^2 = 1$

$x = \pm\dfrac{3}{2}$ $\qquad x = \pm 1$

29. Let $u = x^2$.

$u^2 + 5u - 36 = 0$

$\left(u+9\right)\left(u-4\right) = 0$

$u+9 = 0$ or $u-4 = 0$

$u = -9$ $\qquad u = 4$

$x^2 = -9$ $\qquad x^2 = 4$

$x = \pm 3i$ $\qquad x = \pm 2$

31. Let $u = x+2$.

$u^2 + 6u + 8 = 0$

$\left(u+4\right)\left(u+2\right) = 0$

$u+4 = 0$ or $u+2 = 0$

$u = -4$ $\qquad u = -2$

$x+2 = -4$ $\qquad x+2 = -2$

$x = -6$ $\qquad x = -4$

33. Let $u = x + 3$.

$$2u^2 - 9u - 5 = 0$$

$$(2u + 1)(u - 5) = 0$$

$2u + 1 = 0$ or $u - 5 = 0$

$u = -\dfrac{1}{2}$ $u = 5$

$x + 3 = 5$

$x + 3 = -\dfrac{1}{2}$ $x = 2$

$x = -3\dfrac{1}{2}$

35. Let $u = \dfrac{x - 1}{2}$.

$$u^2 + 8u + 15 = 0$$

$$(u + 3)(u + 5) = 0$$

$u + 3 = 0$ or $u + 5 = 0$

$u = -3$ $u = -5$

$\dfrac{x - 1}{2} = -3$ $\dfrac{x - 1}{2} = -5$

$x - 1 = -6$ $x - 1 = -10$

$x = -5$ $x = -9$

37. Let $u = \dfrac{x + 2}{2}$.

$$2u^2 + u - 3 = 0$$

$$(2u + 3)(u - 1) = 0$$

$2u + 3 = 0$ or $u - 1 = 0$

$u = \dfrac{-3}{2}$ $u = 1$

$\dfrac{x + 2}{2} = \dfrac{-3}{2}$ $\dfrac{x + 2}{2} = 1$

$x + 2 = -3$ $x + 2 = 2$

$x = -5$ $x = 0$

39. Let $u = x^{1/3}$.

$$u^2 - 5u + 6 = 0$$

$$(u - 2)(u - 3) = 0$$

$u - 2 = 0$ or $u - 3 = 0$

$u = 2$ $u = 3$

$x^{1/3} = 2$ $x^{1/3} = 3$

$\left(x^{1/3}\right)^3 = 2^3$ $\left(x^{1/3}\right)^3 = 3^3$

$x = 8$ $x = 27$

41. Let $u = x^{1/3}$.

$$2u^2 - 3u - 2 = 0$$

$$(2u + 1)(u - 2) = 0$$

$2u + 1 = 0$ or $u - 2 = 0$

$u = -\dfrac{1}{2}$ $u = 2$

$x^{1/3} = -\dfrac{1}{2}$ $x^{1/3} = 2$

$\left(x^{1/3}\right)^3 = 2^3$

$\left(x^{1/3}\right)^3 = \left(-\dfrac{1}{2}\right)^3$ $x = 8$

$x = -\dfrac{1}{8}$

43. Let $u = x^{1/4}$.

$$u^2 - 5u + 6 = 0$$

$$(u - 3)(u - 2) = 0$$

$u - 3 = 0$ or $u - 2 = 0$

$u = 3$ $u = 2$

$x^{1/4} = 3$ $x^{1/4} = 2$

$\left(x^{1/4}\right)^4 = 3^4$ $\left(x^{1/4}\right)^4 = 2^4$

$x = 81$ $x = 16$

45. Let $u = x^{1/4}$.

$$5u^2 + 8u - 4 = 0$$

$$(5u - 2)(u + 2) = 0$$

$5u - 2 = 0$ or $u + 2 = 0$

$u = \dfrac{2}{5}$ $u = -2$

$x^{1/4} = \dfrac{2}{5}$ $x^{1/4} = -2$

$\left(x^{1/4}\right)^4 = \left(\dfrac{2}{5}\right)^4$ $\left(x^{1/4}\right)^4 = (-2)^4$

$x = \cancel{16}$

$x = \dfrac{16}{625}$

47. Let x represent the time it takes the bus to travel 180 miles.

Vehicle	d	t	r
bus	180	x	$\dfrac{180}{x}$
truck	180	$x+1$	$\dfrac{180}{x+1}$

Rate of bus = Rate of truck + 15

$$\frac{180}{x} = \frac{180}{x+1} + 15$$

$$180(x+1) = 180x + 15x(x+1)$$

$$180x + 180 = 180x + 15x^2 + 15x$$

$$0 = 15x^2 + 15x - 180$$

$$0 = x^2 + x - 12$$

$$0 = (x+4)(x-3)$$

$$x = \cancel{-4}, 3$$

Because time cannot be negative, it takes the bus 3 hours to travel 180 miles.

49. Let x represent the time it takes the winner to run 26 miles.

Person	d	t	r
winner	26	x	$\dfrac{26}{x}$
other	26	$x+1$	$\dfrac{26}{x+1}$

Rate of winner = Rate of other + 3.66

$$\frac{26}{x} = \frac{26}{x+1} + 3.66$$

$$26(x+1) = 26x + 3.66x(x+1)$$

$$26x + 26 = 26x + 3.66x^2 + 3.66x$$

$$0 = 3.66x^2 + 3.66x - 26$$

$$x = \frac{-3.66 \pm \sqrt{(3.66)^2 - 4(3.66)(-26)}}{2(3.66)}$$

$$= \frac{-3.66 \pm \sqrt{394.0356}}{7.32}$$

$$\approx \frac{-3.66 \pm 19.850}{7.32} \approx 2.21, \cancel{-3.21}$$

Because time cannot be negative, it takes the winner approximately 2.21 hours to run 26 miles.

51. Let x represent the time it takes the motorcycle to travel 300 miles.

Vehicle	d	t	r
motorcycle	300	x	$\dfrac{300}{x}$
bus	360	$x+3$	$\dfrac{360}{x+3}$

Rate of motorcycle = Rate of bus + 15

$$\frac{300}{x} = \frac{360}{x+3} + 15$$

$$300(x+3) = 360x + 15x(x+3)$$

$$300x + 900 = 360x + 15x^2 + 45x$$

$$0 = 15x^2 + 105x - 900$$

$$0 = x^2 + 7x - 60$$

$$0 = (x+12)(x-5)$$

$$x = \cancel{-12}, 5$$

Because time cannot be negative, it takes the motorcycle approximately 5 hours to travel 300 miles. However, we are asked for the rate of the bus. Making the necessary substitution, we find that the bus is traveling at a rate of 45 mph.

$$\text{Rate of bus} = \frac{360}{5+3} = \frac{360}{8} = 45 \text{ mph}$$

53. Let x represent the time it took the cyclist to travel 36 miles a year ago.

	d	t	r
Old	36	x	$\dfrac{36}{x}$
New	36	$x-1$	$\dfrac{36}{x-1}$

New rate = Old rate + 6

$$\frac{36}{x-1} = \frac{36}{x} + 6$$

$$36x = 36(x-1) + 6x(x-1)$$

$$36x = 36x - 36 + 6x^2 - 6x$$

$$0 = 6x^2 - 6x - 36$$

$$0 = x^2 - x - 6$$

$$0 = (x-3)(x+2)$$

$$x = 3, \cancel{-2}$$

Because time cannot be negative, it used to take the cyclist approximately 3 hours to travel 36 miles.

a) Old rate $= \dfrac{36}{3} = 12$ mph

New rate $= \dfrac{36}{2} = 18$ mph

b) Old time is 3 hours, new time is 2 hours

55. Let x represent the number of hours for Jody to run the hoop nets.

Worker	Time to complete alone	Rate of work	Time at work	Portion of job completed
Billy	$x+2$	$\dfrac{1}{x+2}$	$\dfrac{12}{5}$	$\dfrac{12}{5(x+2)}$
Jody	x	$\dfrac{1}{x}$	$\dfrac{12}{5}$	$\dfrac{12}{5x}$

Billy's portion + Jody's portion = 1 (entire job)

$$\frac{12}{5(x+2)} + \frac{12}{5x} = 1$$

$$12x + 12(x+2) = 5x(x+2)$$

$$12x + 12x + 24 = 5x^2 + 10x$$

$$0 = 5x^2 - 14x - 24$$

$$0 = (5x+6)(x-4)$$

$$x = \cancel{-\frac{6}{5}}, 4$$

Since a negative time makes no sense in the context of this problem, it takes Jody 4 hours to run the hoop nets alone and Billy $4 + 2 = 6$ hours working alone.

57. Let x represent the number of hours for the new press to print the copies.

Press	Time to complete alone	Rate of work	Time at work	Portion of job completed
New	x	$\dfrac{1}{x}$	2	$\dfrac{2}{x}$
Old	$x+\dfrac{1}{2}$	$\dfrac{1}{x+\dfrac{1}{2}}$	2	$\dfrac{2}{x+\dfrac{1}{2}} = \dfrac{4}{2x+1}$

New press portion + old press portion = 1 (entire job.)

$$\frac{2}{x} + \frac{4}{2x+1} = 1$$

$$2(2x+1) + 4x = x(2x+1)$$

$$4x + 2 + 4x = 2x^2 + x$$

$$0 = 2x^2 - 7x - 2$$

$$x = \frac{-(-7) \pm \sqrt{(-7)^2 - 4(2)(-2)}}{2(2)}$$

$$= \frac{7 \pm \sqrt{49+16}}{4} = \frac{7 \pm \sqrt{65}}{4}$$

$$\approx 3.77, \cancel{-0.27}$$

Since a negative time makes no sense in the context of this problem, it takes the new press about 3.77 hours to run the copies alone and the old press about $3.77 + 0.5 = 4.27$ hours working alone.

59. String 1: tension = 50 lbs., frequency $= x$
String 2: tension = 60 lbs., frequency $= x + 40$ vps

$$\frac{x^2}{(40+x)^2} = \frac{50}{60}$$

$$60x^2 = 50(40+x)^2$$

$$60x^2 = 50(1600 + 80x + x^2)$$

$$60x^2 = 80,000 + 4000x + 50x^2$$

$$10x^2 - 4000x - 80,000 = 0$$

$$x^2 - 400x - 8000 = 0$$

$$x = \frac{-(-400) \pm \sqrt{(-400)^2 - 4(1)(-8000)}}{2(1)}$$

$$= \frac{400 \pm \sqrt{192,000}}{2} \approx 420, \cancel{-19}$$

Since string frequency must be positive, string 1's frequency is 420 vps, and string 2's frequency is $420 + 40 = 460$ vps.

Review Exercises

1. $-\dfrac{b}{2a} = -\dfrac{8}{2 \cdot 2} = -\dfrac{8}{4} = -2$

2. $\dfrac{4ac - b^2}{4a} = \dfrac{4(2)(5) - (-8)^2}{4(2)}$

$$= \frac{40 - 64}{8}$$

$$= \frac{-24}{8}$$

$$= -3$$

3. $2x + 3y = 6$

   x-intercept: $2x + 3 \cdot 0 = 6$

   $\qquad\qquad\quad 2x = 6$

   $\qquad\qquad\quad\; x = 3$

   $\qquad\qquad\quad\; (3, 0)$

   y-intercept: $2 \cdot 0 + 3y = 6$

   $\qquad\qquad\quad 3y = 6$

   $\qquad\qquad\quad\; y = 2$

   $\qquad\qquad\quad\; (0, 2)$

4. $f(x) = 2x^2 - 3x + 1$

   $f(-1) = 2(-1)^2 - 3(-1) + 1$

   $\qquad\quad = 2 \cdot 1 + 3 + 1$

   $\qquad\quad = 2 + 3 + 1$

   $\qquad\quad = 6$

5. $f(x) = x^2 - 3$

x	$f(x)$	(x, y)
-2	1	$(-2, 1)$
-1	-2	$(-1, -2)$
0	-3	$(0, -3)$
1	-2	$(1, -2)$
2	1	$(2, 1)$

   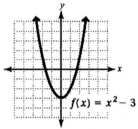

6. $f(x) = 3x^2$

x	$f(x)$	(x, y)
-2	12	$(-2, 12)$
-1	3	$(-1, 3)$
0	0	$(0, 0)$
1	3	$(1, 3)$
2	12	$(2, 12)$

   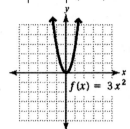

Exercise Set 9.4

1. $f(x) = 5(x - 2)^2 - 3$

   Vertex: $(2, -3)$; axis of symmetry $x = 2$

3. $g(x) = -2(x + 1)^2 - 5$

   Vertex: $(-1, -5)$; axis of symmetry $x = -1$

5. $k(x) = x^2 + 2$

   Vertex: $(0, 2)$; axis of symmetry $x = 0$

7. Can be rewritten as $h(x) = 3(x + 5)^2 + 0$

   Vertex: $(-5, 0)$; axis of symmetry $x = -5$

9. Can be rewritten as $f(x) = -0.5(x - 0)^2 + 0$

   Vertex: $(0, 0)$; axis of symmetry $x = 0$

11. $f(x) = x^2 - 4x + 8$

    x-coordinate: $\dfrac{-(-4)}{2(1)} = \dfrac{4}{2} = 2$

    y-coordinate: $y = (2)^2 - 4(2) + 8$

    $\qquad\qquad\qquad = 4 - 8 + 8$

    $\qquad\qquad\qquad = 4$

    vertex: $(2, 4)$

    axis of symmetry: $x = 2$

13. $k(x) = 2x^2 + 16x + 27$

x-coordinate: $\dfrac{-(16)}{2(2)} = \dfrac{-16}{4} = -4$

y-coordinate: $y = 2(-4)^2 + 16(-4) + 27$
$$= 32 - 64 + 27$$
$$= -5$$

vertex: $(-4, -5)$
axis of symmetry: $x = -4$

15. $g(x) = -3x^2 + 2x + 1$

x-coordinate: $\dfrac{-(2)}{2(-3)} = \dfrac{-2}{-6} = \dfrac{1}{3}$

y-coordinate: $y = -3\left(\dfrac{1}{3}\right)^2 + 2\left(\dfrac{1}{3}\right) + 1$
$$= -3 \cdot \dfrac{1}{9} + \dfrac{2}{3} + 1$$
$$= -\dfrac{1}{3} + \dfrac{2}{3} + 1$$
$$= \dfrac{4}{3}$$

vertex: $\left(\dfrac{1}{3}, \dfrac{4}{3}\right)$

axis of symmetry: $x = \dfrac{1}{3}$

17. $h(x) = -3x^2$
   a) downward
   b) $(0, 0)$
   c) $x = 0$
   d)
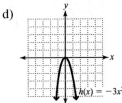

19. $k(x) = \dfrac{1}{4}x^2$
   a) upward
   b) $(0, 0)$
   c) $x = 0$
   d)
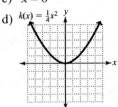

21. $f(x) = 4x^2 - 3$
   a) upward
   b) $(0, -3)$
   c) $x = 0$
   d)
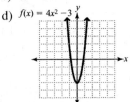

23. $g(x) = -0.5x^2 + 2$
   a) downward
   b) $(0, 2)$
   c) $x = 0$
   d)
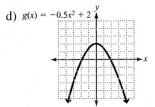

25. $f(x) = (x - 3)^2 + 2$
   a) upward
   b) $(3, 2)$
   c) $x = 3$
   d)
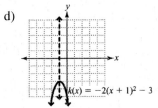

27. $k(x) = -2(x + 1)^2 - 3$
   a) downward
   b) $(-1, -3)$
   c) $x = -1$
   d)

29. $h(x) = \frac{1}{3}(x-2)^2 - 1$

a) upward

b) $(2, -1)$

c) $x = 2$

d)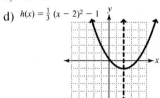

31. $h(x) = x^2 + 6x + 9$

a) x-intercept: $x^2 + 6x + 9 = 0$

$$(x+3)^2 = 0$$
$$x + 3 = \pm\sqrt{0}$$
$$x + 3 = 0$$
$$x = -3$$
$$(-3, 0)$$

y-intercept: $y = 0^2 + 6(0) + 9 = 9$

$(0, 9)$

b) $y = x^2 + 6x + 9$

$$y = (x+3)^2$$
$$h(x) = (x+3)^2$$

c) upward

d) $(-3, 0)$

e) $x = -3$

f)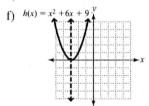

g) Domain: $\{x \mid x \text{ is a real number}\}$ or $(-\infty, \infty)$

Range: $\{y \mid y \geq 0\}$ or $[0, \infty)$

33. $g(x) = -3x^2 + 6x - 5$

a) x-intercepts:

$$-3x^2 + 6x - 5 = 0$$
$$x = \frac{-6 \pm \sqrt{6^2 - 4(-3)(-5)}}{2(-3)}$$
$$= \frac{-6 \pm \sqrt{36 - 60}}{-6}$$
$$= \frac{-6 \pm \sqrt{-24}}{-6}$$

Since the discriminant is negative, there are no x-intercepts.

y-intercept: $y = -3(0)^2 + 6(0) - 5 = -5$

$(0, -5)$

b) $y = -3x^2 + 6x - 5$

$$y + 5 = -3(x^2 - 2x)$$
$$y + 5 - 3 = -3(x^2 - 2x + 1)$$
$$y + 2 = -3(x-1)^2$$
$$y = -3(x-1)^2 - 2$$
$$g(x) = -3(x-1)^2 - 2$$

c) downward

d) $(1, -2)$

e) $x = 1$

f)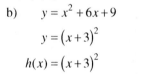

$g(x) = -3x^2 + 6x - 5$

g) Domain: $\{x \mid x \text{ is a real number}\}$ or $(-\infty, \infty)$

Range: $\{y \mid y \leq -2\}$ or $(-\infty, -2]$

35. $f(x) = 2x^2 + 6x + 3$

a) x-intercepts:

$$2x^2 + 6x + 3 = 0$$

$$x = \frac{-6 \pm \sqrt{6^2 - 4(2)(3)}}{2(2)}$$

$$= \frac{-6 \pm \sqrt{36 - 24}}{4}$$

$$= \frac{-6 \pm \sqrt{12}}{4}$$

$$= \frac{-6 \pm 2\sqrt{3}}{4}$$

$$= \frac{-3 \pm \sqrt{3}}{2}$$

$$\left(\frac{-3 + \sqrt{3}}{2}, 0\right), \left(\frac{-3 - \sqrt{3}}{2}, 0\right)$$

y-intercept: $y = 2(0)^2 + 6(0) + 3 = 3$

(0, 3)

b) $$y = 2x^2 + 6x + 3$$

$$y - 3 = 2x^2 + 6x$$

$$y - 3 = 2\left(x^2 + 3x\right)$$

$$y - 3 + \frac{9}{2} = 2\left(x^2 + 3x + \frac{9}{4}\right)$$

$$y + \frac{3}{2} = 2\left(x + \frac{3}{2}\right)^2$$

$$y = 2\left(x + \frac{3}{2}\right)^2 - \frac{3}{2}$$

$$f(x) = 2\left(x + \frac{3}{2}\right)^2 - \frac{3}{2}$$

c) upward

d) $\left(-\dfrac{3}{2}, -\dfrac{3}{2}\right)$

e) $x = -\dfrac{3}{2}$

f) $f(x) = 2x^2 + 6x + 3$

g) Domain: $\{x \mid x \text{ is a real number}\}$ or $(-\infty, \infty)$

Range: $\left\{y \mid y \geq -\dfrac{3}{2}\right\}$ or $\left[-\dfrac{3}{2}, \infty\right)$

37. Finding the vertex will allow us to know the highest point and how many seconds it takes to reach the highest point.

a) t-coordinate: $t = \dfrac{-45}{2(-4.9)} \approx 4.59$ sec is the time it takes to reach the highest point.

b) h-coordinate:

$h = -4.9(4.59)^2 + 45(4.59) \approx 103.32$ meters is the maximum height that the rocket reaches.

c) The t-intercepts describe the times that the rocket is on the ground.

$$h = 0 : 0 = -4.9t^2 + 45t$$

$$t = \frac{-(45) \pm \sqrt{(45)^2 - 4(-4.9)(0)}}{2(-4.9)}$$

$$= \frac{-45 \pm \sqrt{2025 + 0}}{-9.8} = \frac{-45 \pm 45}{-9.8}$$

$$= \frac{-45 + 45}{-9.8} = 0 \text{ (starting point)}$$

or $= \dfrac{-45 - 45}{-9.8} = 9.18$ sec.

39. a) x-coordinate: $-\dfrac{3.2}{2(-0.8)} = 2$

The maximum height will be the y-coordinate of the vertex.

y-coordinate: $y = -0.8(2)^2 + 3.2(2) + 6 = 9.2$ feet is the maximum height that the ball reaches.

b) To find how far the ball travels, we need to find the x-value when it hits the ground, which is where $y = 0$.

$$y = 0 : 0 = -0.8x^2 + 3.2x + 6$$

$$x = \frac{-(3.2) \pm \sqrt{(3.2)^2 - 4(-0.8)(6)}}{2(-0.8)}$$

$$\approx -1.39, 5.39$$

The negative values does not make sense in this context, so the ball travels approximately 5.39 feet.

c)

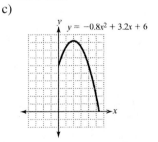

$y = -0.8x^2 + 3.2x + 6$

41. Finding the vertex will allow us to know the highest number of CDs sold and how many weeks it takes to reach that point.

$$n(t) = -200t^2 + 4000t$$

$$n(t) = -200(t^2 - 20t)$$

$$n(t) - 20,000 = -200(t^2 - 20t + 100)$$

$$n(t) = -200(t - 10)^2 + 20,000$$

Vertex: $(10, 20,000)$

a) The tenth week   b) 20,000

43. Because the enclosed space is to be rectangular, the sum of the length and width is $\frac{1}{2}(400) = 200$ feet. Let x represent the width and $200 - x$ represent the length.

$$y = x(200 - x)$$

$$y = 200x - x^2$$

$$y = -x^2 + 200x$$

$$y - 10,000 = -(x^2 - 200x + 10,000)$$

$$y - 10,000 = -(x - 100)^2$$

$$y = -(x - 100)^2 + 10,000$$

The vertex, $(100, 10,000)$ shows the highest point on the graph, or the maximum area to be achieved.
The area is maximized if the width is 100 feet and the length is $200 - 100 = 100$ feet.

45. The vertex would be the lowest point on the graph.

$$C(n) = n^2 - 110n + 5000$$

$$m = n^2 - 110n + 5000$$

$$m - 5000 = n^2 - 110n$$

$$m - 5000 + 3025 = n^2 - 110n + 3025$$

$$m + 1075 = (n - 55)^2$$

$$m = (n - 55)^2 + 1075$$

$$c(n) = (n - 55)^2 + 1075$$

55 units would minimize the cost.

47. Let one of the integers be x. Then the other integer is $x + 12$.

$$y = x(x + 12)$$

$$y = x^2 + 12x$$

$$y + 36 = x^2 + 12x + 36$$

$$y + 36 = (x + 6)^2$$

$$y = (x + 6)^2 - 36$$

The vertex $(-6, -36)$ shows one of the integers is -6 the other is $-6 + 12 = 6$, and the product is -36.

Review Exercises

1. $-|-2 + 3 \cdot 4| - 4^0 = -|-2 + 12| - 1$
$$= -|10| - 1$$
$$= -10 - 1$$
$$= -11$$

2. $x^2 + 2x = 15$
$$x^2 + 2x - 15 = 0$$
$$(x + 5)(x - 3) = 0$$
$$x + 5 = 0 \quad \text{or} \quad x - 3 = 0$$
$$x = -5 \qquad x = 3$$

3. $x^2 = -20$
$$x = \pm\sqrt{-20}$$
$$x = \pm 2i\sqrt{5}$$

4. $4x - 8 \le 2x + 1$
$$2x \le 9$$
$$x \le 4.5$$

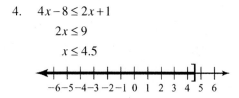

5. $|x - 3| \ge 8$
$$x - 3 \le -8 \quad \text{or} \quad x - 3 \ge 8$$
$$x \le -5 \qquad x \ge 11$$

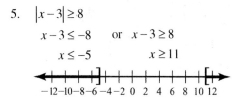

6. $\begin{cases} x+y>2 \\ 2x-3y\le 6 \end{cases}$

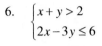
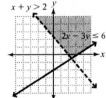

Exercise Set 9.5

1. a) $x=-5,-1$ b) $(-5,-1)$

    c) $(-\infty,-5)\cup(-1,\infty)$

3. a) $x=-1,3$ b) $(-\infty,-1]\cup[3,\infty)$

    c) $[-1,3]$

5. a) $x=-2$ b) $(-\infty,-2)\cup(-2,\infty)$

    c) $\varnothing$

7. a) $\varnothing$ b) $\varnothing$

    c) $\mathbb{R}$ or $(-\infty,\infty)$

9. $(x+4)(x+2)=0$

    $x+4=0$ or $x+2=0$

    $x=-4$ $x=-2$

    For $(-\infty,-4)$, we choose -5.

    $(-5+4)(-5+2)<0$ False

    $-1(-3)<0$

    $3<0$

    For $(-4,-2)$, we choose -3.

    $(-3+4)(-3+2)<0$ True

    $1(-1)<0$

    $-1<0$

    For $(-2,\infty)$, we choose 0.

    $(0+4)(0+2)<0$ False

    $4\cdot 2<0$

    $8<0$

    Solution set: $(-4,-2)$

    ++|+(+|+)+|+|+|+|+|+|+|++
    $-6-5-4-3-2-1\ 0\ 1\ 2\ 3\ 4\ 5\ 6$

11. $(x-2)(x-5)=0$

    $x-2=0$ or $x-5=0$

    $x=2$ $x=5$

    For $(-\infty,2)$, we choose 0.

    $(0-2)(0-5)>0$ True

    $-2(-5)>0$

    $10>0$

    For $(2,5)$, we choose 3.

    $(3-2)(3-5)>0$ False

    $1(-2)>0$

    $-2>0$

    For $(5,\infty)$, we choose 6.

    $(6-2)(6-5)>0$ True

    $4(1)>0$

    $4>0$

    Solution set: $(-\infty,2)\cup(5,\infty)$

    ++|+|+|+|+|+|+(+|+|+|+|+|+)++
    $-6-5-4-3-2-1\ 0\ 1\ 2\ 3\ 4\ 5\ 6$

13. $x^2+5x+4=0$

    $(x+4)(x+1)=0$

    $x+4=0$ or $x+1=0$

    $x=-4$ $x=-1$

    For $(-\infty,-4)$, we choose -5

    $(-5)^2+5(-5)+4<0$ False

    $25-25+4<0$

    $4<0$

    For $(-4,-1)$, we choose -3.

    $(-3)^2+5(-3)+4<0$ True

    $9-15+4<0$

    $-2<0$

    For $(-1,\infty)$, we choose 0.

    $(0)^2+5(0)+4<0$ False

    $4<0$

    Solution set: $(-4,-1)$

    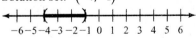

15. $x^2 - 4x + 3 = 0$

$(x-3)(x-1) = 0$

$x - 3 = 0$ or $x - 1 = 0$

$x = 3$ $x = 1$

For $(-\infty, 1)$, we choose 0.

$(0)^2 - 4(0) + 3 > 0$ True

$3 > 0$

For $(1, 3)$, we choose 2.

$(2)^2 - 4(2) + 3 > 0$ False

$4 - 8 + 3 > 0$

$-1 > 0$

For $(3, \infty)$, we choose 4.

$(4)^2 - 4(4) + 3 > 0$ True

$16 - 16 + 3 > 0$

$3 > 0$

Solution set: $(-\infty, 1) \cup (3, \infty)$

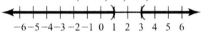

17. $b^2 - 6b + 8 = 0$

$(b-4)(b-2) = 0$

$b - 4 = 0$ or $b - 2 = 0$

$b = 4$ $b = 2$

For $(-\infty, 2)$, we choose 0.

$(0)^2 - 6(0) + 8 \leq 0$ False

$8 \leq 0$

For $(2, 4)$, we choose 3.

$(3)^2 - 6(3) + 8 \leq 0$ True

$9 - 18 + 8 \leq 0$

$-1 \leq 0$

For $(4, \infty)$, we choose 5.

$(5)^2 - 6(5) + 8 \leq 0$ False

$25 - 30 + 8 \leq 0$

$3 \leq 0$

Solution set: $[2, 4]$

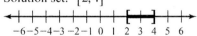

19. $y^2 - 4y - 5 = 0$

$(y+1)(y-5) = 0$

$y + 1 = 0$ or $y - 5 = 0$

$y = -1$ $y = 5$

For $(-\infty, -1)$, we choose –2.

$(-2)^2 - 5 \geq 4(-2)$ True

$4 - 5 \geq -8$

$-1 \geq -8$

For $(-1, 5)$, we choose 3.

$(3)^2 - 5 \geq 4(3)$ False

$9 - 5 \geq 12$

$4 \geq 12$

For $(5, \infty)$, we choose 6.

$(6)^2 - 5 \geq 4(6)$ True

$36 - 5 \geq 24$

$31 \geq 24$

Solution set: $(-\infty, -1] \cup [5, \infty)$

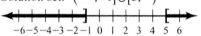

21. $a^2 - 3a - 10 = 0$

$(a-5)(a+2) = 0$

$a - 5 = 0$ or $a + 2 = 0$

$a = 5$ $a = -2$

For $(-\infty, -2)$, we choose –3.

$(-3)^2 - 3(-3) < 10$ False

$9 + 9 < 10$

$18 < 10$

For $(-2, 5)$, we choose 0.

$(0)^2 - 3(0) < 10$ True

$0 < 10$

For $(5, \infty)$, we choose 6.

$(6)^2 - 3(6) < 10$ False

$36 - 18 < 10$

$18 < 10$

Solution set: $(-2, 5)$

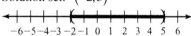

23. $y^2 + 6y + 9 = 0$

$(y+3)^2 = 0$

Since $(y+3)^2 \geq 0$ is always nonnegative, the solution set is $(-\infty, \infty)$ or $\mathbb{R}$.

25. $2c^2 - 4c + 7 = 0$

$$c = \frac{-(-4) \pm \sqrt{(-4)^2 - 4(2)(7)}}{2 \cdot 2}$$

$$= \frac{4 \pm \sqrt{16 - 56}}{4}$$

$$= \frac{4 \pm \sqrt{-40}}{4}$$

$$= \frac{4 \pm 2i\sqrt{10}}{4} = \frac{2 \pm i\sqrt{10}}{2}$$

There are no c-intercepts, so the function is either always positive or always negative. Test the value zero.

$2(0)^2 - 4(0) + 7 = 7 > 0$

Therefore, the function is always positive, and the solution set is $\varnothing$.

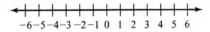

27. $4r^2 + 21r + 5 = 0$

$(4r + 1)(r + 5) = 0$

$4r + 1 = 0 \qquad$ or $\quad r + 5 = 0$

$\qquad r = -0.25 \qquad\qquad r = -5$

For $(-\infty, -5)$, we choose -6.

$4(-6)^2 + 21(-6) + 5 > 0 \qquad$ True

$\qquad 144 - 126 + 5 > 0$

$\qquad\qquad\qquad 23 > 0$

For $(-5, -0.25)$, we choose -1.

$4(-1)^2 + 21(-1) + 5 > 0 \qquad$ False

$\qquad 4 - 21 + 5 > 0$

$\qquad\qquad -12 > 0$

For $(-0.25, \infty)$, we choose 0.

$4(0)^2 + 21(0) + 5 > 0 \qquad$ True

$\qquad\qquad 5 > 0$

Solution set: $(-\infty, -5) \cup (-0.25, \infty)$

29. $x^2 - 5x = 0$

$x(x - 5) = 0$

$x = 0 \quad$ or $\quad x - 5 = 0$

$\qquad\qquad\qquad x = 5$

For $(-\infty, 0)$, we choose -1.

$(-1)^2 - 5(-1) > 0 \qquad$ True

$\qquad 1 + 5 > 0$

$\qquad\qquad 6 > 0$

For $(0, 5)$, we choose 1.

$(1)^2 - 5(1) > 0 \qquad$ False

$\qquad 1 - 5 > 0$

$\qquad\qquad -4 > 0$

For $(5, \infty)$, we choose 6.

$(6)^2 - 5(6) > 0 \qquad$ True

$\qquad 36 - 30 > 0$

$\qquad\qquad 6 > 0$

Solution set: $(-\infty, 0) \cup (5, \infty)$

31. $x^2 - 6x = 0$

$x(x - 6) = 0$

$x = 0 \quad$ or $\quad x - 6 = 0$

$\qquad\qquad\qquad x = 6$

For $(-\infty, 0)$, we choose -1.

$(-1)^2 \leq 6(-1) \qquad$ False

$\qquad 1 \leq -6$

For $(0, 6)$, we choose 1.

$(1)^2 \leq 6(1) \qquad$ True

$\qquad 1 \leq 6$

For $(6, \infty)$, we choose 7.

$(7)^2 \leq 6(7) \qquad$ False

$\qquad 49 \leq 42$

Solution set: $[0, 6]$

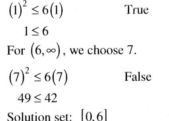

33. Since $(x + 1)^2$ is always 0 or positive, every real number is a solution.

Solution set: $(-\infty, \infty)$ or $\mathbb{R}$

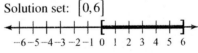

35. $(x-4)(x+2)(x+4)=0$

$x-4=0$ or $x+2=0$ or $x+4=0$

$x=4$ $\qquad x=-2 \qquad x=-4$

Interval	$(-\infty,-4)$	$(-4,-2)$	$(-2,4)$	$(4,\infty)$
Test No.	-5	-3	0	5
Results	-27	7	-32	63
T/F	F	T	F	T

Solution set: $[-4,-2]\cup[4,\infty)$

37. $(x+2)(x+6)(x-1)=0$

$x+2=0$ or $x+6=0$ or $x-1=0$

$x=-2 \qquad x=-6 \qquad x=1$

Interval	$(-\infty,-6)$	$(-6,-2)$	$(-2,1)$	$(1,\infty)$
Test No.	-7	-3	0	2
Results	-40	12	-12	32
T/F	T	F	T	F

Solution set: $(-\infty,-6)\cup(-2,1)$

39. $3c^2+4c-1=0$

$$c=\frac{-4\pm\sqrt{4^2-4(3)(-1)}}{2(3)}=\frac{-2\pm\sqrt{7}}{3}$$

Interval	$\left(-\infty,\dfrac{-2-\sqrt{7}}{3}\right)$	$\left(\dfrac{-2-\sqrt{7}}{3},\dfrac{-2+\sqrt{7}}{3}\right)$
Test No.	-2	0
Results	3	-1
T/F	F	T

Interval	$\left(\dfrac{-2+\sqrt{7}}{3},\infty\right)$
Test No.	1
Results	6
T/F	F

Solution set: $\left(\dfrac{-2-\sqrt{7}}{3},\dfrac{-2+\sqrt{7}}{3}\right)$

41. $4r^2+8r-3=0$

$$r=\frac{-8\pm\sqrt{8^2-4(4)(-3)}}{2(4)}=\frac{-2\pm\sqrt{7}}{2}$$

Interval	$\left(-\infty,\dfrac{-2-\sqrt{7}}{2}\right)$	$\left(\dfrac{-2-\sqrt{7}}{2},\dfrac{-2+\sqrt{7}}{2}\right)$
Test No.	-3	0
Results	9	-3
T/F	T	F

Interval	$\left(\dfrac{-2+\sqrt{7}}{2},\infty\right)$
Test No.	1
Results	9
T/F	T

Solution set: $\left(-\infty,\dfrac{-2-\sqrt{7}}{2}\right]\cup\left[\dfrac{-2+\sqrt{7}}{2},\infty\right)$

43. $-0.2a^2-1.6a-2=0$

$$a=\frac{-(-1.6)\pm\sqrt{(-1.6)^2-4(-0.2)(-2)}}{2(-0.2)}$$

$$=-4\pm\sqrt{6}$$

Interval	$(-\infty,-4-\sqrt{6})$	$(-4-\sqrt{6},-4+\sqrt{6})$
Test No.	-7	-2
Results	-0.6	0.4
T/F	T	F

Interval	$(-4+\sqrt{6},\infty)$
Test No.	0
Results	-2
T/F	T

Solution set: $(-\infty,-4-\sqrt{6}\,]\cup[-4+\sqrt{6},\infty)$

45. $\qquad\dfrac{a+4}{a-1}=0 \qquad\qquad a-1=0$

$\qquad\qquad\qquad\qquad\qquad\qquad a=1$

$(a-1)\dfrac{a+4}{a-1}=0(a-1)$

$\qquad a+4=0$

$\qquad\qquad a=-4$

Interval	$(-\infty,-4)$	$(-4,1)$	$(1,\infty)$
Test No.	-5	0	2
Results	$1/6$	-4	6
T/F	T	F	T

Solution set: $(-\infty,-4)\cup(1,\infty)$

47. $$\frac{n+1}{n+5} = 0 \qquad\qquad n+5 = 0$$
$$n = -5$$
$$(n+5)\frac{n+1}{n+5} = 0(n+5)$$
$$n+1 = 0$$
$$n = -1$$

Interval	$(-\infty,-5)$	$(-5,-1)$	$(-1,\infty)$
Test No.	-6	-2	0
Results	5	$-1/3$	$1/5$
T/F	F	T	F

Solution set: $(-5,-1]$

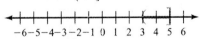

49. $$\frac{6}{x+4} = 0 \qquad\qquad x+4 = 0$$
$$x = -4$$
$$(x+4)\frac{6}{x+4} = 0(x+4)$$
$$6 \neq 0$$

Interval	$(-\infty,-4)$	$(-4,\infty)$
Test No.	-6	0
Results	-3	$3/2$
T/F	F	T

Solution set: $(-4,\infty)$

51. $$\frac{c}{c+3} = 3 \qquad\qquad c+3 = 0$$
$$c = -3$$
$$(c+3)\frac{c}{c+3} = 3(c+3)$$
$$c = 3c + 9$$
$$-2c = 9$$
$$c = -\frac{9}{2}$$

Interval	$(-\infty,-9/2)$	$(-9/2,-3)$
Test No.	-6	-4
Results	$2 < 3$	$4 < 3$
T/F	T	F

Interval	$(-3,\infty)$
Test No.	0
Results	$0 < 3$
T/F	T

Solution set: $\left(-\infty,-\dfrac{9}{2}\right) \cup (-3,\infty)$

53. $$\frac{a+5}{a-4} = 4 \qquad\qquad a-4 = 0$$
$$a = 4$$
$$(a-4)\frac{a+5}{a-4} = 4(a-4)$$
$$a+5 = 4a - 16$$
$$-3a = -21$$
$$a = 7$$

Interval	$(-\infty,4)$	$(4,7)$
Test No.	0	5
Results	$-5/4 > 4$	$10 > 4$
T/F	F	T

Interval	$(7,\infty)$
Test No.	8
Results	$13/4 > 4$
T/F	F

Solution set: $(4,7)$

55. $$\frac{p+3}{p-3} = 4 \qquad\qquad p-3 = 0$$
$$p = 3$$
$$(p-3)\frac{p+3}{p-3} = 4(p-3)$$
$$p+3 = 4p - 12$$
$$-3p = -15$$
$$p = 5$$

Interval	$(-\infty,3)$	$(3,5)$	$(5,\infty)$
Test No.	0	4	6
Results	$-1 \geq 4$	$7 \geq 4$	$3 \geq 4$
T/F	F	T	F

Solution set: $(3,5]$

57.
$$\frac{(k+3)(k-2)}{k-5} = 0$$

$$(k-5)\frac{(k+3)(k-2)}{k-5} = 0(k-5)$$

$$(k+3)(k-2) = 0$$

$$k+3 = 0 \qquad k-2 = 0 \qquad k-5 = 0$$

$$k = -3 \qquad k = 2 \qquad k = 5$$

Interval	$(-\infty,-3)$	$(-3,2)$	$(2,5)$	$(5,\infty)$
Test No.	-4	0	3	6
Results	$-2/3$	$6/5$	-3	36
T/F	T	F	T	F

Solution set: $(-\infty,-3] \cup [2,5)$

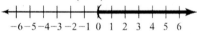

$$-6\ -5\ -4\ -3\ -2\ -1\ \ 0\ \ 1\ \ 2\ \ 3\ \ 4\ \ 5\ \ 6$$

59. Because $(2x-1)^2$ is always nonnegative, we only set the denominator equal to 0; $x = 0$.

Interval	$(-\infty,0)$	$(0,\infty)$
Test No.	-1	1
Results	-9	1
T/F	F	T

Solution set: $(0,\infty)$

$$-6\ -5\ -4\ -3\ -2\ -1\ \ 0\ \ 1\ \ 2\ \ 3\ \ 4\ \ 5\ \ 6$$

61.
$$\frac{(x-5)(x-2)}{x+1} = 0$$

$$(x+1)\frac{(x-5)(x-2)}{x+1} = 0(x+1)$$

$$(x-5)(x-2) = 0$$

$$x-5 = 0 \qquad x-2 = 0 \qquad x+1 = 0$$

$$x = 5 \qquad x = 2 \qquad x = -1$$

Interval	$(-\infty,-1)$	$(-1,2)$	$(2,5)$	$(5,\infty)$
Test No.	-2	0	3	6
Results	-28	10	-0.5	$4/7$
T/F	T	F	T	F

Solution set: $(-\infty,-1) \cup (2,5)$

$$-6\ -5\ -4\ -3\ -2\ -1\ \ 0\ \ 1\ \ 2\ \ 3\ \ 4\ \ 5\ \ 6$$

63. When the ball is on the ground, $h = 0$.

$$-16t^2 + 80t + 96 = 0$$

$$-16(t^2 - 5t - 6) = 0$$

$$t^2 - 5t - 6 = 0$$

$$(t-6)(t+1) = 0$$

$$t = 6, \cancel{-1}$$

a) 6 sec.

b) $-16t^2 + 80t + 96 = 192$

$$-16t^2 + 80t - 96 = 0$$

$$-16(t^2 - 5t + 6) = 0$$

$$t^2 - 5t + 6 = 0$$

$$(t-2)(t-3) = 0$$

$$t = 2,3$$

The ball is 192 ft. above the ground at 2 sec. and then again at 3 sec.

c) $(2,3)$ sec.

d) $[0,2) \cup (3,6]$ sec.

65. Let $h = x - 2$ and $b = x + 6$

$$(x-2)(x+6) \geq 20$$

$$x^2 + 4x - 12 \geq 20$$

$$x^2 + 4x - 32 \geq 0$$

$$(x+8)(x-4) \geq 0$$

$$x \leq -8 \text{ or } x \geq 4$$

a) $x \geq 4$ in.

b) $b \geq 4 + 6$
    $b \geq 10$ in.

$h \geq 4 - 2$

$h \geq 2$ in.

67.
$$\frac{x_2 - 5}{x_2 - 2} \leq \frac{1}{2} \qquad\qquad x_2 - 2 = 0$$

$$\qquad\qquad\qquad x_2 = 2$$

$$\frac{x_2 - 5}{x_2 - 2} = \frac{1}{2}$$

$$x_2 - 2 = 2x_2 - 10$$

$$-x_2 = -8$$

$$x_2 = 8$$

Interval	$(-\infty,2)$	$(2,8)$	$(8,\infty)$
Test No.	0	4	10
Results	$\dfrac{5}{2} \leq \dfrac{1}{2}$	$-\dfrac{1}{2} \leq \dfrac{1}{2}$	$\dfrac{5}{8} \leq \dfrac{1}{2}$
T/F	F	T	F

So, $2 < x \leq 8$

Review Exercises

1. Add 9.
$$x^2 - 6x + 9 = (x-3)^2$$

2. Add $\dfrac{25}{4}$.
$$x^2 + 5x + \frac{25}{4} = \left(x + \frac{5}{2}\right)^2$$

3.
$$\sqrt{x-2} = 4$$
$$\left(\sqrt{x-2}\right)^2 = 4^2$$
$$x - 2 = 16$$
$$x = 18$$

4. Downward, because the coefficient of x^2 is negative.

5.
$$x^2 + 4x - 12 = 0$$
$$(x+6)(x-2) = 0$$
$$x + 6 = 0 \quad \text{or} \quad x - 2 = 0$$
$$x = -6 \qquad\qquad x = 2$$
$$(-6, 0), \ (2, 0)$$

6.
$$y = x^2 - 6x + 5$$
$$y - 5 = x^2 - 6x$$
$$y - 5 + 9 = x^2 - 6x + 9$$
$$y + 4 = (x-3)^2$$
$$y = (x-3)^2 - 4$$
The vertex is (3, –4)

Exercises Set 9.6

1.
$$(f+g)(x) = f(x) + g(x)$$
$$= (2x) + (3x - 1)$$
$$= 5x - 1$$
$$(f-g)(x) = f(x) - g(x)$$
$$= (2x) - (3x - 1)$$
$$= (2x) + (-3x + 1)$$
$$= -x + 1$$

3.
$$(f+g)(x) = f(x) + g(x)$$
$$= (x-5) + (2x - 3)$$
$$= 3x - 8$$
$$(f-g)(x) = f(x) - g(x)$$
$$= (x-5) - (2x - 3)$$
$$= (x-5) + (-2x + 3)$$
$$= -x - 2$$

5.
$$(f+g)(x) = f(x) + g(x)$$
$$= (x^2 - 4x + 7) + (x - 3)$$
$$= x^2 - 3x + 4$$
$$(f-g)(x) = f(x) - g(x)$$
$$= (x^2 - 4x + 7) - (x - 3)$$
$$= (x^2 - 4x + 7) + (-x + 3)$$
$$= x^2 - 5x + 10$$

7.
$$(f+g)(x) = f(x) + g(x)$$
$$= (2x^2 - 5x - 3) + (x^2 + 5)$$
$$= 3x^2 - 5x + 2$$
$$(f-g)(x) = f(x) - g(x)$$
$$= (2x^2 - 5x - 3) - (x^2 + 5)$$
$$= (2x^2 - 5x - 3) + (-x^2 - 5)$$
$$= x^2 - 5x - 8$$

9. $(f+g)(x) = f(x) + g(x)$
$$= (-5x^2 + 4x + 8) + (3x^2 - 2x - 1)$$
$$= -2x^2 + 2x + 7$$
$(f-g)(x) = f(x) - g(x)$
$$= (-5x^2 + 4x + 8) - (3x^2 - 2x - 1)$$
$$= (-5x^2 + 4x + 8) + (-3x^2 + 2x + 1)$$
$$= -8x^2 + 6x + 9$$

11. $(f \cdot g)(x) = f(x)g(x)$
$$= (2x)(x-5)$$
$$= 2x^2 - 10x$$

13. $(f \cdot g)(x) = f(x)g(x)$
$$= (x+1)(x-2)$$
$$= x^2 - 2x + x - 2$$
$$= x^2 - x - 2$$

15. $(f \cdot g)(x) = f(x)g(x)$
$$= (3x+2)(x+2)$$
$$= 3x^2 + 6x + 2x + 4$$
$$= 3x^2 + 8x + 4$$

17. $(f \cdot g)(x) = f(x)g(x)$
$$= (x^2 - 3x + 2)(x^2 + 2x - 7)$$
$$= x^4 + 2x^3 - 7x^2 - 3x^3 - 6x^2 + 21x$$
$$\quad + 2x^2 + 4x - 14$$
$$= x^4 - x^3 - 11x^2 + 25x - 14$$

19. $(f \cdot g)(x) = f(x)g(x)$
$$= (2x^2 - 3x + 1)(3x^2 - x - 1)$$
$$= 6x^4 - 2x^3 - 2x^2 - 9x^3 + 3x^2 + 3x$$
$$\quad + 3x^2 - x - 1$$
$$= 6x^4 - 11x^3 + 4x^2 + 2x - 1$$

21. $(f / g)(x) = \dfrac{f(x)}{g(x)} = \dfrac{2x^2 - 6x}{2x}$
$$= \dfrac{2x^2}{2x} - \dfrac{6x}{2x}$$
$$= x - 3, \ x \neq 0$$

23. $(f / g)(x) = \dfrac{f(x)}{g(x)} = \dfrac{6x^2 - 3x + 6}{3x}$
$$= \dfrac{6x^2}{3x} - \dfrac{3x}{3x} + \dfrac{6}{3x}$$
$$= 2x - 1 + \dfrac{2}{x}, \ x \neq 0$$

25. $(f / g)(x) = \dfrac{f(x)}{g(x)} = \dfrac{x^2 - 14x + 45}{x - 9}$

$$
\begin{array}{r}
x - 5 \\
x-9 \overline{)\, x^2 - 14x + 45} \\
\underline{x^2 - 9x} \\
-5x + 45 \\
\underline{-5x + 45} \\
0
\end{array}
$$

$(f / g)(x) = x - 5, \ x \neq 9$

27. $(f / g)(x) = \dfrac{f(x)}{g(x)} = \dfrac{2x^2 - 3x - 5}{2x - 5}$

$$
\begin{array}{r}
x + 1 \\
2x-5 \overline{)\, 2x^2 - 3x - 5} \\
\underline{2x^2 - 5x} \\
2x - 5 \\
\underline{2x - 5} \\
0
\end{array}
$$

$(f / g)(x) = x + 1, \ x \neq \dfrac{5}{2}$

29. $(f / g)(x) = \dfrac{f(x)}{g(x)} = \dfrac{2x^2 - 5x - 7}{x + 5}$

$$
\begin{array}{r}
2x - 15 \\
x+5 \overline{)\, 2x^2 - 5x - 7} \\
\underline{2x^2 + 10x} \\
-15x - 7 \\
\underline{-15x - 75} \\
68
\end{array}
$$

$(f / g)(x) = 2x - 15 + \dfrac{68}{x + 5}, \ x \neq -5$

31. a) $(p+q)(x) = p(x) + q(x)$
$= (5x-1) + (2x+3)$
$= 7x+2$

b) $(p-q)(x) = p(x) - q(x)$
$= (5x-1) - (2x+3)$
$= (5x-1) + (-2x-3)$
$= 3x-4$

c) $(p \cdot q)(x) = p(x) \cdot q(x)$
$= (5x-1)(2x+3)$
$= 10x^2 + 15x - 2x - 3$
$= 10x^2 + 13x - 3$

d) $(p/q)(x) = \dfrac{p(x)}{q(x)} = \dfrac{5x-1}{2x+3}, \quad x \neq -\dfrac{3}{2}$

33. a) $(p+q)(x) = p(x) + q(x)$
$= (2x^2 - 3x + 5) + (x-1)$
$= 2x^2 - 2x + 4$

b) $(p-q)(x) = p(x) - q(x)$
$= (2x^2 - 3x + 5) - (x-1)$
$= (2x^2 - 3x + 5) + (-x+1)$
$= 2x^2 - 4x + 6$

c) $(p \cdot q)(x) = p(x) \cdot q(x)$
$= (2x^2 - 3x + 5) \cdot (x-1)$
$= 2x^3 - 2x^2 - 3x^2 + 3x + 5x - 5$
$= 2x^3 - 5x^2 + 8x - 5$

d) $(p/q)(x) = \dfrac{p(x)}{q(x)} = \dfrac{2x^2 - 3x + 5}{x-1}$

$$\begin{array}{r} 2x-1 \\ x-1 \overline{)2x^2 - 3x + 5} \\ \underline{2x^2 - 2x} \\ -x+5 \\ \underline{-x+1} \\ 4 \end{array}$$

$(p/q)(x) = 2x - 1 + \dfrac{4}{x-1}, \quad x \neq 1$

35. a) $(p+q)(x) = 7x+2$
$(p+q)(-2) = 7(-2) + 2$
$= -14 + 2$
$= -12$

b) $(p-q)(x) = 3x-4$
$(p-q)(5) = 3(5) - 4$
$= 15 - 4$
$= 11$

c) $(p \cdot q)(x) = 10x^2 + 13x - 3$
$(p \cdot q)(-1) = 10(-1)^2 + 13(-1) - 3$
$= 10(1) - 13 - 3$
$= 10 - 13 - 3$
$= -6$

d) $(p/q)(x) = \dfrac{5x-1}{2x+3}$
$(p/q)(7) = \dfrac{5(7) - 1}{2(7) + 3}$
$= \dfrac{35 - 1}{14 + 3}$
$= \dfrac{34}{17}$
$= 2$

37. a) $(p+q)(x) = 2x^2 - 2x + 4$
$(p+q)(3) = 2(3)^2 - 2(3) + 4$
$= 2(9) - 6 + 4$
$= 18 - 6 + 4$
$= 16$

b) $(p-q)(x) = 2x^2 - 4x + 6$
$(p-q)(-4) = 2(-4)^2 - 4(-4) + 6$
$= 2(16) + 16 + 6$
$= 32 + 16 + 6$
$= 54$

c) $(p \cdot q)(x) = 2x^3 - 5x^2 + 8x - 5$

$(p \cdot q)(2) = 2(2)^3 - 5(2)^2 + 8(2) - 5$

$= 2(8) - 5(4) + 16 - 5$

$= 16 - 20 + 16 - 5$

$= 7$

d) $(p/q)(x) = 2x - 1 + \dfrac{4}{x - 1}$

$(p/q)(-3) = 2(-3) - 1 + \dfrac{4}{(-3) - 1}$

$= -6 - 1 + \dfrac{4}{-4}$

$= -6 - 1 - 1$

$= -8$

39. a) $(f + g)(x) = f(x) + g(x)$

$= (3x + 1) + (x + 5)$

$= 4x + 6$

b)

x	$f(x)$	$g(x)$	$(f + g)(x)$
-4	-11	1	-10
-2	-5	3	-2
0	1	5	6
2	7	7	14
4	13	9	22

c)

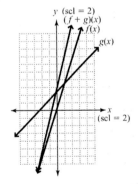

d) The output values for $f(x)$ added to the output values of $g(x)$ are equal to the output values of $(f + g)(x)$.

41. a) $(h - k)(x) = h(x) - k(x)$

$= (2x - 9) - (3x + 5)$

$= (2x - 9) + (-3x - 5)$

$= -x - 14$

b)

x	$h(x)$	$k(x)$	$(h - k)(x)$
-4	-17	-7	-10
-2	-13	-1	-12
0	-9	5	-14
2	-5	11	-16
4	-1	17	-18

c)

d) The output values for $k(x)$ subtracted from the output values of $h(x)$ are equal to the output values of $(h - k)(x)$.

43. a) $t(x) = w(x) + p(x)$

$= (3x + 2) + (x + 2)$

$= 4x + 4$

b) $t(x) = 4x + 4$

$t(200) = 4(200) + 4$

$= 800 + 4$

$= 804$

The total cost is $804.

c)

45. a) $p(x) = r(x) - c(x)$

$p(x) = (x^2 - 3x - 4) - (x + 8)$

$= (x^2 - 3x - 4) + (-x - 8)$

$= x^2 - 4x - 12$

b) $p(x) = x^2 - 4x - 12$

$p(100) = (100)^2 - 4(100) - 12$

$= 10,000 - 400 - 12$

$= 9588$

If $x = 100$, the net profit is \$9588.

c)

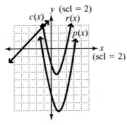

d) $p(x) = 0$

$x^2 - 4x - 12 = 0$

$(x - 6)(x + 2) = 0$

$x - 6 = 0$ or $x + 2 = 0$

$x = 6$ $x = -2$

A total of 6 units must be sold for the company to break even.

47. a) $A(x) = l(x) \cdot h(x)$

$= (3x + 2)(2x)$

$= 6x^2 + 4x$

b) $A(x) = 6x^2 + 4x$

$A(3) = 6(3)^2 + 4(3)$

$= 6(9) + 12$

$= 54 + 12$

$= 66$

The area of the front face is 66 square feet.

49. a) $h(x) = \dfrac{V(x)}{l(x) \cdot w(x)}$

$h(x) = \dfrac{4x^3 - 7x^2 - 14x - 3}{(x - 3)(x + 1)}$

$= \dfrac{4x^3 - 7x^2 - 14x - 3}{x^2 - 2x - 3}$

$$
\begin{array}{r}
4x+1 \\
x^2 - 2x - 3 \overline{)4x^3 - 7x^2 - 14x - 3} \\
\underline{4x^3 - 8x^2 - 12x} \\
x^2 - 2x - 3 \\
\underline{x^2 - 2x - 3} \\
0
\end{array}
$$

$h(x) = 4x + 1$

b) $h(x) = 4x + 1$

$h(4) = 4(4) + 1$

$= 16 + 1$

$= 17$

The height of the box is 17 cm.

Review Exercises

1. $f(x) = x^2 + 1$

$f(2) = (2)^2 + 1$

$= 4 + 1$

$= 5$

2. $f(x) = x^2 + 1$

$f(3a - 2) = (3a - 2)^2 + 1$

$= (9a^2 - 12a + 4) + 1$

$= 9a^2 - 12a + 5$

3. $f(x) = x^2 + 1$

$f(\sqrt{a - 1}) = (\sqrt{a - 1})^2 + 1$

$= (a - 1) + 1$

$= a$ for $a \geq 1$

4. $y = 2x + 4$

$y - 4 = 2x$

$\dfrac{y - 4}{2} = x$

5. $y = x^3 + 2$

$y - 2 = x^3$

$\sqrt[3]{y - 2} = x$

Chapter 9 Review Exercises

1. $x^2 = 16$

$x = \pm\sqrt{16}$

$x = \pm 4$

2. $y^2 = \dfrac{1}{36}$

$y = \pm\sqrt{\dfrac{1}{36}}$

$y = \pm\dfrac{1}{6}$

3. $k^2 + 2 = 30$

$k^2 = 28$

$k = \pm\sqrt{28}$

$k = \pm 2\sqrt{7}$

4. $3x^2 = 42$

$x^2 = \dfrac{42}{3}$

$x^2 = 14$

$x = \pm\sqrt{14}$

5. $5h^2 + 24 = 9$

$5h^2 = -15$

$h^2 = -3$

$h = \pm\sqrt{-3}$

$h = \pm i\sqrt{3}$

6. $(x + 7)^2 = 25$

$(x + 7)^2 = \pm\sqrt{25}$

$x + 7 = \pm 5$

$x = -7 \pm 5$

$= -7 + 5 = -2$

$= -7 - 5 = -12$

7. $(x - 9)^2 = -16$

$x - 9 = \pm\sqrt{-16}$

$x - 9 = \pm 4i$

$x = 9 \pm 4i$

8. $\left(m + \dfrac{3}{5}\right)^2 = \dfrac{16}{25}$

$m + \dfrac{3}{5} = \pm\sqrt{\dfrac{16}{25}}$

$m + \dfrac{3}{5} = \pm\dfrac{4}{5}$

$m = -\dfrac{3}{5} \pm \dfrac{4}{5}$

$= -\dfrac{3}{5} + \dfrac{4}{5} = \dfrac{1}{5}$

$= -\dfrac{3}{5} - \dfrac{4}{5} = -\dfrac{7}{5}$

9. $A = \pi r^2$

$22{,}500\pi = \pi r^2$

$0 = \pi r^2 - 22{,}500\pi$

$0 = \pi\left(r^2 - 22{,}500\right)$

$0 = r^2 - 22{,}500$

$0 = (r + 150)(r - 150)$

$r = \cancel{-150}, 150$

The radius of the circle was 150 feet.

10. $400 = \dfrac{1}{2}50v^2$

$400 = 25v^2$

$16 = v^2$

$4 = v$

The velocity was 4 m/sec.

11. $\pi r^2(4) = \dfrac{4}{3}\pi\left(9^3\right)$

$4\pi r^2 = 972\pi$

$4\pi r^2 - 972\pi = 0$

$4\pi\left(r^2 - 243\right) = 0$

$r^2 - 243 = 0$

$r^2 = 243$

$r = \sqrt{243}$

$r = 9\sqrt{3}$

The radius is $9\sqrt{3}$ inches.

12. $m^2 + 8m = -7$

$$\left(\frac{8}{2}\right)^2 = (4)^2 = 16$$

$m^2 + 8m + 16 = -7 + 16$

$(m+4)^2 = 9$

$m + 4 = \pm\sqrt{9}$

$m + 4 = \pm 3$

$m = -4 \pm 3$

$m = -4 - 3 = -7$

$m = -4 + 3 = -1$

13. $u^2 - 6u - 12 = 100$

$u^2 - 6u = 112$

$$\left(-\frac{6}{2}\right)^2 = (-3)^2 = 9$$

$u^2 - 6u + 9 = 112 + 9$

$(u-3)^2 = 121$

$u - 3 = \pm\sqrt{121}$

$u - 3 = \pm 11$

$u = 3 \pm 11$

$u = 3 + 11 = 14$

$u = 3 - 11 = -8$

14. $2b^2 - 6b + 7 = 0$

$b^2 - 3b = -\frac{7}{2}$

$$\left(\frac{-3}{2}\right)^2 = \frac{9}{4}$$

$b^2 - 3b + \frac{9}{4} = -\frac{7}{2} + \frac{9}{4}$

$\left(b - \frac{3}{2}\right)^2 = -\frac{5}{4}$

$b - \frac{3}{2} = \pm\sqrt{-\frac{5}{4}}$

$b - \frac{3}{2} = \pm\frac{i\sqrt{5}}{2}$

$b = \frac{3 \pm i\sqrt{5}}{2}$

15. $\left(\dfrac{\frac{1}{4}}{2}\right)^2 = \left(\dfrac{1}{8}\right)^2 = \dfrac{1}{64}$

$u^2 + \frac{1}{4}u + \frac{1}{64} = \frac{3}{4} + \frac{1}{64}$

$\left(u + \frac{1}{8}\right)^2 = \frac{49}{64}$

$u + \frac{1}{8} = \pm\sqrt{\frac{49}{64}}$

$u + \frac{1}{8} = \pm\frac{7}{8}$

$u = -\frac{1}{8} \pm \frac{7}{8}$

$u = -\frac{1}{8} - \frac{7}{8} = -1$

$u = -\frac{1}{8} + \frac{7}{8} = \frac{3}{4}$

16. $p^2 + 2p - 5 = 0$

$$p = \frac{-(2) \pm \sqrt{(2)^2 - 4(1)(-5)}}{2(1)}$$

$$= \frac{-2 \pm \sqrt{4 + 20}}{2}$$

$$= \frac{-2 \pm \sqrt{24}}{2}$$

$$= \frac{-2 \pm 2\sqrt{6}}{2}$$

$$= -1 \pm \sqrt{6}$$

17. $3x^2 - 2x + 1 = 0$

$$x = \frac{-(-2) \pm \sqrt{(-2)^2 - 4(3)(1)}}{2(3)}$$

$$= \frac{2 \pm \sqrt{4 - 12}}{6}$$

$$= \frac{2 \pm \sqrt{-8}}{6}$$

$$= \frac{2 \pm 2i\sqrt{2}}{6}$$

$$= \frac{1 \pm i\sqrt{2}}{3}$$

18. $2t^2 + t - 5 = 0$

$$t = \frac{-(1) \pm \sqrt{(1)^2 - 4(2)(-5)}}{2(2)}$$

$$= \frac{-1 \pm \sqrt{1 + 40}}{4}$$

$$= \frac{-1 \pm \sqrt{41}}{4}$$

19. Use $200x^2 + 10x - 3 = 0$.

$$x = \frac{-(10) \pm \sqrt{(10)^2 - 4(200)(-3)}}{2(200)}$$

$$= \frac{-10 \pm \sqrt{100 + 2400}}{400}$$

$$= \frac{-10 \pm \sqrt{2500}}{400}$$

$$= \frac{-10 \pm 50}{400}$$

$$= \frac{-10 - 50}{400} = -\frac{3}{20}$$

$$= \frac{-10 + 50}{400} = \frac{1}{10}$$

20. $x(x+4) = 285$

$x^2 + 4x - 285 = 0$

$(x+19)(x-15) = 0$

$x = \cancel{-19}, 15$

The width is 15 ft. and the length is 15 + 4 = 19 ft.

21. $x^2 + (x+9)^2 = 17^2$

$x^2 + x^2 + 18x + 81 = 289$

$2x^2 + 18x - 208 = 0$

$x^2 + 9x - 104 = 0$

$$x = \frac{-9 \pm \sqrt{9^2 - 4(1)(-104)}}{2(1)}$$

$$= \frac{-9 \pm \sqrt{81 + 416}}{2}$$

$$= \frac{-9 \pm \sqrt{497}}{2} \approx \cancel{-15.65}, 6.65$$

The height is 6.65 ft. and the base is 6.65 + 9 = 15.65 ft.

22. $b^2 - 4b - 12 = 0$

$D = (-4)^2 - 4(1)(-12) = 64$; 2 rational

23. $6z^2 - 7z + 5 = 0$

$D = (-7)^2 - 4(6)(5) = -71$; 2 nonreal complex

24. $k^2 + 6k + 9 = 0$

$D = (6)^2 - 4(1)(9) = 0$; 1 rational

25. $0.8x^2 + 1.2x + 0.3 = 0$

$D = (1.2)^2 - 4(0.8)(0.3) = 0.48$; 2 irrational

26.
$$\frac{1}{y} + \frac{1}{y+3} = \frac{2}{3}$$

$$3y(y+3)\left(\frac{1}{y} + \frac{1}{y+3}\right) = 3y(y+3) \cdot \frac{2}{3}$$

$$3(y+3) + 3y = 2y(y+3)$$

$$3y + 9 + 3y = 2y^2 + 6y$$

$$0 = 2y^2 - 9$$

$$2y^2 = 9$$

$$y^2 = \frac{9}{2}$$

$$y = \pm\sqrt{\frac{9}{2}}$$

$$y = \pm\frac{3}{\sqrt{2}} \cdot \frac{\sqrt{2}}{\sqrt{2}}$$

$$y = \pm\frac{3\sqrt{2}}{2}$$

27.
$$\frac{1}{u} + \frac{1}{u-5} = \frac{10}{u^2 - 25}$$

$$u(u-5)(u+5)\left(\frac{1}{u} + \frac{1}{u-5} = \frac{10}{(u-5)(u+5)}\right)$$

$$(u-5)(u+5) + u(u+5) = 10u$$

$$u^2 - 25 + u^2 + 5u = 10u$$

$$2u^2 - 5u - 25 = 0$$

$$(2u+5)(u-5) = 0$$

$$2u+5 = 0 \quad \text{or} \quad u-5 = 0$$

$$u = -\frac{5}{2} \qquad\qquad u = \cancel{5}$$

28. $6 - 5x^{-1} + x^{-2} = 0$

$$x^2\left(6 - 5x^{-1} + x^{-2}\right) = x^2 \cdot 0$$

$$6x^2 - 5x + 1 = 0$$

$$(3x - 1)(2x - 1) = 0$$

$$3x - 1 = 0 \quad \text{or} \quad 2x - 1 = 0$$

$$x = \frac{1}{3} \qquad\qquad x = \frac{1}{2}$$

29. $2 - 3x^{-1} - x^{-2} = 0$

$$x^2\left(2 - 3x^{-1} - x^{-2}\right) = x^2 \cdot 0$$

$$2x^2 - 3x - 1 = 0$$

$$x = \frac{-(-3) \pm \sqrt{(-3)^2 - 4(2)(-1)}}{2(2)}$$

$$= \frac{3 \pm \sqrt{9 + 8}}{4}$$

$$= \frac{3 \pm \sqrt{17}}{4}$$

30. $14\sqrt{x} + 45 = 0$

$$14\sqrt{x} = -45$$

$$\sqrt{x} = -\frac{45}{14}$$

$$\text{no solution}$$

31. $\sqrt{4m} = 3m - 1$

$$\left(\sqrt{4m}\right)^2 = (3m - 1)^2$$

$$4m = 9m^2 - 6m + 1$$

$$0 = 9m^2 - 10m + 1$$

$$0 = (9m - 1)(m - 1)$$

$$9m - 1 = 0 \quad \text{or} \quad m - 1 = 0$$

$$m = \cancel{\frac{1}{9}} \qquad\qquad m = 1$$

32. $\sqrt{6r + 13} = 2r + 1$

$$\left(\sqrt{6r + 13}\right)^2 = (2r + 1)^2$$

$$6r + 13 = 4r^2 + 4r + 1$$

$$0 = 4r^2 - 2r - 12$$

$$0 = 2r^2 - r - 6$$

$$0 = (2r + 3)(r - 2)$$

$$2r + 3 = 0 \qquad \text{or} \quad r - 2 = 0$$

$$r = \cancel{-\frac{3}{2}} \qquad\qquad r = 2$$

33. $\sqrt{21t + 2} + t = 2 + 4t$

$$\sqrt{21t + 2} = 2 + 3t$$

$$\left(\sqrt{21t + 2}\right)^2 = (2 + 3t)^2$$

$$21t + 2 = 4 + 12t + 9t^2$$

$$0 = 2 - 9t + 9t^2$$

$$0 = (2 - 3t)(1 - 3t)$$

$$2 - 3t = 0 \quad \text{or} \quad 1 - 3t = 0$$

$$t = \frac{2}{3} \qquad\qquad t = \frac{1}{3}$$

34.   Let $u = x^2$

$$u^2 - 5u + 6 = 0$$

$$(u - 2)(u - 3) = 0$$

$$u - 2 = 0 \quad \text{or} \quad u - 3 = 0$$

$$u = 2 \qquad\qquad u = 3$$

$$x^2 = 2 \qquad\qquad x^2 = 3$$

$$x = \pm\sqrt{2} \qquad\qquad x = \pm\sqrt{3}$$

35.   Let $u = m^2$

$$2u^2 - 3u + 1 = 0$$

$$(2u - 1)(u - 1) = 0$$

$$2u - 1 = 0 \qquad\qquad u - 1 = 0$$

$$u = \frac{1}{2} \qquad\qquad u = 1$$

$$\qquad\qquad\qquad m^2 = 1$$

$$m^2 = \frac{1}{2} \qquad\qquad m = \pm 1$$

$$m = \pm\sqrt{\frac{1}{2} \cdot \frac{2}{2}} = \pm\frac{\sqrt{2}}{2}$$

36.   Let $u = x + 5$

$$6u^2 - 5u + 1 = 0$$

$$(2u - 1)(3u - 1) = 0$$

$$2u - 1 = 0 \quad \text{or} \quad 3u - 1 = 0$$

$$u = \frac{1}{2} \qquad\qquad u = \frac{1}{3}$$

$$x + 5 = \frac{1}{2} \qquad\qquad x + 5 = \frac{1}{3}$$

$$x = -\frac{9}{2} \qquad\qquad x = -\frac{14}{3}$$

37. Let $u = \dfrac{x-1}{3}$

$$u^2 + 10u + 9 = 0$$
$$(u+1)(u+9) = 0$$

$u + 1 = 0$ or	$u + 9 = 0$
$u = -1$	$u = -9$
$\dfrac{x-1}{3} = -1$	$\dfrac{x-1}{3} = -9$
$x - 1 = -3$	$x - 1 = -27$
$x = -2$	$x = -26$

38. Let $u = p^{1/3}$

$$u^2 - 11u + 24 = 0$$
$$(u-8)(u-3) = 0$$

$u - 8 = 0$ or	$u - 3 = 0$
$u = 8$	$u = 3$
$p^{1/3} = 8$	$p^{1/3} = 3$
$\left(p^{1/3}\right)^3 = 8^3$	$\left(p^{1/3}\right)^3 = 3^3$
$p = 512$	$p = 27$

39. Let $u = a^{1/4}$

$$5u^2 + 13u - 6 = 0$$
$$(5u-2)(u+3) = 0$$

$5u - 2 = 0$ or	$u + 3 = 0$
$u = \dfrac{2}{5}$	$u = -3$
$a^{1/4} = \dfrac{2}{5}$	$a^{1/4} = -3$
$\left(a^{1/4}\right)^4 = \left(\dfrac{2}{5}\right)^4$	$\left(a^{1/4}\right)^4 = (-3)^4$
$a = \dfrac{16}{625}$	$a = \cancel{81}$

40. $f(x) = -2x^2$

a) x-intercept: $-2x^2 = 0$

$$x^2 = 0$$
$$x = \pm\sqrt{0}$$
$$x = 0$$
$$(0, 0)$$

y-intercept: $y = -2(0)^2 = 0$

(0, 0)

b) downward

c) (0, 0)

d) $x = 0$

e)

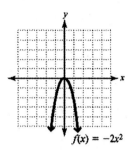

$f(x) = -2x^2$

41. $g(x) = \dfrac{1}{2}x^2 + 1$

a) x-intercept: $\dfrac{1}{2}x^2 + 1 = 0$

$$\frac{1}{2}x^2 = -1$$
$$x^2 = -2$$
$$x = \pm\sqrt{-2}$$
$$x = \pm i\sqrt{2}$$

Because these solutions are not real, there are no x-intercepts.

y-intercept: $y = \dfrac{1}{2}(0)^2 + 1 = 1$

(0, 1)

b) upward

c) (0, 1)

d) $x = 0$

e)

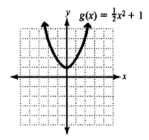

$g(x) = \frac{1}{2}x^2 + 1$

42. $h(x) = -\dfrac{1}{3}(x-2)^2$

 a) x-intercept: $-\dfrac{1}{3}(x-2)^2 = 0$

$$(x-2)^2 = 0$$
$$x - 2 = \pm\sqrt{0}$$
$$x = 2 \pm \sqrt{0}$$
$$x = 2$$
$$(2,0)$$

y-intercept: $y = -\dfrac{1}{3}(0-2)^2 = -\dfrac{1}{3}\cdot 4 = -\dfrac{4}{3}$

$$\left(0, -\dfrac{4}{3}\right)$$

 b) downward
 c) $(2, 0)$
 d) $x = 2$
 e)

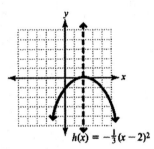

$h(x) = -\frac{1}{3}(x-2)^2$

43. $k(x) = 4(x+3)^2 - 2$

 a) x-intercepts: $0 = 4(x+3)^2 - 2$

$$\dfrac{2}{4} = (x+3)^2$$
$$\pm\dfrac{\sqrt{2}}{2} = x + 3$$
$$-3 \pm \dfrac{\sqrt{2}}{2} = x$$
$$\left(-3 + \dfrac{\sqrt{2}}{2}, 0\right), \left(-3 - \dfrac{\sqrt{2}}{2}, 0\right)$$

y-intercept: $y = 4(0+3)^2 - 2$
$$y = 4\cdot 9 - 2$$
$$y = 34$$
$$(0, 34)$$

 b) upward
 c) $(-3, -2)$
 d) $x = -3$

 e)

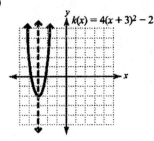

$k(x) = 4(x+3)^2 - 2$

44. a) $y = x^2 + 2x - 1$
$$y + 1 = x^2 + 2x$$
$$y + 1 + 1 = x^2 + 2x + 1$$
$$y + 2 = (x+1)^2$$
$$y = (x+1)^2 - 2$$
$$m(x) = (x+1)^2 - 2$$

 b) x-intercepts: $x = \dfrac{-2 \pm \sqrt{2^2 - 4(1)(-1)}}{2(1)}$

$$= \dfrac{-2 \pm 2\sqrt{2}}{2}$$
$$= -1 \pm \sqrt{2}$$
$$\left(-1 + \sqrt{2}, 0\right), \left(-1 - \sqrt{2}, 0\right)$$

y-intercept: $y = 0 + 2\cdot 0 - 1 = -1$
$$(0, -1)$$

 c) upward
 d) using part (a): $(-1, -2)$
 e) $x = -1$
 f)

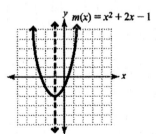

$m(x) = x^2 + 2x - 1$

 g) Domain: $\{x \mid x \text{ is a real number}\}$ or $(-\infty, \infty)$
   Range: $\{y \mid y \geq -2\}$ or $[-2, \infty)$

45. a) $y = -0.5x^2 + 4x - 6$

$y + 6 = -0.5x^2 + 4x$

$y + 6 = -0.5\left(x^2 - 8x\right)$

$y + 6 - 8 = -0.5\left(x^2 - 8x + 16\right)$

$y - 2 = -0.5\left(x - 4\right)^2$

$y = -0.5\left(x - 4\right)^2 + 2$

$p(x) = -0.5\left(x - 4\right)^2 + 2$

b) x-intercepts: $0 = -0.5x^2 + 4x - 6$

$0 = 5x^2 - 40x + 60$

$0 = x^2 - 8x + 12$

$0 = (x - 2)(x - 6)$

$x - 2 = 0 \qquad x - 6 = 0$

$x = 2 \qquad x = 6$

$(2, 0), (6, 0)$

y-intercept: $y = -0.5 \cdot 0 + 4 \cdot 0 - 6 = -6$

$(0, -6)$

c) downward

d) from part (a): (4, 2)

e) $x = 4$

f)

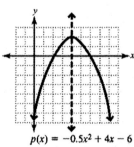

$p(x) = -0.5x^2 + 4x - 6$

g) Domain: $\{x \mid x \text{ is a real number}\}$ or $(-\infty, \infty)$

Range: $\{y \mid y \le 2\}$ or $(-\infty, 2]$

46. Finding the vertex will give the highest point that the acrobat will reach.

$t = \dfrac{-24}{2(-16)} = \dfrac{3}{4} = 0.75$

$h = -16(0.75)^2 + 24(0.75)$

$h = -9 + 18$

$h = 9$

The vertex is at $(0.75, 9)$

a) 0.75 sec.

b) 9 ft.

c) $-16t^2 + 24t = 0$

$-8t(2t - 3) = 0$

$t = \cancel{0}, 1.5 \text{ sec.}$

d)

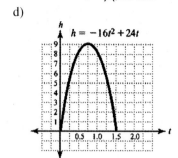

47. a) Find the vertex. $x = \dfrac{-2.16}{2(-0.03)} = 36$

$y = -0.03(36)^2 + 2.16(36)$

$y = 38.88$

The maximum height is 38.88 yd.

b) The distance of the punt is the x-coordinate of the vertex times 2, or 72 yd.

48. $(x + 5)(x - 3) = 0$

$x + 5 = 0 \quad$ or $\quad x - 3 = 0$

$x = -5 \qquad\qquad x = 3$

Interval	$(-\infty, -5)$	$(-5, 3)$	$(3, \infty)$
Test No.	-6	0	4
Results	$9 > 0$	$-15 > 0$	$9 > 0$
T/F	T	F	T

Solution set: $(-\infty, -5) \cup (3, \infty)$

49. $n^2 - 6n + 8 = 0$

$(n - 4)(n - 2) = 0$

$n - 4 = 0 \quad$ or $\quad n - 2 = 0$

$n = 4 \qquad\qquad n = 2$

Interval	$(-\infty, 2)$	$(2, 4)$	$(4, \infty)$
Test No.	0	3	5
Results	$0 \le -8$	$-9 \le -8$	$-5 \le -8$
T/F	F	T	F

Solution set: [2, 4]

50. $x^2 + 9x + 14 = 0$

$(x+7)(x+2) = 0$

$x + 7 = 0$ or $x + 2 = 0$

$x = -7$ $x = -2$

Interval	$(-\infty, -7)$	$(-7, -2)$	$(-2, \infty)$
Test No.	-8	-3	0
Results	$6 < 0$	$-4 < 0$	$14 < 0$
T/F	F	T	F

Solution set: $(-7, -2)$

51. $(x+3)(x-1)(x-2) = 0$

$x + 3 = 0$ or $x - 1 = 0$ or $x - 2 = 0$

$x = -3$ $x = 1$ $x = 2$

Interval	$(-\infty, -3)$	$(-3, 1)$	$(1, 2)$	$(2, \infty)$
Test No.	-4	0	1.5	3
Results	-30	6	-1.125	12
T/F	F	T	F	T

Solution set: $[-3, 1] \cup [2, \infty)$

52. a) $56 \leq \dfrac{1}{2}(x+4)(x-2)$

$112 \leq x^2 + 2x - 8$

$0 \leq x^2 + 2x - 120$

$0 \leq (x+12)(x-10)$

$x \leq -12$ or $x \geq 10$

Since x must be at least 2, $x \geq 10$ in.

b) height ≥ 8 in., base ≥ 14 in.

53. $\dfrac{a+3}{a-1} = 0$ $a - 1 = 0$

$a = 1$

$(a-1)\dfrac{a+3}{a-1} = 0(a-1)$

$a + 3 = 0$

$a = -3$

Interval	$(-\infty, -3)$	$(-3, 1)$	$(1, \infty)$
Test No.	-5	0	2
Results	$1/3 \geq 0$	$-3 \geq 0$	$5 \geq 0$
T/F	T	F	T

Solution set: $(-\infty, -3] \cup (1, \infty)$

54. $\dfrac{r}{r+2} = 2$ $r + 2 = 0$

$r = -2$

$(r+2)\dfrac{r}{r+2} = 2(r+2)$

$r = 2r + 4$

$-4 = r$

Interval	$(-\infty, -4)$	$(-4, -2)$
Test No.	-5	-3
Results	$5/3 < 2$	$3 < 2$
T/F	T	F

Interval	$(-2, \infty)$
Test No.	0
Results	$0 < 2$
T/F	T

Solution set: $(-\infty, -4) \cup (-2, \infty)$

55. $\dfrac{n-3}{n-4} = 5$ $n - 4 = 0$

$n = 4$

$(n-4)\dfrac{n-3}{n-4} = 5(n-4)$

$n - 3 = 5n - 20$

$-4n = -17$

$n = \dfrac{17}{4}$

Interval	$(-\infty, 4)$	$\left(4, \dfrac{17}{4}\right)$
Test No.	0	4.1
Results	$3/4 \leq 5$	$11 \leq 5$
T/F	T	F

Interval	$\left(\dfrac{17}{4}, \infty\right)$
Test No.	5
Results	$2 \leq 5$
T/F	T

Solution set: $(-\infty, 4) \cup \left[\dfrac{17}{4}, \infty\right)$

56. $\dfrac{(k+2)(k-3)}{k-5} = 0$

$(k-5)\dfrac{(k+2)(k-3)}{k-5} = 0(k-5)$

$(k+2)(k-3) = 0$

$k+2 = 0 \qquad k-3 = 0 \qquad k-5 = 0$

$k = -2 \qquad k = 3 \qquad k = 5$

Interval	$(-\infty, -2)$	$(-2, 3)$	$(3, 5)$	$(5, \infty)$
Test No.	-3	0	4	6
Results	$-3/4 < 0$	$\dfrac{6}{5} < 0$	$-6 < 0$	$24 < 0$
T/F	T	F	T	F

Solution set: $(-\infty, -2) \cup (3, 5)$

57. a) $(f+g)(x) = f(x) + g(x)$

$= (-3x+9) + (4x+1)$

$= x + 10$

b) $(f-g)(x) = f(x) - g(x)$

$= (-3x+9) - (4x+1)$

$= (-3x+9) + (-4x-1)$

$= -7x + 8$

c) $(f \cdot g)(x) = f(x) \cdot g(x)$

$= (-3x+9)(4x+1)$

$= -12x^2 - 3x + 36x + 9$

$= -12x^2 + 33x + 9$

58. a) $(f+g)(x) = f(x) + g(x)$

$= (-x-2) + (8x+2)$

$= 7x$

b) $(f-g)(x) = f(x) - g(x)$

$= (-x-2) - (8x+2)$

$= (-x-2) + (-8x-2)$

$= -9x - 4$

c) $(f+g)(x) = f(x) \cdot g(x)$

$= (-x-2)(8x+2)$

$= -8x^2 - 2x - 16x - 4$

$= -8x^2 - 18x - 4$

59. a) $(f+g)(x) = f(x) + g(x)$

$= (3x^2 - x + 5) + (x+2)$

$= 3x^2 + 7$

b) $(f-g)(x) = f(x) - g(x)$

$= (3x^2 - x + 5) - (x+2)$

$= (3x^2 - x + 5) + (-x-2)$

$= 3x^2 - 2x + 3$

c) $(f \cdot g)(x) = f(x) \cdot g(x)$

$= (3x^2 - x + 5)(x+2)$

$= 3x^3 + 6x^2 - x^2 - 2x + 5x + 10$

$= 3x^3 + 5x^2 + 3x + 10$

60. a) $(f+g)(x) = f(x) + g(x)$

$= (x^2 - 2x - 1) + (3x+1)$

$= x^2 + x$

b) $(f-g)(x) = f(x) - g(x)$

$= (x^2 - 2x - 1) - (3x+1)$

$= (x^2 - 2x - 1) + (-3x-1)$

$= x^2 - 5x - 2$

c) $(f \cdot g)(x) = f(x) \cdot g(x)$

$= (x^2 - 2x - 1)(3x+1)$

$= 3x^3 + x^2 - 6x^2 - 2x - 3x - 1$

$= 3x^3 - 5x^2 - 5x - 1$

61. $(f/g)(x) = \dfrac{f(x)}{g(x)} = \dfrac{6x^4 - 15x^3 + 12x^2}{3x^2}$

$= \dfrac{6x^4}{3x^2} - \dfrac{15x^3}{3x^2} + \dfrac{12x^2}{3x^2}$

$= 2x^2 - 5x + 4, \; x \neq 0$

62. $(f / g)(x) = \dfrac{f(x)}{g(x)}$

$= \dfrac{2x^3 - 17x^2 + 36x - 17}{2x - 5}, \; x \neq \dfrac{5}{2}$

$$
\begin{array}{r}
x^2 - 6x + 3 \\
2x-5{\overline{\smash{\big)}\,2x^3 - 17x^2 + 36x - 17}} \\
\underline{2x^3 - 5x^2} \\
-12x^2 + 36x \\
\underline{-12x^2 + 30x} \\
6x - 17 \\
\underline{6x - 15} \\
-2
\end{array}
$$

$(f / g)(x) = x^2 - 6x + 3 + \dfrac{-2}{2x - 5}, \; x \neq \dfrac{5}{2}$

63. a) $c(x) = w(x) + p(x)$

$= (3x + 5) + (x - 1)$

$= 4x + 4$

   b) $c(x) = 4x + 4$

$c(225) = 4(225) + 4$

$= 900 + 4$

$= 904$

The total cost is \$904.

   c)

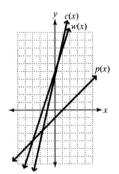

64. a) $h(x) = \dfrac{V(x)}{l(x) \cdot w(x)}$

$h(x) = \dfrac{6x^3 + 41x^2 + 26x - 24}{(3x + 4)(x + 6)}$

$= \dfrac{6x^3 + 41x^2 + 26x - 24}{3x^2 + 22x + 24}$

$$
\begin{array}{r}
2x - 1 \\
3x^2 + 22x + 24{\overline{\smash{\big)}\,6x^3 + 41x^2 + 26x - 24}} \\
\underline{6x^3 + 44x^2 + 48x} \\
-3x^2 - 22x - 24 \\
\underline{-3x^2 - 22x - 24} \\
0
\end{array}
$$

$h(x) = 2x - 1$

   b) $h(x) = 2x - 1$

$h(9) = 2(9) - 1$

$= 18 - 1$

$= 17$

The height of the box is 17 in.

Chapter 9 Practice Test

1. $x^2 = 81$

$x = \pm\sqrt{81}$

$x = \pm 9$

2. $(x - 3)^2 = 20$

$x - 3 = \pm\sqrt{20}$

$x - 3 = \pm\sqrt{4 \cdot 5}$

$x - 3 = \pm 2\sqrt{5}$

$x = 3 \pm 2\sqrt{5}$

3. $x^2 - 8x = -4$

$x^2 - 8x + 16 = -4 + 16$

$(x - 4)^2 = 12$

$x - 4 = \pm\sqrt{12}$

$x - 4 = \pm\sqrt{4 \cdot 3}$

$x - 4 = \pm 2\sqrt{3}$

$x = 4 \pm 2\sqrt{3}$

4. $3m^2 - 6m = 5$

$$\frac{3m^2 - 6m}{3} = \frac{5}{3}$$

$$m^2 - 2m = \frac{5}{3}$$

$$m^2 - 2m + 1 = \frac{5}{3} + 1$$

$$(m-1)^2 = \frac{8}{3}$$

$$m - 1 = \pm\sqrt{\frac{8}{3}}$$

$$m - 1 = \pm\frac{2\sqrt{2}}{\sqrt{3}}$$

$$m - 1 = \pm\frac{2\sqrt{2}}{\sqrt{3}} \cdot \frac{\sqrt{3}}{\sqrt{3}}$$

$$m - 1 = \frac{2\sqrt{6}}{3}$$

$$m = 1 \pm \frac{2\sqrt{6}}{3}$$

$$m = \frac{3 \pm 2\sqrt{6}}{3}$$

5. $2x^2 + x - 6 = 0$

Let $a = 2$, $b = 1$, and $c = -6$.

$$x = \frac{-(1) \pm \sqrt{(1)^2 - 4(2)(-6)}}{2(2)}$$

$$= \frac{-1 \pm \sqrt{1 + 48}}{4}$$

$$= \frac{-1 \pm \sqrt{49}}{4}$$

$$= \frac{-1 \pm 7}{4}$$

$$= -2, \frac{3}{2}$$

6. $x^2 - 8x + 15 = 0$

Let $a = 1$, $b = -8$, and $c = 15$.

$$x = \frac{-(-8) \pm \sqrt{(-8)^2 - 4(1)(15)}}{2(1)}$$

$$= \frac{8 \pm \sqrt{64 - 60}}{2}$$

$$= \frac{8 \pm \sqrt{4}}{2}$$

$$= \frac{8 \pm 2}{2}$$

$$= 3, 5$$

7. $u^2 - 16 = -6u$

$$u^2 + 6u - 16 = 0$$

$$(u+8)(u-2) = 0$$

$$u + 8 = 0 \quad \text{or} \quad u - 2 = 0$$

$$u = -8 \qquad\qquad u = 2$$

8. $4w^2 + 6w + 3 = 0$

Let $a = 4$, $b = 6$, and $c = 3$.

$$x = \frac{-(6) \pm \sqrt{(6)^2 - 4(4)(3)}}{2(4)}$$

$$= \frac{-6 \pm \sqrt{36 - 48}}{8}$$

$$= \frac{-6 \pm \sqrt{-12}}{8}$$

$$= \frac{-6 \pm 2i\sqrt{3}}{8} = \frac{-3 \pm i\sqrt{3}}{4}$$

9. $x^2 + 16 = 0$

$$x^2 = -16$$

$$x = \pm\sqrt{-16}$$

$$x = \pm 4i$$

10. $3k^2 = -5k$

$$3k^2 + 5k = 0$$

$$k(3k + 5) = 0$$

$$k = 0 \quad \text{or} \quad 3k + 5 = 0$$

$$k = -\frac{5}{3}$$

11.
$$\frac{1}{x+2}+\frac{1}{x}=\frac{5}{12}$$

$$12x(x+2)\left(\frac{1}{x+2}+\frac{1}{x}\right)=12x(x+2)\cdot\frac{5}{12}$$

$$12x+12(x+2)=5x(x+2)$$

$$12x+12x+24=5x^2+10x$$

$$0=5x^2-14x-24$$

$$0=(5x+6)(x-4)$$

$$5x+6=0 \quad \text{or} \quad x-4=0$$

$$x=-\frac{6}{5} \qquad\qquad x=4$$

12.
$$3-x^{-1}-2x^{-2}=0$$

$$x^2\left(3-x^{-1}-2x^{-2}\right)=x^2\cdot 0$$

$$3x^2-x-2=0$$

$$(3x+2)(x-1)=0$$

$$3x+2=0 \quad \text{or} \quad x-1=0$$

$$x=-\frac{2}{3} \qquad\qquad x=1$$

13. $9\sqrt{x}+8=0 \qquad \text{No solution}$

$$9\sqrt{x}=-8$$

$$\sqrt{x}=-\frac{8}{9}$$

14. $\sqrt{x+8}-x=2$

$$\sqrt{x+8}=x+2$$

$$\left(\sqrt{x+8}\right)^2=(x+2)^2$$

$$x+8=x^2+4x+4$$

$$0=x^2+3x-4$$

$$0=(x+4)(x-1)$$

$$x+4=0 \quad \text{or} \quad x-1=0$$

$$x=\cancel{-4} \qquad\qquad x=1$$

The solution -4 is extraneous. The only solution is 1.

15. Let $u=a^2$.

$$9u^2+26u-3=0$$

$$(9u-1)(u+3)=0$$

$$9u-1=0 \quad \text{or} \quad u+3=0$$

$$u=\frac{1}{9} \qquad\qquad u=-3$$

$$x^2=\frac{1}{9} \qquad\qquad x^2=-3$$

$$x=\pm\frac{1}{3} \qquad\qquad x=\pm i\sqrt{3}$$

16. Let $u=x+1$.

$$u^2+3u-4=0$$

$$(u+4)(u-1)=0$$

$$u+4=0 \quad \text{or} \quad u-1=0$$

$$u=-4 \qquad\qquad u=1$$

$$x+1=-4 \qquad\qquad x+1=1$$

$$x=-5 \qquad\qquad x=0$$

17. a) x-intercepts:

$$-x^2+6x-4=0$$

$$x^2-6x+4=0$$

$$x=\frac{-(-6)\pm\sqrt{(-6)^2-4(1)(4)}}{2(1)}$$

$$=\frac{6\pm\sqrt{36-16}}{2}$$

$$=\frac{6\pm\sqrt{20}}{2}$$

$$=\frac{6\pm 2\sqrt{5}}{2}$$

$$=3\pm\sqrt{5}$$

The x-intercepts are $\left(3+\sqrt{5},0\right),\left(3-\sqrt{5},0\right)$.

y-intercept: Let $x=0$.

$$y=-x^2+6x-4$$

$$y=-(0)^2+6(0)-4$$

$$y=-4$$

The y-intercept is $(0, -4)$.

b)
$$y=-x^2+6x-4$$

$$y+4=-\left(x^2-6x\right)$$

$$y+4-9=-\left(x^2-6x+9\right)$$

$$y-5=-(x-3)^2$$

$$f(x)=-(x-3)^2+5$$

c) Because $a < 0$, the parabola opens downward.

d) From part (b) we see that $h = 3$ and $k = 5$. Therefore, the vertex is $(h, k) = (3, 5)$.

e) The axis of symmetry is given by $x = h$. Therefore, the axis of symmetry for this parabola is $x = 3$.

f)

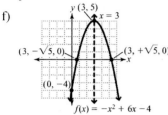

g) The domain is $\{x \mid x \text{ is a real number}\}$ or $(-\infty, \infty)$. The range is $\{y \mid y \leq 5\}$ or $(-\infty, 5]$.

18. $(x+1)(x-4) = 0$

$x + 1 = 0 \quad$ or $\quad x - 4 = 0$

$x = -1 \qquad\qquad x = 4$

Interval	$(-\infty, -1]$	$[-1, 4]$	$[4, \infty)$
Test No.	-2	0	5
Results	$6 \leq 0$	$-4 \leq 0$	$6 \leq 0$
T/F	F	T	F

Solution set: $[-1, 4]$

![number line from -6 to 6 with solid dots at -1 and 4, shaded between]

19. $\dfrac{x+2}{x-1} = 0 \qquad\qquad x - 1 = 0$

$\qquad\qquad\qquad\qquad x = 1$

$(x-1)\dfrac{x+2}{x-1} = 0(x-1)$

$x + 2 = 0$

$x = -2$

Interval	$(-\infty, -2)$	$(-2, 1)$	$(1, \infty)$
Test No.	-5	0	2
Results	$1/2 > 0$	$-2 > 0$	$4 > 0$
T/F	T	F	T

Solution set: $(-\infty, -2) \cup (1, \infty)$

![number line from -6 to 6 with open parenthesis at -2 and 1]

20. $f(x) = x^2$, $g(x) = 3x^2 - 2$

a) $(f + g)(x) = f(x) + g(x)$
$$= \left(x^2\right) + \left(3x^2 - 2\right)$$
$$= 4x^2 - 2$$

b) $(f - g)(x) = f(x) - g(x)$
$$= \left(x^2\right) - \left(3x^2 - 2\right)$$
$$= \left(x^2\right) + \left(-3x^2 + 2\right)$$
$$= -2x^2 + 2$$

c) $(f \cdot g)(x) = f(x)g(x)$
$$= \left(x^2\right)\left(3x^2 - 2\right)$$
$$= 3x^4 - 2x^2$$

21. $f(x) = 20x^4 + 36x^3 - 28x^2$ and $g(x) = 4x^2$

$$(f / g)(x) = \frac{f(x)}{g(x)} = \frac{20x^4 + 36x^3 - 28x^2}{4x^2}$$
$$= \frac{20x^4}{4x^2} + \frac{36x^3}{4x^2} - \frac{28x^2}{4x^2}$$
$$= 5x^2 + 9x - 7$$

22. Let $d = 180$.

$$16t^2 + 32t = d$$
$$16t^2 + 32t = 180$$
$$16t^2 + 32t - 180 = 0$$
$$4\left(4t^2 + 8t - 45\right) = 0$$
$$4t^2 + 8t - 45 = 0$$
$$(2t - 5)(2t + 9) = 0$$

$2t - 5 = 0 \quad$ or $\quad 2t + 9 = 0$

$2t = 5 \qquad\qquad 2t = -9$

$t = \dfrac{5}{2} \qquad\qquad t = -\dfrac{9}{2}$

The solution $-\dfrac{9}{2}$ does not make sense in the context of the problem. Therefore, the ball takes 2.5 seconds to fall 180 feet.

23. Finding the vertex will show the maximum height that the arrow will reach.

$$x = \frac{-1.3}{2(-0.02)} = 32.5$$

$$y = -0.02(32.5)^2 + 1.3(32.5) + 8 = 29.125$$

The vertex is (32.5, 29.125).

a) The maximum height is 29.125 m.

b) $-0.02x^2 + 1.3x + 8 = 0$

$$x = \frac{-1.3 \pm \sqrt{1.3^2 - 4(-0.02)(8)}}{2(-0.02)}$$

$$x = \frac{-1.3 \pm \sqrt{2.33}}{-0.04}$$

$$x \approx 70.66, \cancel{-5.66}$$

The distance can't be negative, so the arrow travels 70.66 m.

c)

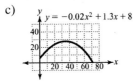

24. a)
$$12w(w + 15) \le 12{,}000$$

$$12w^2 + 180w - 12{,}000 \le 0$$

$$w^2 + 15w - 1000 \le 0$$

$$(w + 40)(w - 25) \le 0$$

$$-40 \le w \le 25$$

Since $w > 0$, $0 < w \le 25$ ft.

b) $15 < l \le 40$ ft.

25. a)
$$A(x) = l(x) \cdot h(x)$$

$$\frac{A(x)}{l(x)} = h(x)$$

$$\frac{36x^2 - x - 2}{4x - 1} = h(x)$$

$$\frac{\cancel{(4x-1)}(9x + 2)}{\cancel{4x-1}} = h(x)$$

$$9x + 2 = h(x)$$

b) Find $h(15)$.

$$h(x) = 9x + 2$$

$$h(15) = 9(15) + 2$$

$$= 135 + 2$$

$$= 137$$

The height of the box is 137 cm

Chapters 1–9 Cumulative Review

1. False; $(45, -12)$ is in quadrant IV.

2. False; a system of equations that has no solution is said to be inconsistent.

3. True

4. parabola, $a > 0$

5. $\sqrt[n]{ab}$

6. $\sqrt{a}, -\sqrt{a}$

7. $(-3x - 4) - (3x + 2) = (-3x - 4) + (-3x - 2)$
$$= -6x - 6$$

8. $(8n^2)(-7mn^3) = -56mn^5$

9. $(x - 3)(4x^2 - 2x + 1)$
$$= 4x^3 - 2x^2 + x - 12x^2 + 6x - 3$$
$$= 4x^3 - 14x^2 + 7x - 3$$

10. $m^3 + 8 = (m + 2)(m^2 - 2m + 4)$

11. $x^2 + 10x + 25 = (x + 5)^2$

12. $\dfrac{4u^2 + 4u + 1}{u + 2u^2} \cdot \dfrac{u}{2u^2 - u - 1}$

$$= \frac{\cancel{(2u+1)}\,(2u+1)}{\cancel{u}\,(1 + 2u)} \cdot \frac{\cancel{u}}{\cancel{(2u+1)}\,(u - 1)}$$

$$= \frac{1}{u - 1}$$

13. $\dfrac{3}{x - 2} - \dfrac{2}{x + 2}$

$$= \frac{3(x + 2)}{(x - 2)(x + 2)} - \frac{2(x - 2)}{(x + 2)(x - 2)}$$

$$= \frac{3x + 6 - 2x + 4}{(x - 2)(x + 2)}$$

$$= \frac{x + 10}{(x - 2)(x + 2)}$$

14. $x^{3/4} \cdot x^{-1/4} = x^{3/4 + (-1/4)} = x^{1/2}$

15. $\dfrac{\sqrt{n}}{\sqrt{m}-\sqrt{n}} = \dfrac{\sqrt{n}}{\sqrt{m}-\sqrt{n}} \cdot \dfrac{\sqrt{m}+\sqrt{n}}{\sqrt{m}+\sqrt{n}}$

$= \dfrac{\sqrt{mn}+\sqrt{n^2}}{\left(\sqrt{m}\right)^2 - \left(\sqrt{n}\right)^2}$

$= \dfrac{\sqrt{mn}+n}{m-n}$

16. $3n-5 = 7n+9$

$-4n-5 = 9$

$-4n = 14$

$n = -\dfrac{7}{2}$

17. $|3x+4|+3 = 7$

$|3x+4| = 4$

$3x+4 = -4 \quad \text{or} \quad 3x+4 = 4$

$3x = -8 \qquad\qquad 3x = 0$

$x = -\dfrac{8}{3} \qquad\qquad x = 0$

18. $2x^2 + 7x = 15$

$2x^2 + 7x - 15 = 0$

$(2x-3)(x+5) = 0$

$2x-3 = 0 \quad \text{or} \quad x+5 = 0$

$x = \dfrac{3}{2} \qquad\qquad x = -5$

19. $\dfrac{5}{x-2} - \dfrac{3}{x} = \dfrac{11}{3x}$

$3x(x-2)\left(\dfrac{5}{x-2} - \dfrac{3}{x}\right) = 3x(x-2)\cdot\dfrac{11}{3x}$

$3x\cdot 5 - 3(x-2)\cdot 3 = 11(x-2)$

$15x - 9x + 18 = 11x - 22$

$6x + 18 = 11x - 22$

$40 = 5x$

$8 = x$

20. $\sqrt{5x-4} = 9$

$\left(\sqrt{5x-4}\right)^2 = 9^2$

$5x - 4 = 81$

$5x = 85$

$x = 17$

21. $x^2 - 6x + 11 = 0$

Let $a = 1$, $b = -6$, and $c = 11$.

$x = \dfrac{-(-6) \pm \sqrt{(-6)^2 - 4(1)(11)}}{2(1)}$

$= \dfrac{6 \pm \sqrt{36-44}}{2}$

$= \dfrac{6 \pm \sqrt{-8}}{2}$

$= \dfrac{6 \pm 2i\sqrt{2}}{2}$

$= 3 \pm i\sqrt{2}$

22. a) $-2 < x+4 < 5$

$-6 < x < 1$

$(-6,1)$

b) $\{x \mid -6 < x < 1\}$ c) $(-6,1)$

23. $x^2 + x = 12$

$x^2 + x - 12 = 0$

$(x+4)(x-3) = 0$

$x+4 = 0 \quad \text{or} \quad x-3 = 0$

$x = -4 \qquad\qquad x = 3$

Interval	$(-\infty,-4)$	$(-4,3)$
Test No.	-5	0
Results	$20 \le 12$	$0 \le 12$
T/F	F	T

Interval	$(3,\infty)$
Test No.	4
Results	$20 \le 12$
T/F	F

a)

b) $\{x \mid -4 \le x \le 3\}$ c) $[-4,3]$

24. $f(x) = -2x$

x	$f(x)$	(x, y)
-1	2	$(-1, 2)$
0	0	$(0, 0)$
1	-2	$(1, -2)$

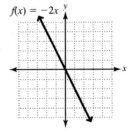

$f(x) = -2x$

25. Solve for y. $3x - 2y = 6$

$$-2y = -3x + 6$$

$$y = \frac{3}{2}x - 3$$

So $m = \frac{3}{2}$. The perpendicular slope is $-\frac{2}{3}$.

Use $m = -\frac{2}{3}$ and $(-2, 4)$.

$$y - 4 = -\frac{2}{3}\left(x - (-2)\right)$$

$$y - 4 = -\frac{2}{3}x - \frac{4}{3}$$

$$3y - 12 = -2x - 4$$

$$2x + 3y = 8$$

26. $\begin{cases} x + y + z = 5 & \text{Eqtn. 1} \\ 2x + y - 2z = -5 & \text{Eqtn. 2} \\ x - 2y + z = 8 & \text{Eqtn. 3} \end{cases}$

Multiply equation 1 by -1 and add to equation 3. This makes equation 4.

$$\begin{array}{r} -x - y - z = -5 \\ \underline{x - 2y + z = 8} \\ -3y = 3 \\ y = -1 \quad \text{Eqtn. 4} \end{array}$$

Multiply equation 1 by -2 and add to equation 2. This makes equation 5.

$$\begin{array}{r} -2x - 2y - 2z = -10 \\ \underline{2x + y - 2z = -5} \\ -y - 4z = -15 \quad \text{Eqtn. 5} \end{array}$$

Substitute equation 4 into equation 5 and solve for z.

$$-(-1) - 4z = -15$$

$$1 - 4z = -15$$

$$-4z = -16$$

$$z = 4$$

Substitute the values for y and z to solve for x.

$$x - 1 + 4 = 5$$

$$x + 3 = 5$$

$$x = 2$$

Solution: $(2, -1, 4)$

27. Complete a table.

	Rate	Time	Distance
not experienced	$\dfrac{300}{t} - 10$	$t + 1$	300
experienced	$\dfrac{300}{t}$	t	300

$$\left(\frac{300}{t} - 10\right)(t + 1) = 300$$

$$300 + \frac{300}{t} - 10t - 10 = 300$$

$$\frac{300}{t} - 10t - 10 = 0$$

$$300 - 10t^2 - 10t = 0$$

$$t^2 + t - 30 = 0$$

$$(t + 6)(t - 5) = 0$$

$$t + 6 = 0 \qquad t - 5 = 0$$

$$t = \cancel{-6} \qquad t = 5$$

It takes the experienced driver 5 hours to drive 300 miles.

28. Create a table.

	Concentrate	Vol. of solution	Vol. of saline
15%	0.15	50	$0.15(50)$
40%	0.40	x	$0.40x$
30%	0.30	$x + 50$	$0.30(x + 50)$

Translate the information in the table to an equation and solve.

$$0.15(50) + 0.40x = 0.3(x + 50)$$

$$7.5 + 0.4x = 0.3x + 15$$

$$0.1x = 7.5$$

$$x = 75$$

75 ml of 40% solution must be added.

29.$\quad v = \dfrac{k}{T} \quad$ So, $\quad v = \dfrac{48}{T}$

$\qquad 6 = \dfrac{k}{8} \qquad\qquad v = \dfrac{48}{15}$

$\qquad 48 = k \qquad\qquad v = 3.2$ cm/sec.

30. a)$\; t = \sqrt{\dfrac{24}{16}} = \dfrac{\sqrt{24}}{4} = \dfrac{\sqrt{4 \cdot 6}}{4} = \dfrac{2\sqrt{6}}{4} = \dfrac{\sqrt{6}}{2}$ sec.

$\quad$ b)$\qquad t = \sqrt{\dfrac{h}{16}}$

$\qquad\qquad 3 = \dfrac{\sqrt{h}}{4}$

$\qquad\qquad 12 = \sqrt{h}$

$\qquad\qquad 12^2 = \left(\sqrt{h}\right)^2$

$\qquad\quad 144$ ft. $= h$

Chapter 10

Exponential and Logarithmic Functions

Exercise Set 10.1

1. $(f \circ g)(0) = f[g(0)]$
   $= f(3)$
   $= 3 \cdot 3 + 5$
   $= 14$

3. $(h \circ f)(1) = h[f(1)]$
   $= h(8)$
   $= \sqrt{8+1}$
   $= 3$

5. $(f \circ g)(-2) = f[g(-2)]$
   $= f(7)$
   $= 3 \cdot 7 + 5$
   $= 26$

7. $(f \circ g)(x) = f[g(x)]$
   $= f(x^2 + 3)$
   $= 3(x^2 + 3) + 5$
   $= 3x^2 + 9 + 5$
   $= 3x^2 + 14$

9. $(f \circ h)(x) = f[h(x)]$
   $= f(\sqrt{x+1})$
   $= 3\sqrt{x+1} + 5$

11. $(h \circ f)(0) = h[f(0)]$
    $= h(5)$
    $= \sqrt{5+1}$
    $= \sqrt{6}$

13. $(f \circ g)(x) = f[g(x)]$
    $= f(3x+4)$
    $= 2(3x+4) - 2$
    $= 6x + 8 - 2$
    $= 6x + 6$

$(g \circ f)(x) = g[f(x)]$
$= g(2x-2)$
$= 3(2x-2) + 4$
$= 6x - 6 + 4$
$= 6x - 2$

15. $(f \circ g)(x) = f[g(x)]$
    $= f(x^2 + 1)$
    $= x^2 + 1 + 2$
    $= x^2 + 3$

    asdf

$(g \circ f)(x) = g[f(x)]$
$= g(x+2)$
$= (x+2)^2 + 1$
$= x^2 + 4x + 4 + 1$
$= x^2 + 4x + 5$

17. $(f \circ g)(x) = f[g(x)]$
    $= f(3x)$
    $= (3x)^2 + 3(3x) - 4$
    $= 9x^2 + 9x - 4$

$(g \circ f)(x) = g[f(x)]$
$= g(x^2 + 3x - 4)$
$= 3(x^2 + 3x - 4)$
$= 3x^2 + 9x - 12$

19. $(f \circ g)(x) = f[g(x)]$
    $= f(2x-5)$
    $= \sqrt{2x-5+2}$
    $= \sqrt{2x-3}$

$(g \circ f)(x) = g[f(x)]$
$= g(\sqrt{x+2})$
$= 2\sqrt{x+2} - 5$

21. $(f \circ g)(x) = f[g(x)]$

$$= f\left(\frac{x-3}{x}\right)$$

$$= \frac{\dfrac{x-3}{x}+1}{\dfrac{x-3}{x}}$$

$$= \frac{2x-3}{x-3}$$

$(g \circ f)(x) = g[f(x)]$

$$= g\left(\frac{x+1}{x}\right)$$

$$= \frac{\dfrac{x+1}{x}-3}{\dfrac{x+1}{x}}$$

$$= \frac{1-2x}{x+1}$$

23. The domain is $[0, \infty)$ and the range is $[3, \infty)$

25. Yes

27. No

29. $(f \circ g)(x) = f[g(x)]$

$$= f(x-5)$$

$$= x-5+5$$

$$= x$$

$(g \circ f)(x) = g[f(x)]$

$$= g(x+5)$$

$$= x+5-5$$

$$= x$$

Yes

31. $(f \circ g)(x) = f[g(x)]$

$$= f\left(\frac{x}{6}\right)$$

$$= 6 \cdot \frac{x}{6}$$

$$= x$$

$(g \circ f)(x) = g[f(x)]$

$$= g(6x)$$

$$= \frac{6x}{6}$$

$$= x$$

Yes

33. $(f \circ g)(x) = f[g(x)]$

$$= f\left(\frac{x+3}{2}\right)$$

$$= 2 \cdot \frac{x+3}{2} - 3$$

$$= x+3-3$$

$$= x$$

$(g \circ f)(x) = g[f(x)]$

$$= g(2x-3)$$

$$= \frac{2x-3+3}{2}$$

$$= \frac{2x}{2}$$

$$= x$$

Yes

35. $(f \circ g)(x) = f[g(x)]$

$$= f\left(\sqrt[3]{x+4}\right)$$

$$= \left(\sqrt[3]{x+4}\right)^3 - 4$$

$$= x+4-4$$

$$= x$$

$(g \circ f)(x) = g[f(x)]$

$$= g(x^3 - 4)$$

$$= \sqrt[3]{x^3 - 4 + 4}$$

$$= \sqrt[3]{x^3}$$

$$= x$$

Yes

37. $(f \circ g)(x) = f[g(x)]$

$$= f\left(\sqrt{x}\right)$$

$$= \left(\sqrt{x}\right)^2$$

$$= x$$

$(g \circ f)(x) = g[f(x)]$

$$= g(x^2)$$

$$= \sqrt{x^2}$$

$$= |x|$$

No

39. $(f \circ g)(x) = f[g(x)]$
$$= f(\sqrt{x})$$
$$= (\sqrt{x})^2$$
$$= x$$
$(g \circ f)(x) = g[f(x)]$
$$= g(x^2)$$
$$= \sqrt{x^2}$$
$$= |x|$$
$$= x, \text{ since } x \geq 0$$
Yes

41. $(f \circ g)(x) = f[g(x)]$
$$= f\left(\frac{3-5x}{x}\right)$$
$$= \frac{3}{\frac{3-5x}{x}+5} = \frac{3x}{3-5x+5x}$$
$$= \frac{3x}{3} = x$$
$(g \circ f)(x) = g[f(x)]$
$$= g\left(\frac{3}{x+5}\right)$$
$$= \frac{3-5\left(\frac{3}{x+5}\right)}{\frac{3}{x+5}} = \frac{3-\frac{15}{x+5}}{\frac{3}{x+5}}$$
$$= \frac{3(x+5)-15}{3} = \frac{3x+15-15}{3}$$
$$= \frac{3x}{3} = x$$
Yes

43. Yes, because $(f \circ g)(x) = (g \circ f)(x) = x$:
$$(f \circ f)(x) = f[f(x)] = f\left(\frac{1}{x}\right) = \frac{1}{\frac{1}{x}} = x$$

45. Yes. Every horizontal line that can intersect this graph does so at one and only one point, so the function is one to one.

47. No. A horizontal line can intersect this graph in more than one point, so the function is not one to one.

49. $f^{-1} = \{(2,-3),(-3,-1),(4,0),(6,4)\}$

51. $f^{-1} = \{(-2,7),(2,9),(1,-4),(3,3)\}$

53. $y = x+6$
$$x = y+6$$
$$x-6 = y$$
$$f^{-1}(x) = x-6$$

55. $y = 2x+3$
$$x = 2y+3$$
$$x-3 = 2y$$
$$\frac{x-3}{2} = y$$
$$f^{-1}(x) = \frac{x-3}{2}$$

57. $y = x^3-1$
$$x = y^3-1$$
$$x+1 = y^3$$
$$\sqrt[3]{x+1} = y$$
$$f^{-1}(x) = \sqrt[3]{x+1}$$

59. $y = \frac{2}{x+2}$
$$x = \frac{2}{y+2}$$
$$x(y+2) = 2$$
$$xy+2x = 2$$
$$xy = 2-2x$$
$$y = \frac{2-2x}{x}$$
$$f^{-1}(x) = \frac{2-2x}{x}$$

61. $y = \frac{x+2}{x-3}$
$$x = \frac{y+2}{y-3}$$
$$x(y-3) = y+2$$
$$xy-3x = y+2$$
$$xy-y = 3x+2$$
$$y(x-1) = 3x+2$$
$$y = \frac{3x+2}{x-1}$$
$$f^{-1}(x) = \frac{3x+2}{x-1}$$

63.
$$y = \sqrt{x-2}$$
$$x = \sqrt{y-2}$$
$$x^2 = y-2$$
$$x^2 + 2 = y$$
$$f^{-1}(x) = x^2 + 2, x \geq 0$$

65.
$$y = 2x^3 + 4$$
$$x = 2y^3 + 4$$
$$x - 4 = 2y^3$$
$$\frac{x-4}{2} = y^3$$
$$\sqrt[3]{\frac{x-4}{2}} = y$$
$$f^{-1}(x) = \sqrt[3]{\frac{x-4}{2}}$$

67.
$$y = \sqrt[3]{x+2}$$
$$x = \sqrt[3]{y+2}$$
$$x^3 = y+2$$
$$x^3 - 2 = y$$
$$f^{-1}(x) = x^3 - 2$$

69.
$$y = 2\sqrt[3]{2x+4}$$
$$x = 2\sqrt[3]{2y+4}$$
$$\frac{x}{2} = \sqrt[3]{2y+4}$$
$$\frac{x^3}{8} = 2y+4$$
$$\frac{x^3}{8} - 4 = 2y$$
$$\frac{1}{2}\left(\frac{x^3}{8} - 4\right) = \frac{1}{2} \cdot 2y$$
$$\frac{x^3}{16} - 2 = y$$
$$f^{-1}(x) = \frac{x^3}{16} - 2$$

71.

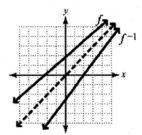

73.

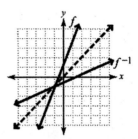

75.

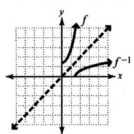

77.

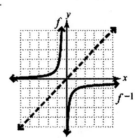

79.

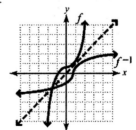

81.

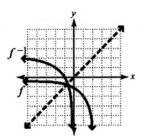

83. a) $y = 0.05x + 100$

$x = 0.05y + 100$

$x - 100 = 0.05y$

$\dfrac{x - 100}{0.05} = y$

b) x represents the salary, y represents the sales

c) $y = \dfrac{x - 100}{0.05}$

$= \dfrac{350 - 100}{0.05}$

$= \dfrac{250}{0.05}$

$= 5000$

If the salary was \$350, sales were \$5000.

85. a) $y = 45x + 65(105 - x)$

$y = 45x + 6825 - 65x$

$y = -20x + 6825$

$x = -20y + 6825$

$x - 6825 = -20y$

$\dfrac{x - 6825}{-20} = y$

$y = \dfrac{6825 - x}{20}$

b) x represents the cost, y represents the number of 36-in. fans.

c) $f^{-1}(x) = \dfrac{6825 - x}{20}$

$f^{-1}(5225) = \dfrac{6825 - 5225}{20}$

$= 80$

The number of 36-inch fans was 80.

87. $(5, -4)$

89. $y = ax + b$

$x = ay + b$

$x - b = ay$

$\dfrac{x - b}{a} = y$

$f^{-1}(x) = \dfrac{x - b}{a}$

Review Exercises

1. $32 = 2^5$ 2. $2^3 = 8$

3. $\left(-\dfrac{1}{3}\right)^3 = -\dfrac{1}{27}$ 4. $4^{-2} = \left(\dfrac{1}{4}\right)^2 = \dfrac{1}{16}$

5. $4^{3/2} = \left(\sqrt{4}\right)^3 = 2^3 = 8$

6. Vertex: $(3, 1)$

Exercise Set 10.2

1. $f(x) = 3^x$

x	$y = f(x)$
-1	$\dfrac{1}{3}$
0	1
1	3
2	9

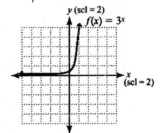

3. $f(x) = 4^x - 3$

x	$y = f(x)$
-1	$4^{-1} - 3 = -\dfrac{11}{4}$
0	$4^0 - 3 = -2$
1	$4^1 - 3 = 1$
2	$4^2 - 3 = 13$

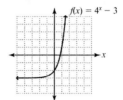

5. $f(x) = \left(\dfrac{1}{3}\right)^x$

x	$y = f(x)$
-1	$\left(\dfrac{1}{3}\right)^{-1} = 3$
0	$\left(\dfrac{1}{3}\right)^0 = 1$
1	$\left(\dfrac{1}{3}\right)^1 = \dfrac{1}{3}$
2	$\left(\dfrac{1}{3}\right)^2 = \dfrac{1}{9}$

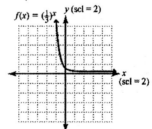

7. $f(x) = \left(\dfrac{2}{3}\right)^x + 2$

x	$y = f(x)$
-1	$\left(\dfrac{2}{3}\right)^{-1} + 2 = \dfrac{7}{2}$
0	$\left(\dfrac{2}{3}\right)^0 + 2 = 3$
1	$\left(\dfrac{2}{3}\right)^1 + 2 = \dfrac{8}{3}$
2	$\left(\dfrac{2}{3}\right)^2 + 2 = \dfrac{22}{9}$

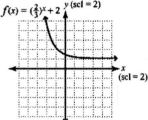

9. $f(x) = -3^x$

x	$y = f(x)$
-1	$-\dfrac{1}{3}$
0	-1
1	-3
2	-9

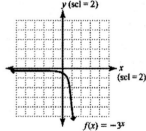

11. $f(x) = 2^{x-2}$

x	$y = f(x)$
-1	$2^{-1-2} = \dfrac{1}{8}$
0	$2^{0-2} = \dfrac{1}{4}$
1	$2^{1-2} = \dfrac{1}{2}$
2	$2^{2-2} = 1$

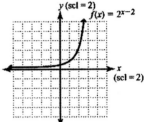

15. $f(x) = 2^{2x-3}$

x	$y = f(x)$
-1	$2^{2(-1)-3} = \dfrac{1}{32}$
0	$2^{2(0)-3} = \dfrac{1}{8}$
1	$2^{2(1)-3} = \dfrac{1}{2}$
2	$2^{2(2)-3} = 2$

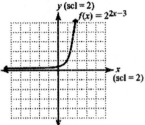

13. $f(x) = 2^{-x}$

x	$y = f(x)$
-1	$2^{-(-1)} = 2$
0	$2^{-(0)} = 1$
1	$2^{-(1)} = \dfrac{1}{2}$
2	$2^{-(2)} = \dfrac{1}{4}$

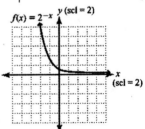

17. $f(x) = 3^{-x+2}$

x	$y = f(x)$
-1	$3^{-(-1)+2} = 27$
0	$3^{-(0)+2} = 9$
1	$3^{-(1)+2} = 3$
2	$3^{-(2)+2} = 1$

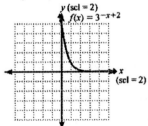

19. $2^x = 8$

$2^x = 2^3$

$x = 3$

21. $\quad 8^x = 32$

$\left(2^3\right)^x = 2^5$

$2^{3x} = 2^5$

$3x = 5$

$x = \dfrac{5}{3}$

23. $16^x = 4$

$\left(2^4\right)^x = 2^2$

$2^{4x} = 2^2$

$4x = 2$

$x = \dfrac{1}{2}$

25. $6^x = \dfrac{1}{36}$

$6^x = 6^{-2}$

$x = -2$

27. $\left(\dfrac{1}{3}\right)^x = 9$

$\left(3^{-1}\right)^x = 3^2$

$3^{-x} = 3^2$

$-x = 2$

$x = -2$

29. $\left(\dfrac{2}{3}\right)^x = \dfrac{8}{27}$

$\left(\dfrac{2}{3}\right)^x = \left(\dfrac{2}{3}\right)^3$

$x = 3$

31. $\left(\dfrac{1}{2}\right)^x = 16$

$\left(2^{-1}\right)^x = 2^4$

$2^{-x} = 4$

$-x = 4$

$x = -4$

33. $25^{x+1} = 125$

$\left(5^2\right)^{x+1} = 5^3$

$5^{2x+2} = 5^3$

$2x + 2 = 3$

$2x = 1$

$x = \dfrac{1}{2}$

35. $8^{2x-1} = 32^{x-3}$

$\left(2^3\right)^{2x-1} = \left(2^5\right)^{x-3}$

$2^{6x-3} = 2^{5x-15}$

$6x - 3 = 5x - 15$

$x = -12$

37. a)

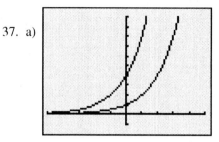

b) The graph of g is the graph of f shifted 2 units to the left.

39. a)

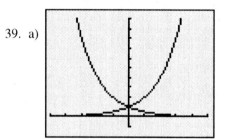

b) The graph of g is the graph of f reflected about the y-axis.

41. $A = A_0 \left(2\right)^{t/20}$

$A = 100 \left(2\right)^{120/20}$

$A = 6400$
After 120 minutes, there will be 6400 cells present.

43. $A = A_0 \left(2\right)^{t/50}$

$A = 6 \left(2\right)^{500/50}$

$A = 6144$
There will be 6144 billion people in 2500 if their growth is uncontrolled.

45. $A = P \left(1 + \dfrac{r}{n}\right)^{nt}$

$A = 10,000 \left(1 + \dfrac{0.08}{4}\right)^{4 \cdot 12}$

$A = 25,870.70$
The account value will be \$25,870.70 after 12 years.

47. $A = A_0 \left(\dfrac{1}{2}\right)^{t/h}$

$A = 5\left(\dfrac{1}{2}\right)^{2160/270}$

$A = 0.020$

After 2160 days, 0.020 gram would remain.

49. $y = 2632.31(1.033)^x$

$y = 2632.31(1.033)^{56}$

$y \approx 16,216.42$

51. $f(x) = 2.5(0.7)^x$

$f(2) = 2.5(0.7)^2$

$= 1.225$

The chlorine level after 2 days is 1.225 parts per million.

53. $T(x) = 1.97(1.503)^x$

$T(27) = 1.97(1.503)^{27}$

$\approx 118,130$

If the trend continues, a total of about 118,130 million transistors could be put on a single chip in 2020.

Review Exercises

1. $5^3 = 125$

2. $\left(\dfrac{1}{3}\right)^{-2} = \left(\dfrac{3}{1}\right)^2 = 9$

3. $4^{-3} = \dfrac{1}{4^3} = \dfrac{1}{64}$

4. $5^{2/3} = \sqrt[3]{5^2} = \left(\sqrt[3]{5}\right)^2$

5. $f(x) = 3x + 4$

$y = 3x + 4$

$x = 3y + 4$

$x - 4 = 3y$

$\dfrac{x-4}{3} = y$

$f^{-1}(x) = \dfrac{x-4}{3}$

6.

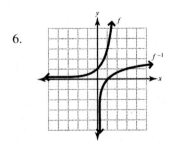

Exercise Set 10.3

1. $\log_2 32 = 5$

3. $\log_{10} 1000 = 3$

5. $\log_e x = 4$

7. $\log_5 \dfrac{1}{125} = -3$

9. $\log_{10} \dfrac{1}{100} = -2$

11. $\log_{625} 5 = \dfrac{1}{4}$

13. $\log_{1/4} \dfrac{1}{16} = 2$

15. $\log_7 \sqrt{7} = \dfrac{1}{2}$

17. $3^4 = 81$

19. $4^{-2} = \dfrac{1}{16}$

21. $10^2 = 100$

23. $e^5 = a$

25. $e^{-4} = \dfrac{1}{e^4}$

27. $\left(\dfrac{1}{8}\right)^2 = \dfrac{1}{64}$

29. $\left(\dfrac{1}{5}\right)^{-2} = 25$

31. $7^{1/2} = \sqrt{7}$

33. $2^5 = x$

$32 = x$

35. $5^{-2} = x$

$\dfrac{1}{5^2} = x$

$\dfrac{1}{25} = x$

37. $3^y = 81$

$3^y = 3^4$

$y = 4$

39. $5^y = \dfrac{1}{25}$

$5^y = 5^{-2}$

$y = -2$

41. $b^3 = 1000$

$b^3 = 10^3$

$b = 10$

43. $m^{-4} = \dfrac{1}{16}$

$m^{-4} = 16^{-1}$

$m^{-4} = \left(2^4\right)^{-1}$

$m^{-4} = 2^{-4}$

$m = 2$

45. $\left(\dfrac{1}{2}\right)^2 = x$

$\dfrac{1}{4} = x$

47. $\left(\dfrac{1}{3}\right)^{-5} = h$

$3^5 = h$

$243 = h$

49. $\left(\dfrac{1}{3}\right)^y = \dfrac{1}{9}$

$\left(\dfrac{1}{3}\right)^y = \left(\dfrac{1}{3}\right)^2$

$y = 2$

51. $\left(\dfrac{1}{2}\right)^t = 64$

$\left(2^{-1}\right)^t = 2^6$

$2^{-t} = 2^6$

$-t = 6$

$t = -6$

53. $\log_2 x = 4$

$x = 2^4$

$x = 16$

55. $\log_{1/4} h = 3$

$h = \left(\dfrac{1}{4}\right)^3$

$h = \dfrac{1}{64}$

57. $\log_b 16 = 4$

$b^4 = 16$

$b^4 = 2^4$

$b = 2$

59. $\dfrac{1}{3}\log_5 c = 1$

$\log_5 c = 3$

$c = 5^3$

$c = 125$

61. $3\log_t 9 = 6$

$\log_t 9 = 2$

$t^2 = 9$

$t^2 = 3^2$

$t = 3$

63. $\dfrac{1}{2}\log_4 m = -1$

$\log_4 m = -2$

$m = 4^{-2}$

$m = \dfrac{1}{4^2}$

$m = \dfrac{1}{16}$

65. $f(x) = \log_4 x$

$y = \log_4 x$

$4^y = x$

y	-1	0	1	2
x	$\dfrac{1}{4}$	1	4	16

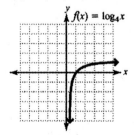

67. $f(x) = \log_{1/3} x$

$y = \log_{1/3} x$

$\left(\dfrac{1}{3}\right)^y = x$

y	-2	-1	0	1
x	9	3	1	$\dfrac{1}{3}$

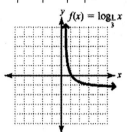

69. $f(x) = 62 + 35\log_{10}(x-4)$

$f(14) = 62 + 35\log_{10}(14-4)$

$= 62 + 35\log_{10} 10$

$= 62 + 35(1)$

$= 97$

At age 14, a girl has reached 97% of her adult height.

71. $35 = 95 - 30\log_2 x$

$-60 = -30\log_2 x$

$2 = \log_2 x$

$x = 2^2$

$x = 4$

After 4 days, 35% of the students recall important features of the lecture.

73. $\log_b b = 1$ because $b^1 = b$.

75. $\log_b 1 = 0$ because $b^0 = 1$.

Review Exercises

1. x^{-6}

2. $x^4 \cdot x^2 = x^{4+2} = x^6$

3. $\left(x^3\right)^5 = x^{3 \cdot 5} = x^{15}$

4. $\dfrac{x^6}{x^3} = x^{6-3} = x^3$

5. $\dfrac{\left(x^3\right)^2 \cdot x^4}{x^5} = \dfrac{x^6 \cdot x^4}{x^5} = \dfrac{x^{10}}{x^5} = x^{10-5} = x^5$

6. $\sqrt[4]{x^3} = x^{3/4}$

Exercise Set 10.4

1. $8^{\log_8 2} = 2$　　　　　　3. $a^{\log_a r} = r$

5. $a^{\log_a 4x} = 4x$　　　　7. $\log_3 3^5 = 5$

9. $\log_e e^y = y$　　　　　11. $\log_a a^{7x} = 7x$

13. $\log_2 5y = \log_2 5 + \log_2 y$

15. $\log_a pq = \log_a p + \log_a q$

17. $\log_4 mnp = \log_4 m + \log_4 n + \log_4 p$

19. $\log_a x(x-5) = \log_a x + \log_a (x-5)$

21. $\log_3 5 + \log_3 8 = \log_3 (5 \cdot 8) = \log_3 40$

23. $\log_4 3 + \log_4 9 = \log_4 (3 \cdot 9) = \log_4 27$

25. $\log_a 7 + \log_a m = \log_a 7m$

27. $\log_4 a + \log_4 b = \log_4 ab$

29. $\log_a 2 + \log_a x + \log_a (x+5) = \log_a 2x(x+5)$
$$= \log_a (2x^2 + 10x)$$

31. $\log_4 (x+1) + \log_4 (x+3) = \log_4 (x+1)(x+3)$
$$= \log_4 (x^2 + 4x + 3)$$

33. $\log_2 \dfrac{7}{9} = \log_2 7 - \log_2 9$

35. $\log_a \dfrac{x}{5} = \log_a x - \log_a 5$

37. $\log_a \dfrac{a}{b} = \log_a a - \log_a b = 1 - \log_a b$

39. $\log_a \dfrac{x}{x-3} = \log_a x - \log_a (x-3)$

41. $\log_4 \dfrac{2x-3}{4x+5} = \log_4 (2x-3) - \log_4 (4x+5)$

43. $\log_6 24 - \log_6 3 = \log_6 \dfrac{24}{3} = \log_6 8$

45. $\log_2 24 - \log_2 12 = \log_2 \dfrac{24}{12} = \log_2 2 = 1$

47. $\log_a x - \log_a 3 = \log_a \dfrac{x}{3}$

49. $\log_4 p - \log_4 q = \log_4 \dfrac{p}{q}$

51. $\log_b x - \log_b (x-4) = \log_b \dfrac{x}{x-4}$

53. $\log_x (x^2 - x) - \log_x (x-1) = \log_x \dfrac{x^2 - x}{x-1}$
$$= \log_x \dfrac{x(x-1)}{x-1}$$
$$= \log_x x$$
$$= 1$$

55. $\log_4 3^6 = 6\log_4 3$

57. $\log_a x^7 = 7\log_a x$

59. $\log_a \sqrt{3} = \log_a 3^{1/2} = \dfrac{1}{2}\log_a 3$

61. $\log_3 \sqrt[3]{x^2} = \log_3 x^{2/3} = \dfrac{2}{3}\log_3 x$

63. $\log_a \dfrac{1}{6^2} = \log_a 6^{-2} = -2\log_a 6$

65. $\log_a \dfrac{1}{y^2} = \log_a y^{-2} = -2\log_a y$

67. $4\log_3 5 = \log_3 5^4$

69. $-3\log_2 x = \log_2 x^{-3} = \log_2 \dfrac{1}{x^3}$

71. $\dfrac{1}{2}\log_7 64 = \log_7 64^{1/2} = \log_7 \sqrt{64} = \log_7 8$

73. $\dfrac{3}{4}\log_a x = \log_a x^{3/4} = \log_a \sqrt[4]{x^3}$

75. $\dfrac{2}{3}\log_a 8 = \log_a 8^{2/3}$
$$= \log_a \sqrt[3]{8^2}$$
$$= \log_a \sqrt[3]{64}$$
$$= \log_a 4$$

77. $-\dfrac{1}{2}\log_3 x = \log_3 x^{-1/2} = \log_3 \dfrac{1}{x^{1/2}} = \log_3 \dfrac{1}{\sqrt{x}}$

79. $\log_a \dfrac{x^3}{y^4} = \log_a x^3 - \log_a y^4 = 3\log_a x - 4\log_a y$

81. $\log_3 a^4 b^2 = \log_3 a^4 + \log_3 b^2 = 4\log_3 a + 2\log_3 b$

83. $\log_a \dfrac{xy}{z} = \log_a xy - \log_a z$

$= \log_a x + \log_a y - \log_a z$

85. $\log_x \dfrac{a^2}{bc^3} = \log_x a^2 - \log_x bc^3$

$= \log_x a^2 - \left(\log_x b + \log_x c^3\right)$

$= 2\log_x a - \left(\log_x b + 3\log_x c\right)$

$= 2\log_x a - \log_x b - 3\log_x c$

87. $\log_4 \sqrt[4]{\dfrac{x^3}{y}} = \log_4 \left(\dfrac{x^3}{y}\right)^{1/4}$

$= \log_4 \left(\dfrac{x^{3/4}}{y^{1/4}}\right)$

$= \log_4 x^{3/4} - \log_4 y^{1/4}$

$= \dfrac{3}{4}\log_4 x - \dfrac{1}{4}\log_4 y$

89. $\log_a \sqrt[3]{\dfrac{x^2 y}{z^3}} = \log_a \left(\dfrac{x^2 y}{z^3}\right)^{1/3}$

$= \log_a \left(\dfrac{x^{2/3} y^{1/3}}{z^{3/3}}\right)$

$= \log_a x^{2/3} y^{1/3} - \log_a z$

$= \log_a x^{2/3} + \log_a y^{1/3} - \log_a z$

$= \dfrac{2}{3}\log_a x + \dfrac{1}{3}\log_a y - \log_a z$

91. $3\log_3 2 - 2\log_3 4 = \log_3 2^3 - \log_3 4^2$

$= \log_3 8 - \log_3 16$

$= \log_3 \dfrac{8}{16}$

$= \log_3 \dfrac{1}{2}$

93. $4\log_b x + 3\log_b y = \log_b x^4 + \log_b y^3$

$= \log_b x^4 y^3$

95. $\dfrac{1}{2}\left(\log_a 5 - \log_a 7\right) = \dfrac{1}{2}\log_a \left(\dfrac{5}{7}\right)$

$= \log_a \left(\dfrac{5}{7}\right)^{1/2}$

$= \log_a \sqrt{\dfrac{5}{7}}$

97. $\dfrac{2}{3}\left(\log_a x^2 + \log_a y^3\right) = \dfrac{2}{3}\log_a \left(x^2 y^3\right)$

$= \log_a \left(x^2 y^3\right)^{2/3}$

$= \log_a \sqrt[3]{\left(x^2 y^3\right)^2}$

99. $\log_b x + \log_b \left(3x - 2\right) = \log_b x\left(3x - 2\right)$

$= \log_b \left(3x^2 - 2x\right)$

101. $3\log_a \left(x - 2\right) - 4\log_a \left(x + 1\right)$

$= \log_a \left(x - 2\right)^3 - \log_a \left(x + 1\right)^4$

$= \log_a \dfrac{\left(x - 2\right)^3}{\left(x + 1\right)^4}$

103. $2\log_a x + 4\log_a z - 3\log_a w - 6\log_a u$

$= \log_a x^2 + \log_a z^4 - \log_a w^3 - \log_a u^6$

$= \log_a x^2 z^4 - \log_a w^3 - \log_a u^6$

$= \log_a \dfrac{x^2 z^4}{w^3} - \log_a u^6$

$= \log_a \dfrac{x^2 z^4}{w^3 u^6}$

Review Exercises

1. $\left(10^{9.5}\right)\left(10^{-12}\right) = 10^{9.5 + (-12)} = 10^{-2.5}$

2. $\dfrac{10^{-3}}{10^{-12}} = 10^{-3 - (-12)} = 10^9$

3. $A = 10,000\left(1 + \dfrac{0.06}{4}\right)^{5(4)}$

$A = \$13,468.55$

4. $\log_{10} 50 = 1.6990$

5. $10^{1.6532} = 45$

6. $e^{-1.3863} = 0.25$

Exercise Set 10.5

1. $\log 64 = 1.8062$

3. $\log 0.0067 = -2.1739$

5. $\log 435.6 = 2.6391$

7. $\log\left(1.5 \times 10^4\right) = 4.1761$

9. $\log\left(1.6\times10^{-6}\right) = -5.7959$

11. $\ln 9.34 = 2.2343$

13. $\ln 79.2 = 4.3720$

15. $\ln 0.034 = -3.3814$

17. $\ln\left(5.4\times e^{4}\right) = 5.6864$

19. $\log e = 0.4343$

21. Error results because the domain of $\log_a x$ is $(0,\infty)$; so log 0 is undefined.

23. $\log 100 = \log_{10} 10^{2} = 2$

25. $\log \dfrac{1}{100} = \log_{10} 10^{-2} = -2$

27. $\log \sqrt[3]{10} = \log_{10} 10^{1/3} = \dfrac{1}{3}$

29. $\log 0.001 = \log_{10} 10^{-3} = -3$

31. $\ln e^{3} = \log_e e^{3} = 3$

33. $\ln \sqrt{e} = \log_e e^{1/2} = \dfrac{1}{2}$

35. $d = 10\log \dfrac{I}{I_0}$

$d = 10\log \dfrac{10^{-3}}{10^{-12}}$

$d = 10\log 10^{9}$

$d = 10\cdot 9$

$d = 90$

The reading for a firecracker is 90dB.

37. $d = 10\log \dfrac{I}{I_0}$

$60 = 10\log \dfrac{I}{10^{-12}}$

$6 = \log \dfrac{I}{10^{-12}}$

$6 = \log I - \log 10^{-12}$

$6 = \log I - \left(-12\right)$

$6 = \log I + 12$

$-6 = \log_{10} I$

$I = 10^{-6}$

The sound intensity of a noisy office is 10^{-6} watts/m^2.

39. $\text{pH} = -\log\left[\text{H}_3\text{O}^{+}\right]$

$\text{pH} = -\log\left[1.6\times10^{-3}\right]$

$\text{pH} = 2.796$

41. $3.5 = -\log\left[\text{H}_3\text{O}^{+}\right]$

$\log\left[\text{H}_3\text{O}^{+}\right] = -3.5$

$10^{-3.5} = \left[\text{H}_3\text{O}^{+}\right]$

The hydronium ion concentration is $10^{-3.5}$ moles/L .

43. $t = \dfrac{1}{r}\ln\dfrac{A}{P}$

$t = \dfrac{1}{0.04}\ln\dfrac{5000}{2000}$

≈ 22.9

It will take approximately 23 years.

45. San Francisco: $7.8 = \log \dfrac{I}{I_0}$

$7.8 = \log I - \log I_0$

$7.8 + \log I_0 = \log I$

$10^{7.8+\log I_0} = I$

Alaska: $8.4 = \log \dfrac{I}{I_0}$

$8.4 = \log I - \log I_0$

$8.4 + \log I_0 = \log i$

$10^{8.4+\log I_0} = I$

$\dfrac{10^{8.4+\log I_0}}{10^{7.8+\log I_0}} = 10^{\left(8.4+\log I_0\right)-\left(7.8+\log I_0\right)}$

$= 10^{0.6}$

≈ 4

The 1964 Alaska earthquake was about four times as severe as the 1906 San Francisco earthquake.

47. $\log E_s = 11.8 + 1.5\left(7.3\right)$

$\log_{10} E_s = 22.75$

$E_s = 10^{22.75}$

The energy released was $10^{22.75}$ ergs.

49. $y = -69.45 + 22.7 \ln x$

$y = -69.45 + 22.7 \ln(40)$

$y \approx -69.45 + 84$

$y \approx 14.55$

The purchasing power of $1.00 was approximately 40 cents in 1985.

51. $y = 3\log 110$

$y \approx 6.124$

and

$6.124(1000) = 6124$ ft.

Review Exercises

1. $\log_3(2x+1) - \log_3(x-1) = \log_3 \dfrac{2x+1}{x-1}$

2. $3^2 = 2x + 5$

3. $3x - (7x+2) = 12 - 2(x-4)$

$3x - 7x - 2 = 12 - 2x + 8$

$-4x - 2 = 20 - 2x$

$-2x - 2 = 20$

$-2x = 22$

$x = -11$

4. $x^2 + 2x = 15$

$x^2 + 2x - 15 = 0$

$(x+5)(x-3) = 0$

$x + 5 = 0 \ \text{ or } \ x - 3 = 0$

$x = -5 \qquad\quad x = 3$

5. $\dfrac{5}{x} + \dfrac{3}{x+1} = \dfrac{23}{3x}$

$3x(x+1)\left[\dfrac{5}{x} + \dfrac{3}{x+1}\right] = 3x(x+1)\left[\dfrac{23}{3x}\right]$

$3(x+1)(5) + 3x(3) = (x+1)23$

$15x + 15 + 9x = 23x + 23$

$24x + 15 = 23x + 23$

$x = 8$

6. $\sqrt{5x-1} = 7$

$\left(\sqrt{5x-1}\right)^2 = 7^2$

$5x - 1 = 49$

$5x = 50$

$x = 10$

Exercise Set 10.6

1. $2^x = 9$

$\log 2^x = \log 9$

$x \log 2 = \log 9$

$x = \dfrac{\log 9}{\log 2} \approx 3.1699$

3. $5^{2x} = 32$

$\log 5^{2x} = \log 32$

$2x \log 5 = \log 32$

$2x = \dfrac{\log 32}{\log 5}$

$x = \dfrac{\log 32}{\log 5} \div 2 \approx 1.0767$

5. $5^{x+3} = 10$

$\log 5^{x+3} = \log 10$

$(x+3)\log 5 = 1$

$x \log 5 + 3 \log 5 = 1$

$x \log 5 = 1 - 3 \log 5$

$x = \dfrac{1 - 3 \log 5}{\log 5} \approx -1.5693$

7. $8^{x-2} = 6$

$\log 8^{x-2} = \log 6$

$(x-2)\log 8 = \log 6$

$x \log 8 - 2 \log 8 = \log 6$

$x \log 8 = \log 6 + 2 \log 8$

$x = \dfrac{\log 6 + 2 \log 8}{\log 8} \approx 2.8617$

9. $4^{x+2} = 5^x$

$\log 4^{x+2} = \log 5^x$

$(x+2)\log 4 = x \log 5$

$x \log 4 + 2 \log 4 = x \log 5$

$x \log 4 - x \log 5 = -2 \log 4$

$x(\log 4 - \log 5) = -2 \log 4$

$x = \dfrac{-2 \log 4}{\log 4 - \log 5} \approx 12.4251$

11.
$$2^{x+1} = 3^{x-2}$$
$$\log 2^{x+1} = \log 3^{x-2}$$
$$(x+1)\log 2 = (x-2)\log 3$$
$$x\log 2 + \log 2 = x\log 3 - 2\log 3$$
$$x\log 2 - x\log 3 = -2\log 3 - \log 2$$
$$x(\log 2 - \log 3) = -2\log 3 - \log 2$$
$$x = \frac{-2\log 3 - \log 2}{\log 2 - \log 3} \approx 7.1285$$

13.
$$e^{3x} = 5$$
$$\ln e^{3x} = \ln 5$$
$$3x = \ln 5$$
$$x = \frac{\ln 5}{3} \approx 0.5365$$

15.
$$e^{0.03x} = 25$$
$$\ln e^{0.03x} = \ln 25$$
$$0.03x = \ln 25$$
$$x = \frac{\ln 25}{0.03} \approx 107.2959$$

17.
$$e^{-0.022x} = 5$$
$$\ln e^{-0.022x} = \ln 5$$
$$-0.022x = \ln 5$$
$$x = -\frac{\ln 5}{0.022} \approx -73.1563$$

19.
$$\ln e^{4x} = 24$$
$$4x = 24$$
$$x = 6$$

21.
$$\log_4 (x+5) = 2$$
$$4^2 = x+5$$
$$16 = x+5$$
$$11 = x$$

23.
$$\log_4 (4x-8) = 2$$
$$4^2 = 4x-8$$
$$16 = 4x-8$$
$$24 = 4x$$
$$6 = x$$

25.
$$\log_4 x^2 = 2$$
$$4^2 = x^2$$
$$16 = x^2$$
$$\pm\sqrt{16} = x$$
$$\pm 4 = x$$

27.
$$\log_6 (x^2 + 5x) = 2$$
$$6^2 = x^2 + 5x$$
$$36 = x^2 + 5x$$
$$0 = x^2 + 5x - 36$$
$$0 = (x+9)(x-4)$$
$$x+9 = 0 \quad \text{or} \quad x-4 = 0$$
$$x = -9 \qquad\qquad x = 4$$

29.
$$\log(4x-3) = \log(3x+4)$$
$$4x-3 = 3x+4$$
$$x = 7$$

31.
$$\ln(3x+4) = \ln(x-6)$$
$$3x+4 = x-6$$
$$2x = -10$$
$$x = \cancel{-5}$$
No solution

33.
$$\log_9 (x^2 + 4x) = \log_9 12$$
$$x^2 + 4x = 12$$
$$x^2 + 4x - 12 = 0$$
$$(x+6)(x-2) = 0$$
$$x+6 = 0 \quad \text{or} \quad x-2 = 0$$
$$x = -6 \qquad\qquad x = 2$$

35.
$$\log_4 x + \log_4 8 = 2$$
$$\log_4 8x = 2$$
$$4^2 = 8x$$
$$16 = 8x$$
$$2 = x$$

37.
$$\log_2 x - \log_2 5 = 1$$
$$\log_2 \frac{x}{5} = 1$$
$$2^1 = \frac{x}{5}$$
$$10 = x$$

39. $\log_3 x + \log_3 (x+6) = 3$

$\qquad \log_3 x(x+6) = 3$

$\qquad \log_3 (x^2 + 6x) = 3$

$\qquad\qquad 3^3 = x^2 + 6x$

$\qquad\qquad 27 = x^2 + 6x$

$\qquad\qquad 0 = x^2 + 6x - 27$

$\qquad\qquad 0 = (x+9)(x-3)$

$x + 9 = 0 \quad$ or $\quad x - 3 = 0$

$\quad x = \cancel{-9} \qquad\qquad x = 3$

41. $\log_3 (2x+15) + \log_3 x = 3$

$\qquad \log_3 x(2x+15) = 3$

$\qquad \log_3 (2x^2 + 15x) = 3$

$\qquad\qquad 3^3 = 2x^2 + 15x$

$\qquad\qquad 27 = 2x^2 + 15x$

$\qquad\qquad 0 = 2x^2 + 15x - 27$

$\qquad\qquad 0 = (2x-3)(x+9)$

$2x - 3 = 0 \quad$ or $\quad x + 9 = 0$

$\quad x = \dfrac{3}{2} \qquad\qquad x = \cancel{-9}$

43. $\log_2 (7x+3) - \log_2 (2x-3) = 3$

$\qquad \log_2 \dfrac{7x+3}{2x-3} = 3$

$\qquad\qquad 2^3 = \dfrac{7x+3}{2x-3}$

$\qquad\qquad 8(2x-3) = 7x+3$

$\qquad\qquad 16x - 24 = 7x + 3$

$\qquad\qquad 9x = 27$

$\qquad\qquad x = 3$

45. $\log_2 (3x+8) - \log_2 (x+1) = 2$

$\qquad \log_2 \dfrac{3x+8}{x+1} = 2$

$\qquad\qquad 2^2 = \dfrac{3x+8}{x+1}$

$\qquad\qquad 4 = \dfrac{3x+8}{x+1}$

$\qquad\qquad 4x + 4 = 3x + 8$

$\qquad\qquad x = 4$

47. $\log_8 2x + \log_8 6 = \log_8 10$

$\qquad \log_8 12x = \log_8 10$

$\qquad\qquad 12x = 10$

$\qquad\qquad x = \dfrac{5}{6}$

49. $\ln x + \ln (2x-1) = \ln 10$

$\qquad \ln x(2x-1) = \ln 10$

$\qquad \ln (2x^2 - x) = \ln 10$

$\qquad\qquad 2x^2 - x = 10$

$\qquad\qquad 2x^2 - x - 10 = 0$

$\qquad\qquad (2x-5)(x+2) = 0$

$2x - 5 = 0 \quad$ or $\quad x + 2 = 0$

$\quad x = \dfrac{5}{2} \qquad\qquad x = \cancel{-2}$

51. $\log x - \log (x-5) = \log 6$

$\qquad \log \dfrac{x}{x-5} = \log 6$

$\qquad\qquad \dfrac{x}{x-5} = 6$

$\qquad\qquad x = 6x - 30$

$\qquad\qquad -5x = -30$

$\qquad\qquad x = 6$

53. $\log_6 (3x+4) - \log_6 (x-2) = \log_6 8$

$\qquad \log_6 \dfrac{3x+4}{x-2} = \log_6 8$

$\qquad\qquad \dfrac{3x+4}{x-2} = 8$

$\qquad\qquad 3x + 4 = 8x - 16$

$\qquad\qquad 20 = 5x$

$\qquad\qquad 4 = x$

55. $\qquad 8000 = 5000\left(1 + \dfrac{.05}{4}\right)^{4t}$

$\qquad\qquad \dfrac{8}{5} = (1.0125)^{4t}$

$\qquad\qquad \log \dfrac{8}{5} = (4t)\log 1.0125$

$\qquad\qquad \dfrac{\log \dfrac{8}{5}}{4\log 1.0125} = t$

$\qquad\qquad 9.5 \approx t$

It will take about 9.5 years.

57. a) $A = 8000e^{0.06(15)}$

    $A = 19,676.82$

    The value of the account will be \$19,676.82.

    b) $14000 = 8000e^{0.06t}$

    $\dfrac{14}{8} = e^{0.06t}$

    $\ln \dfrac{7}{4} = \ln e^{0.06t}$

    $\ln \dfrac{7}{4} = 0.06t$

    $\dfrac{\ln \dfrac{7}{4}}{0.06} = t$

    $9.3 \approx t$

    It will take about 9.3 years.

59. a) $P(0.08) = e^{21.5(0.08)}$

    $P(0.08) = 5.58\%$

    b) $50 = e^{21.5b}$

    $\ln 50 = \ln e^{21.5b}$

    $\ln 50 = 21.5b$

    $\dfrac{\ln 50}{21.5} = b$

    $0.18 = b$

61. a) $A = 500e^{0.04(14)}$

    $A \approx 875$

    There will be 875 mosquitoes.

    b) $10,000 = 500e^{0.04t}$

    $20 = e^{0.04t}$

    $\ln 20 = \ln e^{0.04t}$

    $\ln 20 = 0.04t$

    $\dfrac{\ln 20}{0.04} = t$

    $75 \approx t$

    It will take about 75 days.

63. a) $A = 6.8e^{0.011(10)}$

    $A \approx 7.59$

    The world population will be about 7.59 billion in 2020.

    b) $7.0 = 6.8e^{0.011t}$

    $\dfrac{7.0}{6.8} = e^{0.011t}$

    $\ln \dfrac{7.0}{6.8} = \ln e^{0.011t}$

    $\ln \dfrac{7.0}{6.8} = 0.011t$

    $2.6 \approx t$

    Year: $2010 + 2.6 = 2012$

65. a) $P = 0.48 \ln(50 + 1)$

    $P = 0.48 \ln 51$

    $P \approx 1.89$

    The barometric pressure is 1.89 inches of mercury.

    b) $1.5 = 0.48 \ln(x + 1)$

    $3.125 = \ln(x + 1)$

    $e^{3.125} = x + 1$

    $e^{3.125} - 1 = x$

    $21.8 \approx x$

    The distance is about 21.8 miles.

67. a) $f(10) = 29 + 48.8 \log(10 + 1)$

    $f(10) = 29 + 48.8 \log(11)$

    $f(10) \approx 79.8$

    At age 10, a boy has reached 79.8% of his adult height.

    b) $75 = 29 + 48.8 \log(x + 1)$

    $46 = 48.8 \log(x + 1)$

    $0.9426 = \log(x + 1)$

    $10^{0.9426} = x + 1$

    $10^{0.9426} - 1 = x$

    $7.76 = x$

    A boy will attain 75% of his adult height at about age 8.

69. $y = 1.244(1.043)^x$

    $y = 1.244(1.043)^{50}$

    $y \approx 10.21$

    In 2020, it will take \$10.21 to have the same purchasing power as \$1.00 in 1970.

71. $y = 78.62(1.092)^x$

$y = 78.62(1.092)^{29}$

$y \approx 1009.2$

Approximately \$1009.2 billion will be spent on recreation in 2014.

73. $\log_4 12 = \dfrac{\log 12}{\log 4} \approx 1.7925$

75. $\log_8 3 = \dfrac{\log 3}{\log 8} \approx 0.5283$

77. $\dfrac{\log 5}{\log 0.5} \approx -2.3219$

79. $\log_{1/4} \dfrac{3}{5} = \dfrac{\log \dfrac{3}{5}}{\log \dfrac{1}{4}} \approx 0.3685$

81. $P = 95 - 30\log_2 x$

$P = 95 - 30\log_2 3$

$P = 95 - 30\dfrac{\log 3}{\log 2}$

$P \approx 47$

After 3 days, about 47% of the lecture is retained.

Review Exercises

1. $(3x+2) - (2x-1) = (3x+2) + (-2x+1)$

$\qquad = x+3$

2. $\left(3\sqrt{5}+2\right)\left(2\sqrt{5}-1\right)$

$= 3\sqrt{5} \cdot 2\sqrt{5} + 3\sqrt{5} \cdot (-1) + 2 \cdot 2\sqrt{5} + 2 \cdot (-1)$

$= 30 - 3\sqrt{5} + 4\sqrt{5} - 2$

$= 28 + \sqrt{5}$

3. $f(-5) = (-5)^2 + 4 = 25 + 4 = 29$

4. $(f+g)(x) = x^2 + 4 + 2x - 1 = x^2 + 2x + 3$

5. $f \circ g = f\left[g(x)\right]$

$= (2x-1)^2 + 4$

$= 4x^2 - 4x + 1 + 4$

$= 4x^2 - 4x + 5$

6. Using the answer from #5:

$(f \circ g)(2) = 4(2)^2 - 4(2) + 5$

$= 16 - 8 + 5$

$= 13$

Chapter 10 Review Exercises

1. $(f \circ g)(3) = f\left[g(3)\right]$

$= f(7)$

$= 3(7) + 4$

$= 21 + 4$

$= 25$

2. $(g \circ f)(3) = g\left[f(3)\right]$

$= g(13)$

$= 13^2 - 2$

$= 169 - 2$

$= 167$

3. $f\left[g(0)\right] = f(-2)$

$= 3(-2) + 4$

$= -6 + 4$

$= -2$

4. $g\left[f(0)\right] = g(4)$

$= 4^2 - 2$

$= 16 - 2$

$= 14$

5. $(f \circ g)(x) = f\left[g(x)\right]$

$= f(2x+3)$

$= 3(2x+3) - 6$

$= 6x + 9 - 6$

$= 6x + 3$

$(g \circ f)(x) = g\left[f(x)\right]$

$= g(3x-6)$

$= 2(3x-6) + 3$

$= 6x - 12 + 3$

$= 6x - 9$

6. $(f \circ g)(x) = f[g(x)]$
$= f(3x - 7)$
$= (3x - 7)^2 + 4$
$= 9x^2 - 42x + 49 + 4$
$= 9x^2 - 42x + 53$
$(g \circ f)(x) = g[f(x)]$
$= g(x^2 + 4)$
$= 3(x^2 + 4) - 7$
$= 3x^2 + 12 - 7$
$= 3x^2 + 5$

7. $(f \circ g)(x) = f[g(x)]$
$= f(2x - 1)$
$= \sqrt{2x - 1 - 3}$
$= \sqrt{2x - 4}$
$(g \circ f)(x) = g[f(x)]$
$= g(\sqrt{x - 3})$
$= 2\sqrt{x - 3} - 1$

8. $(f \circ g)(x) = f[g(x)]$
$= f\left(\dfrac{x - 4}{x}\right)$
$= \dfrac{\dfrac{x - 4}{x} + 3}{\dfrac{x - 4}{x}}$
$= \dfrac{x - 4 + 3x}{x - 4} = \dfrac{4x - 4}{x - 4}$
$(g \circ f)(x) = g[f(x)]$
$= g\left(\dfrac{x + 3}{x}\right)$
$= \dfrac{\dfrac{x + 3}{x} - 4}{\dfrac{x + 3}{x}}$
$= \dfrac{x + 3 - 4x}{x + 3}$
$= \dfrac{3 - 3x}{x + 3}$

9. $(f \circ g)(x) = f[g(x)]$
$= f\left(\dfrac{x - 2}{3}\right)$
$= 3\left(\dfrac{x - 2}{3}\right) + 2$
$= x - 2 + 2$
$= x$
$(g \circ f)(x) = g[f(x)]$
$= g(3x + 2)$
$= \dfrac{3x + 2 - 2}{3}$
$= \dfrac{3x}{3}$
$= x$
Yes, f and g are inverse functions.

10. $(f \circ g)(x) = f[g(x)]$
$= f(\sqrt[3]{x - 6})$
$= (\sqrt[3]{x - 6})^3 + 6$
$= x - 6 + 6$
$= x$
$(g \circ f)(x) = g[f(x)]$
$= g(x^3 + 6)$
$= \sqrt[3]{x^3 + 6 - 6}$
$= x$
Yes, f and g are inverse functions.

11. $(f \circ g)(x) = f[g(x)]$
$= f(\sqrt{x + 2})$
$= (\sqrt{x + 2})^2 - 3$
$= x + 2 - 3$
$= x - 1$
No, f and g are not inverse functions.

12. $(f \circ g)(x) = f\big[g(x)\big]$

$= f\left(\dfrac{-4x}{x+1}\right)$

$= \dfrac{\dfrac{-4x}{x+1}}{\dfrac{-4x}{x+1}+4}$

$= \dfrac{(x+1)\left(\dfrac{-4x}{x+1}\right)}{(x+1)\left(\dfrac{-4x}{x+1}+4\right)}$

$= \dfrac{-4x}{-4x+4x+4}$

$= \dfrac{-4x}{4}$

$= -x$

No, f and g are not inverse functions.

13. No, not a one-to-one function

14. Yes, it is a one-to-one function.

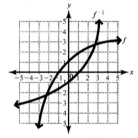

15. $\quad y = 5x+4$

$\quad x = 5y+4$

$x-4 = 5y$

$\dfrac{x-4}{5} = y$

$f^{-1}(x) = \dfrac{x-4}{5}$

16. $\quad y = x^3+6$

$\quad x = y^3+6$

$x-6 = y^3$

$\sqrt[3]{x-6} = y$

$f^{-1}(x) = \sqrt[3]{x-6}$

17. $\quad y = \dfrac{4}{x+5}$

$\quad x = \dfrac{4}{y+5}$

$xy+5x = 4$

$xy = 4-5x$

$y = \dfrac{4-5x}{x}$

$f^{-1}(x) = \dfrac{4-5x}{x}$

18. $\quad y = \sqrt[3]{3x+2}$

$\quad x = \sqrt[3]{3y+2}$

$x^3 = 3y+2$

$x^3-2 = 3y$

$\dfrac{x^3-2}{3} = y$

$f^{-1}(x) = \dfrac{x^3-2}{3}$

19. $(-6, 4)$

20. $f^{-1}(b) = a$

21. $f(x) = 3^x$

x	$y = f(x)$
-1	$\dfrac{1}{3}$
0	1
1	3
2	9

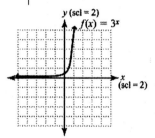

22. $f(x) = \left(\dfrac{1}{2}\right)^x$

x	$y = f(x)$
-1	$\left(\dfrac{1}{2}\right)^{(-1)} = 2$
0	$\left(\dfrac{1}{2}\right)^{(0)} = 1$
1	$\left(\dfrac{1}{2}\right)^{(1)} = \dfrac{1}{2}$
2	$\left(\dfrac{1}{2}\right)^{(2)} = \dfrac{1}{4}$

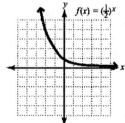

23. $f(x) = 2^{x-3}$

x	$y = f(x)$
-1	$2^{-1-3} = \dfrac{1}{16}$
0	$2^{0-3} = \dfrac{1}{8}$
1	$2^{1-3} = \dfrac{1}{4}$
2	$2^{2-3} = \dfrac{1}{2}$

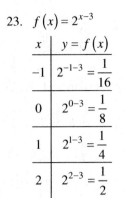

24. $f(x) = 3^{-x+2}$

x	$y = f(x)$
-1	$3^{-(-1)+2} = 27$
0	$3^{-(0)+2} = 9$
1	$3^{-(1)+2} = 3$
2	$3^{-(2)+2} = 1$

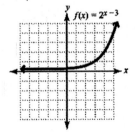

25. $5^x = 625$

$5^x = 5^4$

$x = 4$

26. $16^x = 64$

$\left(4^2\right)^x = 4^3$

$4^{2x} = 4^3$

$2x = 3$

$x = \dfrac{3}{2}$

27. $6^x = \dfrac{1}{36}$

$6^x = \dfrac{1}{6^2}$

$6^x = 6^{-2}$

$x = -2$

28. $\left(\dfrac{3}{4}\right)^x = \dfrac{16}{9}$

$\left(\dfrac{3}{4}\right)^x = \left(\dfrac{4}{3}\right)^2$

$\left(\dfrac{3}{4}\right)^x = \left(\dfrac{3}{4}\right)^{-2}$

$x = -2$

29. $4^{x-1} = 64$

$4^{x-1} = 4^3$

$x - 1 = 3$

$x = 4$

30. $5^{x+2} = 25^x$

$5^{x+2} = \left(5^2\right)^x$

$5^{x+2} = 5^{2x}$

$x + 2 = 2x$

$2 = x$

31. $\left(\dfrac{1}{3}\right)^{-x} = 27$

$\left(3^{-1}\right)^{-x} = 3^3$

$3^x = 3^3$

$x = 3$

32. $8^{3x-2} = 16^{4x}$

$\left(2^3\right)^{3x-2} = \left(2^4\right)^{4x}$

$2^{9x-6} = 2^{16x}$

$9x - 6 = 16x$

$-6 = 7x$

$-\dfrac{6}{7} = x$

33. $A = A_0 \cdot 2^{t/100}$

$A = 500 \cdot 2^{365/100}$

$A \approx 6277$

There are 6277 cells after one year.

34. $A = 25000\left(1 + \dfrac{0.06}{12}\right)^{12 \cdot 8}$

$A = 40{,}353.57$

The account is worth \$40,353.57 after eight years.

35. $A = A_0\left(\dfrac{1}{2}\right)^{t/h}$

$A = 50\left(\dfrac{1}{2}\right)^{300/107}$

$A \approx 7.16$

After 300 days, 7.16 grams remain.

36. $y = 4.0144(1.028)^x$

 $y = 4.0144(1.028)^{25}$

 $y \approx 8.01$

 In 2015, the minimum wage will be approximately \$8.01.

37. $\log_7 343 = 3$

38. $\log_4 \dfrac{1}{64} = -3$

39. $\log_{3/2} \dfrac{81}{16} = 4$

40. $\log_{11} \sqrt[3]{11} = \dfrac{1}{3}$

41. $9^2 = 81$

42. $\left(\dfrac{1}{5}\right)^{-3} = 125$

43. $a^4 = 16$

44. $e^b = c$

45. $3^{-4} = x$

 $\dfrac{1}{3^4} = x$

 $\dfrac{1}{81} = x$

46. $\left(\dfrac{1}{2}\right)^{-2} = x$

 $2^2 = x$

 $4 = x$

47. $2^x = 32$

 $2^x = 2^5$

 $x = 5$

48. $\left(\dfrac{1}{4}\right)^x = 16$

 $\left(4^{-1}\right)^x = 4^2$

 $4^{-x} = 4^2$

 $-x = 2$

 $x = -2$

49. $x^4 = 81$

 $x^4 = 3^4$

 $x = 3$

50. $x^3 = \dfrac{1}{1000}$

 $x = \sqrt[3]{\dfrac{1}{1000}}$

 $x = \dfrac{1}{10}$

51. $\left(\dfrac{3}{4}\right)^x = \dfrac{9}{16}$

 $\left(\dfrac{3}{4}\right)^x = \left(\dfrac{3}{4}\right)^2$

 $x = 2$

52. $81^{1/4} = x$

 $\sqrt[4]{81} = x$

 $3 = x$

53. $f(x) = \log_4 x$

 $y = \log_4 x$

 $4^y = x$

y	-1	0	1	2
x	$\dfrac{1}{4}$	1	4	16

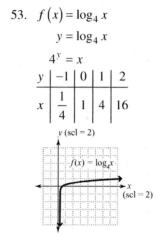

54. $f(x) = \log_{1/3} x$

 $y = \log_{1/3} x$

 $\left(\dfrac{1}{3}\right)^y = x$

y	-2	-1	0	1
x	9	3	1	$\dfrac{1}{3}$

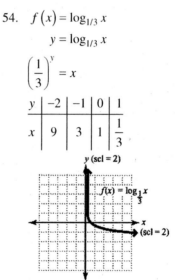

55. Domain is $(-\infty, \infty)$ and range is $(0, \infty)$ because $f(x) = \log_b x$ and $g(x) = b^x$ are inverses.

56. $d = 10\log\dfrac{I}{I_0}$

    $d = 10\log\dfrac{1000 I_0}{I_0}$

    $d = 10\log 1000$

    $d = 10\log 10^3$

    $d = 10 \cdot 3$

    $d = 30$

    The decibel reading is 30 dB.

57. $3^{\log_3 8} = 8$

58. $\log_4 4^6 = 6$

59. $\log_6 6x = \log_6 6 + \log_6 x = 1 + \log_6 x$

60. $\log_4 x(2x-5) = \log_4 x + \log_4 (2x-5)$

61. $\log_3 4 + \log_3 8 = \log_3 (4 \cdot 8) = \log_3 32$

62. $\log_5 3 + \log_5 x + \log_5 (x-2) = \log_5 3x(x-2)$
    $$= \log_5 (3x^2 - 6x)$$

63. $\log_b \dfrac{x}{5} = \log_b x - \log_b 5$

64. $\log_a \dfrac{3x-2}{4x+3} = \log_a (3x-2) - \log_a (4x+3)$

65. $\log_8 32 - \log_8 16 = \log_8 \dfrac{32}{16} = \log_8 2$

66. $\log_2 (x+5) - \log_2 (2x-3) = \log_2 \dfrac{x+5}{2x-3}$

67. $\log_3 7^4 = 4\log_3 7$

68. $\log_a \sqrt[3]{x} = \log_a x^{1/3} = \dfrac{1}{3}\log_a x$

69. $\log_4 \dfrac{1}{a^4} = \log_4 a^{-4} = -4\log_4 a$

70. $\log_a \sqrt[5]{a^4} = \log_a a^{4/5} = \dfrac{4}{5}\log_a a = \dfrac{4}{5}$

71. $4\log_a x = \log_a x^4$

72. $\dfrac{3}{5}\log_a y = \log_a y^{3/5} = \log_a \sqrt[5]{y^3}$

73. $\log_a x^2 y^3 = \log_a x^2 + \log_a y^3$
    $$= 2\log_a x + 3\log_a y$$

74. $\log_a \dfrac{c^4}{d^3} = \log_a c^4 - \log_a d^3$
    $$= 4\log_a c - 3\log_a d$$

75. $\log_a \dfrac{x^2 y^3}{z^4} = \log_a x^2 + \log_a y^3 - \log_a z^4$
    $$= 2\log_a x + 3\log_a y - 4\log_a z$$

76. $\log_a \sqrt{\dfrac{a^3}{b^4}} = \log_a \left(\dfrac{a^3}{b^4}\right)^{1/2}$
    $$= \log_a \left(\dfrac{a^{3/2}}{b^{4/2}}\right)$$
    $$= \log_a a^{3/2} - \log_a b^2$$
    $$= \dfrac{3}{2}\log_a a - 2\log_a b$$
    $$= \dfrac{3}{2} - 2\log_a b$$

77. $3\log_x y + 5\log_x z = \log_x y^3 + \log_x z^5$
    $$= \log_x y^3 z^5$$

78. $3\log_a 4 - 2\log_a 3 = \log_a 4^3 - \log_a 3^2$
    $$= \log_a \dfrac{4^3}{3^2}$$

79. $\dfrac{1}{4}\left(2\log_a x + 3\log_a y\right) = \dfrac{2}{4}\log_a x + \dfrac{3}{4}\log_a y$
    $$= \log_a x^{2/4} + \log_a y^{3/4}$$
    $$= \log_a x^{2/4} y^{3/4}$$
    $$= \log_a \left(x^2 y^3\right)^{1/4}$$
    $$= \log_a \sqrt[4]{x^2 y^3}$$

80. $4\log_a (x+5) + 2\log_a (x-3)$
    $$= \log_a (x+5)^4 + \log_a (x-3)^2$$
    $$= \log_a (x+5)^4 (x-3)^2$$

81. $\log 326 = 2.5132$

82. $\log 0.0035 = -2.4559$

83. $\ln 0.043 = -3.1466$

84. $\ln 92 = 4.5218$

85. $\log 0.00001 = \log_{10} 10^{-5} = -5$

86. $\ln \sqrt[4]{e} = \ln e^{1/4} = \dfrac{1}{4}$

87. $d = 10 \log \dfrac{I}{I_0}$

$d = 10 \log \dfrac{10^{-3.5}}{10^{-12}}$

$d = 10 \log 10^{8.5}$

$d = 10(8.5)$

$d = 85$

The decibel reading of the thunder was 85 dB.

88. $\text{pH} = -\log \left[H_3O^+ \right]$

$\text{pH} = -\log 0.001259$

$\text{pH} \approx 2.9$

89. $A = Pe^{rt}$

$10,000 = 6000e^{0.03t}$

$\dfrac{10}{6} = e^{0.03t}$

$\ln \dfrac{5}{3} = \ln e^{0.03t}$

$\ln \dfrac{5}{3} = 0.03t$

$\dfrac{\ln \dfrac{5}{3}}{0.03} = t$

$17 \approx t$

It will take approximately 17 years.

90. Earthquake 1: $\quad 7.8 = \log \dfrac{I}{I_0}$

$7.8 = \log I - \log I_0$

$7.8 + \log I_0 = \log I$

$10^{7.8 + \log I_0} = I$

Earthquake 2: $\quad 6.8 = \log \dfrac{I}{I_0}$

$6.8 = \log I - \log I_0$

$6.8 + \log I_0 = \log I$

$10^{6.8 + \log I_0} = I$

$\dfrac{10^{7.8 + \log I_0}}{10^{6.8 + \log I_0}} = 10^{(7.8 + \log I_0) - (6.8 + \log + I_0)}$

$= 10^{7.8 - 6.8}$

$= 10^1$

$= 10$

Earthquake 1 was ten times as severe as earthquake 2.

91. $9^x = 32$

$\ln 9^x = \ln 32$

$x \ln 9 = \ln 32$

$x = \dfrac{\ln 32}{\ln 9} \approx 1.5773$

92. $3^{5x} = 19$

$\ln 3^{5x} = \ln 19$

$5x \ln 3 = \ln 19$

$5x = \dfrac{\ln 19}{\ln 3}$

$x = \dfrac{\dfrac{\ln 19}{\ln 3}}{5} \approx 0.5360$

93. $6^{2x-1} = 22$

$\ln 6^{2x-1} = \ln 22$

$(2x-1)\ln 6 = \ln 22$

$2x \ln 6 - \ln 6 = \ln 22$

$2x \ln 6 = \ln 22 + \ln 6$

$x = \dfrac{\ln 22 + \ln 6}{2 \ln 6} \approx 1.3626$

94. $4^{2x-3} = 5^{x+1}$

$\ln 4^{2x-3} = \ln 5^{x+1}$

$(2x-3)\ln 4 = (x+1)\ln 5$

$2x \ln 4 - 3 \ln 4 = x \ln 5 + \ln 5$

$2x \ln 4 - x \ln 5 = \ln 5 + 3 \ln 4$

$x(2 \ln 4 - \ln 5) = \ln 5 + 3 \ln 4$

$x = \dfrac{\ln 5 + 3 \ln 4}{2 \ln 4 - \ln 5} \approx 4.9592$

95. $e^{4x} = 11$

$\ln e^{4x} = \ln 11$

$4x = \ln 11$

$x = \dfrac{\ln 11}{4} \approx 0.5995$

96. $e^{-0.003x} = 5$

$\ln e^{-0.003x} = \ln 5$

$-0.003x = \ln 5$

$x = -\dfrac{\ln 5}{0.003} \approx -536.4793$

97.
$$A = P\left(1 + \frac{r}{n}\right)^{nt}$$

$$18000 = 15000\left(1 + \frac{0.03}{12}\right)^{12 \cdot t}$$

$$1.2 = (1.0025)^{12t}$$

$$\ln 1.2 = \ln(1.0025)^{12t}$$

$$\ln 1.2 = 12t \ln(1.0025)$$

$$\frac{\ln 1.2}{12 \ln(1.0025)} = t$$

$$6.1 \approx t$$

It will take approximately 6.1 years.

98. a) $A = 7000e^{0.07(8)}$

$A = \$12,254.71$

b) $12,000 = 7000e^{0.07t}$

$$\frac{12}{7} = e^{0.07t}$$

$$\ln \frac{12}{7} = \ln e^{0.07t}$$

$$\ln \frac{12}{7} = 0.07t$$

$$\frac{\ln \frac{12}{7}}{0.07} = t \approx 7.7$$

It will take about 7.7 years.

99. a) $A = 400e^{0.03(12)}$

$A \approx 573$

The population of the colony will be about 573 ants after one year.

b) $800 = 400e^{0.03t}$

$$2 = e^{0.03t}$$

$$\ln 2 = \ln e^{0.03t}$$

$$\ln 2 = 0.03t$$

$$\frac{\ln 2}{0.03} = t$$

$$23 \approx t$$

The population will be 800 ants after about 23 months.

100. $G = \dfrac{t}{3.3 \log \dfrac{b}{B}}$

$$G = \frac{240}{3.3 \log \dfrac{1,000,000}{1000}}$$

$$G = \frac{240}{3.3 \log 1000}$$

$$G = \frac{240}{3.3 \log 10^3}$$

$$G = \frac{240}{3.3(3)} \approx 24$$

The generation time is about 24 minutes.

101. $\log_4(x+8) = 2$

$$4^2 = x + 8$$

$$16 = x + 8$$

$$8 = x$$

102. $\log_2(x^2 + 2x) = 3$

$$2^3 = x^2 + 2x$$

$$0 = x^2 + 2x - 8$$

$$0 = (x+4)(x-2)$$

$$x + 4 = 0 \quad \text{or} \quad x - 2 = 0$$

$$x = -4 \qquad\qquad x = 2$$

103. $\log(3x - 8) = \log(x - 2)$

$$3x - 8 = x - 2$$

$$2x = 6$$

$$x = 3$$

104. $\log 5 + \log x = 2$

$$\log 5x = 2$$

$$10^2 = 5x$$

$$100 = 5x$$

$$20 = x$$

105. $\log_3 x - \log_3 4 = 2$

$$\log_3 \frac{x}{4} = 2$$

$$3^2 = \frac{x}{4}$$

$$9 = \frac{x}{4}$$

$$36 = x$$

106. $\log_2 x + \log_2 (x-6) = 4$

$\log_2 x(x-6) = 4$

$\log_2 (x^2 - 6x) = 4$

$2^4 = x^2 - 6x$

$0 = x^2 - 6x - 16$

$0 = (x-8)(x+2)$

$x - 8 = 0$ or $x + 2 = 0$

$x = 8$ $\qquad$ $x = \cancel{-2}$

107. $\log_4 x + \log_4 (x+2) = \log_4 8$

$\log_4 x(x+2) = \log_4 8$

$x(x+2) = 8$

$x^2 + 2x = 8$

$x^2 + 2x - 8 = 0$

$(x+4)(x-2) = 0$

$x + 4 = 0$ $\quad$ or $\quad$ $x - 2 = 0$

$x = \cancel{-4}$ $\qquad$ $x = 2$

108. $\log_3 (5x+2) - \log_3 (x-2) = 2$

$\log_3 \dfrac{5x+2}{x-2} = 2$

$3^2 = \dfrac{5x+2}{x-2}$

$9(x-2) = 5x+2$

$9x - 18 = 5x + 2$

$4x = 20$

$x = 5$

109. $\log_4 15 = \dfrac{\log 15}{\log 4} \approx 1.9534$

110. $\log_{1/2} 6 = \dfrac{\log 6}{\log 0.5} \approx -2.5850$

Chapter 10 Practice Test

1. Let $f(x) = x^2 - 6$ and $g(x) = 3x - 5$.

$f[g(x)] = f(3x - 5)$

$= (3x-5)^2 - 6$

$= 9x^2 - 30x + 25 - 6$

$= 9x^2 - 30x + 19$

2. $f(x) = 4x - 3$

a) $\qquad y = 4x - 3$

$x = 4y - 3$

$x + 3 = 4y$

$\dfrac{x+3}{4} = y$

$f^{-1}(x) = \dfrac{x+3}{4}$

b) $f\left[f^{-1}(x)\right] = f\left(\dfrac{x+3}{4}\right)$

$= 4\left(\dfrac{x+3}{4}\right) - 3$

$= x + 3 - 3$

$= x$

$f^{-1}\left[f(x)\right] = f^{-1}(4x - 3)$

$= \dfrac{4x - 3 + 3}{4}$

$= \dfrac{4x}{4}$

$= x$

c) The graphs are symmetric about the graph of $y = x$.

3.

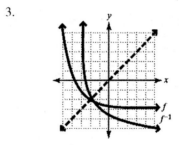

4. $f(x) = 2^{x-1}$

x	$y = f(x)$
-1	$2^{-1-1} = \dfrac{1}{4}$
0	$2^{0-1} = \dfrac{1}{2}$
1	$2^{1-1} = 1$
2	$2^{2-1} = 2$

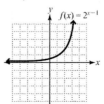

5. $32^{x-2} = 8^{2x}$

$\left(2^5\right)^{x-2} = \left(2^3\right)^{2x}$

$2^{5x-10} = 2^{6x}$

$5x - 10 = 6x$

$-10 = x$

6.   a) $A = 20{,}000\left(1 + \dfrac{0.05}{4}\right)^{4(15)}$

$A = \$42{,}143.63$

b) $30{,}000 = 20{,}000\left(1 + \dfrac{0.05}{4}\right)^{4\cdot t}$

$1.5 = \left(1.0125\right)^{4t}$

$\ln 1.5 = \ln\left(1.0125\right)^{4t}$

$\ln 1.5 = 4t \ln\left(1.0125\right)$

$\dfrac{\ln 1.5}{4\ln\left(1.0125\right)} = t$

$8.2 \approx t$

It will take about 8.2 years.

7.   a) $A = 100\left(\dfrac{1}{2}\right)^{216/30}$

$A = 0.68$

About 0.68 gram remains.

b) $20 = 100\left(\dfrac{1}{2}\right)^{t/30}$

$0.2 = \left(0.5\right)^{t/30}$

$\ln 0.2 = \ln\left(0.5\right)^{t/30}$

$\ln 0.2 = \dfrac{t}{30}\ln 0.5$

$\dfrac{30\ln 0.2}{\ln 0.5} = t$

$69.66 \approx t$

It will take about 69.66 minutes, or 1.16 hours.

8.   a)   In 1960: $y = 3.17\left(1.026\right)^{60}$

$y \approx 14.8$

There were about 14.8 million people.

In 2012: $y = 3.17\left(1.026\right)^{112}$

$y \approx 56.2$

There were about 56.2 million people.

b) $31.4 = 3.17\left(1.026\right)^{x}$

$\dfrac{31.4}{3.17} = 1.026^{x}$

$\ln\left(\dfrac{31.4}{3.17}\right) = \ln 1.026^{x}$

$\ln\left(\dfrac{31.4}{3.17}\right) = x\ln 1.026$

$\dfrac{\ln\left(\dfrac{31.4}{3.17}\right)}{\ln 1.026} = x$

$89 \approx x$

89 years after 1900, in 1989.

9. $\left(\dfrac{1}{3}\right)^{-4} = 81$

10. $\log_6 \dfrac{1}{216} = x$

$6^{x} = \dfrac{1}{216}$

$6^{x} = 216^{-1}$

$6^{x} = \left(6^3\right)^{-1}$

$6^{x} = 6^{-3}$

$x = -3$

11. $\log_x 625 = 4$

$x^4 = 625$

$x^4 = 5^4$

$x = 5$

12. $f\left(x\right) = \log_2 x$

$y = \log_2 x$

$2^{y} = x$

y	-1	0	1	2
x	$\dfrac{1}{2}$	1	2	4

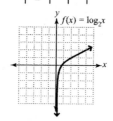

13. $\log_b \dfrac{x^4 y^2}{z} = \log_b x^4 y^2 - \log_b z$

$\qquad = \log_b x^4 + \log_b y^2 - \log_b z$

$\qquad = 4\log_b x + 2\log_b y - \log_b z$

14. $\log_b \sqrt[4]{\dfrac{x^5}{y^7}} = \log_b \left(\dfrac{x^5}{y^7} \right)^{1/4}$

$\qquad = \log_b \dfrac{x^{5/4}}{y^{7/4}}$

$\qquad = \log_b x^{5/4} - \log_b y^{7/4}$

$\qquad = \dfrac{5}{4}\log_b x - \dfrac{7}{4}\log_b y$

15. $\dfrac{3}{4}\left(2\log_b x + 3\log_b y \right) = \dfrac{6}{4}\log_b x + \dfrac{9}{4}\log_b y$

$\qquad = \log_b x^{6/4} + \log_b y^{9/4}$

$\qquad = \log_b x^{6/4} y^{9/4}$

$\qquad = \log_b \left(x^6 y^9 \right)^{1/4}$

$\qquad = \log_b \sqrt[4]{x^6 y^9}$

16. a) $A = 300e^{-0.0028(500)}$

$\qquad A \approx 74$

About 74 grams remain.

b) $\qquad A = 300e^{-0.0028t}$

$\qquad 150 = 300e^{-0.0028t}$

$\qquad 0.5 = e^{-0.0028t}$

$\qquad \ln 0.5 = \ln e^{-0.0028t}$

$\qquad \ln 0.5 = -0.0028t$

$\qquad \dfrac{\ln 0.5}{-0.0028} = t$

$\qquad 247.6 \approx t$

The half-life is about 247.6 days.

17. $\qquad 6^{x-3} = 19$

$\qquad \ln 6^{x-3} = \ln 19$

$\qquad (x-3)\ln 6 = \ln 19$

$\qquad x\ln 6 - 3\ln 6 = \ln 19$

$\qquad x\ln 6 = \ln 19 + 3\ln 6$

$\qquad x = \dfrac{\ln 19 + 3\ln 6}{\ln 6}$

$\qquad x \approx 4.6433$

18. $\log_3 x + \log_3 (x+6) = 3$

$\qquad \log_3 x(x+6) = 3$

$\qquad \log_3 \left(x^2 + 6x \right) = 3$

$\qquad 3^3 = x^2 + 6x$

$\qquad 0 = x^2 + 6x - 27$

$\qquad 0 = (x+9)(x-3)$

$x+9 = 0 \qquad$ or $\quad x-3 = 0$

$\qquad x = \cancel{-9} \qquad\qquad x = 3$

19. $\log(4x+2) - \log(3x-2) = \log 2$

$\qquad \log \dfrac{4x+2}{3x-2} = \log 2$

$\qquad \dfrac{4x+2}{3x-2} = 2$

$\qquad 4x+2 = 6x-4$

$\qquad 6 = 2x$

$\qquad 3 = x$

20. a) $n = 3.3\log \dfrac{b}{B}$

$\qquad n = 3.3\log \dfrac{10{,}000{,}000}{100}$

$\qquad n = 3.3\log (100{,}000)$

$\qquad n = 16.5$

There were 16.5 generations.

b) $n = 3.3\log \dfrac{b}{B}$

$\qquad 9.9 = 3.3\log \dfrac{b}{100}$

$\qquad 3 = \log \dfrac{b}{100}$

$\qquad 10^3 = \dfrac{b}{100}$

$\qquad 1000 = \dfrac{b}{100}$

$\qquad 100{,}000 = b$

There will be 100,000 bacteria.

Chapters 1–10 Cumulative Review

1. True

2. True

3. False; the range of $f(x) = x^2 + 4$ is $[4, \infty)$.

4. -8

5. x-intercepts

6. $x - 3 = 4$; $x - 3 = -4$

7. $6 - 8 \div 2 \cdot 3^2 - 4\left(3 - 4 \cdot 2^3\right)$

$= 6 - 8 \div 2 \cdot 3^2 - 4\left(3 - 4 \cdot 8\right)$

$= 6 - 8 \div 2 \cdot 3^2 - 4\left(3 - 32\right)$

$= 6 - 8 \div 2 \cdot 3^2 - 4\left(-29\right)$

$= 6 - 8 \div 2 \cdot 9 - 4\left(-29\right)$

$= 6 - 4 \cdot 9 - \left(-116\right)$

$= 6 - 36 + 116$

$= 86$

8. $\dfrac{\left(4a^{-4}\right)^3 \left(2a^2\right)^{-3}}{\left(2a^{-2}\right)^3} = \dfrac{64a^{-12} \cdot 2^{-3} a^{-6}}{2^3 a^{-6}}$

$= \dfrac{2^6 a^{-12} \cdot 2^{-3} a^{-6}}{2^3 a^{-6}}$

$= \dfrac{2^3 a^{-18}}{2^3 a^{-6}}$

$= a^{-18 - (-6)}$

$= a^{-12}$

$= \dfrac{1}{a^{12}}$

9. $\dfrac{21p^2 q^4 - 14 p^5 q^3 + 6 p^3 q^7}{7 p^3 q^5}$

$= \dfrac{21 p^2 q^4}{7 p^3 q^5} - \dfrac{14 p^5 q^3}{7 p^3 q^5} + \dfrac{6 p^3 q^7}{7 p^3 q^5}$

$= 3 p^{-1} q^{-1} - 2 p^2 q^{-2} + \dfrac{6}{7} q^2$

$= \dfrac{3}{pq} - \dfrac{2p^2}{q^2} + \dfrac{6q^2}{7}$

10. $\dfrac{2x+3}{x^2 + 6x + 9} - \dfrac{x+4}{x+3} = \dfrac{2x+3}{(x+3)^2} - \dfrac{(x+4)(x+3)}{(x+3)^2}$

$= \dfrac{2x+3}{(x+3)^2} - \dfrac{x^2 + 7x + 12}{(x+3)^2}$

$= \dfrac{2x+3 - x^2 - 7x - 12}{(x+3)^2}$

$= \dfrac{-x^2 - 5x - 9}{(x+3)^2}$

11. $4\sqrt{6c^3} \cdot 3\sqrt{10c^5} = 12\sqrt{6c^3 \cdot 10c^5}$

$= 12\sqrt{60c^8}$

$= 12\sqrt{4c^8 \cdot 15}$

$= 12 \cdot 2c^4 \sqrt{15}$

$= 24c^4 \sqrt{15}$

12. $(4+3i)(2-4i) = 8 - 16i + 6i - 12i^2$

$= 8 - 10i - 12(-1)$

$= 8 - 10i + 12$

$= 20 - 10i$

13. $\log_b b^{2x} = 2x$

14. $\dfrac{d^2 - d - 12}{2d^2} \div \dfrac{3d^2 + 13d + 12}{d}$

$= \dfrac{d^2 - d - 12}{2d^2} \cdot \dfrac{d}{3d^2 + 13d + 12}$

$= \dfrac{(d-4)\,\cancel{(d+3)}}{2 \cdot d \cdot \cancel{d}} \cdot \dfrac{\cancel{d}}{(3d+4)\,\cancel{(d+3)}}$

$= \dfrac{d-4}{2d(3d+4)}$

15. $x^4 - 9x^2 - 4x^2 y^2 + 36 y^2$

$= \left(x^4 - 9x^2\right) - \left(4x^2 y^2 - 36 y^2\right)$

$= x^2\left(x^2 - 9\right) - 4y^2\left(x^2 - 9\right)$

$= \left(x^2 - 9\right)\left(x^2 - 4y^2\right)$

$= (x+3)(x-3)(x+2y)(x-2y)$

16. $24x^4 y - 30x^3 y^2 + 9x^2 y^3$

$= 3x^2 y\left(8x^2 - 10xy + 3y^2\right)$

$= 3x^2 y(4x - 3y)(2x - y)$

17. $|2x - 3| - 5 = -4$

$|2x - 3| = 1$

$2x - 3 = 1$ or $2x - 3 = -1$

$2x = 4$ $2x = 2$

$x = 2$ $x = 1$

18. $2|6x - 10| > -8$

$|6x - 10| > -4$

Because an absolute value is always nonnegative, this statement is true for all real numbers.

19. $\begin{cases} 4x - 5y = -22 \\ 3x + 2y = -5 \end{cases}$

Solve the first equation for x: $x = \dfrac{5}{4}y - \dfrac{22}{4}$

Substitute $\dfrac{5}{4}y - \dfrac{22}{4}$ for x in the second equation
and solve for y.

$$3x + 2y = -5$$
$$3\left(\frac{5}{4}y - \frac{22}{4}\right) + 2y = -5$$
$$\frac{15}{4}y - \frac{66}{4} + \frac{8}{4}y = -5$$
$$\frac{23}{4}y - \frac{66}{4} = -5$$
$$4\left(\frac{23}{4}y - \frac{66}{4}\right) = 4(-5)$$
$$23y - 66 = -20$$
$$23y = 46$$
$$y = 2$$

$$x = \frac{5}{4}y - \frac{22}{4}$$
$$x = \frac{5}{4}(2) - \frac{22}{4}$$
$$x = \frac{10}{4} - \frac{22}{4}$$
$$x = -\frac{12}{4}$$
$$x = -3$$

Solution: $(-3, 2)$

20. Let $u = 2x - 3$.

$$u^2 - 2u = 8$$
$$u^2 - 2u - 8 = 0$$
$$(u - 4)(u + 2) = 0$$

$u - 4 = 0$ or $u + 2 = 0$

$u = 4$ $u = -2$

$2x - 3 = 4$ $2x - 3 = -2$

$2x = 7$ $2x = 1$

$x = \dfrac{7}{2}$ $x = \dfrac{1}{2}$

21. $\dfrac{x}{x-2} - \dfrac{4}{x-1} = \dfrac{2}{x^2 - 3x + 2}$

$$\frac{x}{x-2} - \frac{4}{x-1} = \frac{2}{(x-2)(x-1)}$$
$$(x-2)(x-1)\left(\frac{x}{x-2} - \frac{4}{x-1}\right)$$
$$= (x-2)(x-1)\left(\frac{2}{(x-2)(x-1)}\right)$$
$$x(x-1) - 4(x-2) = 2$$
$$x^2 - x - 4x + 8 = 2$$
$$x^2 - 5x + 6 = 0$$
$$(x-3)(x-2) = 0$$

$x - 3 = 0$ or $x - 2 = 0$

$x = 3$ $x = 2$

Notice that if $x = 2$, then $\dfrac{x}{x-2}$ and $\dfrac{2}{x^2 - 3x + 2}$
are undefined. Therefore, $x = 2$ is extraneous.
The only solution is $x = 3$.

22. $\sqrt{x+4} - \sqrt{x-4} = 4$

$$\sqrt{x+4} = \sqrt{x-4} + 4$$
$$\left(\sqrt{x+4}\right)^2 = \left(\sqrt{x-4} + 4\right)^2$$
$$x+4 = x-4 + 8\sqrt{x-4} + 16$$
$$x+4 = x + 12 + 8\sqrt{x-4}$$
$$-8 = 8\sqrt{x-4}$$
$$-1 = \sqrt{x-4}$$
$$(-1)^2 = \left(\sqrt{x-4}\right)^2$$
$$1 = x - 4$$
$$5 = x$$

Check:

$$\sqrt{5+4} - \sqrt{5-4} \overset{?}{=} 4$$
$$\sqrt{9} - \sqrt{1} \overset{?}{=} 4$$
$$3 - 1 \overset{?}{=} 4$$
$$2 \neq 4$$

Because 5 is an extraneous solution, this
equation has no solution.

23. $4^{x+2} = 8$

$$\left(2^2\right)^{x+2} = 2^3$$

$$2^{2x+4} = 2^3$$

$$2x + 4 = 3$$

$$2x = -1$$

$$x = -\frac{1}{2}$$

24. $\frac{1}{2}\log_2 x = 2$

$$\log_2 x^{1/2} = 2$$

$$2^2 = x^{1/2}$$

$$4 = \sqrt{x}$$

$$4^2 = \left(\sqrt{x}\right)^2$$

$$16 = x$$

25. $\log(3x-5) + \log x = \log 12$

$$\log\left[(3x-5)x\right] = \log 12$$

$$\log\left(3x^2 - 5x\right) = \log 12$$

$$3x^2 - 5x = 12$$

$$3x^2 - 5x - 12 = 0$$

$$(3x+4)(x-3) = 0$$

$$3x + 4 = 0 \quad \text{or} \quad x - 3 = 0$$

$$x = -\frac{4}{3} \quad\quad\quad x = 3$$

26. $2x - 5y = 10$

$$-5y = -2x + 10$$

$$y = \frac{2}{5}x - 2$$

a) $m = \frac{2}{5}$

b) y-intercept: $(0, -2)$

   x-intercept:

   $$2x - 5(0) = 10$$

   $$2x = 10$$

   $$x = 5$$

   $$(5, 0)$$

c)

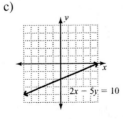

27. $a^2 + b^2 = c^2$

$$n^2 + (n-1)^2 = (n+8)^2$$

$$n^2 + n^2 - 2n + 1 = n^2 + 16n + 64$$

$$n^2 - 18n - 63 = 0$$

$$(n-21)(n+3) = 0$$

$$n - 21 = 0 \quad \text{or} \quad n + 3 = 0$$

$$n = 21 \quad\quad\quad n = -3$$

Lengths cannot be negative, so disregard the negative solution. One leg has a length of 21. The other leg has length $21 - 1 = 20$, and the hypotenuse has length $21 + 8 = 29$.

28. $A = P\left(1 + \dfrac{r}{n}\right)^{nt}$

$$A = 15,000\left(1 + \frac{0.04}{12}\right)^{12\cdot3}$$

$$A = 15,000\left(1.00\overline{3}\right)^{36}$$

$$A \approx 16,909.08$$

After three years, there will be $16,909.08 in the account.

29. Complete a table.

Categories	Rate of Work	Time at Work	Amt of Task Completed
Ethan	$\dfrac{1}{2}$	t	$\dfrac{t}{2}$
Pam	$\dfrac{1}{1.5} = \dfrac{1}{\frac{3}{2}} = \dfrac{2}{3}$	t	$\dfrac{2t}{3}$

$$\frac{t}{2} + \frac{2t}{3} = 1$$

$$6 \cdot \left(\frac{t}{2} + \frac{2t}{3}\right) = 6 \cdot (1)$$

$$3t + 4t = 6$$

$$7t = 6$$

$$t = \frac{6}{7} \approx 0.857$$

Working together, it takes Ethan and Pam 0.857 hour, or about 51 minutes, to complete the job.

30. Create a table.

	rate	time	distance
car	$x+10$	$\dfrac{300}{x+10}$	300
bus	x	$\dfrac{300}{x}$	300

Translate the information in the table to an equation and solve.

$$\frac{300}{x+10}+1=\frac{300}{x}$$

$$x(x+10)\left(\frac{300}{x+10}+1\right)=x(x+10)\left(\frac{300}{x}\right)$$

$$300x+x(x+10)=300(x+10)$$

$$300x+x^2+10x=300x+3000$$

$$x^2+10x-3000=0$$

$$(x+60)(x-50)=0$$

$$x+60=0 \quad \text{or} \quad x-50=0$$

$$x=\cancel{-60} \qquad\qquad x=50$$

Disregard the negative rate. If the bus travels at a rate of 50 miles per hour, then it takes

$$\frac{300}{50+10}=\frac{300}{60}=5 \text{ hours for the car to travel 300}$$

miles.

Chapter 11
Conic Sections

Exercise Set 11.1

1. $y = (x-1)^2 + 2$

   opens upward

   vertex is $(1,2)$

   axis of symmetry is $x = 1$

   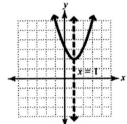

3. $$y = -x^2 - 2x + 3$$
   $$y - 3 = -1(x^2 + 2x)$$
   $$y - 3 - 1 = -1(x^2 + 2x + 1)$$
   $$y - 4 = -1(x+1)^2$$
   $$y = -1(x+1)^2 + 4$$

   opens downward

   vertex is $(-1, 4)$

   axis of symmetry is $x = -1$

   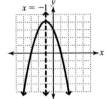

5. $x = (y+2)^2 - 2$

   opens right

   vertex is $(-2,-2)$

   axis of symmetry is $y = -2$

   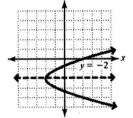

7. $x = -(y-1)^2 + 3$

   opens left

   vertex is $(3,1)$

   axis of symmetry is $y = 1$

   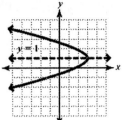

9. $x = 2(y+2)^2 - 4$

   opens right

   vertex is $(-4,-2)$

   axis of symmetry is $y = -2$

   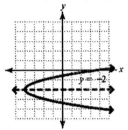

11. $x = -3(y+2)^2 - 5$

    opens left

    vertex is $(-5,-2)$

    axis of symmetry is $y = -2$

    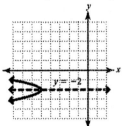

13. $x = y^2 + 4y + 3$
       $x - 3 = y^2 + 4y$
       $x - 3 + 4 = y^2 + 4y + 4$
       $x + 1 = (y + 2)^2$
       $x = (y + 2)^2 - 1$

opens right

vertex is $(-1, -2)$

axis of symmetry is $y = -2$

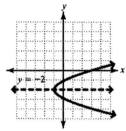

15. $x = -y^2 + 6y - 5$
       $x + 5 = -(y^2 - 6y)$
       $x + 5 - 9 = -(y^2 - 6y + 9)$
       $x - 4 = -(y - 3)^2$
       $x = -(y - 3)^2 + 4$

opens left

vertex is $(4, 3)$

axis of symmetry is $y = 3$

17. $x = 2y^2 + 8y + 3$
       $x - 3 = 2(y^2 + 4y)$
       $x - 3 + 8 = 2(y^2 + 4y + 4)$
       $x + 5 = 2(y + 2)^2$
       $x = 2(y + 2)^2 - 5$

opens right

vertex is $(-5, -2)$

axis of symmetry is $y = -2$

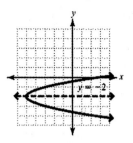

19. $x = 3y^2 - 6y + 1$
       $x - 1 = 3(y^2 - 2y)$
       $x - 1 + 3 = 3(y^2 - 2y + 1)$
       $x = (y - 1)^2 - 2$

opens right

vertex is $(-2, 1)$

axis of symmetry is $y = 1$

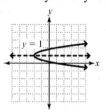

21. $x = -2y^2 + 4y + 5$
       $x - 5 = -2(y^2 - 2y)$
       $x - 5 - 2 = -2(y^2 - 2y + 1)$
       $x - 7 = -2(y - 1)^2$
       $x = -2(y - 1)^2 + 7$

opens left

vertex is $(7, 1)$

axis of symmetry is $y = 1$

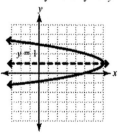

23. c   25. a

27. $(-4,2)$ and $(-1,6)$

$$d = \sqrt{(-4-(-1))^2 + (2-6)^2}$$
$$= \sqrt{(-3)^2 + (-4)^2}$$
$$= \sqrt{9+16}$$
$$= \sqrt{25}$$
$$= 5$$

$$\text{Midpoint} = \left(\frac{-4+(-1)}{2}, \frac{2+6}{2}\right)$$
$$= \left(\frac{-5}{2}, \frac{8}{2}\right)$$
$$= \left(-\frac{5}{2}, 4\right)$$

29. $(-8,-4)$ and $(-3,8)$

$$d = \sqrt{(-8-(-3))^2 + (-4-8)^2}$$
$$= \sqrt{(-5)^2 + (-12)^2}$$
$$= \sqrt{25+144}$$
$$= \sqrt{169}$$
$$= 13$$

$$\text{Midpoint} = \left(\frac{-8+(-3)}{2}, \frac{-4+8}{2}\right)$$
$$= \left(\frac{-11}{2}, \frac{4}{2}\right)$$
$$= \left(-\frac{11}{2}, 2\right)$$

31. $(-8,-10)$ and $(4,-5)$

$$d = \sqrt{(-8-4)^2 + (-10-(-5))^2}$$
$$= \sqrt{(-12)^2 + (-5)^2}$$
$$= \sqrt{144+25}$$
$$= \sqrt{169}$$
$$= 13$$

$$\text{Midpoint} = \left(\frac{-8+4}{2}, \frac{-10+(-5)}{2}\right)$$
$$= \left(\frac{-4}{2}, \frac{-15}{2}\right)$$
$$= \left(-2, -\frac{15}{2}\right)$$

33. $(2,4)$ and $(4,8)$

$$d = \sqrt{(2-4)^2 + (4-8)^2}$$
$$= \sqrt{(-2)^2 + (-4)^2}$$
$$= \sqrt{4+16}$$
$$= \sqrt{20}$$
$$= 2\sqrt{5}$$

$$\text{Midpoint} = \left(\frac{2+4}{2}, \frac{4+8}{2}\right)$$
$$= \left(\frac{6}{2}, \frac{12}{2}\right)$$
$$= (3,6)$$

35. $(-5,2)$ and $(3,-2)$

$$d = \sqrt{(-5-3)^2 + (2-(-2))^2}$$
$$= \sqrt{(-8)^2 + (4)^2}$$
$$= \sqrt{64+16}$$
$$= \sqrt{80}$$
$$= 4\sqrt{5}$$

$$\text{Midpoint} = \left(\frac{-5+3}{2}, \frac{2+(-2)}{2}\right)$$
$$= \left(\frac{-2}{2}, \frac{0}{2}\right)$$
$$= (-1,0)$$

37. $(6,-2)$ and $(1,-5)$

$$d = \sqrt{(6-1)^2 + (-2-(-5))^2}$$
$$= \sqrt{(5)^2 + (3)^2}$$
$$= \sqrt{25+9}$$
$$= \sqrt{34}$$

$$\text{Midpoint} = \left(\frac{6+1}{2}, \frac{-2+(-5)}{2}\right)$$
$$= \left(\frac{7}{2}, \frac{-7}{2}\right)$$
$$= \left(\frac{7}{2}, -\frac{7}{2}\right)$$

39. Find the distance from the center to a point on the circle. This is the radius.

$(4, 2)$ and $(8, -1)$

$$r = \sqrt{(8-4)^2 + (-1-2)^2}$$
$$= \sqrt{(4)^2 + (-3)^2}$$
$$= \sqrt{16 + 9}$$
$$= \sqrt{25}$$
$$= 5$$

41. Find the distance from the center to a point on the circle. This is the radius.

$(2, -6)$ and $(10, -1)$

$$r = \sqrt{(10-2)^2 + (-1-(-6))^2}$$
$$= \sqrt{(8)^2 + (5)^2}$$
$$= \sqrt{64 + 25}$$
$$= \sqrt{89}$$

43. $(x-2)^2 + (y-1)^2 = 4$

Center $(2, 1)$; radius 2

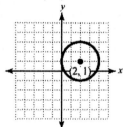

45. $(x+3)^2 + (y+2)^2 = 81$

Center $(-3, -2)$; radius 9

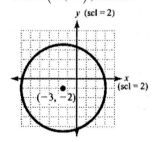

47. $(x-5)^2 + (y+3)^2 = 49$

Center $(5, -3)$; radius 7

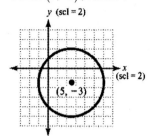

49. $(x-1)^2 + (y+1)^2 = 12$

Center $(1, -1)$; radius $\sqrt{12} = 2\sqrt{3}$

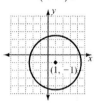

51. $(x+4)^2 + (y+2)^2 = 32$

Center $(-4, -2)$; radius $\sqrt{32} = 4\sqrt{2}$

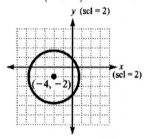

53.
$$x^2 + y^2 + 8x - 6y + 16 = 0$$
$$(x^2 + 8x) + (y^2 - 6y) = -16$$
$$(x^2 + 8x + 16) + (y^2 - 6y + 9) = -16 + 16 + 9$$
$$(x+4)^2 + (y-3)^2 = 9$$

Center $(-4, 3)$; radius 3

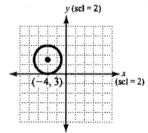

55.
$$x^2 + y^2 + 10x - 4y - 35 = 0$$
$$\left(x^2 + 10x\right) + \left(y^2 - 4y\right) = 35$$
$$\left(x^2 + 10x + 25\right) + \left(y^2 - 4y + 4\right) = 35 + 25 + 4$$
$$\left(x + 5\right)^2 + \left(y - 2\right)^2 = 64$$

Center $\left(-5, 2\right)$; radius 8

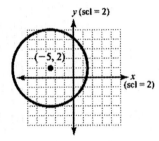

57.
$$x^2 + y^2 + 14x - 4y + 49 = 0$$
$$\left(x^2 + 14x\right) + \left(y^2 - 4y\right) = -49$$
$$\left(x^2 + 14x + 49\right) + \left(y^2 - 4y + 4\right) = -49 + 49 + 4$$
$$\left(x + 7\right)^2 + \left(y - 2\right)^2 = 4$$

Center $\left(-7, 2\right)$; radius 2

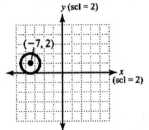

59. b   **61.** c

63. $x^2 + y^2 = 49$

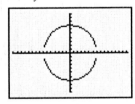

65. $\left(x - 2\right)^2 + \left(y + 3\right)^2 = 25$

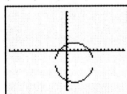

67. Center: $\left(4, 2\right) = \left(h, k\right)$; radius: 4
$$\left(x - h\right)^2 + \left(y - k\right)^2 = r^2$$
$$\left(x - 4\right)^2 + \left(y - 2\right)^2 = 4^2$$
$$\left(x - 4\right)^2 + \left(y - 2\right)^2 = 16$$

69. Center: $\left(-4, -3\right) = \left(h, k\right)$; radius: 5
$$\left(x - h\right)^2 + \left(y - k\right)^2 = r^2$$
$$\left(x - \left(-4\right)\right)^2 + \left(y - \left(-3\right)\right)^2 = 5^2$$
$$\left(x + 4\right)^2 + \left(y + 3\right)^2 = 25$$

71. Center: $\left(6, -2\right) = \left(h, k\right)$; radius: $\sqrt{14}$
$$\left(x - h\right)^2 + \left(y - k\right)^2 = r^2$$
$$\left(x - 6\right)^2 + \left(y - \left(-2\right)\right)^2 = \left(\sqrt{14}\right)^2$$
$$\left(x - 6\right)^2 + \left(y + 2\right)^2 = 14$$

73. Center: $\left(-5, 2\right) = \left(h, k\right)$; radius: $3\sqrt{5}$
$$\left(x - h\right)^2 + \left(y - k\right)^2 = r^2$$
$$\left(x - \left(-5\right)\right)^2 + \left(y - 2\right)^2 = \left(3\sqrt{5}\right)^2$$
$$\left(x + 5\right)^2 + \left(y - 2\right)^2 = 45$$

75. Find the distance from the center to a point on the circle. This is the radius.
$\left(2, 4\right)$ and $\left(5, 8\right)$
$$r = \sqrt{\left(2 - 5\right)^2 + \left(4 - 8\right)^2}$$
$$= \sqrt{\left(-3\right)^2 + \left(-4\right)^2}$$
$$= \sqrt{9 + 16}$$
$$= \sqrt{25}$$
$$= 5$$
The equation in standard form is:
$$\left(x - 2\right)^2 + \left(y - 4\right)^2 = 25$$

77. Find the distance from the center to a point on the circle. This is the radius.
$\left(2, 4\right)$ and $\left(7, 16\right)$
$$r = \sqrt{\left(2 - 7\right)^2 + \left(4 - 16\right)^2}$$
$$= \sqrt{\left(-5\right)^2 + \left(-12\right)^2}$$
$$= \sqrt{25 + 144}$$
$$= \sqrt{169}$$
$$= 13$$
The equation in standard form is
$$\left(x - 2\right)^2 + \left(y - 4\right)^2 = 169$$

79. The center is $(2, -5)$ since all points are an equal distance from that point. Also, since all points are 8 units from the center, the radius is 8.

The equation in standard form is:
$(x-2)^2 + (y+5)^2 = 64$

81. $(4, -2)$ and $(-2, 6)$

$\text{diameter} = \sqrt{(4-(-2))^2 + (-2-6)^2}$
$= \sqrt{(4+2)^2 + (-8)^2}$
$= \sqrt{36+64}$
$= \sqrt{100}$
$= 10$

The radius is $10 \div 2 = 5$.

$\text{Midpoint} = \left(\dfrac{4+(-2)}{2}, \dfrac{-2+6}{2}\right)$
$= \left(\dfrac{2}{2}, \dfrac{4}{2}\right)$
$= (1, 2)$

The center of the circle is $(1, 2)$.
The equation in standard form is:
$(x-1)^2 + (y-2)^2 = 25$

83. $(6, -2)$ and $(-2, 8)$

$\text{diameter} = \sqrt{(6-(-2))^2 + (-2-8)^2}$
$= \sqrt{(6+2)^2 + (-10)^2}$
$= \sqrt{64+100}$
$= \sqrt{164}$
$= 2\sqrt{41}$

The radius is $2\sqrt{41} \div 2 = \sqrt{41}$.

$\text{Midpoint} = \left(\dfrac{6+(-2)}{2}, \dfrac{-2+8}{2}\right)$
$= \left(\dfrac{4}{2}, \dfrac{6}{2}\right)$
$= (2, 3)$

The center of the circle is $(2, 3)$.
The equation in standard form is:
$(x-2)^2 + (y-3)^2 = 41$

85. a)
$h = -16t^2 + 96t + 112$
$h - 112 = -16(t^2 - 6t)$
$h - 112 - 144 = -16(t^2 - 6t + 9)$
$h - 256 = -16(t-3)^2$
$h = -16(t-3)^2 + 256$

The vertex for the parabola is (3, 256).
The maximum height for the rock will be 256 ft.

b) It will take 3 seconds to reach the maximum height.

c)
$0 = -16t^2 + 96t + 112$
$0 = -16(t^2 - 6t - 7)$
$0 = t^2 - 6t - 7$
$0 = (t-7)(t+1)$
$t - 7 = 0 \qquad t + 1 = 0$
$t = 7 \qquad\quad t = \cancel{-1}$

The rock will hit the ground after 7 seconds.

87. We must find a to use the standard form $y = a(x-h)^2 + k$. The parabola has vertex $(0, 18)$. When $y = 0$, $x = \dfrac{30}{2} = 15$. Then

$y = a(x-h)^2 + k$
$0 = a(15-0)^2 + 18$
$-18 = a \cdot 225$
$\dfrac{-18}{225} = a$
$a = -\dfrac{2}{25} = -0.08$

Substitute into the general equation.
$y = a(x-h)^2 + k$
$y = -0.08(x-0)^2 + 18$
$y = -0.08x^2 + 18$
or
$y = -\dfrac{2}{25}x^2 + 18$

89. $y = 0.0038x^2 - 0.3475x + 8.316$
$y = 0.0038(17)^2 - 0.3475(17) + 8.316$
$y \approx 3.5\%$

91. $x^2 + y^2 = 20^2$
$x^2 + y^2 = 400$

93. $x^2 + (y-110)^2 = 100^2$
$x^2 + (y-110)^2 = 10,000$

95. $x^2 + y^2 = (6370 + 230)^2$

    $x^2 + y^2 = 6600^2$

    $x^2 + y^2 = 43,560,000$

Review Exercises

1.

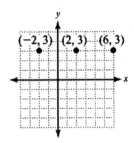

2. $25x^2 - 9y^2 = (5x + 3y)(5x - 3y)$

3.

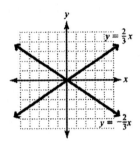

4. $\dfrac{x^2}{16} = 1$

    $x^2 = 16$

    $x = \pm\sqrt{16}$

    $x = \pm 4$

5. $9x^2 + 16 \cdot 0 = 144$

    $9x^2 = 144$

    $x^2 = 16$

    $x = \pm 4$

    $(-4, 0), (4, 0)$

    $9 \cdot 0 + 16y^2 = 144$

    $16y^2 = 144$

    $y^2 = \dfrac{144}{16}$

    $y^2 = 9$

    $y = \pm 3$

    $(0, -3), (0, 3)$

6. $25x^2 + 9 \cdot 0 = 225$

    $25x^2 = 225$

    $x^2 = 9$

    $x = \pm 3$

    $(-3, 0), (3, 0)$

$25 \cdot 0 + 9y^2 = 225$

    $9y^2 = 225$

    $y^2 = \dfrac{225}{9}$

    $y^2 = 25$

    $y = \pm 5$

    $(0, -5), (0, 5)$

Exercise Set 11.2

1. $\dfrac{x^2}{81} + \dfrac{y^2}{64} = 1$

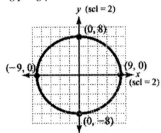

3. $\dfrac{x^2}{36} + y^2 = 1$

5. $4x^2 + 9y^2 = 36$

    $\dfrac{4x^2}{36} + \dfrac{9y^2}{36} = \dfrac{36}{36}$

    $\dfrac{x^2}{9} + \dfrac{y^2}{4} = 1$

7. $36x^2 + 4y^2 = 144$

$\dfrac{36x^2}{144} + \dfrac{4y^2}{144} = \dfrac{144}{144}$

$\dfrac{x^2}{4} + \dfrac{y^2}{36} = 1$

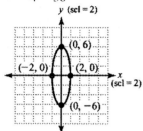

9. $\dfrac{(x-1)^2}{49} + \dfrac{(y+3)^2}{25} = 1$

11. $\dfrac{(x-4)^2}{4} + \dfrac{(y+3)^2}{36} = 1$

13. $\dfrac{x^2}{25} + \dfrac{y^2}{4} = 1$

15. $\dfrac{(y-2)^2}{36} + \dfrac{(x+4)^2}{25} = 1$

17. $\dfrac{x^2}{9} + \dfrac{y^2}{16} = 1$

19. $\dfrac{x^2}{9} - \dfrac{y^2}{4} = 1$

21. $\dfrac{y^2}{36} - \dfrac{x^2}{9} = 1$

23. $9x^2 - y^2 = 36$

$\dfrac{9x^2}{36} - \dfrac{y^2}{36} = \dfrac{36}{36}$

$\dfrac{x^2}{4} - \dfrac{y^2}{36} = 1$

25. $9y^2 - 25x^2 = 225$

$\dfrac{9y^2}{225} - \dfrac{25x^2}{225} = \dfrac{225}{225}$

$\dfrac{y^2}{25} - \dfrac{x^2}{9} = 1$

27. $\dfrac{x^2}{36} - \dfrac{y^2}{4} = 1$

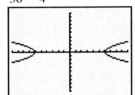

29. $\dfrac{x^2}{9} - \dfrac{y^2}{4} = 1$

31. c   33. b

35. $\begin{aligned} 9x^2 + 16y^2 &= 144 \\ \dfrac{9x^2}{144} + \dfrac{16y^2}{144} &= \dfrac{144}{144} \\ \dfrac{x^2}{16} + \dfrac{y^2}{9} &= 1 \end{aligned}$

ellipse

37. $\begin{aligned} x^2 + y^2 - 6x + 8y - 75 &= 0 \\ x^2 - 6x + y^2 + 8y &= 75 \\ x^2 - 6x + 9 + y^2 + 8y + 16 &= 75 + 9 + 16 \\ (x-3)^2 + (y+4)^2 &= 100 \end{aligned}$

circle

39. $(x-2)^2 + (y+2)^2 = 49$

circle

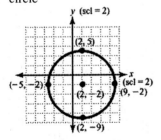

41. $y = 2(x+1)^2 + 3$

parabola

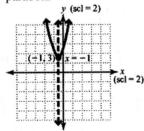

43. $\dfrac{x^2}{36} + \dfrac{y^2}{81} = 1$

ellipse

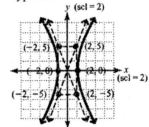

45. $\dfrac{x^2}{4} - \dfrac{y^2}{25} = 1$

hyperbola

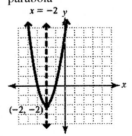

47. $\begin{aligned} y &= 2x^2 + 8x + 6 \\ y - 6 &= 2(x^2 + 4x) \\ y - 6 + 8 &= 2(x^2 + 4x + 4) \\ y &= 2(x+2)^2 - 2 \end{aligned}$

parabola

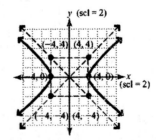

49. $\dfrac{x^2}{16} - \dfrac{y^2}{16} = 1$

hyperbola

51. $\dfrac{(x+3)^2}{9}+\dfrac{(y+2)^2}{36}=1$

ellipse

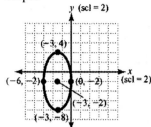

53. a) $400x^2+256y^2=102{,}400$

$$\dfrac{400}{102{,}400}x^2+\dfrac{256}{102{,}400}y^2=\dfrac{102{,}400}{102{,}400}$$

$$\dfrac{x^2}{256}+\dfrac{y^2}{400}=1$$

Yes, the sailboat will clear the bridge. The height of the bridge at the center is $\sqrt{400}=20$ ft., and the boat's mast is only 18 ft. above the water.

b) $\sqrt{256}=16$

The bridge is $2(16)=32$ ft. at the base of the arch.

55. $\dfrac{x^2}{3.6^2}+\dfrac{y^2}{2.88^2}=1$

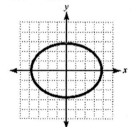

57. $\dfrac{x^2}{1.5^2}+\dfrac{y^2}{2^2}=1$

$$\dfrac{x^2}{2.25}+\dfrac{y^2}{4}=1$$

59. Rewrite as: $\dfrac{x^2}{5^2}+\dfrac{y^2}{4^2}=1$

Now, use substitution: $c^2=5^2-4^2$

$$c^2=25-16$$
$$c^2=9$$
$$c=\sqrt{9}$$
$$c=3\text{ units}$$

Review Exercises

1. $3x^2+y=6$

$$y=-3x^2+6$$

2. $\begin{cases}x+y=3\\2x+y=4\end{cases}$

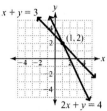

Solution: $(1,2)$

3. $\begin{cases}2x+y=1\\3x+4y=-6\end{cases}$

Solve the first equation for y: $y=1-2x$

$3x+4(1-2x)=-6y=1-2\cdot2$

$3x+4-8x=-6y=-3$

$-5x=-10$

$x=2$

Solution: $(2,-3)$

4. $4(2x+3y=6)3(-3)-4y=-25$

$\dfrac{3(3x-4y=-25)}{}-9-4y=-25$

$8x+12y=24-4y=-16$

$\dfrac{9x-12y=-75}{}y=4$

$17x=-51$

$x=-3$

Solution: $(-3,4)$

5. $3x^2+10x-8=0$

$(3x-2)(x+4)=0$

$3x-2=0$ or $x+4=0$

$x=\dfrac{2}{3}x=-4$

6. $4x^2=36$

$x^2=9$

$x=\pm3$

Exercise Set 11.3

1. $\begin{cases} x^2 + 2y = 1 \\ 2x + y = 2 \end{cases}$

   Solve the second equation for y: $y = -2x + 2$

   $$x^2 + 2y = 1$$
   $$x^2 + 2(-2x + 2) = 1$$
   $$x^2 - 4x + 4 = 1$$
   $$x^2 - 4x + 3 = 0$$
   $$(x - 3)(x - 1) = 0$$
   $$x - 3 = 0 \quad \text{or} \quad x - 1 = 0$$
   $$x = 3 \qquad\quad x = 1$$
   $$x = 3 \quad y = -2(3) + 2 = -6 + 2 = -4 \qquad (3, -4)$$
   $$x = 1 \quad y = -2(1) + 2 = -2 + 2 = 0 \qquad (1, 0)$$

3. $\begin{cases} y = x^2 + 4x + 4 \\ 3x - y = -6 \end{cases}$

   The first equation is solved for y.
   $$3x - (x^2 + 4x + 4) = -6$$
   $$3x - x^2 - 4x - 4 = -6$$
   $$0 = x^2 + x - 2$$
   $$0 = (x + 2)(x - 1)$$
   $$x + 2 = 0 \quad \text{or} \quad x - 1 = 0$$
   $$x = -2 \qquad\quad x = 1$$
   $$x = -2: \quad y = (-2)^2 + 4(-2) + 4$$
   $$y = 4 - 8 + 4$$
   $$y = 0$$
   $$(-2, 0)$$
   $$x = 1: \quad y = (1)^2 + 4(1) + 4$$
   $$y = 1 + 4 + 4$$
   $$y = 9$$
   $$(1, 9)$$

5. $\begin{cases} x^2 + y^2 = 25 \\ x - y = -1 \end{cases}$

   Solve the second equation for x. $x = y - 1$

   $$(y - 1)^2 + y^2 = 25$$
   $$y^2 - 2y + 1 + y^2 = 25$$
   $$2y^2 - 2y - 24 = 0$$
   $$y^2 - y - 12 = 0$$
   $$(y - 4)(y + 3) = 0$$
   $$y - 4 = 0 \quad \text{or} \quad y + 3 = 0$$
   $$y = 4 \qquad\quad y = -3$$
   $$y = 4: \quad x = 4 - 1$$
   $$x = 3$$
   $$(3, 4)$$

$$y = -3: \quad x = -3 - 1$$
$$x = -4$$
$$(-4, -3)$$

7. $\begin{cases} x^2 + y^2 = 10 \\ 3x + y = 6 \end{cases}$

   Solve the second equation for y. $y = 6 - 3x$

   $$x^2 + y^2 = 10$$
   $$x^2 + (6 - 3x)^2 = 10$$
   $$x^2 + 36 - 36x + 9x^2 = 10$$
   $$10x^2 - 36x + 26 = 0$$
   $$2(5x^2 - 18x + 13) = 0$$
   $$2(5x - 13)(x - 1) = 0$$
   $$5x - 13 = 0 \qquad\qquad \text{or} \quad x - 1 = 0$$
   $$x = \frac{13}{5} = 2.6 \qquad\qquad\qquad x = 1$$
   $$x = 2.6: \quad y = 6 - 3(2.6)$$
   $$y = 6 - 7.8$$
   $$y = -1.8$$
   $$(2.6, -1.8)$$
   $$x = 1: \quad y = 6 - 3(1)$$
   $$y = 6 - 3$$
   $$y = 3$$
   $$(1, 3)$$

9. $\begin{cases} x^2 + 2y^2 = 4 \\ x + y = 5 \end{cases}$

   Solve the second equation for x. $x = 5 - y$

   $$(5 - y)^2 + 2y^2 = 4$$
   $$25 - 10y + y^2 + 2y^2 - 4 = 0$$
   $$3y^2 - 10y + 21 = 0$$
   $$y = \frac{10 \pm \sqrt{10^2 - 4(3)(21)}}{2(3)}$$
   $$y = \frac{10 \pm \sqrt{-152}}{6}$$

   Because the discriminant of the quadratic equation is negative, we know that the equation cannot be solved using real numbers. Therefore there is no real solution.

11. $\begin{cases} y = 2x^2 + 3 \\ 2x - y = -3 \end{cases}$

The first equation is solved for y.

$2x - (2x^2 + 3) = -3$

$2x - 2x^2 - 3 = -3$

$0 = 2x^2 - 2x$

$0 = 2x(x - 1)$

$2x = 0$ or $x - 1 = 0$

$x = 0$ $x = 1$

$x = 0$: $y = 2 \cdot 0 + 3$

        $y = 3$

        $(0, 3)$

$x = 1$: $y = 2 \cdot 1 + 3$

        $y = 2 + 3$

        $y = 5$

        $(1, 5)$

13. $\begin{cases} y = (x - 3)^2 + 2 \\ y = -(x - 2)^2 + 3 \end{cases}$

$(x - 3)^2 + 2 = -(x - 2)^2 + 3$

$x^2 - 6x + 9 + 2 = -x^2 + 4x - 4 + 3$

$2x^2 - 10x + 12 = 0$

$x^2 - 5x + 6 = 0$

$(x - 2)(x - 3) = 0$

$x - 2 = 0$ or $x - 3 = 0$

$x = 2$ $x = 3$

$x = 2$: $y = -(2 - 2)^2 + 3 = 0 + 3 = 3$ $(2, 3)$

$x = 3$: $y = (3 - 3)^2 + 2 = 0 + 2 = 2$ $(3, 2)$

15. $\begin{cases} x^2 + y^2 = 20 \\ x^2 - y^2 = 12 \end{cases}$

$\begin{array}{l} x^2 + y^2 = 20 \\ \underline{x^2 - y^2 = 12} \\ 2x^2 \quad\;\; = 32 \\ x^2 \quad\;\;\; = 16 \\ x \quad\;\;\;\; = \pm 4 \end{array}$

$x = -4$: $(-4)^2 + y^2 = 20$

          $16 + y^2 = 20$

             $y^2 = 4$

              $y = \pm 2$

              $(-4, -2), (-4, 2)$

$x = 4$: $(4)^2 + y^2 = 20$

         $16 + y^2 = 20$

            $y^2 = 4$

             $y = \pm 2$

             $(4, -2), (4, 2)$

17. $\begin{cases} 4x^2 + 9y^2 = 72 \\ x^2 + y^2 = 13 \end{cases}$

$\begin{array}{l} 4x^2 + 9y^2 = 72 \\ \underline{-4(x^2 + y^2 = 13)} \\ 4x^2 + 9y^2 = 72 \\ \underline{-4x^2 - 4y^2 = -52} \\ 5y^2 = 20 \\ y^2 = 4 \\ y = \pm 2 \end{array}$

$y = -2$: $x^2 + (-2)^2 = 13$

          $x^2 + 4 = 13$

             $x^2 = 9$

             $x = \pm 3$

             $(-3, -2), (3, -2)$

$y = 2$: $x^2 + (2)^2 = 13$

         $x^2 + 4 = 13$

            $x^2 = 9$

            $x = \pm 3$

            $(-3, 2), (3, 2)$

19. $\begin{cases} 4x^2 - y^2 = 15 \\ x^2 + y^2 = 5 \end{cases}$

$\begin{array}{l} 4x^2 - y^2 = 15 \\ \underline{x^2 + y^2 \;= 5} \\ 5x^2 \quad\;\;\; = 20 \\ x^2 \quad\;\;\;\; = 4 \\ x \quad\;\;\;\;\; = \pm 2 \end{array}$

$x = -2$: $(-2)^2 + y^2 = 5$

          $4 + y^2 = 5$

             $y^2 = 1$

             $y = \pm 1$

             $(-2, -1), (-2, 1)$

$x = 2$: $(2)^2 + y^2 = 5$

         $4 + y^2 = 5$

            $y^2 = 1$

            $y = \pm 1$

            $(2, -1), (2, 1)$

21. $\begin{cases} x^2 + 3y^2 = 36 \\ x = y^2 - 6 \end{cases}$

The second equation is solved for x.

$$\left(y^2 - 6\right)^2 + 3y^2 = 36$$
$$y^4 - 12y^2 + 36 + 3y^2 - 36 = 0$$
$$y^4 - 9y^2 = 0$$
$$y^2\left(y^2 - 9\right) = 0$$
$$y^2\left(y + 3\right)\left(y - 3\right) = 0$$
$$y = 0, -3, 3$$

$y = 0: \quad x = 0^2 - 6$
$\qquad x = -6$
$\qquad \left(-6, 0\right)$

$y = -3: \quad x = \left(-3\right)^2 - 6$
$\qquad x = 9 - 6$
$\qquad x = 3$
$\qquad \left(3, -3\right)$

$y = 3: \quad x = \left(3\right)^2 - 6$
$\qquad x = 9 - 6$
$\qquad x = 3$
$\qquad \left(3, 3\right)$

23. $\begin{cases} 9x^2 + 4y^2 = 36 \\ 4x^2 - 9y^2 = 36 \end{cases}$

$$4\left(9x^2 + 4y^2 = 36\right)$$
$$-9\left(4x^2 - 9y^2 = 36\right)$$
$$\overline{\rule{0pt}{1.2em}}$$
$$36x^2 + 16y^2 = 144$$
$$-36x^2 + 81y^2 = -324$$
$$\overline{\rule{0pt}{1.2em}}$$
$$97y^2 = -180$$
$$y^2 = -1.856$$

y is imaginary, no solution

25. $\begin{cases} y = x^2 \\ x^2 + y^2 = 20 \end{cases}$

The first equation is already solved for y.

$$x^2 + \left(x^2\right)^2 = 20$$
$$x^2 + x^4 = 20$$
$$x^4 + x^2 - 20 = 0$$
$$\left(x^2 - 4\right)\left(x^2 + 5\right) = 0$$
$$x^2 = 4$$
$$x = \pm 2$$

$x^2 + 5 = 0$ would have yielded an imaginary answer.

$x = -2: \quad y = \left(-2\right)^2$
$\qquad y = 4$
$\qquad \left(-2, 4\right)$

$x = 2: \quad y = \left(2\right)^2$
$\qquad y = 4$
$\qquad \left(2, 4\right)$

27. $\begin{cases} 25x^2 - 16y^2 = 400 \\ x^2 + 4y^2 = 16 \end{cases}$

Use elimination.

$$25x^2 - 16y^2 = 400$$
$$4\left(x^2 + 4y^2 = 16\right)$$
$$\overline{\rule{0pt}{1.2em}}$$
$$25x^2 - 16y^2 = 400$$
$$4x^2 + 16y^2 = 64$$
$$\overline{\rule{0pt}{1.2em}}$$
$$29x^2 = 464$$
$$x^2 = 16$$
$$x = \pm 4$$

$x = -4: \quad \left(-4\right)^2 + 4y^2 = 16$
$\qquad 16 + 4y^2 = 16$
$\qquad 4y^2 = 0$
$\qquad y^2 = 0$
$\qquad y = 0$
$\qquad \left(-4, 0\right)$

$x = 4: \quad 4^2 + 4y^2 = 16$
$\qquad 16 + 4y^2 = 16$
$\qquad 4y^2 = 0$
$\qquad y^2 = 0$
$\qquad y = 0$
$\qquad \left(4, 0\right)$

29. $\begin{cases} xy = 4 \\ 2x^2 - y^2 = 4 \end{cases}$

Solve the first equation for y. $y = \dfrac{4}{x}$

$$2x^2 - \left(\dfrac{4}{x}\right)^2 = 4$$

$$2x^2 - \dfrac{16}{x^2} = 4$$

$$2x^4 - 16 = 4x^2$$

$$x^4 - 2x^2 - 8 = 0$$

$$\left(x^2 - 4\right)\left(x^2 + 2\right) = 0$$

$x^2 - 4 = 0$ or $x^2 + 2 = 0$

    $x^2 = 4$ $x^2 = -2$

      $x = \pm 2$ Leads to

                           imaginary

                           solutions

$x = -2:$ $y = \dfrac{4}{-2}$

            $y = -2$

               $(-2, -2)$

$x = 2:$ $y = \dfrac{4}{2}$

          $y = 2$

             $(2, 2)$

31. $\begin{cases} y = x^2 - 2x - 3 \\ y = -x^2 + 6x + 7 \end{cases}$

They are both solved for y, so use substitution.

$$-x^2 + 6x + 7 = x^2 - 2x - 3$$

$$0 = 2x^2 - 8x - 10$$

$$0 = x^2 - 4x - 5$$

$$0 = (x - 5)(x + 1)$$

$$x = 5, -1$$

$x = 5:$ $y = (5)^2 - 2(5) - 3$

               $= 25 - 10 - 3$

               $= 12$

                $(5, 12)$

$x = -1:$ $y = (-1)^2 - 2(-1) - 3$

                 $= 1 + 2 - 3$

                 $= 0$

                  $(-1, 0)$

33. $\begin{cases} 4x^2 + 5y^2 = 36 \\ 4x^2 - 3y^2 = 4 \end{cases}$

$$\begin{aligned} 4x^2 + 5y^2 &= 36 \\ -\left(4x^2 - 3y^2 = 4\right) \\ \hline 4x^2 + 5y^2 &= 36 \\ -4x^2 + 3y^2 &= -4 \\ \hline 8y^2 &= 32 \\ y^2 &= 4 \\ y &= \pm 2 \end{aligned}$$

$y = -2:$ $4x^2 + 5(-2)^2 = 36$

            $4x^2 + 5 \cdot 4 = 36$

            $4x^2 + 20 = 36$

               $4x^2 = 16$

                 $x^2 = 4$

                  $x = \pm 2$

                $(-2, -2), (2, -2)$

$y = 2:$ $4x^2 + 5(2)^2 = 36$

            $4x^2 + 5 \cdot 4 = 36$

            $4x^2 + 20 = 36$

               $4x^2 = 16$

                 $x^2 = 4$

                  $x = \pm 2$

                $(-2, 2), (2, 2)$

35. $\begin{cases} x = -y^2 + 2 \\ x^2 - 5y^2 = 4 \end{cases}$

The first equation is already solved for x.

$$\left(-y^2 + 2\right)^2 - 5y^2 = 4$$

$$y^4 - 4y^2 + 4 - 5y^2 = 4$$

$$y^4 - 9y^2 = 0$$

$$y^2\left(y^2 - 9\right) = 0$$

$$y = 0, \pm 3$$

$y = 0:$ $x = -0^2 + 2$

            $x = 2$

             $(2, 0)$

$y = -3:$ $x = -(-3)^2 + 2$

            $x = -9 + 2$

            $x = -7$

             $(-7, -3)$

$y = 3:$ $x = -(3)^2 + 2$

            $x = -9 + 2$

            $x = -7$

             $(-7, 3)$

37. Answers may vary, but one possible system is
$$\begin{cases} x + y = 4 \\ x^2 + y^2 = 1 \end{cases}.$$

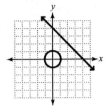

39. Let x be one of the integers and y be the other integer.

$x^2 + y^2 = 34$

$\underline{x^2 - y^2 = 16}$

$2x^2 \quad = 50$

$x^2 \quad = 25$

$x \quad = \pm 5$

$x = -5 : (-5)^2 + y^2 = 34$

$\qquad 25 + y^2 = 34$

$\qquad y^2 = 9$

$\qquad y = \pm 3$

$\qquad (-5, -3), (-5, 3)$

$x = 5 : \quad (5)^2 + y^2 = 34$

$\qquad 25 + y^2 = 34$

$\qquad y^2 = 9$

$\qquad y = \pm 3$

$\qquad (5, -3), (5, 3)$

41. Let x be one dimension and let y be the other.

$xy = 144$

$2x + 2y = 52$

Solve the second equation for x. $x = 26 - y$

$(26 - y)y = 144$

$26y - y^2 = 144$

$0 = y^2 - 26y + 144$

$0 = (y - 8)(y - 18)$

$y = 8, 18$

If $y = 8$: $xy = 144$ and $x = 18$.

If $y = 18$: $xy = 144$ and $x = 8$.

The computer keyboard is 8 in. by 18 in.

43. $-3x^2 + 120 = 11x + 28$

$0 = 3x^2 + 11x - 92$

$0 = \cancel{(3x + 23)}(x - 4)$

$x = 4$

$3x + 23 = 0$ would have yielded a negative answer.

So, because $x = 4$ and x is in hundreds of units, at equilibrium there would be 400 chairs.

The price per chair would be $p = 11(4) + 28 = 44 + 28 = \$72$.

45.

47.

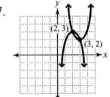

Review Exercises

1. $x + 3y \le 6$ Yes

$0 + 3 \cdot 0 \le 6$

$0 \le 6$

2. $x + 3y \le 6$ No

$3 + 3 \cdot 4 \le 6$

$3 + 12 \le 6$

$15 \le 6$

3. $y \ge 3$

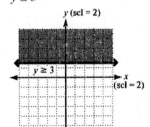

4. $y < 2x + 3$

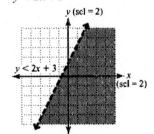

5. $2x + 3y < -6$

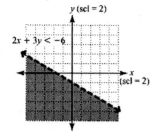

6. $x \geq -2$

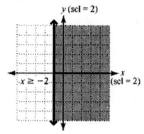

Exercise Set 11.4

1. $x^2 + y^2 \leq 9$

    First graph the related equation $x^2 + y^2 = 9$ with a solid line.
    Let region 1 be the area inside the circle, and let region 2 be the area outside the circle.
    Region 1: Choose (0, 0).
    $0^2 + 0^2 \leq 9$
    $0 + 0 \leq 9$
    $0 \leq 9$
    True, so region 1 is in the solution set.
    Region 2: Choose (4, 0).
    $4^2 + 0^2 \leq 9$
    $16 + 0 \leq 9$
    $16 \leq 9$
    False, so region 2 is not in the solution set.

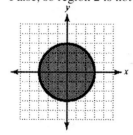

3. $y > x^2$

    First graph the related equation $y = x^2$ with a dotted line.
    Let region 1 be the area above the boundary, and let region 2 be the area below the boundary.
    Region 1: Choose (0, 1).
    $1 > 0^2$
    $1 > 0$
    True, so region 1 is in the solution set.
    Region 2: Choose (2, 0).
    $0 > 2^2$
    $0 > 4$
    False, so region 2 is not in the solution set.

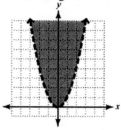

5. $y > 2(x + 2)^2 - 3$

    First graph the related equation $y = 2(x + 2)^2 - 3$ with a dotted line.
    Let region 1 be the area above the boundary, and let region 2 be the area below the boundary.
    Region 1: Choose $(-2, 0)$.
    $0 > 2(-2 + 2)^2 - 3$
    $0 > 2(0)^2 - 3$
    $0 > 0 - 3$
    $0 > -3$
    True, so region 1 is in the solution set.
    Region 2: Choose (0, 0).
    $0 > 2(0 + 2)^2 - 3$
    $0 > 2(2)^2 - 3$
    $0 > 2(4) - 3$
    $0 > 8 - 3$
    $0 > 5$
    False, so region 2 is not in the solution set.

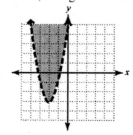

7. $\dfrac{x^2}{16}+\dfrac{y^2}{9}\le 1$

First graph the related equation $\dfrac{x^2}{16}+\dfrac{y^2}{9}=1$ with

a solid line.

Let region 1 be the area inside the ellipse, and let region 2 be the area outside the ellipse.

Region 1: Choose $(0, 0)$.

$$\dfrac{0^2}{16}+\dfrac{0^2}{9}\le 1$$

$$0+0\le 1$$

$$0\le 1$$

True, so region 1 is in the solution set.

Region 2: Choose $(5, 0)$.

$$\dfrac{5^2}{16}+\dfrac{0^2}{9}\le 1$$

$$\dfrac{25}{16}+0\le 1$$

$$\dfrac{25}{16}\le 1$$

False, so region 2 is not in the solution set.

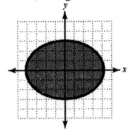

9. $\dfrac{x^2}{25}-\dfrac{y^2}{4}>1$

First graph the related equation $\dfrac{x^2}{25}-\dfrac{y^2}{4}=1$ with

a dotted line.

Let region 1 be the area to the left of the left branch of the hyperbola, let region 3 be the area to the right of the right branch of the hyperbola, and let region 2 be the area between the two branches.

Region 1: Choose $(-8,0)$.

$$\dfrac{(-8)^2}{25}-\dfrac{0^2}{4}>1$$

$$\dfrac{64}{25}-\dfrac{0}{4}>1$$

$$\dfrac{64}{25}>1$$

True, so region 1 is in the solution set.

Region 2: Choose $(0,0)$.

$$\dfrac{0^2}{25}-\dfrac{0^2}{4}>1$$

$$\dfrac{0}{25}-\dfrac{0}{4}>1$$

$$0>1$$

False, so region 2 is not in the solution set.

Region 3: Choose $(8,0)$.

$$\dfrac{8^2}{25}-\dfrac{0^2}{4}>1$$

$$\dfrac{64}{25}-\dfrac{0}{4}>1$$

$$\dfrac{64}{25}>1$$

True, so region 3 is in the solution set.

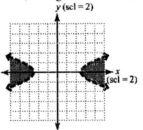

11. $x^2+y^2<16$

First graph the related equation $x^2+y^2=16$

with a dotted line.

Let region 1 be the area inside the circle, and let region 2 be the area outside the circle.

Region 1: Choose $(0, 0)$.

$$0^2+0^2<16$$

$$0<16$$

True, so region 1 is in the solution set.

Region 2: Choose $(5, 0)$.

$$5^2+0^2<16$$

$$25+0<16$$

$$25<16$$

False, so region 2 is not in the solution set.

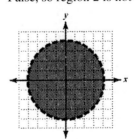

13. $y \geq -2(x-3)^2 + 3$

First graph the related equation

$y = -2(x-3)^2 + 3$ with a solid line.

Let region 1 be the area above the boundary, and let region 2 be the area below the boundary.

Region 1: Choose $(0,0)$.

$0 \geq -2(0-3)^2 + 3$

$0 \geq -2(-3)^2 + 3$

$0 \geq -2(9) + 3$

$0 \geq -18 + 3$

$0 \geq -15$

True, so region 1 is in the solution set.

Region 2: Choose $(4,-2)$.

$-2 \geq -2(4-3)^2 + 3$

$-2 \geq -2(1)^2 + 3$

$-2 \geq -2(1) + 3$

$-2 \geq -2 + 3$

$-2 \geq 1$

False, so region 2 is not in the solution set.

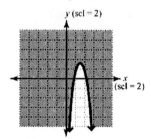

15. $\dfrac{x^2}{16} + \dfrac{y^2}{4} \leq 1$

First graph the related equation $\dfrac{x^2}{16} + \dfrac{y^2}{4} = 1$ with a solid line.

Let region 1 be the area inside the ellipse, and let region 2 be the area outside the ellipse.

Region 1: Choose $(0, 0)$.

$\dfrac{0^2}{16} + \dfrac{0^2}{4} \leq 1$

$0 + 0 \leq 1$

$0 \leq 1$

True, so region 1 is in the solution set.

Region 2: Choose $(5, 0)$.

$\dfrac{5^2}{16} + \dfrac{0^2}{4} \leq 1$

$\dfrac{25}{16} + 0 \leq 1$

$\dfrac{25}{16} \leq 1$

False, so region 2 is not in the solution set.

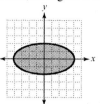

17. $\dfrac{y^2}{25} - \dfrac{x^2}{4} \leq 1$

First graph the related equation $\dfrac{y^2}{25} - \dfrac{x^2}{4} = 1$ with a solid line.

Let region 1 be the area above the top branch of the hyperbola, let region 3 be the area below the bottom branch of the hyperbola, and let region 2 be the area between the two branches.

Region 1: Choose $(0,8)$.

$\dfrac{8^2}{25} - \dfrac{0^2}{4} \leq 1$

$\dfrac{64}{25} - \dfrac{0}{4} \leq 1$

$\dfrac{64}{25} \leq 1$

False, so region 1 is not in the solution set.

Region 2: Choose $(0,0)$.

$\dfrac{0^2}{25} - \dfrac{0^2}{4} \leq 1$

$\dfrac{0}{25} - \dfrac{0}{4} \leq 1$

$0 - 0 \leq 1$

$0 \leq 1$

True, so region 2 is in the solution set.

Region 3: Choose $(0,-8)$.

$\dfrac{(-8)^2}{25} - \dfrac{0^2}{4} \leq 1$

$\dfrac{64}{25} - \dfrac{0}{4} \leq 1$

$\dfrac{64}{25} \leq 1$

False, so region 3 is not in the solution set.

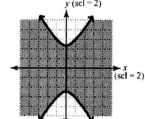

19. $y < x^2 + 4x - 5$

First graph the related equation $y = x^2 + 4x - 5$
with a dotted line.
Let region 1 be the area above the boundary, and
let region 2 be the area below the boundary.
Region 1: Choose $(0, 0)$.

$0 < 0^2 + 4(0) - 5$
$0 < 0 + 0 - 5$
$0 < -5$
False, so region 1 is not in the solution set.
Region 2: Choose $(4, 0)$.
$0 < 4^2 + 4(4) - 5$
$0 < 16 + 16 - 5$
$0 < 27$
True, so region 2 is in the solution set.

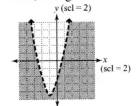

21. $\begin{cases} y < -x^2 \\ 2x - y < 4 \end{cases}$

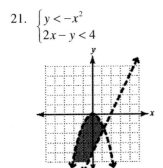

23. $\begin{cases} 3x + 2y \geq -6 \\ x^2 + y^2 \leq 25 \end{cases}$

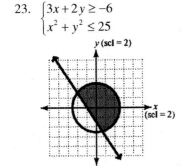

25. $\begin{cases} x^2 + y^2 > 4 \\ x^2 + y^2 > 9 \end{cases}$

27. $\begin{cases} y > x^2 + 1 \\ 2x + y < 3 \end{cases}$

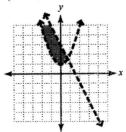

29. $\begin{cases} y < -x^2 + 3 \\ y > x^2 - 2 \end{cases}$

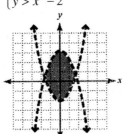

31. $\begin{cases} \dfrac{x^2}{25} + \dfrac{y^2}{9} \leq 1 \\ x^2 + y^2 \geq 4 \end{cases}$

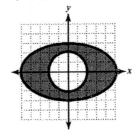

33. $\begin{cases} \dfrac{x^2}{25} + \dfrac{y^2}{9} < 1 \\ y > x^2 + 1 \end{cases}$

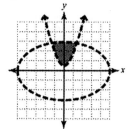

35. $\begin{cases} \dfrac{x^2}{9} - \dfrac{y^2}{4} \le 1 \\ \dfrac{x^2}{25} + \dfrac{y^2}{9} \le 1 \end{cases}$

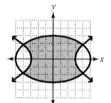

37. $\begin{cases} \dfrac{x^2}{4} - \dfrac{y^2}{4} > 1 \\ y > 2 \end{cases}$

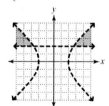

39. $\begin{cases} 3x + 2y \le 6 \\ x - y > -3 \\ x + 6y \ge 2 \end{cases}$

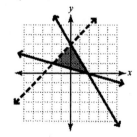

41. $\begin{cases} \dfrac{x^2}{16} + \dfrac{y^2}{4} \le 1 \\ x^2 + y^2 \le 9 \\ y \le x \end{cases}$

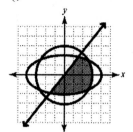

43. $\begin{cases} \dfrac{x^2}{49} - \dfrac{y^2}{16} \le 1 \\ \dfrac{x^2}{64} + \dfrac{y^2}{36} \le 1 \\ 2x - y \le -3 \end{cases}$

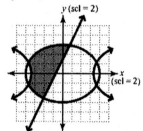

45. $\begin{cases} y < 2x^2 + 8 \\ 2x + y > 3 \\ x - y < 4 \end{cases}$

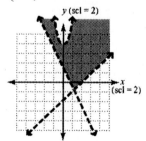

Review Exercises

1. $(3 \cdot 1 + 2) + (3 \cdot 2 + 2) + (3 \cdot 3 + 2) + (3 \cdot 4 + 2)$
   $= (3 + 2) + (6 + 2) + (9 + 2) + (12 + 2)$
   $= 5 + 8 + 11 + 14$
   $= 38$

2. $\dfrac{n}{2}(a_1 + a_n) = \dfrac{12}{2}(-8 + 60)$
   $= 6(52)$
   $= 312$

3. $a_1 + (n-1)d = -12 + (25-1)(-3)$
$$= -12 + (-72)$$
$$= -84$$

4. $n = 1$ $2(1^2) - 3 = 2 - 3 = -1$

    $n = 2$ $2(2^2) - 3 = 8 - 3 = 5$

    $n = 3$ $2(3^2) - 3 = 18 - 3 = 15$

    $n = 4$ $2(4^2) - 3 = 32 - 3 = 29$

5. $n = 1$ $\dfrac{(-1)^1}{3(1) - 2} = \dfrac{-1}{3 - 2} = -1$

    $n = 2$ $\dfrac{(-1)^2}{3(2) - 2} = \dfrac{1}{6 - 2} = \dfrac{1}{4}$

    $n = 3$ $\dfrac{(-1)^3}{3(3) - 2} = \dfrac{-1}{9 - 2} = -\dfrac{1}{7}$

    $n = 4$ $\dfrac{(-1)^4}{3(4) - 2} = \dfrac{1}{12 - 2} = \dfrac{1}{10}$

6. Let n, $(n + 2)$, and $(n + 4)$ be the three consecutive odd integers.
$$n + (n + 2) + (n + 4) = 207$$
$$3n + 6 = 207$$
$$3n = 201$$
$$n = 67$$
The three integers are 67, 69, and 71.

Chapter 11 Review Exercises

1. $y = 2(x - 3)^2 - 5$

    opens upward

    vertex is $(3, -5)$

    axis of symmetry is $x = 3$

    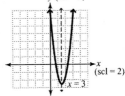

2. $y = -2(x + 2)^2 + 3$

    opens downward

    vertex is $(-2, 3)$

    axis of symmetry is $x = -2$

    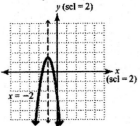

3. $x = -(y - 2)^2 + 4$

    opens left

    vertex is $(4, 2)$

    axis of symmetry is $y = 2$

    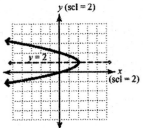

4. $x = 2(y + 3)^2 - 2$

    opens right

    vertex is $(-2, -3)$

    axis of symmetry is $y = -3$

    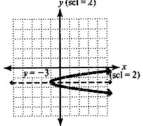

5.
$$x = y^2 + 6y + 8$$
$$x - 8 = y^2 + 6y$$
$$x - 8 + 9 = y^2 + 6y + 9$$
$$x + 1 = (y + 3)^2$$
$$x = (y + 3)^2 - 1$$

opens right

vertex is $(-1, -3)$

axis of symmetry is $y = -3$

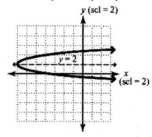

6.
$$x = 2y^2 - 8y - 6$$
$$x + 6 = 2(y^2 - 4y)$$
$$x + 6 + 8 = 2(y^2 - 4y + 4)$$
$$x + 14 = 2(y - 2)^2$$
$$x = 2(y - 2)^2 - 14$$

opens right

vertex is $(-14, 2)$

axis of symmetry is $y = 2$

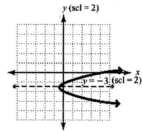

7.
$$x = -y^2 - 2y + 3$$
$$x - 3 = -(y^2 + 2y)$$
$$x - 3 - 1 = -(y^2 + 2y + 1)$$
$$x - 4 = -(y + 1)^2$$
$$x = -(y + 1)^2 + 4$$

opens left

vertex is $(4, -1)$

axis of symmetry is $y = -1$

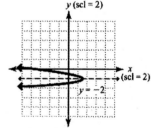

8.
$$x = -3y^2 - 12y - 9$$
$$x + 9 = -3(y^2 + 4y)$$
$$x + 9 - 12 = -3(y^2 + 4y + 4)$$
$$x - 3 = -3(y + 2)^2$$
$$x = -3(y + 2)^2 + 3$$

opens left

vertex is $(3, -2)$

axis of symmetry is $y = -2$

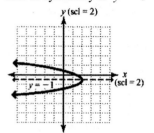

9. $(-1, -2)$ and $(-5, 1)$

$$d = \sqrt{(-1 - (-5))^2 + (-2 - 1)^2}$$
$$= \sqrt{(4)^2 + (-3)^2}$$
$$= \sqrt{16 + 9}$$
$$= \sqrt{25}$$
$$= 5$$

$$\text{Midpoint} = \left(\frac{-1 + (-5)}{2}, \frac{-2 + 1}{2} \right)$$
$$= \left(\frac{-6}{2}, \frac{-1}{2} \right)$$
$$= \left(-3, -\frac{1}{2} \right)$$

10. $(2,-5)$ and $(-2,3)$

$$d = \sqrt{(2-(-2))^2 + (-5-3)^2}$$
$$= \sqrt{(4)^2 + (-8)^2}$$
$$= \sqrt{16+64}$$
$$= \sqrt{80}$$
$$= 4\sqrt{5}$$

$$\text{Midpoint} = \left(\frac{2+(-2)}{2}, \frac{-5+3}{2}\right)$$
$$= \left(\frac{0}{2}, \frac{-2}{2}\right)$$
$$= (0,-1)$$

11. Find the distance from the center to a point on the circle. This is the radius.

$$r = \sqrt{(6-(-2))^2 + (-4-2)^2}$$
$$= \sqrt{(8)^2 + (-6)^2}$$
$$= \sqrt{64+36}$$
$$= \sqrt{100}$$
$$= 10$$

12. Find the distance from the center to a point on the circle. This is the radius.

$$r = \sqrt{(-6-3)^2 + (8-(-4))^2}$$
$$= \sqrt{(-9)^2 + (12)^2}$$
$$= \sqrt{81+144}$$
$$= \sqrt{225}$$
$$= 15$$

The equation in standard form is:
$$(x+6)^2 + (y-8)^2 = 225$$

13. a) $$h = -16t^2 + 32t + 128$$
$$h - 128 = -16(t^2 - 2t)$$
$$h - 128 - 16 = -16(t^2 - 2t + 1)$$
$$h - 144 = -16(t-1)^2$$
$$h = -16(t-1)^2 + 144$$

The vertex for the parabola is (1, 144). The maximum height for the rock will be 144 ft.

b) It will take 1 second to reach the maximum height.

c) $$0 = -16t^2 + 32t + 128$$
$$0 = -16(t^2 - 2t - 8)$$
$$0 = t^2 - 2t - 8$$
$$0 = (t-4)(t+2)$$
$$t - 4 = 0 \quad \text{or} \quad t + 2 = 0$$
$$t = 4 \qquad\qquad t = \cancel{-2}$$

The object will strike the ground after 4 seconds.

14. $(x-3)^2 + (y+2)^2 = 25$

Center $(3,-2)$; radius 5

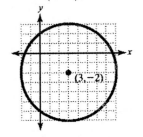

15. $(x+5)^2 + (y-1)^2 = 4$

Center $(-5,1)$; radius 2

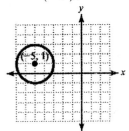

16. $$x^2 + y^2 - 4x + 8y + 11 = 0$$
$$(x^2 - 4x) + (y^2 + 8y) = -11$$
$$(x^2 - 4x + 4) + (y^2 + 8y + 16) = -11 + 4 + 16$$
$$(x-2)^2 + (y+4)^2 = 9$$

Center $(2,-4)$; radius 3

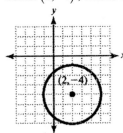

17.
$$x^2 + y^2 + 10x + 2y + 22 = 0$$
$$\left(x^2 + 10x\right) + \left(y^2 + 2y\right) = -22$$
$$\left(x^2 + 10x + 25\right) + \left(y^2 + 2y + 1\right) = -22 + 25 + 1$$
$$\left(x + 5\right)^2 + \left(y + 1\right)^2 = 4$$

Center $\left(-5, -1\right)$; radius 2

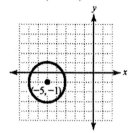

18. $\left(x - 6\right)^2 + \left(y + 8\right)^2 = 9^2$
$\left(x - 6\right)^2 + \left(y + 8\right)^2 = 81$

19. $\left(x - \left(-3\right)\right)^2 + \left(y - \left(-5\right)\right)^2 = 10^2$
$\left(x + 3\right)^2 + \left(y + 5\right)^2 = 100$

20. Since the rope is 20 ft. long and that is the distance from the center to any point on the edge of the circle, the radius is 20.
The equation in standard form is:
$x^2 + y^2 = 20^2$
$x^2 + y^2 = 400$

21. $\dfrac{x^2}{49} + \dfrac{y^2}{25} = 1$

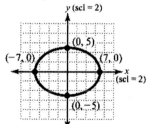

22. $\dfrac{x^2}{9} + \dfrac{y^2}{25} = 1$

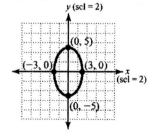

23. $4x^2 + 9y^2 = 36$
$$\frac{4x^2}{36} + \frac{9y^2}{36} = \frac{36}{36}$$
$$\frac{x^2}{9} + \frac{y^2}{4} = 1$$

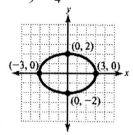

24. $25x^2 + 9y^2 = 225$
$$\frac{25x^2}{225} + \frac{9y^2}{225} = \frac{225}{225}$$
$$\frac{x^2}{9} + \frac{y^2}{25} = 1$$

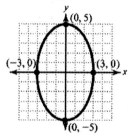

25. $\dfrac{\left(x + 2\right)^2}{16} + \dfrac{\left(y - 3\right)^2}{4} = 1$

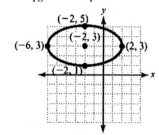

26. $\dfrac{\left(x - 1\right)^2}{9} + \dfrac{\left(y + 4\right)^2}{25} = 1$

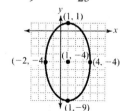

27. $\dfrac{x^2}{25^2} + \dfrac{y^2}{15^2} = 1$

$\dfrac{x^2}{625} + \dfrac{y^2}{225} = 1$

28. $\dfrac{x^2}{1.625^2} + \dfrac{y^2}{2^2} = 1$

$\dfrac{x^2}{1.625^2} + \dfrac{y^2}{4} = 1$

29. $\dfrac{x^2}{25} - \dfrac{y^2}{16} = 1$

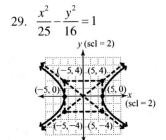

30. $\dfrac{x^2}{36} - \dfrac{y^2}{9} = 1$

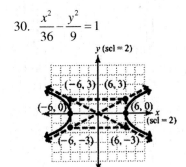

31. $y^2 - 9x^2 = 36$

$\dfrac{y^2}{36} - \dfrac{9x^2}{36} = \dfrac{36}{36}$

$\dfrac{y^2}{36} - \dfrac{x^2}{4} = 1$

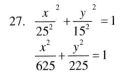

32. $25y^2 - 9x^2 = 225$

$\dfrac{25y^2}{225} - \dfrac{9x^2}{225} = \dfrac{225}{225}$

$\dfrac{y^2}{9} - \dfrac{x^2}{25} = 1$

33. $\begin{cases} y = 2x^2 - 3 \\ 2x - y = -1 \end{cases}$

The first equation is already solved for y.

$2x - \left(2x^2 - 3\right) = -1$

$2x - 2x^2 + 3 = -1$

$0 = 2x^2 - 2x - 4$

$0 = x^2 - x - 2$

$0 = (x - 2)(x + 1)$

$x - 2 = 0$ or $x + 1 = 0$

$x = 2 \qquad\qquad x = -1$

$x = 2:\qquad\quad y = 2 \cdot 2^2 - 3$

$\qquad\qquad\qquad y = 2 \cdot 4 - 3$

$\qquad\qquad\qquad y = 8 - 3$

$\qquad\qquad\qquad y = 5$

$\qquad\qquad\qquad (2, 5)$

$x = -1:\qquad\quad y = 2(-1)^2 - 3$

$\qquad\qquad\qquad y = 2 \cdot 1 - 3$

$\qquad\qquad\qquad y = 2 - 3$

$\qquad\qquad\qquad y = -1$

$\qquad\qquad\qquad (-1, -1)$

34. $\begin{cases} x^2 + y^2 = 17 \\ x - y = 3 \end{cases}$

Solve the second equation for x. $x = y + 3$

$$(y+3)^2 + y^2 = 17$$
$$y^2 + 6y + 9 + y^2 = 17$$
$$2y^2 + 6y - 8 = 0$$
$$y^2 + 3y - 4 = 0$$
$$(y+4)(y-1) = 0$$
$$y + 4 = 0 \quad \text{or} \quad y - 1 = 0$$
$$y = -4 \qquad\qquad y = 1$$
$$y = -4: \qquad x = -4 + 3$$
$$x = -1$$
$$(-1, -4)$$
$$y = 1: \qquad x = 1 + 3$$
$$x = 4$$
$$(4, 1)$$

35. $\begin{cases} 25x^2 + 3y^2 = 100 \\ 2x - y = -3 \end{cases}$

Solve the second equation for y. $y = 2x + 3$

$$25x^2 + 3(2x+3)^2 = 100$$
$$25x^2 + 3(4x^2 + 12x + 9) = 100$$
$$25x^2 + 12x^2 + 36x + 27 = 100$$
$$37x^2 + 36x - 73 = 0$$
$$(37x + 73)(x - 1) = 0$$
$$37x + 73 = 0 \qquad \text{or} \quad x - 1 = 0$$
$$37x = -73 \qquad\qquad x = 1$$
$$x = -\frac{73}{37}$$
$$x = -\frac{73}{37}: \qquad y = 2\left(-\frac{73}{37}\right) + 3$$
$$y = -\frac{35}{37}$$
$$\left(-\frac{73}{37}, -\frac{35}{37}\right)$$
$$x = 1: \qquad y = 2 \cdot 1 + 3$$
$$y = 5$$
$$(1, 5)$$

36. $\begin{cases} x^2 + y^2 = 64 \\ x^2 - y^2 = 64 \end{cases}$

$$x^2 + y^2 = 64$$
$$\underline{x^2 - y^2 = 64}$$
$$2x^2 \quad\;\; = 128$$
$$x^2 \qquad = 64$$
$$x \qquad\;\; = \pm 8$$

$$x = -8: \qquad (-8)^2 + y^2 = 64$$
$$64 + y^2 = 64$$
$$y^2 = 0$$
$$y = 0$$
$$(-8, 0)$$

$$x = 8: \qquad (8)^2 + y^2 = 64$$
$$64 + y^2 = 64$$
$$y^2 = 0$$
$$y = 0$$
$$(8, 0)$$

37. $\begin{cases} x^2 + y^2 = 25 \\ 25y^2 - 16x^2 = 256 \end{cases}$

$$16(x^2 + y^2 = 25)$$
$$\underline{-16x^2 + 25y^2 = 256}$$
$$16x^2 + 16y^2 \quad\; = 400$$
$$\underline{-16x^2 + 25y^2 = 256}$$
$$41y^2 = 656$$
$$y^2 = 16$$
$$y = \pm 4$$

$$y = -4: \qquad x^2 + (-4)^2 = 25$$
$$x^2 + 16 = 25$$
$$x^2 = 9$$
$$x = \pm 3$$
$$(-3, -4), (3, -4)$$

$$x = 4: \qquad x^2 + (4)^2 = 25$$
$$x^2 + 16 = 25$$
$$x^2 = 9$$
$$x = \pm 3$$
$$(-3, 4), (3, 4)$$

38. $\begin{cases} x^2 + 5y^2 = 36 \\ 4x^2 - 7y^2 = 36 \end{cases}$

$-4\left(x^2 + 5y^2 = 36\right)$

$\underline{4x^2 - 7y^2 = 36}$

$\underline{-4x^2 - 20y^2 = -144}$

$\underline{4x^2 - 7y^2\quad = 36}$

$\qquad -27y^2 = -108$

$\qquad\quad y^2 = 4$

$\qquad\quad y\ = \pm 2$

$y = 2:\qquad x^2 + 5(2)^2 = 36$

$\qquad\qquad\quad x^2 + 5\cdot 4 = 36$

$\qquad\qquad\quad x^2 + 20 = 36$

$\qquad\qquad\qquad\quad x^2 = 16$

$\qquad\qquad\qquad\quad x = \pm 4$

$\qquad\qquad\qquad (-4, 2), (4, 2)$

$y = -2:\qquad x^2 + 5(-2)^2 = 36$

$\qquad\qquad\quad\ x^2 + 5\cdot 4 = 36$

$\qquad\qquad\quad\ x^2 + 20 = 36$

$\qquad\qquad\qquad\quad x^2 = 16$

$\qquad\qquad\qquad\quad x = \pm 4$

$\qquad\qquad\qquad (-4, -2), (4, -2)$

39. $\begin{cases} y = x^2 - 2 \\ 4x^2 + 5y^2 = 36 \end{cases}$

The first equation is already solved for y.

$4x^2 + 5\left(x^2 - 2\right)^2 = 36$

$4x^2 + 5\left(x^4 - 4x^2 + 4\right) = 36$

$4x^2 + 5x^4 - 20x^2 + 20 = 36$

$5x^4 - 16x^2 - 16 = 0$

$\cancel{\left(5x^2 + 4\right)}\left(x^2 - 4\right) = 0$

$\qquad\qquad\qquad x^2 = 4$

$\qquad\qquad\qquad x = \pm 2$

$5x^2 + 4 = 0$ would have yielded an imaginary solution.

$x = -2:\ y = (-2)^2 - 2$

$\qquad\qquad y = 4 - 2$

$\qquad\qquad y = 2$

$\qquad\qquad (-2, 2)$

$x = 2:\quad y = (2)^2 - 2$

$\qquad\qquad y = 4 - 2$

$\qquad\qquad y = 2$

$\qquad\qquad (2, 2)$

40. $\begin{cases} y = x^2 - 1 \\ 4y^2 - 5x^2 = 16 \end{cases}$

The first equation is already solved for y.

$4\left(x^2 - 1\right)^2 - 5x^2 = 16$

$4\left(x^4 - 2x^2 + 1\right) - 5x^2 = 16$

$4x^4 - 8x^2 + 4 - 5x^2 = 16$

$4x^4 - 13x^2 - 12 = 0$

$\cancel{\left(4x^2 + 3\right)}\left(x^2 - 4\right) = 0$

$\qquad\qquad\qquad x^2 = 4$

$\qquad\qquad\qquad x = \pm 2$

$4x^2 + 3 = 0$ would have yielded an imaginary solution.

$x = -2:\ y = (-2)^2 - 1$

$\qquad\qquad y = 4 - 1$

$\qquad\qquad y = 3$

$\qquad\qquad (-2, 3)$

$x = 2:\quad y = (2)^2 - 1$

$\qquad\qquad y = 4 - 1$

$\qquad\qquad y = 3$

$\qquad\qquad (2, 3)$

41. Let x be one of the integers and y be the other integer.

$\begin{cases} x^2 + y^2 = 89 \\ x^2 - y^2 = 39 \end{cases}$

$x^2 + y^2 = 89$

$\underline{x^2 - y^2 = 39}$

$2x^2\quad\ = 128$

$x^2\quad\ = 64$

$x\qquad = \pm 8$

$x = -8:\ (-8)^2 + y^2 = 89$

$\qquad\qquad\ 64 + y^2 = 89$

$\qquad\qquad\qquad y^2 = 25$

$\qquad\qquad\qquad y = \pm 5$

$\qquad\qquad\qquad (-8, -5), (-8, 5)$

$x = 8:\quad (8)^2 + y^2 = 89$

$\qquad\qquad\ 64 + y^2 = 89$

$\qquad\qquad\qquad y^2 = 25$

$\qquad\qquad\qquad y = \pm 5$

$\qquad\qquad\qquad (8, -5), (8, 5)$

42. Let x be one dimension and let y be the other.

$$\begin{cases} xy = 80 \\ 2x + 2y = 36 \end{cases}$$

Solve the second equation for x. $x = 18 - y$

$$(18 - y) y = 80$$
$$18y - y^2 = 80$$
$$0 = y^2 - 18y + 80$$
$$0 = (y - 10)(y - 8)$$
$$y - 10 = 0 \quad \text{or} \quad y - 8 = 0$$
$$y = 10 \qquad\qquad y = 8$$

If $y = 10$, then $xy = 80$ and $x = 8$.

If $y = 8$, then $xy = 80$ and $x = 10$.

The rug is 8 ft. by 10 ft.

43.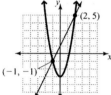

44.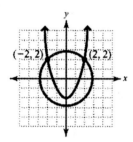

45. $x^2 + y^2 \le 64$

First graph the related equation $x^2 + y^2 = 64$ with a solid line.

Let region 1 be the area inside the circle, and let region 2 be the area outside the circle.

Region 1: Choose (0, 0).

$$0^2 + 0^2 \le 64$$
$$0 + 0 \le 64$$
$$0 \le 64$$

True, so region 1 is in the solution set.

Region 2: Choose (10, 0).

$$10^2 + 0^2 \le 64$$
$$100 + 0 \le 64$$
$$100 \le 64$$

False, so region 2 is not in the solution set.

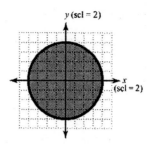

46. $y < 2(x - 3)^2 + 4$

First graph the related equation $y = 2(x - 3)^2 + 4$ with a dotted line.

Let region 1 be the area above the boundary, and let region 2 be the area below the boundary.

Region 1: Choose (2, 10).

$$10 < 2(2 - 3)^2 + 4$$
$$10 < 2(-1)^2 + 4$$
$$10 < 2(1) + 4$$
$$10 < 2 + 4$$
$$10 < 6$$

False, so region 1 is not in the solution set.

Region 2: Choose $(0, 0)$.

$$0 < 2(0 - 3)^2 + 4$$
$$0 < 2(-3)^2 + 4$$
$$0 < 2(9) + 4$$
$$0 < 18 + 4$$
$$0 < 22$$

True, so region 2 is in the solution set.

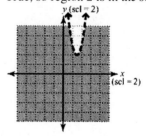

47. $\dfrac{y^2}{25}+\dfrac{x^2}{49}>1$

First graph the related equation $\dfrac{y^2}{25}+\dfrac{x^2}{49}=1$ with

a dotted line.
Let region 1 be the area inside the ellipse, and let
region 2 be the area outside the ellipse.
Region 1: Choose $(0, 0)$.

$\dfrac{0^2}{25}+\dfrac{0^2}{49}>1$

$0+0>1$

$0>1$

False, so region 1 is not in the solution set.
Region 2: Choose $(0, 6)$.

$\dfrac{6^2}{25}+\dfrac{0^2}{49}>1$

$\dfrac{36}{25}+0>1$

$\dfrac{36}{25}>1$

True, so region 2 is in the solution set.

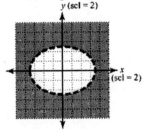

48. $\dfrac{x^2}{25}-\dfrac{y^2}{36}<1$

First graph the related equation $\dfrac{x^2}{25}-\dfrac{y^2}{36}=1$ with

a dotted line.
Let region 1 be the area to the left of the left
branch of the hyperbola, let region 3 be the area
to the right of the right branch of the hyperbola,
and let region 2 be the area between the two
branches.
Region 1: Choose $(-8,0)$.

$\dfrac{(-8)^2}{25}-\dfrac{0^2}{36}<1$

$\dfrac{64}{25}-\dfrac{0}{36}<1$

$\dfrac{64}{25}<1$

False, so region 1 is not in the solution set.

Region 2: Choose $(0,0)$.

$\dfrac{0^2}{25}-\dfrac{0^2}{36}<1$

$\dfrac{0}{25}-\dfrac{0}{36}<1$

$0<1$

True, so region 2 is in the solution set.
Region 3: Choose $(8,0)$.

$\dfrac{(8)^2}{25}-\dfrac{0^2}{36}<1$

$\dfrac{64}{25}-\dfrac{0}{36}<1$

$\dfrac{64}{25}<1$

False, so region 3 is not in the solution set.

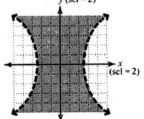

49. $\begin{cases} y \geq x^2-3 \\ 2x+y<2 \end{cases}$

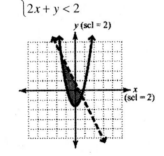

50. $\begin{cases} x+2y<4 \\ x^2+y^2 \leq 25 \end{cases}$

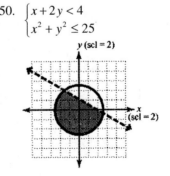

51. $\begin{cases} y > x^2 - 2 \\ y < -x^2 + 1 \end{cases}$

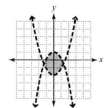

52. $\begin{cases} \dfrac{x^2}{9} + \dfrac{y^2}{25} \le 1 \\ x^2 + y^2 \ge 9 \end{cases}$

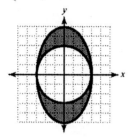

53. $\begin{cases} \dfrac{y^2}{4} - \dfrac{x^2}{9} \le 1 \\ \dfrac{x^2}{9} + \dfrac{y^2}{25} \le 1 \end{cases}$

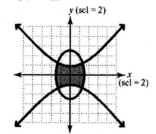

54. $\begin{cases} \dfrac{y^2}{4} + \dfrac{x^2}{16} \le 1 \\ x^2 + y^2 \le 9 \\ y \le x + 1 \end{cases}$

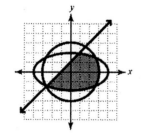

Chapter 11 Practice Test

1. $y = 2(x+1)^2 - 4$; $a = 2$, $h = -1$; and $k = -4$

The parabola opens upward, has its vertex at $(-1, -4)$, and has its axis of symmetry at $x = -1$.

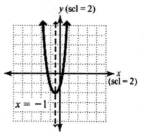

2. $x = -2(y-3)^2 + 1$; $a = -2$, $k = 3$; and $h = 1$

The parabola opens left, has its vertex at $(1, 3)$, and has its axis of symmetry at $y = 3$.

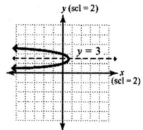

3. $$x = y^2 + 4y - 3$$
$$x + 3 = y^2 + 4y$$
$$x + 3 + 4 = y^2 + 4y + 4$$
$$x + 7 = (y+2)^2$$
$$x = (y+2)^2 - 7$$

$a = 1$, $k = -2$; and $h = -7$

The parabola opens right, has its vertex at $(-7, -2)$, and has its axis of symmetry at $y = -2$.

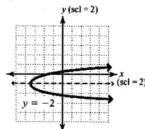

4. $(x_2, y_2) = (2, -3)$ and $(x_1, y_1) = (6, -5)$

$$d = \sqrt{(x_2 - x_1)^2 + (y_2 - y_1)^2}$$
$$= \sqrt{(2-6)^2 + (-3-(-5))^2}$$
$$= \sqrt{(-4)^2 + (2)^2}$$
$$= \sqrt{16 + 4}$$
$$= \sqrt{20}$$
$$= 2\sqrt{5}$$

$$\text{Midpoint} = \left(\frac{x_1 + x_2}{2}, \frac{y_1 + y_2}{2} \right)$$
$$= \left(\frac{6+2}{2}, \frac{-5 + (-3)}{2} \right)$$
$$= \left(\frac{8}{2}, \frac{-8}{2} \right)$$
$$= (4, -4)$$

5. $(x+4)^2 + (y-3)^2 = 36$

Because $h = -4$, $k = 3$, and $r = 6$, the center is $(-4, 3)$ and the radius is 6.

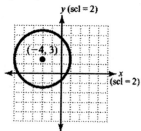

6.
$$x^2 + y^2 + 4x - 10y + 20 = 0$$
$$(x^2 + 4x) + (y^2 - 10y) = -20$$
$$(x^2 + 4x + 4) + (y^2 - 10y + 25) = -20 + 4 + 25$$
$$(x+2)^2 + (y-5)^2 = 9$$

Because $h = -2$, $k = 5$, and $r = 3$, the center is $(-2, 5)$ and the radius is 3.

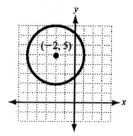

7. The center is $(2, -4)$, so $h = 2$ and $k = -4$. Find the distance from the center to a point on the circle $(-4, 4)$. This is the radius.

$$d = \sqrt{(2 - (-4))^2 + (-4 - 4)^2}$$
$$= \sqrt{(6)^2 + (-8)^2}$$
$$= \sqrt{36 + 64}$$
$$= \sqrt{100}$$
$$= 10$$

The equation in standard form is
$$(x-2)^2 + (y+4)^2 = 100$$

8. $16x^2 + 36y^2 = 576$
$$\frac{16x^2}{576} + \frac{36y^2}{576} = \frac{576}{576}$$
$$\frac{x^2}{36} + \frac{y^2}{16} = 1$$

$a = 6$, $b = 4$

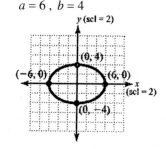

9. $\dfrac{(x+2)^2}{4} + \dfrac{(y-1)^2}{25} = 1$

$a = 2$, $b = 5$, $h = -2$, $k = 1$

10. $\dfrac{x^2}{49} - \dfrac{y^2}{25} = 1$

$a = 7$, $b = 5$

11. $\dfrac{y^2}{9} - \dfrac{x^2}{16} = 1$

$a = 4$, $b = 3$

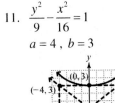

12. $\begin{cases} y = (x+1)^2 + 2 \\ 2x + y = 8 \end{cases}$

Solve the second equation for y. $y = 8 - 2x$.

$$y = (x+1)^2 + 2$$
$$8 - 2x = (x+1)^2 + 2$$
$$8 - 2x = x^2 + 2x + 1 + 2$$
$$0 = x^2 + 4x - 5$$
$$0 = (x+5)(x-1)$$
$$x = -5, 1$$

$x = -5:$ $y = 8 - 2(-5)$
 $y = 8 + 10$
 $y = 18$
 $(-5, 18)$

$x = 1:$ $y = 8 - 2(1)$
 $y = 8 - 2$
 $y = 6$
 $(1, 6)$

13. $\begin{cases} 3x - y = 4 \\ x^2 + y^2 = 34 \end{cases}$

Solve the first equation for y. $y = 3x - 4$.

$$x^2 + y^2 = 34$$
$$x^2 + (3x-4)^2 = 34$$
$$x^2 + 9x^2 - 24x + 16 = 34$$
$$10x^2 - 24x - 18 = 0$$
$$2(5x^2 - 12x - 9) = 0$$
$$5x^2 - 12x - 9 = 0$$
$$(5x+3)(x-3) = 0$$
$$x = -\dfrac{3}{5}, 3$$

$x = -\dfrac{3}{5}:$ $y = 3\left(-\dfrac{3}{5}\right) - 4$

$\phantom{x = -\dfrac{3}{5}:}$ $y = -\dfrac{29}{5}$

$\phantom{x = -\dfrac{3}{5}:}$ $\left(-\dfrac{3}{5}, -\dfrac{29}{5}\right)$

$x = 3:$ $y = 3(3) - 4$
 $y = 9 - 4$
 $y = 5$
 $(3, 5)$

14. $\begin{cases} x^2 + y^2 = 13 \\ 3x^2 + 4y^2 = 48 \end{cases}$

$$-3(x^2 + y^2 = 13)$$
$$3x^2 + 4y^2 = 48$$

$$\begin{aligned} -3x^2 - 3y^2 &= -39 \\ 3x^2 + 4y^2 &= 48 \\ \hline y^2 &= 9 \\ y &= \pm 3 \end{aligned}$$

$y = -3:$ $x^2 + (-3)^2 = 13$
 $x^2 + 9 = 13$
 $x^2 = 4$
 $x = \pm 2$
 $(-2, -3), (2, -3)$

$y = 3:$ $x^2 + (3)^2 = 13$
 $x^2 + 9 = 13$
 $x^2 = 4$
 $x = \pm 2$
 $(2, 3), (-2, 3)$

15. $\begin{cases} x^2 - 2y^2 = 1 \\ 4x^2 + 7y^2 = 64 \end{cases}$

$$-4(x^2 - 2y^2 = 1)$$
$$4x^2 + 7y^2 = 64$$

$$\begin{aligned} -4x^2 + 8y^2 &= -4 \\ 4x^2 + 7y^2 &= 64 \\ \hline 15y^2 &= 60 \\ y^2 &= 4 \\ y &= \pm 2 \end{aligned}$$

$y = -2:$ $x^2 - 2(-2)^2 = 1$
 $x^2 - 8 = 1$
 $x^2 = 9$
 $x = \pm 3$
 $(3, -2), (-3, -2)$

$y = 2:$ $x^2 - 2(2)^2 = 1$
 $x^2 - 8 = 1$
 $x^2 = 9$
 $x = \pm 3$
 $(3, 2), (-3, 2)$

16. $y \le -2(x+3)^2 + 2$

First graph the related equation

$y = -2(x+3)^2 + 2$ with a solid line.

Let region 1 be the area above the boundary, and let region 2 be the area below the boundary.

Region 1: Choose $(0,0)$.

$0 \le -2(0+3)^2 + 2$

$0 \le -2(3)^2 + 2$

$0 \le -2(9) + 2$

$0 \le -18 + 2$

$0 \le -16$

False, so region 1 is not in the solution set.

Region 2: Choose $(-2,-6)$.

$-6 < -2(-2+3)^2 + 2$

$-6 < -2(1)^2 + 2$

$-6 < -2 + 2$

$-6 < 0$

True, so region 2 is in the solution set.

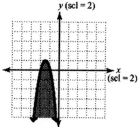

17. $\dfrac{x^2}{4} + \dfrac{y^2}{9} > 1$

First graph the related equation $\dfrac{x^2}{4} + \dfrac{y^2}{9} = 1$ with a dotted line.

Let region 1 be the area inside the ellipse, and let region 2 be the area outside the ellipse.

Region 1: Choose $(0, 0)$.

$\dfrac{0^2}{4} + \dfrac{0^2}{9} > 1$

$0 + 0 > 1$

$0 > 1$

False, so region 1 is not in the solution set.

Region 2: Choose $(0, 4)$.

$\dfrac{0^2}{4} + \dfrac{4^2}{9} > 1$

$0 + \dfrac{16}{9} > 1$

$\dfrac{16}{9} > 1$

True, so region 2 is in the solution set.

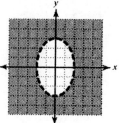

18. $\begin{cases} y \ge x^2 - 4 \\ \dfrac{x^2}{9} + \dfrac{y^2}{16} \le 1 \end{cases}$

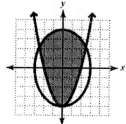

19. $x = 0:\ y = -\dfrac{1}{2}\cdot 0 + 18$

$y = 18$

The height is 18.

$y = 0: \quad 0 = -\dfrac{1}{2}x^2 + 18$

$ -18 = -\dfrac{1}{2}x^2$

$ 36 = x^2$

$ \pm 6 = x$

The distance across the base is 12.

20. $a = \dfrac{19}{2} = 9.5$ and $b = \dfrac{2}{2} = 1$

$\dfrac{x^2}{a^2} + \dfrac{y^2}{b^2} = 1$

$\dfrac{x^2}{9.5^2} + y^2 = 1$

Chapters 1–11 Cumulative Review

1. False; a positive base raised to a negative exponent yields a reciprocal.

2. True

3. True

4. $a_1 b_2 - a_2 b_1$

5. i

6. $d = \sqrt{\left(x_2 - x_1\right)^2 + \left(y_2 - y_1\right)^2}$

7. $6 - (-3) + 2^{-4} = 6 + 3 + \dfrac{1}{16} = 9\dfrac{1}{16}$

8. $\left(5m^3 n^{-2}\right)^{-3} = 5^{-3} m^{3(-3)} n^{-2(-3)}$

$$= \dfrac{1}{5^3} m^{-9} n^6$$

$$= \dfrac{n^6}{125 m^9}$$

9. $\sqrt{-20} = \sqrt{-4 \cdot 5} = 2i\sqrt{5}$

10. $ax + bx + ay + by = x(a+b) + y(a+b)$
$$= (a+b)(x+y)$$

11. $k^4 - 81 = \left(k^2 + 9\right)\left(k^2 - 9\right)$
$$= \left(k^2 + 9\right)(k+3)(k-3)$$

12. $|2x+1| = |x-3|$

$2x+1 = x-3 \quad$ or $\quad 2x+1 = -(x-3)$
$\qquad x = -4 \qquad\qquad 2x+1 = -x+3$
$\qquad\qquad\qquad\qquad\qquad 3x = 2$
$$x = \dfrac{2}{3}$$

13. $\sqrt{x-3} = 7$

$\left(\sqrt{x-3}\right)^2 = 7^2$

$\qquad x - 3 = 49$

$\qquad\quad x = 52$

14. $4x^2 - 2x + 1 = 0$

$$x = \dfrac{-(-2) \pm \sqrt{(-2)^2 - 4(4)(1)}}{2(4)} = \dfrac{2 \pm \sqrt{4-16}}{8}$$

$$= \dfrac{2 \pm \sqrt{-12}}{8} = \dfrac{2 \pm 2i\sqrt{3}}{8} = \dfrac{1 \pm i\sqrt{3}}{4}$$

15. $9 + 24x^{-1} + 16x^{-2} = 0$

$x^2\left(9 + 24x^{-1} + 16x^{-2}\right) = x^2 \cdot 0$

$\qquad 9x^2 + 24x + 16 = 0$

$\qquad\qquad (3x+4)^2 = 0$

$\qquad\qquad \sqrt{(3x+4)^2} = 0$

$\qquad\qquad\qquad 3x + 4 = 0$

$$x = -\dfrac{4}{3}$$

16. $9^x = 27$

$\left(3^2\right)^x = 3^3$

$\qquad 3^{2x} = 3^3$

$\qquad 2x = 3$

$$x = \dfrac{3}{2} = 1.5$$

17. $\log_3(x+1) - \log_3 x = 2$

$$\log_3 \dfrac{x+1}{x} = 2$$

$$3^2 = \dfrac{x+1}{x}$$

$\qquad 9x = x + 1$

$\qquad 8x = 1$

$$x = \dfrac{1}{8}$$

18. $5m^2 - 3m = 0$

$m(5m-3) = 0$

$m = 0 \quad$ or $\quad 5m - 3 = 0$

$$m = \dfrac{3}{5}$$

For $(-\infty, 0)$, we choose -1.

$5(-1)^2 - 3(-1) < 0 \qquad$ False
$\qquad 5 + 3 < 0$
$\qquad\qquad 8 < 0$

For $\left(0, \dfrac{3}{5}\right)$, we choose $\dfrac{1}{5}$.

$5\left(\dfrac{1}{5}\right)^2 - 3\left(\dfrac{1}{5}\right) < 0 \qquad$ True

$$\dfrac{1}{5} - \dfrac{3}{5} < 0$$

$$-\dfrac{2}{5} < 0$$

For $\left(\dfrac{3}{5}, \infty\right)$, we choose 2.

$5(2)^2 - 3(2) < 0 \qquad$ False
$\qquad 20 - 6 < 0$
$\qquad\qquad 14 < 0$

a)

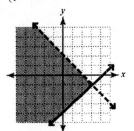

b) $\left\{ m \mid 0 < m < \dfrac{3}{5} \right\}$

c) $\left(0, \dfrac{3}{5} \right)$

19. a) $(f+g)(x) = (2x+1) + (x^2 - 1)$
$\qquad = x^2 + 2x$

b) $(f-g)(x) = (2x+1) - (x^2 - 1)$
$\qquad = (2x+1) + (-x^2 + 1)$
$\qquad = -x^2 + 2x + 2$

c) $(f \cdot g)(x) = (2x+1)(x^2 - 1)$
$\qquad = 2x^3 - 2x + x^2 - 1$
$\qquad = 2x^3 + x^2 - 2x - 1$

d) $(f / g)(x) = \dfrac{2x+1}{x^2 - 1}; x \neq \pm 1$

e) $(f \circ g)(x) = f\left[g(x) \right]$
$\qquad = f\left(x^2 - 1 \right)$
$\qquad = 2\left(x^2 - 1 \right) + 1$
$\qquad = 2x^2 - 2 + 1$
$\qquad = 2x^2 - 1$

f) $(g \circ f)(x) = g\left[f(x) \right]$
$\qquad = g(2x+1)$
$\qquad = (2x+1)^2 - 1$
$\qquad = 4x^2 + 4x + 1 - 1$
$\qquad = 4x^2 + 4x$

20. $\log_5 125 = 3$

21. $\log_5 x^5 y = \log_5 x^5 + \log_5 y = 5\log_5 x + \log_5 y$

22. Find the distance from the center to a point on the circle. This is the radius.
$r = \sqrt{(2-7)^2 + (4-16)^2}$
$\quad = \sqrt{(-5)^2 + (-12)^2}$
$\quad = \sqrt{25 + 144}$
$\quad = \sqrt{169}$
$\quad = 13$
The equation in standard form is:
$(x-2)^2 + (y-4)^2 = 13^2$
$(x-2)^2 + (y-4)^2 = 169$

23. $\begin{cases} y < -x + 2 \\ y \geq x - 4 \end{cases}$

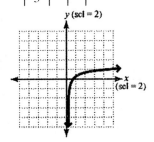

24. $f(x) = \log_3 x$
$\quad y = \log_3 x$
$\quad 3^y = x$

y	-1	0	1	2
x	$\dfrac{1}{3}$	1	3	9

25. $\dfrac{x^2}{4} + \dfrac{y^2}{9} = 1$

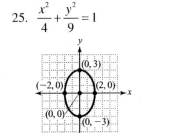

26. a) $\quad V = \pi r^2 h$
$\qquad \dfrac{V}{\pi r^2} = \dfrac{\pi r^2 h}{\pi r^2}$
$\qquad\qquad h = \dfrac{V}{\pi r^2}$

b) $h = \dfrac{V}{\pi r^2}$
$\quad h = \dfrac{15 \cancel{\pi}}{\cancel{\pi} r^2}$
$\quad h = \dfrac{15}{r^2}$

27. Let s be pounds of shrimp and t be pounds of tuna consumed.

    Translate to a system of equations. $\begin{cases} s+t = 6.3 \\ s = 0.5+t \end{cases}$

    $\begin{array}{ll} 0.5+t+t = 6.3 & s = 0.5+2.9 \\ \quad\quad 2t = 5.8 & s = 3.4 \\ \quad\quad\quad t = 2.9 \end{array}$

    2.9 lbs. of tuna and 3.4 lbs. of shrimp

28. Translate to a system of equations.

    Let x be the money market fund, y be the income fund, and z be the growth fund.

    $\begin{cases} x+y+z = 5000 \\ 0.05x+0.06y+0.03z = 255 \\ z = x-500 \end{cases}$

    $\begin{array}{ll} z = 1500-500 & 1500+y+1000 = 5000 \\ z = 1000 & \quad\quad 2500+y = 5000 \\ & \quad\quad\quad\quad\quad y = 2500 \end{array}$

    $1500 at 5%, $2500 at 6%, and $1000 at 3%

29. $M = \log \dfrac{I}{I_0}$

    $M = \log \dfrac{10^{2.9}}{10^{-4}}$

    $M = \log 10^{6.9}$

    $M = 6.9$

30. $\dfrac{x^2}{75^2} + \dfrac{y^2}{20^2} = 1$

    $\dfrac{x^2}{5625} + \dfrac{y^2}{400} = 1$

FAFSA
Save Key

E-mail
andresmontalvo432gmail.com

Villarica 57